Progress in Mathematics
Volume 101

Harmonic Analysis on Reductive Groups

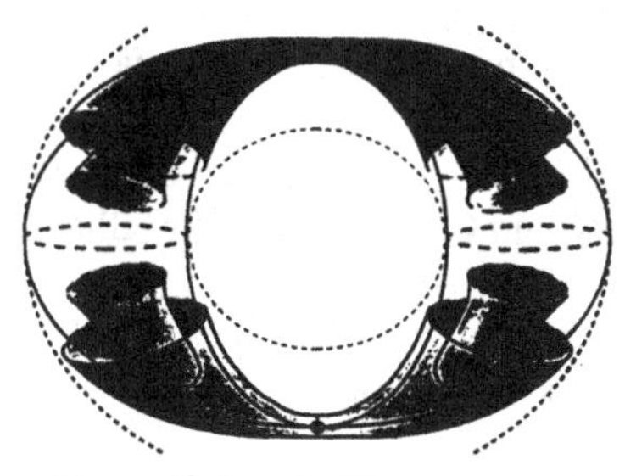

Bowdoin College 1989

William Barker
Paul Sally
Editors

Birkhäuser 1991
Boston • Basel • Berlin

William H. Barker
Department of Mathematics
Bowdoin College
Brunswick, ME 04011

Paul J. Sally, Jr.
Department of Mathematics
University of Chicago
Chicago, IL 60637

Library of Congress Cataloging-in-Publication Data

Harmonic analysis on reductive groups : proceedings of the Bowdoin conference 1989, Bowdoin College, 1989 / William Barker, Paul Sally, editors.
p. cm. -- (Progress in mathematics : vol. 101)
Includes bibliographical references.
ISBN 0-8176-3514-9
1. P-adic groups--Congresses. 2. Representations of groups--Congresses. 3. Harmonic analysis--Congresses. I. Barker, William, 1946- . II. Sally, Paul. III. Series.
QA 171.H32 1991 91-33801
512'.74--dc20 CIP

Printed on acid-free paper

ISBN 0-8176-3514-9
ISBN 3-7643-3514-9

Typeset in AMS-TEX.
Printed and bound by Edward Bros., Ann Arbor, MI

9 8 7 6 5 4 3 2 1

CONTENTS

PREFACE

A conference on Harmonic Analysis on Reductive Groups was held at Bowdoin College in Brunswick, Maine from July 31 to August 11, 1989. The stated goal of the conference was to explore recent advances in harmonic analysis on both real and p-adic groups. It was the first conference since the AMS Summer Symposium on Harmonic Analysis on Homogeneous Spaces, held at Williamstown, Massachusetts in 1972, to cover local harmonic analysis on reductive groups in such detail and to such an extent. While the Williamstown conference was longer (three weeks) and somewhat broader (nilpotent groups, solvable groups, as well as semisimple and reductive groups), the structure and timeliness of the two meetings was remarkably similar.

The program of the Bowdoin Conference consisted of two parts. First, there were six major lecture series, each consisting of several talks addressing those topics in harmonic analysis on real and p-adic groups which were the focus of intensive research during the previous decade. These lectures began at an introductory level and advanced to the current state of research. Second, there was a series of single lectures in which the speakers presented an overview of their latest research. The principal speakers and their topics for the lecture series were: James Arthur, *Some problems in local harmonic analysis*; Colin Bushnell, *The admissible dual of GL(n) via restriction to compact, open subgroups*; Laurent Clozel, *Invariant harmonic analysis on the Schwartz space of a reductive p-adic group*; Lawrence Corwin, *Constructing the supercuspidal representations of GL(n,F), F p-adic*; Wilfried Schmid, *Construction and classification of irreducible Harish-Chandra modules;* and David Vogan, *Associated varieties and unipotent representations.* The content of these lectures reflected accurately most of the major developments in the field of harmonic analysis on reductive groups since the Williamstown Conference. Those developments of importance which were not covered in the principal lecture series, such as harmonic analysis on semisimple symmetric spaces, were discussed in the individual lectures.

By design, the conference did not treat the extensive applications of local representation theory to the theory of automorphic forms. This is, without doubt, one of the most active and interesting areas of modern mathematics, and conferences on this topic have been held regularly over the past fifteen years. Two of the more comprehensive were held at Corvallis in 1977 (AMS Summer Symposium on Automorphic Forms, Representations, and L-functions, A. Borel and W. Casselman, eds.) and Ann Arbor in 1988 (Automorphic forms, Shimura varieties, and L-functions, L. Clozel and J. Milne, eds.). It should be mentioned, however, that much of the current research in local harmonic analysis, especially for p-adic groups, is motivated by considerations related to various aspects of the Langlands program. Moreover, it appears that some deep results in local harmonic analysis are necessary to achieve further progress in some parts of this program.

Returning to the Williamstown Conference, we note that a number of participants at Williamstown also attended the Bowdoin conference, and, in fact, several people spoke at both conferences. On the other hand, most of the speakers at Bowdoin had not started their research in mathematics in 1972. These facts illustrate both the continuity and the enduring vitality of the research in harmonic analysis on reductive groups. At Williamstown, the principal lecture series related directly to reductive groups were given by Harish-Chandra (Harmonic analysis on reductive p-adic groups); S. Helgason (Functions on symmetric spaces); and V. S. Varadarajan (The theory of characters and the discrete series for semisimple Lie groups). A quick glance at the papers in this volume shows that these lecture series are still directly connected to present day research. For example, Helgason's lectures are related to the papers of Anker and Helgason I, Varadarajan's to the papers of Adams and Herb, and Harish-Chandra's to most of the p-adic papers. It is no surprise that the deep and penetrating work of Harish-Chandra on both real and p-adic groups is reflected in the majority of the papers appearing here. Furthermore, it was at the Williamstown Conference that Roger Howe stated his famous conjecture about the finite-dimensionality of certain spaces of invariant distributions on p-adic groups. Harish-Chandra regarded this conjecture as the key to the study of invariant distributions on these groups. The Howe conjecture was proved by Clozel in the mid-1980s, and many of the consequences are contained in his paper in this volume. At the same time, the papers presented at the Bowdoin Conference contained much mathematics that was developed entirely within the past twenty years. Instances of this are the work on unipotent representations in the unitary dual for real groups, applications of the local trace formula to harmonic analysis, the introduction of D-modules into the study of representation theory, along with many others.

This volume is intended to serve as a reference for both graduate students and researchers working in representation theory and harmonic analysis on reductive groups. While the papers included here represent to a large extent the material covered in the talks at the Conference, they actually contain much more. The principal speakers, and, to some degree, the individual speakers have made a serious effort to give a complete exposition of their topics. We expect that these proceedings will provide a valuable resource for many years.

Paul Sally, Jr.

ACKNOWLEDGMENTS

The Organizing Committee for the conference consisted of William Barker, Rebecca Herb, Paul Sally, and Joseph Wolf. Joseph Bernstein also served on the committee until illness forced him to withdraw. The editors of these proceedings wish to thank their fellow committee members for the time and effort they expended in making the conference a success.

The editors also wish to thank those colleagues who delivered lectures at the conference, and especially those who subsequently contributed manuscripts to the Proceedings. Our thanks further go to the reviewers of the papers—many valuable improvements resulted from their careful reading.

We gratefully acknowledge the support provided by the National Science Foundation in grant number DMS-8804695. Additional support—specifically designated to aid graduate students— was generously supplied by Bowdoin College. These grants permitted a large and diverse group of mathematicians to participate in the conference.

The editors wish to extend a special word of thank to the staff of Birkhäuser Boston for their support and patience during the preparation of this volume. The delays they had to endure may have made lesser persons change professions. We further thank Ann Kostant, also of Birkhäuser Boston, who did an excellent job of converting a number of the longer manuscripts into TEX.

During and prior to the conference, the Organizing Committee was fortunate to have the services of Pam Ohlman as Administrative Assistant. Pam ably coordinated all the daily functions of the conference, and displayed confidence and good humor even under the most trying of circumstances. Sue Theberge, the Academic Coordinator for the Bowdoin Mathematics Department, also helped with the administrative work; in addition, she played an important role in polishing and completing the TEXfiles for this volume.

Finally, we wish to thank Donald Knuth for inventing the marvelous computer typesetting system TEX, and the AMS for the development of the TEXmacro package AMS-TEX. All the papers in this volume were typeset by AMS-TEX, version 2.0, preprint style.

William Barker

Harmonic Analysis on Reductive Groups
Participant List

Prof. Jeffrey Adams	University of Maryland
Dr. Jean-Philippe Anker	Princeton University
Prof. Susumu Ariki	University of Tokyo, Japan
Prof. James G. Arthur	University of Toronto, Canada
Prof. Magdy Ahmed Assem	Purdue University
Prof. Dan Barbasch	Cornell University
Dr. Leticia Barchini	SUNY at Stony Brook
Prof. William H. Barker	Bowdoin College
Dr. Robert Baston	Oxford University, Great Britain
Prof. Robert Bédard	Université du Québec à Montréal, Canada
Prof. Birne T. Binegar	Oklahoma State University
Mr. Juergen Blume	Mathematisches Institut, Bonn, W. Germany
Prof. Brian D. Boe	University of Georgia
Dr. Mladen Bozicevic	Mathematical Sciences Research Institute
Dr. Colin Bushnell	King's College, London, Great Britain
Prof. Jen-Tseh Chang	Oklahoma State
Prof. Laurent Clozel	Université de Paris-Sud, France
Prof. David Collingwood	University of Washington
Prof. Mark Copper	Florida International University
Prof. Lawrence J. Corwin	Rutgers University
Prof. James E. Daly	University of Colorado at Colorado Springs
Dr. Anton Deitmar	University of Muenster, W. Germany
Dr. Hideo Doi	Hiroshima University, Japan
Dr. Fokko du Cloux	Ecole Polytechnique, Paris, France
Dr. Edward Dunne	Oklahoma State University
Prof. Masaaki Eguchi	Hiroshima University, Japan
Prof. Thomas Enright	University of California, San Diego
Prof. Sam R. Evens	Rutgers University
Prof. Mogens Flensted-Jensen	Royal Vet. and Agricultural Un., Denmark
Prof. Jeffrey Fox	University of Colorado
Mr. Masaaki Furusawa	Johns Hopkins University
Prof. Pierre Y. Gaillard	SUNY at Buffalo
Dr. Devra Garfinkle	Rutgers University, Newark
Prof. Paul Gerardin	Penn State University
Mr. David Goldberg	University of Maryland
Mr. Daniel Goldstein	University of Chicago
Prof. Roe Goodman	Rutgers University
Mr. William Graham	Massachusetts Institute of Technology
Prof. Kenneth I. Gross	University of Vermont
Mr. Jeff Hakim	Columbia University
Prof. Henryk Hecht	University of Utah
Prof. Gerrit Heckman	Katholieke Universiteit, The Netherlands
Prof. Sigurdur Helgason	Massachusetts Institute of Technology
Prof. Aloysius G. Helminck	North Carolina State University
Prof. Rebecca Herb	University of Maryland
Prof. Michael J. Heumos	Columbia University
Prof. Roger Howe	Yale University
Mr. Jing-song Huang	Massachusetts Institute of Technology

Prof. James Humphreys	University of Massachusetts, Amherst
Mr. Chris Jantzen	University of Chicago
Prof. Joseph F. Johnson	Rutgers University, Camden
Prof. David Joyner	U. S. Naval Academy
Prof. Takeshi Kawazoe	Keio University, Japan
Prof. David Keys	Rutgers University, Newark
Mr. Donggyun Kim	University of Iowa
Mr. Moritz D. Klingholz	Mathematisches Institut, Bonn, W. Germany
Mr. Toshiyuki Kobayashi	University of Tokyo, Japan
Prof. Kazuhiko Koike	Aoyama Gahiun University, Tokyo, Japan
Prof. Ray A. Kunze	University of Georgia
Prof. Philip C. Kutzko	University of Iowa
Dr. Jian-Shu Li	University of Maryland
Prof. Ron Lipsman	University of Maryland
Prof. Jonathan D. Lubin	National Science Foundation
Prof. George Lusztig	Massachusetts Institute of Technology
Prof. David Manderscheid	University of Iowa
Prof. Lisa Mantini	Oklahoma State University
Prof. Shigeki Matsumoto	Konan University, Japan
Prof. Hisayosi Matumoto	The Institute For Advanced Study
Prof. William McGovern	Yale University
Prof. Hisaichi Midorikawa	Tsuda College, Tokyo, Japan
Prof. Dragan Milicic	University of Utah
Prof. Lawrence Morris	Clark University
Prof. Allen Moy	University of Washington
Prof. Fiona A. Murnaghan	University of Toronto, Canada
Mr. Hiroyuki Nakamura	University of Tokyo, Japan
Dr. Kyo Nishiyama	Tokyo Denki University, Japan
Prof. Hiroyuki Ochiai	Rikkyo University, Tokyo, Japan
Prof. Jeremy Orloff	Tufts University
Prof. Claude Pichet	Université du Québec à Montréal, Canada
Prof. Cary Rader	Ohio State University, Newark
Dr. Félix Recillas-Juárez	Instituto de Matemáticas - UNAM, Mexico
Mr. Zeév Rudnick	Yale University
Prof. Susana A. Salamanca-Riba	University of California, Davis
Prof. Paul J. Sally, Jr.	University of Chicago
Mr. Guillaume Sanje-Mpacko	Rutgers University
Dr. Gordan Savin	Massachusetts Institute of Technology
Prof. Roberto Scaramuzzi	Louisiana State University
Prof. Henrik Schlichtkrull	Royal Vet. and Agricultural Un., Denmark
Prof. Wilfried Schmid	Harvard University
Prof. Jiro Sekiguchi	The Johns Hopkins University
Prof. Freydoon Shahidi	Purdue University
Prof. Diana Shelstad	Rutgers University
Prof. Tom Sherman	Northeastern University
Mr. Nobukazu Shimeno	University of Tokyo, Japan
Prof. Robert Stanton	Ohio State University
Prof. Tim J. Steger	University of Chicago
Prof. Terje Sund	University of Oslo, Norway
Mr. W. Jay Sweet, Jr.	University of Maryland
Prof. Marko Tadic	University of Zagreb, Yugoslavia
Mr. Eng Chye Tan	Yale University

Dr. Helgi Thorleifsson	University of Göttingen, W. Germany
Prof. Tôru Umeda	Yale University
Mr. Tohru Uzawa	Yale University
Dr. Erik P. van den Ban	University of Utrecht, The Netherlands
Prof. Marie-France Vigneras	University of Paris VII, France
Prof. David A. Vogan	Massachusetts Institute of Technology
Prof. Masato Wakayama	Tottori University, Japan
Prof. Antoni Wawrzyñczyk	Universidad Autónoma Metropolitana, Mexico
Prof. Joseph A. Wolf	University of California, Berkeley
Mr. Hon-Wai Wong	Harvard University
Dr. Hiroshi Yamashita	Kyoto University, Japan
Prof. Jae-Hyun Yang	Harvard University
Prof. Yangbo Ye	Cornell University
Prof. Akihiko Yukie	Oklahoma State University
Mr. Chen-bo Zhu	Yale University
Prof. Roger Zierau	Oklahoma State University

LIFTING OF CHARACTERS

JEFFREY ADAMS

University of Maryland

INTRODUCTION

In *Lifting of Characters and Harish-Chandra's Method of Descent* [**1**] we discussed lifting of characters from endoscopic groups in terms which we need for Arthur's conjectures [**4**]. This paper is an expository version of [**1**], written with some important special cases in mind and illustrated by numerous examples. We refer the reader to the introduction to [**1**] for motivation; here we limit ourselves to a summary of some essential points and a discussion of how this paper differs from [**1**].

Let G be a reductive algebraic group defined over $\mathbf{R}$, and let ${}^\vee G$ be the (connected) dual group [**5**]. In [**13**] Vogan defined duality between characters of real forms of G and those of real forms of ${}^\vee G$. An endoscopic group H for G has as its dual group ${}^\vee H$ the centralizer of a semisimple element s of ${}^\vee G$. The main result of [**1**, Theorem 9.7] is that lifting of stable characters from the endoscopic group H, as defined in [**10**], is dual (in the sense of [**13**]) to Harish-Chandra's method of descent from ${}^\vee G$ to ${}^\vee H$.

There are a number of technical issues, all closely related to one another, which arise to obscure the main arguments. These are:

(1) infinitesimal characters:
 (a) non-integral,
 (b) singular,
 (c) the translation principle,
(2) covering groups,
(3) E-groups (failure of subgroups of L-groups to be L-groups),
(4) splittings of L-groups.

In this paper we restrict ourselves to **regular integral infinitesimal character**. We proceed to describe the resulting simplifications.

We need to be more precise about the duality of [**1**]. Fix a semisimple element λ of a Cartan subalgebra of the Lie algebra ${}^\vee\mathfrak{g}$ of ${}^\vee G$. This may be identified with the dual $\mathfrak{t}^*$ of a Cartan subalgebra $\mathfrak{t}$ of $\mathfrak{g}$. Then,

---

Partially supported by NSF grant DMS-8802586.

via the Harish-Chandra homomorphism, λ is identified with an infinitesimal character $\chi(\lambda)$ of G. Now the representations of real forms of G with infinitesimal character λ are dual to representations of real forms of the centralizer ${}^\vee G_\lambda$ of $e^{2\pi i\lambda}$ in ${}^\vee G$. (This is the group ${}^\vee G_{{}^\vee c}$ of [**1**, 2.2].) The difference between ${}^\vee G$ and ${}^\vee G_\lambda$ leads to a number of difficulties with respect to conjugacy and uniqueness of data. For example, the fact that two elements of ${}^\vee G_\lambda$ in distinct ${}^\vee G_\lambda$ conjugacy classes may be conjugate by ${}^\vee G$ causes endless headaches.

Therefore in this paper we assume ${}^\vee G_\lambda = {}^\vee G$, i.e. $e^{2\pi i\lambda}$ is in the center $Z({}^\vee G)$ of ${}^\vee G$. Equivalently, λ is in the lattice of weights of finite dimensional representations of $\mathfrak{g}$, and therefore we are considering representations with *integral* infinitesimal character.

An important ingredient of [**1**] is the definition of representations in terms of "L-data" [**1**, Definition 3.8]. Let ${}^\vee G^\Gamma$ be the L-group of G. Fix an infinitesimal character χ, and let Π be an L-packet of representations of G with this infinitesimal character. Then Π is defined by a map ϕ of the Weil group of $\mathbf{R}$ into ${}^\vee G^\Gamma$, which may be described in terms of (χ and) a triple $(y, {}^dT^\Gamma, {}^dB)$ (cf. (2-2)). Here y is an element of ${}^\vee G^\Gamma - {}^\vee G$, and ${}^dT^\Gamma$ (resp. ${}^dB$) is a Cartan (resp. Borel) subgroup of ${}^\vee G^\Gamma$, satisfying certain conditions. By extending this data we obtain *L-data* S which defines an irreducible representation of a real form of G. That is ([**1**, Definition 3.8]) a set of L-data for G is a 7-tuple:

$$S = (x,\ T^\Gamma,\ P,\ y, {}^dT^\Gamma,\ {}^dP,\ \zeta).$$

Here y and ${}^dT^\Gamma$ are as above, and ${}^dB$ has been replaced by a set ${}^dP$ of positive roots for a certain subset of the roots of ${}^\vee G$. The elements x, T^Γ and P are defined analagously for G. Now the L-group ${}^\vee T^\Gamma$ of T is isomorphic to ${}^dT^\Gamma$, and ζ is a choice of such an isomorphism (cf. (2-8) and [**1**, §3]). (More precisely we need *E-groups* here; see below.) This isomorphism is similar to the pseudo-diagonalization of [**10**].

This data is considerably simplified in the case of integral infinitesimal character. In this case P becomes a set of positive roots of T in G and is identified with a Borel subgroup B of G (similarly for ${}^dP$). Furthermore, in this case the isomorphism ζ is determined completely by B and ${}^dB$. Therefore we define a set of L-data to be a 6-tuple

$$S = (x,\ T^\Gamma,\ B,\ y, {}^dT^\Gamma,\ {}^dB).$$

Note that this data is obtained by symmetrizing (with respect to G and ${}^\vee G$) the data defining an L-packet.

Another issue involving infinitesimal characters is the translation principle. Note that $e^{2\pi i\lambda}$ is unchanged if we replace λ by $\lambda+\nu$ when $e^{2\pi i\nu}=1$, i.e., when ν is contained in the lattice $X_*({}^dT)$. This lattice may be identified with the lattice of weights of finite dimensional representations of

G, and this phenomenon corresponds to the translation principle [**14**]: the categories of representations with these two infinitesimal characters are equivalent (assuming they are both regular).

Thus, given data $(y, {}^dT^\Gamma, {}^dB)$ as above, to obtain a map ϕ of the Weil group into ${}^\vee G^\Gamma$ we need to choose an infinitesimal character χ. This may be chosen such that $\chi = \chi(\lambda)$ where λ satisfies $e^{2\pi i\lambda} = y^2$. Therefore $(y, {}^dT^\Gamma, {}^dB)$ determines a collection of maps ϕ (given by varying λ) and hence a *translation family* of L-packets. Strictly speaking one defines duality and lifting of translation families rather then individual representations (or even virtual representations). However, this is quite cumbersome, and in the case that y^2 is central, can be avoided. In this case the choice of λ is precisely that of an infinitesimal character, which we therefore fix, and we specialize all the translation families to this infinitesimal character. (In general the parameter set $\mathfrak{P}$ of [**1**,§2] is necessary. There is a map from $\mathfrak{P}$ to infinitesimal characters; however, this may fail to be injective in the non-integral case.)

In [**1**] the passage to singular infinitesimal character via the translation principle was carried out in §10. Here we confine ourselves to regular infinitesimal characters.

Mention was made above of *E-groups*. For example, the group ${}^dT^\Gamma$ generated by a Cartan subgroup ${}^dT$ of ${}^\vee G$ and an element $y \in {}^\vee G^\Gamma - {}^\vee G$ may fail to be isomorphic to the L-group of a torus T. Such a group is an example of an E-group [**2**]. Another example is the E-group ${}^\vee H^\Gamma$ of an endoscopic group H, which is generated by ${}^\vee H$ and y. Maps of the Weil group of $\mathbf{R}$ into ${}^\vee T^\Gamma$ or ${}^\vee H^\Gamma$ parametrize representations of (linear) covering groups of T or H. The nature of these coverings is more complicated if the infinitesimal character is not integral, by virtue of the fact that there is a further covering coming from the difference between ${}^\vee G$ and ${}^\vee G_\lambda$. By restricting ourselves to integral infinitesimal character we avoid these complications. In particular, the character τ of [**1**, 4.7] is somewhat simpler in this case.

Finally we note that a significant subtlety which is often overlooked is the need to choose a splitting of the L-group [**8**, §2]. This takes on a somewhat different look in our setup (cf. §1). This has not been eliminated; rather, we have tried to make the role of this choice clear at each step.

We conclude with an outline of the contents of this paper, and a guide to the examples. Section 1 discusses the structural aspects of L-groups and E-groups. Section 2 defines L-packets, representations, and blocks in terms of L-data, and presents duality of [**13**] in these terms. We also introduce the notions of super L-packets and blocks. The main result of section 3 is Theorem (3-2) which describes a super L-packet in terms of data on the dual group. Section 4 discusses endoscopic groups, and section 5 the method of descent. Stability and super-stability are discussed in section 6. Lifting is defined (Definition (7-10)) and discussed (Theorems (7-2,7,16, and 24)) in section 7.

There are no complete proofs and only a few sketches; the reader is referred to [1] for details. In a few places the terminology is in conflict with [1]. The main example is that L-data herein refers to what was called *strong integral* L-data in [1].

The examples discussed are the following:

(1) tori: (1-6, 9, 13, 16), (2-4, 13, 42, 45, 47), (3-8), (4-6, 12);
(2) SL(2) and PGL(2): (1-7, 10, 14), (2-5, 14, 17, 22, 27, 33, 37, 43), (3-14), (4-7, 13, 15, 17), (5-4), (6-3, 12), (7-6, 12);
(3) inner forms of compact groups: (1-17, 19), (2-48), (4-8);
(4) discrete series: (2-18, 29), (3-9, 13), (6-4, 13);
(5) $H = G$: (4-4), (5-6), (7-5, 11);
(6) other: (1-8), (2-7, 20, 21, 30, 31, 32), (3-16).

The author thanks David Vogan who, as joint author of [1], is responsible for much of this material (but is absolved of the responsibility for any errors). We also thank William Barker, Bowdoin College, and the NSF for the conference at which this was presented.

1. L-Groups and E-Groups

Let G be a connected reductive algebraic group. We begin by describing the L-group of G in terms suitable for our use. At the same time we define the more general notion of an *E-group* of G.

Let $\mathrm{Aut}(G)$, $\mathrm{Int}(G)$, and $\mathrm{Out}(G)$ be the (holomorphic) automorphisms, inner automorphisms, and outer automorphisms of G, respectively. There is a split exact sequence:

$$1 \rightarrow \mathrm{Int}(G) \rightarrow \mathrm{Aut}(G) \rightarrow \mathrm{Out}(G) \rightarrow 1. \tag{1-1}$$

We say two elements of $\mathrm{Aut}(G)$ are *inner* to each other, or define the same *inner class*, if they have the same image in $\mathrm{Out}(G)$. The L-group of G is attached to an inner class of involutions, so fix $\gamma \in \mathrm{Out}(G)$ of order two. Let ${}^\vee G$ be the connected complex dual group of G [5]. Let ${}^\vee\gamma$ be the element of $\mathrm{Out}(G)$ obtained from γ via the isomorphism $\mathrm{Out}(G) \simeq \mathrm{Out}({}^\vee G)$ given by *-transpose* on root data (cf. [2,Prop. 3.24]). The L-group of G (and γ) as defined in [5,8] consists primarily of an extension

$$1 \rightarrow {}^\vee G \rightarrow {}^L G \rightarrow \Gamma \rightarrow 1 \tag{1-2}$$

where Γ is the Galois group of $\mathbf{C}/\mathbf{R}$. The action of the non-trivial element σ of Γ is by a quasi-compact involution which is inner to γ. (We say an involution is quasi-compact if there is a Borel subgroup for which every simple imaginary root is compact. This is the involution in the given inner class closest to being the Cartan involution of a compact group.) In addition, the data defining an L-group includes a (conjugacy class of a) splitting of the exact sequence (1-2), so ${}^L G \simeq {}^\vee G \rtimes \Gamma$.

We seek to give a restatement of this definition, which at once generalizes it and is symmetric with respect to G and ${}^\vee G$. We start by noting that an L-group is an example of a group given in the following (preliminary) definition (the full definition is (1-11)). Since we deal with G and ${}^\vee G$ symmetrically, we state this in terms of G.

(1-3) Preliminary Definition. *Let G be a connected reductive algebraic group.*

(1) *An extended group containing G is an algebraic group G^Γ containing G as a subgroup of index two, i.e., the group G^Γ is an extension of G by Γ.*

The L-group of G is distinguished among extended groups containing ${}^\vee G$ by the condition that there is an element $\sigma \in {}^L G - {}^\vee G$ such that conjugation by σ (denoted $\mathrm{int}(\sigma)$) is a quasi-compact involution. On the other hand we say an involution of ${}^\vee G$ is *principal* (or *quasi-split*) if it is the Cartan involution of a quasi-split group (cf. [**2**, Def. 6.13]). Any inner class of order 2 contains a principal involution, which is unique up to conjugation. This involution carries more information than a quasi-compact one—for example it "sees" all Cartan subgroups, not just a fundamental one. For this reason we single out a principal involution in a given inner class.

Turning to the general extended group G^Γ containing G, from the exact sequence $1 \to G \to G^\Gamma \to \Gamma \to 1$, we obtain a homomorphism $\Gamma \to \mathrm{Out}(G)$. Composing this with a splitting $\mathrm{Out}(G) \hookrightarrow \mathrm{Aut}(G)$, we see there is an element $\delta_0 \in G^\Gamma - G$ such that δ_0^2 is contained in the center $Z(G)$ of G. In particular, this says that $\mathrm{int}(\delta_0)$ is an involution. Choosing a principal involution in the inner class defined by $\mathrm{int}(\delta_0)$, we obtain an element $\delta = g\delta_0$ ($g \in G$) with $\delta^2 \in Z(G)$ and such that $\mathrm{int}(\delta)$ is a principal involution. Therefore we concentrate on δ, and we see that any extended group containing G may be realized as follows.

(1-3) (cont.)

(2) *Fix an involution $\gamma \in \mathrm{Out}(G)$. Choose $\theta \in \mathrm{Aut}(G)$ to be a principal involution in the inner class defined by γ, and fix $z \in Z(G)^\theta$ (this set is independent of the choice of θ). The extended group containing G defined by γ and z is the group G^Γ defined as follows:*

$$G^\Gamma = \langle G, \delta \rangle,$$

the group generated by G and a formal element δ. It is subject to the relations $\delta^2 = z$, and $\delta g \delta^{-1} = \theta(g)$ ($g \in G$). This defines the group uniquely up to isomorphism, and is independent of the choice of θ.

Now the L-group of G, and more general E-groups, may be conveniently described. An E-group of G is an extended group containing ${}^\vee G$, so it is defined by the involution ${}^\vee\gamma$ (dual to a given involution γ as above) and z as in (2). In the case of the L-group we find ${}^\vee\delta$ and z as follows. First we need some notation.

Fix a Cartan subgroup ${}^dT$ of ${}^\vee G$ and let ρ be one-half the sum of a set of positive coroots of ${}^dT$ in ${}^\vee G$. Let $z_\rho = \exp(2\pi i\rho) \in Z({}^\vee G)$. Then z_ρ is fixed by all automorphisms of ${}^\vee G$ and is independent of the choices. Let $m_\rho = \exp(\pi i\rho) \in {}^dT$, which is a square-root of z_ρ depending on the choices. By symmetry we also define elements ${}^\vee z_\rho$ and ${}^\vee m_\rho$ contained in G. Now an involution θ of G is principal if and only if there exists a θ-stable Borel subgroup B of G such that, fixing a θ-stable Cartan subgroup $T \subset B$, every root of T in B is either complex or non-compact imaginary ([**2**, Prop, 6.24]). Such a Borel subgroup is said to be *large*, as T is a fundamental Cartan subgroup, and B corresponds to a "large" fundamental series representation of a quasi-split group. Applying this to ${}^\vee G$, choose ${}^dT$ and ${}^dB$ as indicated, and let m_ρ be the corresponding element of ${}^dT$.

Returning to our L-group ${}^LG$ of G, we see that conjugation by $m_\rho\sigma$ is a principal involution (as may be seen by considering the action on root spaces), and therefore we may take ${}^\vee\delta = m_\rho\sigma$. Then we have ${}^\vee\delta^2 = z_\rho$, so in this case $z = z_\rho$.

(1-3) (cont.)

(3) *Fix γ as in (2). An <u>E-group of G</u> (and γ) is an extended group ${}^\vee G^\Gamma$ containing ${}^\vee G$ defined by ${}^\vee\gamma$ and any element of $Z({}^\vee G)^{{}^\vee\theta}$. The <u>L-group</u> of G (and γ) is the E-group ${}^\vee G^\Gamma$ of G defined by ${}^\vee\gamma$ and z_ρ. We say an E-group is <u>genuine</u> if it is not an L-group.*

(4) *Suppose G is defined over* $\mathbf{R}$, *and let θ be a Cartan involution of G. Let γ be the image of θ in* $\mathrm{Out}(G)$; *this is independent of the choice of θ. Then E-groups and L-groups of G are defined by (3). We say two real forms of G are inner to each other if their Cartan involutions are inner, i.e. they have isomorphic L-groups.*

By symmetry we have the L-group of ${}^\vee G$ is the E-group G^Γ of ${}^\vee G$ defined by ${}^\vee z_\rho$. By construction the L-group ${}^\vee G^\Gamma$ just defined (i.e. the E-group defined by z_ρ) is isomorphic to the usual L-group ${}^LG$. To be explicit, choose m_ρ and δ as in (1-3)(3). Then the isomorphism is given by:

(1-4)
$$g \to g \quad (g \in {}^\vee G)$$
$${}^\vee\delta \to m_\rho\sigma$$

(cf. [**2**, 9.7]).

It is often convenient to pass back and forth between these two realizations of the L-group. Since the actions of ${}^\vee\delta$ and σ on ${}^\vee G^\Gamma$ differ by $\mathrm{int}({}^\vee m_\rho)$, these actions agree on a fundamental Cartan subgroup of ${}^\vee G$, and on the center of ${}^\vee G$.

It is not hard to see that two real forms of the same group are inner to each other if and only they they share (i.e. contain isomorphic) fundamental Cartan subgroups. The following Lemma is immediate.

(1-5) Lemma. *Let G be defined over* **R**, *and let $\gamma \in \mathrm{Out}(G)$ be as in (1-3(4)). The following conditions are equivalent:*

(1) *γ is trivial,*
(2) *The action of σ on G is trivial,*
(3) *G is inner to a compact group,*
(4) *${}^\vee G$ is inner to a split group,*
(5) *G contains a compact Cartan subgroup T,*
(6) *The action of σ is trivial on some Cartan subgroup of G,*
(7) *The action of δ is trivial on some Cartan subgroup of G.*

(1-6) **Example:** *L-groups for Tori.* Suppose G is a one-dimensional torus defined over **R**. We write $\times$ for direct product, and $\rtimes$ for semi-direct product with non-trivial action. If G is split, then ${}^\vee G$ is compact, ${}^\vee G^\Gamma \simeq G \times \Gamma$, and $G \simeq G \rtimes \Gamma$, where the action of δ (and σ) is by $g \to g^{-1}$. Note that $G^\Gamma \simeq O(2, \mathbf{C})$. If G is compact the situation is the same with the roles of G and ${}^\vee G$ reversed. □

(1-7) **Example:** *L-groups for SL(2).* Suppose G is a real form of $SL(2)$, so ${}^\vee G = PGL(2)$. Then σ acts trivially on G, δ acts by conjugation by $m_\rho = \mathrm{diag}(i, -i)$, and $\delta^2 = z_\rho = -I$. Note that $\mathrm{int}(\delta)$ is the Cartan involution of $SU(1,1)$.

If G is a real form of $PGL(2)$, the actions of σ and δ are the same as for $SL(2)$, but $\delta^2 = z_\rho = 1$. This defines L-groups G^Γ and ${}^\vee G^\Gamma$ for both G and ${}^\vee G$ which we continue to use in subsequent examples. □

(1-8) **Example:** *L-groups for GL(n).* Suppose G is (the split real form of) $GL(n)$. Then ${}^\vee G = U(n)$, σ acts trivially on ${}^\vee G$, and by $g \to^t g^{-1}$ on G. The action of δ on G is the same as that of σ, and ${}^\vee\delta$ acts on ${}^\vee G$ by conjugation by ${}^\vee m_\rho = (i)^{(n-1)}\mathrm{diag}(1, -1, 1, \ldots, (-1)^{n+1})$. Furthermore ${}^\vee\delta^2 = \delta^2 = (-1)^{n+1} I$. □

(1-9) **Example:** *E-groups for Tori.* If G is a split one-dimensional torus, there is an E-group of G defined by any element ${}^\vee z \in {}^\vee G$. Note that if ${}^\vee z = -1$ this is isomorphic to the subgroup of the L-group of $PGL(2)$ generated by the diagonal torus and ${}^\vee\delta$. This is a typical example of how more general E-groups arise. If G is compact, there is one genuine E-group of G, defined by ${}^\vee z = -1$. Again this is isomorphic to a subgroup of the L-group of $PGL(2)$. □

(1-10) **Example:** *E-groups for SL(2).* If G is $PGL(2)$ there is one additional E-group of G, where the action of δ is as before, but $\delta^2 = -1$. On the other hand the L-group is the unique E-group of $SL(2)$. □

To parametrize the representations of $G(\mathbf{R})$ it is not enough to have a map of the Weil group into the group ${}^LG$ of (1-2); it is necessary to choose an isomorphism ${}^LG \simeq {}^\vee G \rtimes \Gamma$ via a conjugacy class of (admissible) splittings of the exact sequence (1-2) [8]. The necessity of such a choice may be seen by taking G to be a one-dimensional split torus (cf. Example (1-13)). Definition (1-3) is preliminary because it does not take this splitting into account, which in our situation becomes the following data.

(1-11) Definition. *An extended group G^Γ is a group given by Definition (1-3), together with a G-conjugacy class of pairs (δ, B), for $\delta \in G^\Gamma - G$, and B a large Borel subgroup with respect to* $\operatorname{int}(\delta)$.

Given G^Γ we refer to a conjugacy class $\{(\delta, B)\}$ as a choice of *extended-group structure* on G^Γ. Similarly if ${}^\vee G^\Gamma$ is an L-group or an E-group of G we refer to the choice of $\{({}^\vee\delta, {}^dB)\}$ as *L-group* or *E-group* structure on ${}^\vee G^\Gamma$. Note that the choice of L-group structure on ${}^\vee G$ is precisely what was needed for the isomorphism (1-4). Now the parametrization of representations of G via the L-group ${}^\vee G^\Gamma$ depends on the choice of an L-group structure on ${}^\vee G^\Gamma$; two such parametrizations differ by tensoring with a one-dimensional representation of G, trivial on the identity component. Thus we note that given such a structure, then the other possible such choices are parametrized by

$$\{z \epsilon Z(G) | z\theta(z) = 1\} / \{w\theta(w^{-1}) | w \epsilon Z(G)\}. \tag{1-12}$$

We also note that the choice of L-group structure amounts to choosing a large fundamental series representation of a quasisplit form of ${}^\vee G$, which in the duality of §2 will correspond to the trivial representation of a quasisplit form of G. Equivalently the choice is that of a one-dimensional representation of a quasisplit form of G, trivial on the identity component.

(1-13) **Example:** *Tori.*

(1) If G is a compact torus, and ${}^\vee G^\Gamma$ is given, then there is a unique L-group structure on ${}^\vee G^\Gamma$, and no choice is necessary. Note that ${}^\vee\delta$ is conjugate to $-{}^\vee\delta$.
(2) If G is a split torus, then there are two choices for the L-group structure on ${}^\vee G^\Gamma$, given by $\pm {}^\vee\delta$. The corresponding parametrizations of representations of $\mathbf{R}^*$ differ by tensoring with sgn. □

(1-14) **Example:** *SL(2).* For now and for later use we make some choices in $SL(2)$ and $PGL(2)$. Let B^+ (resp. B^-) denote the upper (resp. lower) triangular matrices in $SL(2)$. Let T_c be the diagonal matrices, and let T_s be $SO(2, \mathbf{C})$ embedded in the usual way. Fix any Borel subgroup B_s containing T_s. We use similar notation for $PGL(2)$, which by abuse of notation we think of as a set 2×2 matrices, and append the prefix ^{d} for the same objects on the dual side.

(1) If G is $SL(2)$, then $\{g \cdot ({}^\vee\delta, {}^dB^+) | g \in G\}$ is the unique choice

of L-group structure on ${}^\vee G^\Gamma$. Note that $({}^\vee\delta, {}^dB^+)$ is conjugate to $({}^\vee\delta, {}^dB^-)$.

(2) If G is $PGL(2)$ then the two choices of L-group structure on ${}^\vee G$ are $\{({}^\vee\delta, {}^dB^+)\}$ and $\{({}^\vee\delta, {}^dB^-)\} = \{(-{}^\vee\delta, {}^dB^+)\}$. The resulting parametrizations of representations of $PU(1,1)$ differ by tensoring with sgn. Note that $SL(2,\mathbf{R})$ has two large discrete series representations, and $PGL(2,\mathbf{R})$ has two one-dimensional representations (trivial on the identity component). □

We next discuss real forms of a given group, which we parametrize via their Cartan involutions. Thus given a real form of G, we let θ be a corresponding Cartan involution, which is a holomorphic involution of G; any such involution is the Cartan involution of some real form [7, §VI].

Let $K = G^\theta$. Given θ we recover $G(\mathbf{R})$ as a real form of G which satisfies: $G(\mathbf{R}) \cap K$ is a maximal compact subgroup of $G(\mathbf{R})$. Then the subgroup $G(\mathbf{R})$ of G is determined up to conjugation by K.

Let G^Γ be an extended group containing G. Let x be an element of $G^\Gamma - G$ such that x^2 is contained in $Z(G)$, and let $\overline{x}$ denote the coset $xZ(G)$. Define an involution θ_x of G by $\theta_x(g) = xgx^{-1}$. This defines a map from pairs $(G^\Gamma, \overline{x})$ to involutions of G. This is a bijection between equivalence classes of pairs $(G^\Gamma, \overline{x})$ and equivalence classes of real forms. Here equivalence is defined by conjugation by G, i.e. the action of $\mathrm{Int}(G)$ on G. (This is slightly weaker than the usual notion, which allows the action of $\mathrm{Aut}(G)$ (cf. Example (1-19(4c)).)

(1-15) Definition. *A strong real form of G is a pair (G^Γ, x), for G^Γ an extended-group containing G, and $x \epsilon G^\Gamma - G$, $x^2 \epsilon Z(G)$. Two strong real forms (G^Γ, x) and (G^Γ, y) are said to be equivalent if y is conjugate to x by an element of G.*

If G^Γ is given, we will refer to x as a strong real form.

Note that if $Z(G) = 1$, then the notions of equivalence classes of strong real forms and real forms coincide.

(1-16) **Example:** *Tori.* Let G be a one-dimensional torus defined over $\mathbf{R}$. If G is compact, then G has infinitely many strong real forms, parametrized by $\{z \in G \simeq \mathbf{C}^*\}$. However, if G is split, then the elements of $G^\Gamma - G$ are all conjugate and G has only one strong real form (up to equivalence). Note that in this case $x^2 = 1$ for all $x \in G^\Gamma - G$ (cf. Example (1-6)). □

(1-17) **Example:** *Inner forms of a compact group.* Suppose G is defined over $\mathbf{R}$, and is inner to a compact group, so σ acts trivially on G (Lemma (1-5)). Then (using the Langlands version of G^Γ) we see that the set of equivalence classes of strong real forms is parametrized by $\{g \in G | g^2 \in Z(G)\}/G$. Given a Cartan subgroup T of G, let $X_* = X_*(T)$, and let P_* be the co-weights of T in G (cf. the end of this section). Choosing T to be compact, we may conjugate g into T, and we see the above set is in bijection with:

$$\left(\tfrac{1}{2}P_*/X_*(T)\cap\tfrac{1}{2}P_*\right)/W \tag{1-18}$$

(where W is the Weyl group of T in G) which is computable. Note that G has $|Z(G)|$ strong real forms which are compact, represented by the elements of $Z(G)$. The elements zm_ρ ($z\in Z(G)$) give strong real forms which are quasi-split, some of which may be equivalent to each other. In particular, suppose G is split and consider $Z(G)$ as contained in a split Cartan T_s. Since θ acts by $t\to t^{-1}$ on T_s, it follows that the elements $zm_\rho\delta$ are all conjugate, and there is precisely one split strong real form. Of course if G is adjoint then strong real forms are the same as real forms.

For example if G is $SL(2)$, then G has two real forms, but three strong real forms corresponding to the elements $I,\mathrm{diag}(i,-i)$, and $-I$ of $SL(2)$; or $0,\alpha/4$, and $\alpha/2$ in $\frac{1}{2}P_*(G,T)$ respectively, where α is a root. These correspond to the real forms $SU(2),SU(1,1)$, and $SU(2)$ respectively. It is convenient to think of these as $SU(2,0),SU(1,1)$, and $SU(0,2)$. If G is $PGL(2)$ then G has two strong real forms corresponding to $PU(1,1)\simeq PGL(2,\mathbf{R})$ and $PU(2)\simeq SO(3)$ respectively. □

(1-19) **Example:** *(continued)*. We compute the strong real forms of most simple groups which are inner to a compact group (the only groups not covered are intermediate covering groups of $PU(n)$ and $PO(2n)$).

As is the case above for $SL(2)$ we somewhat casually count strong real forms of $SU(n+1)$ by distinguishing between $SU(p,q)$ and $SU(q,p)$. Thus, to be precise, case (1a) should read: " … $SU(n+1-k,k)$ counted $n+1$ times ($0\le k<\frac{n+1}{2}$) and $SU(\frac{n+1}{2},\frac{n+1}{2})$ counted $\frac{n+1}{2}$ times (if n is odd)". Similar comments hold for the orthogonal and symplectic groups.

(1) Type A_n:
   (a) $SU(n+1)$ has $(n+1)(n+2)/2$ (strong real) forms, corresponding to the real forms $SU(n+1-k,k)$ ($0\le k\le n+1$) counted $\frac{n+1}{2}$ times.
   (b) $PU(n+1)$ has $[\frac{n+1}{2}]+1$ forms $PU(n+1-k,k)$ ($0\le k\le\frac{n+1}{2}$).
(2) Type B_n:
   (a) $Spin(2n+1)$ has $n+2$ forms, given by $Spin(2n+1-2k,2k)$ ($0\le k\le n$), with $Spin(2n+1,0)$ counted twice,
   (b) $SO(2n+1)$ has $n+1$ forms $SO(2n+1-2k,2k)$ ($0\le k\le n$).
(3) Type C_n:
   (a) $Sp(2n)$ has $n+2$ forms, given by $Sp(n-k,k)$ ($0\le k\le n$) and $Sp(2n,\mathbf{R})$,
   (b) $PSp(2n)$ has $[\frac{n}{2}]+2$ forms $PSp(n-k,k)$ ($0\le k\le n/2$) and $PSp(2n,\mathbf{R})$.
(4) Type D_n ($n>1$):
   (a) $Spin(2n)$ has $n+7$ forms $Spin(2n-2k,2k)$ ($0\le k\le n$) with $Spin(2n,0)$ and $Spin(0,2n)$ each counted twice, and $Spin^*(2n)$ counted four times,

(b) $SO(2n)$ has $n+3$ forms $SO(2n-2k,2k)$ $(0 \leq k \leq n)$, and $SO^*(2n)$ counted twice,
(c) $PO(2n)$ has $\left[\frac{n}{2}\right]+3$ forms $PO(2n-2k,2k)$ $(0 \leq k \leq n/2)$, and $PO^*(2n)$ counted twice.

(5) F_4 has 3 forms as usual, one split, one compact, and one of rank one.
(6) G_2 has two forms as usual, one split and one compact.
(7) E_6(adjoint) has 3 forms, and E_6(simply connected) has 9 forms, with each form counted 3 times each.
(8) E_7(adjoint) has 4 forms, and E_7(simply connected) has 6 forms; the split, compact, Hermitian symmetric, and remaining forms are each counted 1,2,1, and 2 times respectively.
(9) E_8 has 3 forms as usual. □

Given (G^Γ, x) let $\theta = \theta_x$ be the Cartan involution $\theta(g) = xgx^{-1}$ of G, and let $K_x = G^\theta$. Write $G(\mathbf{R})_x$ for a real form of G with Cartan involution θ_x (i.e. $G(\mathbf{R})_x \cap K_x$ is a maximal compact subgroup of $G(\mathbf{R})_x$); this is defined up to conjugation by K_x. This defines a map, not necessarily injective, from equivalence classes of strong real forms to equivalence classes of real forms. If x is fixed we write θ, K and $G(\mathbf{R})$. Note that the inner class of G is determined by G^Γ and the equivalence class of its real form by the coset $xZ(G)$.

We recall some facts about Cartan subgroups of G^Γ ([**2**, §9]). By definition a Cartan subgroup of an extended group G^Γ is a subgroup T^Γ generated by a Cartan subgroup T of G, and an element $x \in G^\Gamma - G$, with $x^2 \in T$. If x is a strong real form of G, T^Γ is said to be θ_x-stable. Let ${}^dT^\Gamma$ be a ${}^\vee\theta$-stable Cartan subgroup of ${}^\vee G^\Gamma$, where ${}^\vee\theta = \mathrm{int}({}^\vee\delta)$ for ${}^\vee\delta \in {}^\vee\mathfrak{D}$. Let ${}^\vee T^\Gamma$ be the L-group of a Cartan subgroup T of G (with real form given by $\theta|_T$), with ${}^\vee\theta_T$ its Cartan involution. Then choosing Borel subgroups B and ${}^dB$ of G and ${}^\vee G$, containing T and ${}^dT$, we obtain an isomorphism

$$\zeta = \zeta_{B,{}^dB} : {}^\vee T \rightarrow {}^dT \tag{1-20}$$

([**2**, 9.7]). By changing Borel subgroups this may be modified arbitrarily by the Weyl groups of T or ${}^dT$. We say T^Γ is dual to ${}^dT^\Gamma$ (via $(B, {}^dB)$) if ζ takes ${}^\vee\theta_T$ to $({}^\vee\theta)|_{{}^dT}$. Given x we say ${}^dT^\Gamma$ is relevant if there exists a θ_x-stable Cartan subgroup T^Γ such that T^Γ and ${}^dT^\Gamma$ are dual.

Let $Q^*(G,T) \subset X^*(T) \subset P^*(G,T)$ denote the root lattice, the character lattice, and the weights respectively. Let $Q_*(G,T) \subset X_*(T) \subset P_*(G,T)$ denote the co-root lattice, the lattice of one-parameter subgroups, and the co-weights respectively. Given T^Γ dual to ${}^dT^\Gamma$ we obtain isomorphisms $X^*(T) \simeq X_*({}^dT)$, $W(G,T) \simeq W({}^\vee G, {}^dT)$, and so on, which we also label ζ or $\zeta_{B,{}^dB}$. As the above notation indicates, we use the superscript ^{d} to indicate an object having to do with ${}^\vee G$, not necessarily originating from G in any specified way. Given ζ, we write ${}^\vee$ to indicate objects for ${}^\vee G^\Gamma$

coming from G via ζ. Thus ${}^d\alpha$ denotes a typical root of ${}^dT$ in ${}^\vee G$; whereas for α a root of T in G, $\alpha^\vee$ is the co-root of T in G, and ${}^\vee\alpha$ is the co-root $\zeta(\alpha^\vee)$ of ${}^dT$ in ${}^\vee G$.

Given T^Γ and ${}^dT^\Gamma$ dual via $(B, {}^dB)$, we obtain an isomorphism ${}^\vee T \simeq {}^dT$ as above. This induces isomorphisms $X^*(T) \simeq X_*({}^dT)$, $W(G,T) \simeq W({}^\vee G, {}^dT)$, etc. We denote these isomorphisms ζ, or $\zeta_{B,{}^dB}$ if it is necessary to specify the Borel subgroups.

2. Character Duality

In this section we reformulate the results of [**13**]. Given G defined over $\mathbf{R}$, let ${}^\vee G^\Gamma$ and G^Γ be L- groups for G and ${}^\vee G$ as in §1. We begin by describing maps of the Weil group into ${}^\vee G^\Gamma$. Let $W_{\mathbf{R}}$ be the Weil group of $\mathbf{C}$ over $\mathbf{R}$; it is generated by $\mathbf{C}^*$ and a distinguished element j. These are subject to the relations

(2-1)
$$j^2 = -1 \in \mathbf{C}^*$$
$$jzj^{-1} = \overline{z}.$$

We consider triples $(y, {}^dT^\Gamma, \lambda)$ satisfying

(2-2)

(1) $y \in {}^\vee G^\Gamma - {}^\vee G$, $y^2 \in Z({}^\vee G)$,
(2) ${}^dT^\Gamma$ is a Cartan subgroup of ${}^\vee G^\Gamma$ containing y,
(3) $\lambda \in {}^d\mathfrak{t} \simeq X_*({}^dT) \otimes \mathbf{C}$ (where ${}^d\mathfrak{t} = Lie({}^dT)$), and
(4) $\exp(2\pi i\lambda) = y^2$.

Define a homomorphism $\phi = \phi(y, {}^dT^\Gamma, \lambda)$ from $W_{\mathbf{R}}$ to ${}^\vee G^\Gamma$ by

(2-3)

(1) $\phi(z) = z^\lambda \overline{z}^{Ad(y)\lambda}$ for $z \epsilon \mathbf{C}^* \subseteq W_{\mathbf{R}}$, (where $z^\lambda = \exp(\lambda \log(z))$),
(2) $\phi(j) = e^{-\pi i \lambda} y$.

From the relations (2-1) defining $W_{\mathbf{R}}$ it is straightforward to see that $\phi(y, {}^dT^\Gamma, \lambda)$ is a group homomorphism. It is quasi-admissible ([**2**, Definition 9.8])— it is continuous, commutes with projection on Γ, and the image of $\mathbf{C}^*$ consists of semi-simple elements.

If x is a strong real form of G, we let Π_ϕ be the L-packet of $(\mathfrak{g}, K_x)$-modules defined by (the conjugacy class of) ϕ (cf. [**2,5,8**]). This is a finite set of irreducible $(\mathfrak{g}, K_x)$-modules, all having the same infinitesimal character. We adopt the convention that Π_ϕ is empty if ϕ is not admissible for the given real form.

If $\lambda \in X^*(T) \otimes \mathbf{C}$ we let $\chi(\lambda)$, or simply λ, denote the corresponding infinitesimal character for G via the Harish-Chandra homomorphism. We use the same notation for $\lambda \in X_*({}^dT) \otimes \mathbf{C}$ via an isomorphism $X^*(T) \simeq X_*({}^dT)$. If $\phi = \phi(y, {}^dT^\Gamma, \lambda)$, then the representations in Π_ϕ have infinitesimal character $\chi(\lambda)$.

(2-4) **Example:** *Tori.*

(1) If G is a one-dimensional compact torus, then up to conjugation we may take $y = {}^\vee\delta$, and to specify ϕ it is enough to specify λ satisfying $e^{2\pi i\lambda} = 1$. Thus $\lambda \in \mathbf{Z}$ (identifying $X_*({}^dT)$ with $\mathbf{Z}$), $\phi(z) = (z/\bar{z})^\lambda$, and $\phi(j) = {}^\vee\delta$.

(2) Suppose that G is split. Then we need to specify y and λ satisfying $e^{2\pi i\lambda} = y^2$. Writing $y = y_0\,{}^\vee\delta$ this says $\varepsilon = e^{-\pi i\lambda}y_0 = \pm 1$, so it is enough to choose λ and ε. Note this depends on the choice of ${}^\vee\delta \in {}^\vee G^\Gamma - {}^\vee G$ (cf. (1-13(2))). Then $\phi(z) = (z\bar{z})^\lambda$, and $\phi(j) = \varepsilon\,{}^\vee\delta$. $\square$

(2-5) **Example:** *SL(2).* Let G be $SL(2)$, and use the notation of Example (1-14). Up to conjugation the choices of $(y, {}^dT^\Gamma, \lambda)$ are given as follows. Fix $({}^\vee\delta, {}^dB) \in {}^\vee\mathfrak{D}$.

(1) $({}^\vee\delta, {}^dT_s^\Gamma, k)$ with $k \in X_*({}^dT_s) \simeq \mathbf{Z}$.

(2) $(y_0\,{}^\vee\delta, {}^dT_c^\Gamma, z)$ with $y_0 \in {}^dT_c \simeq \mathbf{C}^*$, $z \in {}^d\mathfrak{t}_c \simeq \mathbf{C}$, and $e^{2\pi iz} = y_0^2$. $\square$

(2-6) Definition. *Given ${}^\vee z \in Z({}^\vee G)^{{}^\vee\theta}$, we say an infinitesimal character χ is associated to ${}^\vee z$ if $\chi = \chi(\lambda)$ for some $\lambda \in X_*({}^dT) \otimes \mathbf{C}$ satisfying* $\exp(2\pi i\lambda) = {}^\vee z$.

We say that χ is integral if $\lambda \in X_*({}^dT)$. If χ is regular, this is equivalent to the condition that χ is the infinitesimal character of a finite dimensional representation of $\mathfrak{g}$. Note that χ is integral if and only if $y^2 \in Z({}^\vee G)$—i.e., y is a strong real form of ${}^\vee G$. This is an important observation: the representation theory of G with infinitesimal character χ associated to y^2 is controlled by the centralizer of y^2 in ${}^\vee G$ (and the strong real forms of this centralizer). By limiting ourselves to integral infinitesimal character this centralizer is always simply ${}^\vee G$.

Furthermore note that if λ satisfies (2-2(4)) then λ may be replaced by $\lambda' = \lambda + \gamma$ for any $\gamma \in X_*({}^dT)$. This corresponds to the translation principle applied to representations: the categories of representations with infinitesimal characters $\chi(\lambda)$ and $\chi(\lambda')$ are isomorphic (assuming they are both regular). Thus we will be working "modulo the translation principle". Therefore in place of the data $(y, {}^dT^\Gamma, \lambda)$ we will use $(y, {}^dT^\Gamma, {}^dB)$ for ${}^dB$ a Cartan subgroup containing ${}^dT$. This data specifies a "translation family" of representations (cf. [1, §2]) as follows. If an infinitesimal character χ associated to y^2 has been specified, then we recover λ as the unique ${}^dB$-dominant element of ${}^d\mathfrak{t}$ satisfying $\chi = \chi(\lambda)$. Varying over all such χ we obtain a translation family of representations. Conversely, given a translation family, we may specialize it to a given infinitesimal character.

Therefore, for the remainder of the paper, we will assume without further mention that given y, we have chosen a regular infinitesimal character χ associated to y^2, and that all representations of G have this infinitesimal character.

The integral infinitesimal characters for G are therefore parametrized (up to translation) by $Z({}^\vee G)^{{}^\vee\theta}$ where ${}^\vee\theta$ is any involution in the given inner class. Given ${}^\vee z \in Z(G)^{{}^\vee\theta}$, we associate to ${}^\vee z$ the strong real forms y of ${}^\vee G$ with $y^2 = {}^\vee z$, which by the discussion above control the representation theory of G with this infinitesimal character.

(2-7) **Example:** *SL(2), SO(2n).*

(1) If G is $SL(2)$, then the only infinitesimal character is ρ, which corresponds to the (strong real) forms $PGL(2)$ and $PU(2)$ of ${}^\vee G$. However, if G is $PGL(2)$, then $-I \in {}^\vee G = SL(2)$ corresponds to infinitesimal character ρ for G and the form $SU(1,1)$, whereas I corresponds to 2ρ and the forms $SU(2,0)$ and $SU(0,2)$.
(2) If G is $SO(2n+1)$, then infinitesimal character ρ corresponds to the form $Sp(2n,\mathbf{R})$ of ${}^\vee G$, and 2ρ corresponds to the forms $Sp(k,n-k)$.
(3) If G is $SO(2n)$, then ρ corresponds the the form $SO^*(2n)$, and 2ρ to the forms $SO(2n-2k,2k)$. If G is $PSO(2n)$, then there are four infinitesimal characters for G. These correspond to four classes of forms of ${}^\vee G = Spin(2n)$:
$\{Spin(2n-2k,2k) | k \in 2\mathbf{Z}\}$, $\{Spin(2n-2k,2k) | k \in 2\mathbf{Z}+1\}$,
$\{Spin^*(2n)$ (2 copies)$\}$, and $\{Spin^*(2n)$ (2 copies)$\}$
respectively.

□

We now wish to extend the data $(y, {}^dT^\Gamma, {}^dB)$ in such a way as to specify a unique representation within an L-packet. We recall the construction of representations of ([**2**, §3]). Given $\phi : W_{\mathbf{R}} \to {}^\vee G^\Gamma$, we consider ϕ as a map into a Cartan subgroup ${}^dT^\Gamma$ of ${}^\vee G^\Gamma$. Now ${}^dT^\Gamma$ is isomorphic to an E-group ${}^\vee T^\Gamma$ of a Cartan subgroup T of G. Choosing such an isomorphism ι, we obtain

$$\iota \circ \phi : W_{\mathbf{R}} \to {}^\vee T^\Gamma, \tag{2-8}$$

associated to which is a character Λ of a certain two-fold cover of T. Using cohomological induction we obtain a standard $(\mathfrak{g}, K_x)$-module I, with unique irreduible submodule J. These modules depend on the choice of the isomorphism ι; the resulting set of irreducible representations $\{J\}$ forms the L-packet attached to ϕ. The following definition incorporates the choice necessary to specify ι, and hence determines a single representation within the given L-packet (cf. [**1**, Def. 3.8]).

(2-9) Definition. *A set of L-data (for G^Γ and ${}^\vee G^\Gamma$) is a 6-tuple*

$$S = (x,\ T^\Gamma,\ B,\ y, {}^dT^\Gamma,\ {}^dB)$$

subject to the following conditions:

a. *T^Γ is a Cartan subgroup of G^Γ, and x is an element of $T^\Gamma - T$,*

b. *${}^dT^\Gamma$ is a Cartan subgroup of ${}^\vee G^\Gamma$, and y is an element of ${}^dT^\Gamma - {}^dT$,*
d. *$B \supseteq T$ is a Borel subgroup of G,*
e. *${}^dB \supseteq {}^dT$ is a Borel subgroup of ${}^\vee G$,*
f. *Let $\zeta = \zeta_{B,{}^dB}$ be the isomorphism ${}^dT \simeq {}^\vee T$ (cf. §1). Then we assume that ζ takes the inverse transpose of $\mathrm{int}(x)|_T$ to $\mathrm{int}(y)|_{{}^dT}$,*
g. *$x^2 \epsilon Z(G)$,*
h. *$y^2 \epsilon Z({}^\vee G)$.*

Two sets S and S' are said to be equivalent if S is conjugate to S' under the obvious action of $G \times G'$.

We say an infinitesimal character χ is associated to S if it is associated to y^2.

Given S as in Definition (2-9), fix an infinitesimal character χ associated to S. Let $\phi = \phi(y, {}^dT^\Gamma, \lambda) : W_{\mathbf{R}} \to {}^\vee G^\Gamma$, where λ is defined as in the discussion preceding Example (2-7). Choose a set Ψ of positive real roots of T in G. Sending ${}^\vee\delta$ to an element ${}^\vee\delta_T$ of ${}^\vee T^\Gamma - {}^\vee T$, which makes $\zeta(\Psi)$ a special set of positive imaginary roots of ${}^dT$ in ${}^\vee G$ (cf. [**2**, 6.29]), we obtain an isomorphism $\iota : {}^dT^\Gamma \simeq {}^\vee T^\Gamma$. As in the discussion preceding Definition (2-9), we obtain a character Λ, with differential λ, of the ρ two-fold cover of the real torus T (with real structure determined by $\mathrm{int}(x)|_T$).

Define

$$I(S) = I(\Psi, \Lambda), \tag{2-10}$$

a standard representation ([**2**, Definition 8.27]). This is defined via the reader's favorite construction of standard modules, i.e. parabolic or cohomological induction. (We assume that devotees of D-modules are no longer reading.) Define

$$J(S) = \text{unique irreducible submodule of } I(S). \tag{2-11}$$

This is independent of the choice of Ψ ([**1**, Lemma 2.10]).

Attached to a set of L-data is a $(\mathfrak{g}, K_x)$-module. Equivalence of L-data corresponds to equivalence of modules as follows. We consider pairs (x, π), where x is a strong real form of G, and π is a $(\mathfrak{g}, K_x)$-module. We define equivalence of such pairs by the natural action of conjugation by G.

(2-12) Theorem ([2, Theorem 9.11]).

Let ${}^\vee G^\Gamma$ be an L-group for G. Fix an element ${}^\vee z \in Z({}^\vee G)^{{}^\vee\theta}$, and fix a regular (integral) infinitesimal character χ associated to ${}^\vee z$ (Definition (2-6)). Then the map from L-data S to the pair $(x, J(S))$ defines a bijection between equivalence classes of L-data with $y^2 = {}^\vee z$, and equivalence classes of pairs (x, π), where x is a strong real form of G and π is an irreducible $(\mathfrak{g}, K_x)$-module of infinitesimal character χ.

(2-13) **Example:** *Tori.*

(1) Let G be a compact one-dimensional torus, so that ${}^\vee G$ is a split torus. In this case L-data reduces to a pair of elements (x, y) and,

since y is necessarily conjugate to ${}^\vee\delta$, we take $y = {}^\vee\delta$. Then x simply determines the strong real form of G. The corresponding representation π is determined by its infinitesimal character, which we may take to be trivial, and hence π is the trivial representation (here $G(\mathbf{R}) \simeq S^1$).

(2) On the other hand suppose G is split and ${}^\vee G$ is compact. Then we take $x = \delta$. G has a unique strong real form and $G(\mathbf{R}) \simeq \mathbf{R}^*$. Choose infinitesimal character $\chi(\lambda)$ associated to y^2. This condition says $e^{2\pi i\lambda} = y^2$, or (writing $y = y_0\,{}^\vee\delta$ as in (2-4)) $e^{-\pi i\lambda}y_0 = \pm 1$. The representation π is then determined by its infinitesimal character χ, and $\pi(-1) = e^{-\pi i\lambda}y_0$. Note that replacing λ by $\lambda + k$ for $k \in \mathbf{Z}$ replaces $\pi(-1)$ with $\pi(-1)(-1)^k$, which corresponds to tensoring π with the restriction of the holomorphic representation $z \to z^k$ of $\mathbf{C}^*$ to $\mathbf{R}^*$; this confirms that this construction behaves correctly under the translation principle. □

(2-14) **Example:** *SL(2).* Fix the strong real form δ of $G = SL(2)$. We use the notation of Examples (1-14) and (2-5). For use here and later we let $t_0 = \mathrm{diag}(i, -i)$ in either $SL(2)$ or $PGL(2)$. Since we are considering only integral infinitesimal character, up to conjugation we may take $y = {}^\vee\delta$ or $y = t_0\,{}^\vee\delta$, and fix $\chi = \rho$. Then (up to conjugation) the choices of L-data are given as follows.

(1) $(\delta, T_c^\Gamma, B^+, {}^\vee\delta, {}^dT_s^\Gamma, {}^dB_s)$ corresponds to the holomorphic discrete series of the strong real form of G given by δ.
(2) $(\delta, T_c^\Gamma, B^-, {}^\vee\delta, {}^dT_s^\Gamma, {}^dB_s)$ corresponds to the anti-holomorphic discrete series.
(3) $(\delta, T_s^\Gamma, B_s, {}^\vee\delta, {}^dT_c^\Gamma, {}^dB_s)$ corresponds to the trivial representation.
(4) $(\delta, T_s^\Gamma, B_s, t_0\,{}^\vee\delta, {}^dT_c^\Gamma, {}^dB_s)$ corresponds to the irreducible principal series (with odd K-types). (If we were to take χ to be 0, this would be the irreducible spherical principal series.)

Note that the L-data

$$S = (\delta, T_c^\Gamma, B^+, {}^\vee\delta, {}^dT_s^\Gamma, {}^dB_s)$$

is conjugate to

$$S' = (-\delta, T_c^\Gamma, B^-, {}^\vee\delta, {}^dT_s^\Gamma, {}^dB_s).$$

Thus it is important that we are keeping track of the strong real forms δ and $-\delta$, and not just their coset mod $Z(G)$; without this distinction Theorem (2-12) would be false in this example. □

We can now conveniently define "super" L-packets. Given a quasi-admissible homomorphism $\phi : W_\mathbf{R} \to {}^\vee G^\Gamma$ and a strong real form x of G, we a obtain a (possibly empty) L-packet Π_ϕ of $(\mathfrak{g}, K_x)$-modules. We write $\Pi_{\phi,x}$ to indicate the dependence on x.

(2-15) Definition. *The super L-packet $\mathbf{\Pi}_\phi$ associated to ϕ is the union of the sets $\Pi_{\phi,x}$ as x varies over a set of representatives of equivalence classes of strong real forms of G.*

Suppose now that $\phi = \phi(y, {}^dT^\Gamma, {}^dB)$. Let $S = (x, T^\Gamma, B, y, {}^dT^\Gamma, {}^dB)$ be a set of L-data, so that $J(S)$ belongs to $\mathbf{\Pi}_\phi$. Up to conjugation any such data is given by some fixed T^Γ and B. It is therefore determined entirely by x, and x in turn is determined by S up to conjugation by T. The following Lemma is an immediate consequence.

(2-16) Lemma. *In the setting just described,*

(1) $\mathbf{\Pi}_\phi$ *is in bijection with*

$$\{x \in T^\Gamma | x^2 \in Z(G)\}/(\textit{conjugation by } T).$$

(2) *Fix δ in T^Γ–T with $\delta^2 \epsilon Z(G)$; for example, take δ in $\mathfrak{D} \cap T^\Gamma$, and write $\theta = \mathrm{int}(\delta)$. Then $\mathbf{\Pi}_\phi$ is in bijection with*

$$\{t \epsilon T | t\theta t \epsilon Z(G)\}/\{s\theta(s^{-1}) | s \epsilon T\}.$$

Notice that the second parameter set is a group.

(2-17) **Example:** *SL(2).* Let G be $SL(2)$, and $\mathbf{\Pi}$ be the super L-packet of discrete series (say with infinitesimal character ρ). Then $\mathbf{\Pi}$ contans four elements: two discrete series of $SU(1,1)$, and the trivial representations of $SU(2,0)$ and $SU(0,2)$.

On the other hand if G is $PGL(2)$, then $\mathbf{\Pi}$ has just two elements, the discrete series of $PU(1,1)$ and the trivial representation of $PU(2)$. □

(2-18) **Example:** *Super-Packets of Discrete Series.* Let $\mathbf{\Pi}$ be a super-packet of discrete series representations. Then ${}^\vee\theta$ acts on ${}^dT$ by $t \to t^{-1}$. It follows from (2-16(2)) that $\mathbf{\Pi}$ contains $|Z(G)|2^n$ elements, where n is the rank of G. In the formulas below we have terms of the form $\binom{n}{n/2}$ which are to be ignored if n is odd.

(A_n) $SU(n+1)$: Since $SU(n+1-k,k)$ has $\binom{n+1}{k}$ discrete series representations, $\mathbf{\Pi}$ has $\sum_0^{n+1} \frac{n+1}{2}\binom{n+1}{k} = \frac{n+1}{2}2^{n+1} = (n+1)2^n$ discrete series representations. However if G is $PU(n)$ then each term in the sum is divided by $n+1$ and $\mathbf{\Pi}$ has 2^n elements.

(B_n) $Spin(2n+1)$: $\mathbf{\Pi}$ has $1+1+\sum_1^n 2\binom{n}{k} = 2^{n+1}$ elements, which for $SO(2n+1)$ becomes $\sum_0^n \binom{n}{k} = 2^n$.

(C_n) $Sp(2n)$: $\mathbf{\Pi}$ has $\sum_0^n \binom{n}{k} + 2^n = 2^{n+1}$ elements. If G is $PSp(2n)$ this becomes $\sum_0^{[\frac{n-1}{2}]} \binom{n}{k} + \frac{1}{2}\binom{n}{n/2} + 2^{n-1} = 2^n$.

(D_n) $Spin(2n)$: $\mathbf{\Pi}$ has $1+1+\sum_1^{n-1} 2\binom{n}{k} + 4 \times 2^{n-1} = 2^{n+2}$ elements. For $SO(2n)$ this becomes $\sum_0^n \binom{n}{k} + 2 \times 2^{n-1} = 2^{n+1}$, whereas for $PO(2n)$ we get $\sum_0^{[\frac{n-1}{2}]} \binom{n}{k} + [\frac{1}{2}\binom{n}{n/2}] + 2 \times 2^{n-2} = 2^n$ elements.

(F_4) Π has $12+3+1=2^4$ elements.
(G_2) Π has $3+1=2^2$ elements.
(E_6) Π has $36+27+1=2^6$ elements in the adjoint case, and $3\times 36+3\times 27+3\times 1=3\times 2^6$ elements in the simply connected case.
(E_7) For the adjoint group Π has $36+1+28+63=2^7$ elements, and $72\times 1+1\times 2+56\times 1+63\times 2=2^8$ in the simply connected case.
(E_8) Π contains $1+135+120=2^8$ elements. □

We turn next to the consideration of blocks of $(\mathfrak{g},K_x)$-modules. Write $\mathfrak{V}(\mathfrak{g},K_x)$ for the Grothendieck group of finite length $(\mathfrak{g},K_x)$-modules, or *virtual modules.* This is a free $\mathbf{Z}$-module having as basis the set of irreducible $(\mathfrak{g},K_x)$-modules. The set of standard modules that admit a unique irreducible submodule also form a basis of $\mathfrak{V}(\mathfrak{g},K_x)$. Passage to this submodule defines a bijection from the standard modules to irreducible modules. For the next definition, fix a regular infinitesimal character for G, and let x be a strong real form of G. A block is minimal space of virtual modules on which the Kazhdan-Lusztig algorithm makes sense:

(2-19) Definition.

(1) *A block $\mathfrak{B}$ of $(\mathfrak{g},K_x)$-modules is a minimal $\mathbf{Z}$-module of virtual modules (with the given infinitesimal character), with the following properties:*
  (a) *If a standard module I is contained in $\mathfrak{B}$, then the irreducible submodule of I is contained in $\mathfrak{B}$,*
  (b) *If an irreducible module π is contained in $\mathfrak{B}$, then the standard module containing π is contained in $\mathfrak{B}$.*

(2) *Fix a block $\mathfrak{B}$. We let $\mathfrak{B}_{irr}$ (respectively $\mathfrak{B}_{std}$) be the set of irreducible (resp. standard) representations contained in $\mathfrak{B}$.*

Note that $\mathfrak{B}_{irr}$ and $\mathfrak{B}_{std}$ are each a basis of $\mathfrak{B}$. We will often think of a block $\mathfrak{B}$ as being the set of irreducible (or standard) modules in a basis of $\mathfrak{B}$.

We may think of an L-packet Π as being a $\mathbf{Z}$-module, given by the $\mathbf{Z}$-span of the representations in Π. Each L-packet is contained in a single block.

(2-20) **Example:** *Finite dimensional representations.* Let $\mathfrak{B}$ be the block containing the finite dimensional representations of a quasi-split (strong) real form of G (so if the infinitesimal character is ρ this contains the trivial representation of G). Then $\mathfrak{B}$ also contains the fundamental series of G. This is the largest block for (any real form of) G. □

(2-21) **Example:** *Singletons.* An irreducible standard module may be a block by itself. For example this holds for a finite-dimensional representation of a compact group, and also for an irreducible minimal principal series of a split group. □

(2-22) Example: *SL(2).* We establish some notation to be used repeatedly. Fix infinitesimal character ρ for $SL(2)$ and $PGL(2)$.

$SL(2)$: Let π_+ (resp. π_-) denote the holomorphic (resp. antiholomorphic) discrete series representation of $SL(2,\mathbf{R}) = SU(1,1)$. Let $\pi^{2,0}_{trivial}$ (resp. $\pi^{0,2}_{trivial}$, $\pi^{1,1}_{trivial}$) denote the trivial representation of $SU(2,0)$ (resp. SU(0,2), $SU(1,1)$). Let ps_e (resp. ps_o) denote the even (resp. odd) principal series representation of $SU(1,1)$. Then (in the Grothendieck group) $ps_e = \pi_+ + \pi_- + \pi^{1,1}_{trivial}$, and ps_o is irreducible.

For a basis of the block $\mathfrak{B}$ of the trivial representation of $SU(1,1)$ we may take $\mathfrak{B}_{irr} = \{\pi_+, \pi_-, \pi^{1,1}_{trivial}\}$ or $\mathfrak{B}_{std} = \{\pi_+, \pi_-, ps_e\}$. These two bases are related via the formulas:

$$\begin{aligned} \pi_+ &= \pi_+ & \pi_+ &= \pi_+ \\ \pi_- &= \pi_- & \pi_- &= \pi_- \\ ps_e &= \pi^{1,1}_{trivial} + \pi_+ + \pi_- & \pi^{1,1}_{trivial} &= ps_e - \pi_+ - \pi_-. \end{aligned}$$

The singleton $\{ps_o\}$ itself is a block, as are $\{\pi^{2,0}_{trivial}\}$ and $\{\pi^{0,2}_{trivial}\}$.

$PGL(2)$: Let π_d be the discrete series representation of $PU(1,1)$; let $\pi^{1,1}_{trivial}$ and $\pi^{1,1}_{sgn}$ denote the trivial and sgn representations respectively. Let $\pi^{2,0}_{trivial}$ denote the trivial representation of $PU(2,0)$. Let $ps_{trivial}$ (resp. ps_{sgn}) be the (reducible) principal series representations of $PU(1,1)$ which are given by $ps_{trivial} = \pi_d + \pi^{1,1}_{trivial}$, and $ps_{sgn} = \pi_d + \pi^{1,1}_{sgn}$ respectively. Finally let $ps_\pm(2\rho)$ be the two (irreducible) principal series representations of $PU(1,1)$ with infinitesimal character 2ρ. These are distinguished by declaring $ps_+(2\rho)$ (resp. $ps_-(2\rho)$) contains the trivial (resp. sgn) representation when restricted to K.

Let $\mathfrak{B}$ be the block of the trivial representation of $PU(1,1)$. Then the irreducible and standard bases of $\mathfrak{B}$ are $\mathfrak{B}_{irr} = \{\pi_d, \pi^{1,1}_{trivial}, \pi^{1,1}_{sgn}\}$ and $\mathfrak{B}_{std} = \{\pi_d, ps_{trivial}, ps_{sgn}\}$. In this case the relations are:

$$\begin{aligned} \pi_d &= \pi_d & \pi_d &= \pi_d \\ ps_{trivial} &= \pi^{1,1}_{trivial} + \pi_d & \pi^{1,1}_{trivial} &= ps_{trivial} - \pi_d \\ ps_{sgn} &= \pi^{1,1}_{sgn} + \pi_d & \pi^{1,1}_{sgn} &= ps_{sgn} - \pi_d \end{aligned}$$

On the other hand $\{ps_+(2\rho)\}$ is a block, as is $\{ps_-(2\rho)\}$. □

(2-23) Theorem ([1, Theorem 3.16]). *Suppose*

$$S = (x,\ T^\Gamma,\ B,\ y, {}^dT^\Gamma,\ {}^dB), \text{ and}$$
$$S' = (x,\ (T^\Gamma)',\ B',\ y',\ ({}^dT^\Gamma)',\ {}^dB')$$

are two sets of strong L-data for the strong real form x of G^Γ. Then $J(S)$ is block equivalent to $J(S')$ if and only if y is conjugate to y'.

Recall that by strengthening the the data $(y, {}^dT^\Gamma, {}^dB)$ giving a map of $W_\mathbf{R}$ into ${}^\vee G^\Gamma$, we obtained a parametrization of irreducible representations in terms of L-data (Theorem (2-12)) . We now *weaken* this data to parametrize blocks.

(2-24) Definition. *Suppose $x \in G^\Gamma - G$ and $y \in {}^\vee G^\Gamma - {}^\vee G$, with x^2 and y^2 central in G and ${}^\vee G$ respectively. We say that the pair (x, y) is admissible if it can be extended to a set of L-data. Equivalently we require that there exist dual Cartan subgroups T^Γ and ${}^dT^\Gamma$ of G^Γ and ${}^\vee G^\Gamma$ containing x and y respectively.*

Two pairs are said to be equivalent if they are conjugate by $G \times {}^\vee G$.

(2-25) Corollary. *Equivalence classes of admissible pairs (x, y) are naturally in bijection with equivalence classes of pairs $(x, \mathfrak{B})$, with x a strong real form of G and $\mathfrak{B}$ a block of $(\mathfrak{g}, K_x)$-modules.*

We write

$$\mathfrak{B}(x, y) \tag{2-26}$$

for the block corresponding to the admissible pair (x, y).

(2-27) **Example:** *SL(2).* The blocks of Example (2-22) are obtained as follows:

(1) $G = SL(2)$:
  (a) $\mathfrak{B}(\delta, {}^\vee\delta)$ is the block $\{ps_e, \pi_+, \pi_-\}$ of the trivial representation of $SU(1,1)$,
  (b) $\mathfrak{B}(t_0\delta, {}^\vee\delta) = \{\pi^{2,0}_{trivial}\}$ for $SU(2,0)$,
  (c) $\mathfrak{B}(-t_0\delta, {}^\vee\delta) = \{\pi^{0,2}_{trivial}\}$ for $SU(0,2)$,
  (d) $\mathfrak{B}(\delta, t_0\, {}^\vee\delta) = \{ps_o\}$ for $SU(1,1)$.

(2) $G = PGL(2)$:
  (a) $\mathfrak{B}(\delta, {}^\vee\delta)$ is the block $\{\pi_d, ps_{trivial}, ps_{sgn}\}$ containing the trivial representation of $PU(1,1)$,
  (b) $\mathfrak{B}(\delta, t_o\, {}^\vee\delta) = \{ps_+(2\rho)\}$ for $PU(1,1)$,
  (c) $\mathfrak{B}(\delta, -t_o\, {}^\vee\delta) = \{ps_-(2\rho)\}$ for $PU(1,1)$,
  (d) $\mathfrak{B}(t_0\delta, {}^\vee\delta) = \{\pi^{2,0}_{trivial}\}$ for $PU(2,0)$. □

Now the data defining a representation (Theorem (2-12)) and that defining a block (Corollary (2-25) are symmetric in G and ${}^\vee G$. We use this to define duality.

(2-28) Definition.

(1) *Suppose $S = (x,\ T^\Gamma,\ B,\ y,\ {}^dT^\Gamma,\ {}^dB)$ is a set of L-data. The set of dual L-data is*

$$ {}^\vee S = (y,\ {}^dT^\Gamma,\ {}^dB,\ x,\ T^\Gamma,\ B).$$

Fix regular infinitesimal characters for G and ${}^\vee G$ associated to y^2 and x^2 respectively. Recall $I(S)$ and $J(S)$ are respectively the *standard and irreducible modules associated to S. We define the dual standard and irreducible modules to be $I({}^\vee S)$ and $J({}^\vee S)$.*

(2) *Suppose (x,y) is an admissible pair (Definition 2-24), corresponding to a block $\mathfrak{B}$ for G (Corollary 2-25). The dual block ${}^\vee\mathfrak{B}$ for ${}^\vee G$ is defined to be the block corresponding to the pair (y,x).*

Note that data $(x, y, {}^dT^\Gamma, {}^dB)$ defining an L-packet is *not* symmetric.

In accordance with the discussion at the end of §1, we will write ${}^d\pi$ for a typical irreducible module in ${}^\vee\mathfrak{B}$. On the other hand, for $\pi \in \mathfrak{B}$ we let ${}^\vee\pi \in {}^\vee\mathfrak{B}$ be the dual module of Definition (2-28). Similarly ${}^dI$ and ${}^\vee I$ will denote a typical standard module in ${}^\vee\mathfrak{B}$ and the dual to a standard module $I \in \mathfrak{B}$ respectively.

(2-29) **Example:** *Discrete Series.* Suppose G is an adjoint group of rank n, inner to a compact group. Then the strong forms of G have 2^n discrete series (cf. Example (2-18)). These are dual to the 2^n minimal principal series of the split form of ${}^\vee G$ obtained by fixing the infinitesimal character, and varying the inducing representation on ${}^\vee M \simeq (\mathbf{Z}/2\mathbf{Z})^n$. If G is not adjoint, we have $|Z(G)|$ such families, corresponding to different (translation families of) infinitesimal characters for ${}^\vee G$. □

(2-30) **Example:** *General Remarks.* Suppose π is a standard module obtained from a Cartan subgroup of G (with strong real form x) with compact (resp. split) part of dimension m (resp. n). Then ${}^\vee\pi$ is obtained from a Cartan subgroup of ${}^\vee G$ having compact (resp. split) part of dimension n (resp. m). In particular, if π is a discrete series representation, then ${}^\vee G$ contains a split Cartan subgroup and ${}^\vee\pi$ is a quotient of a minimal principal series. Note that the quotients of minimal principal series of ${}^\vee G$ lie in a number of different blocks, corresponding to discrete series of different strong forms x of G (all of which have discrete series). More generally, a similar statement holds relating the fundamental series of forms of G to quotients of minimal principal series of quasisplit form(s) of ${}^\vee G$. □

(2-31) **Example:** *The block of finite dimensional representations.* The block defined by $(\delta, {}^\vee\delta)$ is always the block of the finite dimensional representations of the quasisplit (strong) form of G. If the infinitesimal character is ρ, this is the block of the trivial representation. Similar statements hold by symmetry for ${}^\vee G$. □

(2-32) **Example:** *Singletons–cf. Example (2-32).* An irreducible minimal principal series of a split group (which is a block by itself) is dual to a finite dimensional representation of a compact group. (If we allow non-integral infinitesimal character this behavior becomes more common and is dual to representations of subgroups of ${}^\vee G$. For example, a minimal principal series of a split group G with generic infinitesimal character is dual to a character of a compact torus ${}^\vee T$.) □

(2-33) Example: *SL(2).* In Example (2-27), the blocks (a-d) for $SL(2)$ are dual to the corresponding blocks (a-d) for $PGL(2)$. □

Suppose $S = (x, T^\Gamma, B, y, {}^dT^\Gamma, {}^dB)$ is a set of L-data, $I = I(S)$, and $J = J(S)$. Let A be the split part of T (with respect to $\theta = \theta_x$). Let A_f be the split part of a fundamental Cartan subgroup of G, and let $K = K_x = G^\theta$ as usual. Fix a constant $c \in \mathbf{Z}$. Then the length of I is defined to be

$$\begin{aligned} \ell(I) = \frac{1}{2}|\{\alpha \text{ a root of T in B}|\theta\alpha \text{ is not a root in B}\}| \\ + \frac{1}{2}\dim A - \dim(A_f) + \frac{1}{2}\dim(G/K) + c, \end{aligned} \tag{2-34}$$

which is an integer. We define $\ell(S) = \ell(J) = \ell(I)$.

The results of this chapter are all independent of the normalization constant c. For definiteness and later use, we define $\ell_0(\pi)$ by (2-34) with $c = 0$. With this normalization the discrete series has length $\frac{1}{2}\dim(G/K)$, which is minimal.

For every pair (π, I) of an irreducible and a standard module in $\mathfrak{B}$, we define integers $M(I, \pi)$ and $m(\pi, I)$ so that (in the Grothendieck group)

$$\begin{aligned} I = \sum_{\pi \in \mathfrak{B}} m(\pi, I)\pi \\ \pi = \sum_{I \in \mathfrak{B}} M(I, \pi) I. \end{aligned} \tag{2-35}$$

The matrix m is the multiplicity matrix, and M is its inverse; since standard modules have relatively simple characters, the entries of M can be interpreted as coefficients in character formulas for irreducible modules. We define $m({}^d\pi, {}^dI)$ and $M({}^dI, {}^d\pi)$ on the dual block ${}^\vee\mathfrak{B}$ similarly. The next Theorem, Kazhdan-Lusztig duality for Harish-Chandra modules, relates these matrices for $\mathfrak{B}$ and ${}^\vee\mathfrak{B}$. It is the main result (Theorem 13.13) of [**13**] (cf. [**1**, Theorem 3.29]).

(2-36) Theorem. *Fix an admisible pair (x, y) with the corresponding pair of dual blocks $\mathfrak{B}$ and ${}^\vee\mathfrak{B}$. Suppose π is an irreducible module and I is a standard module in $\mathfrak{B}$. Let ${}^\vee\pi$ and ${}^\vee I$ be the dual irreducible and standard modules (Definition (2-28)). Then:*
1. $M(I, \pi) = (-1)^{\ell(I)-\ell(\pi)} m({}^\vee\pi, {}^\vee I)$,
2. $m(\pi, I) = (-1)^{\ell(I)-\ell(\pi)} M({}^\vee I, {}^\vee\pi)$.

The Theorem says that the transpose of the matrix $M(I, \pi)$ is the matrix $\epsilon m({}^\vee\pi, {}^\vee I)\epsilon^{-1}$, where $\epsilon = \mathrm{diag}(\dots, (-1)^{\ell(\pi)}, \cdots)$.

(2-37) Example: $SL(2)$. Let $G = SL(2)$, and consider the dual blocks given by $(\delta, {}^\vee\delta)$, i.e. the blocks containing the trivial representations of $SU(1,1)$ and $PU(1,1)$ (cf. Example (2-27)). Write m and M for the multiplicity matrices for $SL(2)$ with respect to the bases defined in Example (2-22), and ${}^\vee m$ and ${}^\vee M$ the corresponding matrices for $PGL(2)$. Then:

$$m = \begin{pmatrix} 1 & 0 & 0 \\ 0 & 1 & 0 \\ -1 & -1 & 1 \end{pmatrix} \quad {}^\vee M = \begin{pmatrix} 1 & 0 & 1 \\ 0 & 1 & 1 \\ 0 & 0 & 1 \end{pmatrix}.$$

Thus $m = \epsilon({}^\vee M)^t\epsilon$ with $\epsilon = \text{diag}(-1,-1,1)$. Taking inverses we obtain $M = m^{-1} = \epsilon({}^\vee M)^{t^{-1}}\epsilon = \epsilon({}^\vee m)^t\epsilon$. □

(2-38) Definition. *In the setting of Theorem (2-36), define a perfect bilinear pairing* $\mathfrak{B} \times {}^\vee\mathfrak{B} \rightarrow \mathbf{Z}$ *by defining it on* $\mathfrak{B}_{irr} \times {}^\vee\mathfrak{B}_{irr}$ *as follows. Suppose* $\pi \in \mathfrak{B}_{irr}$ *is irreducible, with dual module* ${}^\vee\pi \in {}^\vee\mathfrak{B}_{irr}$. *Define* $< \pi, {}^\vee\pi >= (-1)^{\ell(\pi)}$, *and let* $< \pi, {}^d\sigma >= 0$ *for* ${}^d\sigma$ *not isomorphic to* ${}^\vee\pi$.

Now the following Corollary to Theorem (2-36) is immediate (and is, in fact, easily seen to be equivalent to the Theorem).

(2-39) Corollary. $< I, {}^\vee I >= (-1)^{\ell(I)}$, *and* $< I, {}^dI' >= 0$ *for* ${}^dI'$ *not isomorphic to* ${}^\vee I$.

We now replace our pair of L-groups by E-groups (cf. §1). Recall (§1) we are given an inner class of involutions of G and ${}^\vee G$. Fix elements $z \in Z(G)^\theta$, ${}^\vee z \in Z({}^\vee G)^{{}^\vee\theta}$ (cf. Definition (1-3)). Here θ and ${}^\vee\theta$ are any involutions in the given inner classes. Let G^Γ and ${}^\vee G^\Gamma$ be E-groups associated to these elements (Definition (1-3)). As explained in [**2**], maps of the Weil group into an E-group for G are related to representations of a certain covering of G, as we now describe.

Here is the main Definition. We simply use Definition (2-9) with general E-groups in place of L-groups:

(2-40) Definition. *A set of L-data for E-groups* G^Γ *and* ${}^\vee G^\Gamma$, *defined by* z *and* ${}^\vee z$, *is a 6-tuple* $S = (x, T^\Gamma, B, y, {}^dT^\Gamma, {}^dB)$ *satisfying the conditions of Definition (2-9).*

In most situations we will have $z^2 = 1$ and ${}^\vee z^2 = 1$, so we focus on this case. Let ξ be an element of $P^*({}^\vee G, T)$ corresponding to ${}^\vee z\,{}^\vee z_\rho$ under the isomorphism $Z({}^\vee G) \simeq P^*(G,T)/X^*(T)$ (cf. §1). Then since ${}^\vee z^2 = 1$, we have $2\xi \in X^*(T)$. We briefly recall the construction of a two-fold cover $\tilde{G}$ of G defined by ξ ([**2**, Proposition 7.12]).

Let F^ξ be the finite-dimensional representation of the simply connected covering group of G with extremal weight ξ. Then $F^\xi \otimes F^\xi$ factors to a representation of G. Let $\phi : G \rightarrow \text{End}(F^\xi \otimes F^\xi)$ be the corresponding map, and let $\psi : \mathrm{End}(F^\xi) \to \mathrm{End}(F^\xi \otimes F^\xi)$ be the map $X \to X \otimes X$. Let $\tilde{G}$ be the pullback of ψ via ϕ. Thus

$$\tilde{G} = \{(g, X) | g \in G, X \in \mathrm{End}(F^\xi), \phi(g) = \psi(X)\}.$$

We have a commutative diagram

$$\text{(2-41)} \qquad \begin{array}{ccc} \tilde{G} & \xrightarrow{\pi} & \mathrm{End}(F^\xi) \\ {\scriptstyle\mu}\downarrow & & \downarrow{\scriptstyle\psi} \\ G & \xrightarrow{\phi} & \mathrm{End}(F^\xi \otimes F^\xi) \end{array}$$

where μ and π are projection on the first and second factors respectively.

If $\xi \in X^*(T)$ then $\tilde{G} \simeq G \times \mathbf{Z}/2\mathbf{Z}$.

Given L-data S, let $\theta = \mathrm{int}(x)$ and K_x be as usual. Let $\tilde{K}_x$ denote the inverse image of K_x in $\tilde{G}$. Then $(\mathfrak{g}, \tilde{K}_x)$-modules correspond to representations of $\tilde{G}(\mathbf{R})_x$, the inverse image of $G(\mathbf{R})_x$ in $\tilde{G}$. We say such a module is *genuine* if it is non-trivial on the kernel of the covering map $\tilde{G} \to G$.

(2-42) **Example:** *Tori.* Let G be a split one-dimensional torus, and take $z = -1$. Then $\tilde{G} \simeq \mathbf{C}^*$ is the covering of $G \simeq \mathbf{C}^*$ given by $z \to z^2$. Then, since $G(\mathbf{R}) \simeq \mathbf{R}^*$, we have $\tilde{G}(\mathbf{R}) \simeq \mathbf{R}^* \cup i\mathbf{R}^*$. Note that this E-group arises naturally as the subgroup of the L-group $< SL(2), {}^\vee\delta >$ of $PGL(2)$ generated by the diagonal torus ${}^dT_c$ and ${}^\vee\delta$.

Now take G to be compact, and consider the E-group defined by $z = -1$. Then $\tilde{G}$ is the two-fold cover of G, $\tilde{G}(\mathbf{R}) \simeq S^1$ is the twofold cover of $G(\mathbf{R}) = S^1$, and the E-group is isomorphic to $< {}^dT_s, {}^\vee\delta > \subset < SL(2), {}^\vee\delta >$. □

(2-43) **Example:** *SL(2).* If G is $PGL(2)$ and $z = \mathrm{diag}(i, -i)$, then $\tilde{G} = SL(2)$. Then, if $x = \delta$, we have $G(\mathbf{R})_x \simeq PU(1,1)$ and $\tilde{G}(\mathbf{R})$ is the subgroup of $SL(2, \mathbf{C})$ generated by $SL(2, \mathbf{R})$ and $\mathrm{diag}(i, -i)$. This is a non-split extension of $SL(2, \mathbf{R})$ by $\mathbf{Z}/2\mathbf{Z}$ with action given by the outer automorphism of $SL(2, \mathbf{R})$. We denote this group $\widetilde{PU}(1,1)$. □

(2-44) Theorem (cf. [13, Proposition 15.7]).

Fix an E-group ${}^\vee G^\Gamma$ for G and an element ${}^\vee z \in Z({}^\vee G)^{{}^\vee\theta}$. Let χ be a regular (integral) infinitesimal character χ associated to ${}^\vee z$.

Suppose we are given a set of L-data S with $y^2 = {}^\vee z$. Then there is a standard $(\mathfrak{g}, \tilde{K}_x)$-module $I(S)$ associated to S with infinitesimal character χ having unique irreducible submodule $J(S)$. This correspondence defines a bijection from the set of equivalence classes of L-data with $y^2 = {}^\vee z$ to the set of equivalence classes of pairs (x, π), where π is a genuine irreducible $(\mathfrak{g}, \tilde{K}_x)$-module of infinitesimal character χ.

If (x, y) is admissible (Definition (2-24); that is, if there are any sets of L-data of this form) then the set of classes of L-data containing x and y corresponds to a single block of $(\mathfrak{g}, \tilde{K}_x)$-modules (Definition (2-19)).

(2-45) **Example:** *Tori.* Take the genuine E-group for the split one-dimensional torus G of Example (1-9). Then the data $(\delta, {}^\vee\delta)$ corresponds to the genuine representation of $\tilde{G}(\mathbf{R}) \simeq \mathbf{R}^* \cup i\mathbf{R}^*$ which is trivial on the identity component and sends i to i. On the other hand, if G is compact, this data yields the representation $e^{i\theta/2}$ of the two-fold cover of the circle. □

Suppose $S = (x, T^\Gamma, B, y, {}^dT^\Gamma, {}^dB)$ is a set of L-data. Then we define the dual data $(y, {}^dT^\Gamma, {}^dB, x, T^\Gamma, B)$ as before (Definition (2-28)). We obtain a duality mapping from $(\mathfrak{g}, \tilde{K}_x)$ modules to $({}^\vee\mathfrak{g}, {}^\vee\tilde{K}_y)$ modules, and similarly dual blocks. Then Theorem (2-36) holds as stated.

Note that we allow the E-groups G^Γ and ${}^\vee G^\Gamma$ to vary independently. For example, let G^Γ be an L-group, but let ${}^\vee G^\Gamma$ be a more general E-group. Then we see a block $\mathfrak{B}$ of $(\mathfrak{g}, K_x)$ modules is dual to a block of $({}^\vee g, {}^\vee\tilde{K}_y)$ modules for various covering groups of ${}^\vee G$, including the trivial one. This flexibility will be important in §6.

(2-46) **Example:** *SL(2).* The block containing the trivial representation of $SU(1,1)$ is dual to the block of the trivial representation of $PU(1,1)$. It is also dual to a similar block for the group $\widetilde{PU}(1,1)$ of Example (2-43).

More significantly, consider the singleton consisting of the trivial representation $\pi^{2,0}$ of the strong real form $SU(2,0)$ of $SL(2)$. Since $x^2 = (t_0\delta)^2 = 1$, the corresponding infinitesimal character for ${}^\vee G$ is 2ρ, and $\pi^{2,0}$ is dual to an irreducible principal series of $PU(1,1)$ with this infinitesimal character. Note that $PU(1,1)$ does *not* have an irreducible principal series with infinitesimal character ρ, so this is forced. However $\pi^{2,0}$ is also dual to an irreducible principal series of $\widetilde{PU}(1,1)$ which may be taken to be of infinitesimal character ρ. This is due to the fact that for $\widetilde{PU}(1,1)$ (i.e. $SL(2)$) we may translate by ρ, while this is not allowed for $PGL(2)$.This phenomenon will be important in the definition of super-stability (cf §6). □

There are certain situations in which we need higher order covering groups of G. Let G^{sc} denote the simply connected covering group of G; so the kernel of the covering map $G^{sc} \to G$ is $\pi_1 = \pi_1(G)$. Now the elements of $G^\Gamma - G$ all have the same action on π_1, which we denote θ. Let G^{can} be the covering group of G with kernel $\pi_1(\mathbf{R}) = \pi_1/(1+\theta)\pi_1$ (cf. [**2**, §7]). If x is a strong real form of G, let K_x^{can} denote the inverse image of K_x in G^{can}.

(2-47) **Example:** *Tori.* If G is a compact torus, then θ acts by 1 on $\mathfrak{t}$, and G^{can} is a two-fold cover of G. If G is split, however, then θ acts by -1, G^{can} is the universal cover $\mathbf{C}$ of G, and the covering map is the exponential. □

(2-48) **Example:** *inner forms of compact or split groups.* If G is inner to a split group, then θ acts by -1 on π_1 (since $\pi_1 \subset \mathfrak{t}$ for T a split Cartan subgroup), so $G^{can} = G^{sc}$. Similarly if G is inner to a compact group, the kernel of the covering map $G^{can} \to G$ is a 2-group. □

Now the group of characters of $\pi_1(\mathbf{R})$ is naturally isomorphic to $Z({}^\vee G)^{{}^\vee\theta}$.

(2-49) Definition ([1, 4.7]). *For ${}^\vee z \in Z({}^\vee G)^{{}^\vee\theta}$, let $\tau({}^\vee z)$ denote the corresponding character of $\pi_1(\mathbf{R})$.*

We say a $(\mathfrak{g}, K_x^{can})$ module is of type τ if its restriction to $\pi_1(\mathbf{R})$ is a multiple of a character τ.

(2-50) Proposition. *Let ${}^\vee z$ be the element defining the E-group ${}^\vee G$. Theorem (2-44) holds as stated with $(\mathfrak{g}, K_x^{can})$-modules of type $\tau({}^\vee z\,{}^\vee z_\rho)$ in place of genuine $(\mathfrak{g}, \tilde{K}_x)$-modules.*

The generality of this result over that of Theorem (2-44) is only important when $\pi_1(\mathbf{R})$ is not a 2-group. This can only occur if there is a factor of type A_n, D_{2n+1}, or if the center of G contains a split torus.

3. Parametrization of L-Packets

Let ϕ be an admissible map of the Weil group into the L-group. We now define a perfect pairing

$$\mathbf{\Pi}_\phi \times \tilde{\mathbf{S}}_\phi \to \mathbf{C}^*, \tag{3-1}$$

where $\mathbf{\Pi}_\phi$ is the super L-packet defined by ϕ, and $\tilde{\mathbf{S}}_\phi$ is a certain component group on the dual side. Using $\tilde{\mathbf{S}}_\phi$ we obtain the coefficents that occur in the lift of a stable character of an endoscopic group H.

Fix G and an inner class of real forms of G, with L-group ${}^\vee G^\Gamma$. Let $\phi = \phi(y, {}^dT^\Gamma, \lambda) : W_\mathbf{R} \to {}^\vee G^\Gamma$ be a quasi-admissible homomorphism, with λ regular. (Equivalently, choose ${}^dB$ and a regular infinitesimal character for G associated to y^2, and let $\phi = \phi(y, {}^dT^\Gamma, {}^dB)$.) Let S_ϕ denote the centralizer of ϕ in ${}^\vee G$. Then $S_\phi = ({}^dT)^{{}^\vee\theta}$, where ${}^\vee\theta = {}^\vee\theta_y = \mathrm{int}(y)$. Let $\tilde{S}_\phi$ be the inverse image of S_ϕ in ${}^\vee G^{can}$ and let $\tilde{\mathbf{S}}_\phi = \tilde{S}_\phi/\tilde{S}_\phi^0$, where $\tilde{S}_\phi^0$ is the identity component of $\tilde{S}_\phi$. Let $(\tilde{\mathbf{S}}_\phi)^\wedge$ denote the group of characters of $\tilde{\mathbf{S}}_\phi$. Let $\mathbf{\Pi}_\phi$ be the super L-packet (Definition (2-15)) defined by ϕ, which we consider here as a set of irreducible modules.

(3-2) Theorem ([1,Theorem 5.1]). *There is a canonical bijection:*

$$\mathbf{\Pi}_\phi \leftrightarrow (\tilde{\mathbf{S}}_\phi)^\wedge. \tag{3-3}$$

Equivalently there is a canonical perfect pairing

$$\mathbf{\Pi}_\phi \times \tilde{\mathbf{S}}_\phi \to \mathbf{C}^*. \tag{3-4}$$

Sketch of proof.

Choose B, T^Γ and $\delta \in T^\Gamma - T$ as in Lemma (2-16), so that the standard modules in $\mathbf{\Pi}$ are parametrized by the group

(3-5) $$\mathfrak{F} = \{t \in T | t\theta(t) \in Z(G)\}/\{s\theta(s^{-1})| s \in T\}$$

(where $\theta = \theta_\delta = \mathrm{int}(\delta)$). We compute $\mathfrak{F}\hat{}$, the group of (holomorphic) characters of $\mathfrak{F}$. It follows from a calculation in T and ${}^dT$ that

(3-6)
$$\mathfrak{F}\hat{} \simeq \{\gamma \in X_*({}^dT) | \gamma + {}^\vee\theta(\gamma) = 0\}/\{\mu - {}^\vee\theta\,\mu | \mu \in Q_*({}^\vee G, {}^dT)\} = L_1/L_2.$$

Given $\gamma \in L_1$, let $\tilde{s} = \exp(\frac{1}{2}\gamma) \in {}^dT^{can}$. We obtain a map $L_1 \to \tilde{S}_\phi$; composing with projection gives a map $L_1 \to \tilde{\mathbf{S}}_\phi$. The kernel is L_2 so we have an isomorphism $\mathfrak{F}\hat{} \simeq \tilde{\mathbf{S}}_\phi$. By Pontryagin duality we have $\mathfrak{F} \simeq (\tilde{\mathbf{S}}_\phi)\hat{}$; define $\mathbf{\Pi}_\phi \to (\tilde{\mathbf{S}}_\phi)\hat{}$ by the sequence of maps:

(3-7) $$\mathbf{\Pi}_\phi \leftrightarrow \mathfrak{F} \simeq (\tilde{\mathbf{S}}_\phi)\hat{}.$$

This bijection is independent of the choices. □

(3-8) **Example:** *Tori.* If G is a split one-dimensional torus, then ${}^\vee G$ is compact, and ${}^\vee G^{can}$ is the two-fold cover of ${}^\vee G$. Then $\tilde{S}_\phi = {}^\vee G^{can}$ is connected, and $\tilde{\mathbf{S}}_\phi$ is trivial. In this case there is only one strong real form of G, and each super L-packet is a singleton.

If G is a compact one-dimensional torus, then ${}^\vee G^{can} \simeq \mathbf{C}$ is the simply connected covering group of ${}^\vee G$, $S_\phi = \{\pm 1\}$, and $\tilde{\mathbf{S}}_\phi = \{\tilde{s}_k = \pi i k | k \in \mathbf{Z}\}$. Now the strong real forms of G are parametrized by $z \in \mathbf{C}^*$, and in this case Theorem (3-2) says $\hat{\mathbf{Z}} \simeq \mathbf{C}^*$. Thus for $z \in \mathbf{C}^*$, let π_z be the trivial representation of the strong real form z. Then the pairing of (3-4) is

$$< \pi_z, \tilde{s}_k >= z^k.$$

□

(3-9) **Example:** *Discrete Series.* If $\mathbf{\Pi}_\phi$ is a super L-packet of discrete series, then T is compact, ${}^dT$ is split, and ${}^\vee\theta|_{{}^dT}$ acts by $t \to t^{-1}$. Thus $\mathbf{\Pi}_\phi$ is parametrized by $\mathfrak{F}$, which in this case becomes:

(3-10) $$P_*(G,T)/2X_*(T).$$

Now $S_\phi = {}^dT^{{}^\vee\theta}$ is the elements of order two in ${}^dT$, and has order 2^n ($n = \mathrm{rank}(G)$). Now $G^{can} = G^{sc}$ (cf. Example (2-48)), and $\tilde{\mathbf{S}}_\phi$ has order $|{}^\vee\pi_1| 2^n = |Z(G)| 2^n$; $\tilde{\mathbf{S}}_\phi$ is isomorphic to:

(3-11) $$X_*({}^dT)/2Q_*({}^\vee G, {}^dT).$$

The pairing $<,>$: $\mathbf{\Pi}_\phi \times \tilde{\mathbf{S}}_\phi$ may be realized as follows. Write $\pi(\gamma)$ for the element of $\mathbf{\Pi}_\phi$ corresponding to (the coset of) $\gamma \in P_*(T)$ via (3-10). Similarly suppose $\tilde{s}(\lambda) \in \tilde{\mathbf{S}}_\phi$ corresponds to $\lambda \in X_*({}^dT)$ using (3-11). We have

(3-12) $$< \pi(\gamma), \tilde{s}(\lambda) >= e^{\pi i <\zeta(\gamma), \lambda>}.$$

□

Note that in general $\mathbf{\Pi}_\phi$ contains a distinguished representation π_0, corresponding to the trivial character of $\tilde{\mathbf{S}}_\phi$, so that $< \pi_0, \tilde{s} >= 1$ for all $\tilde{s} \in \tilde{\mathbf{S}}_\phi$.

(3-13) **Example:** *Discrete Series.* For a super L-packet of discrete series $\mathbf{\Pi}$, π_0 is the distinguished large discrete series representation of the quasi-split form of G which was chosen in defining the L-group (cf discussion following (1-12)). □

The numbers $< I, \tilde{s} >$ may be computed using roots and weights in the Lie algebras of T and ${}^dT$ respectively.

(3-14) **Example:** *Discrete Series for SL(2).* Following Example (3-9) (and using the notation of Example (2-22)) we see

$$\mathbf{\Pi}_\phi = \{\pi^+ = \pi(0), \pi_- = \pi(\alpha^\vee), \pi^{2,0} = \pi(\frac{1}{2}\alpha^\vee), \pi^{0,2} = \pi(\frac{3}{2}\alpha^\vee)\}.$$

Similarly $\tilde{\mathbf{S}}_\phi$ is parametrized by $\{0, \frac{1}{2}\beta^\vee, \beta^\vee, \frac{3}{2}\beta^\vee\}$. We identify these with the elements $1, i, -1, -i \in \mathbf{C}^*$ under the isomorphism ${}^dT \simeq \mathbf{C}^*$. Then

$$\begin{aligned} &< \pi^+, z > = 1, \ < \pi^{2,0}, z > = z, \\ &< \pi^-, z > = z^2, < \pi^{0,2}, z > = z^3. \end{aligned} \tag{3-15}$$

□

(3-16) **Example:** *Sp(2n).* We continue with the method of the previous example. Fix L-data $(\delta, B) \in \mathfrak{D}$, and a Cartan subgroup $T \subset B$. Then T is compact (with respect to $\theta = \text{int}(\delta)$), and B is large. Fix ${}^\vee\delta \in {}^\vee\mathfrak{D}$, and a split Cartan subgroup ${}^dT$ (with respect to ${}^\vee\delta = \text{int}({}^\vee\delta)$). Choose any Borel subgroup ${}^dB$ containing ${}^dT$.

Choose the usual coordinates $(a_1, a_2, \ldots, a_n)$ on $\mathfrak{t}$. Then $P_*(G,T)$ is given by the condition $a_i \in \mathbf{Z}$ (for all i) or $a_i \in \mathbf{Z} + \frac{1}{2}$ (for all i), and $X_*(T) = Q_*(G,T)$ is the subset of elements with integral entries. Making similar choices for the dual group $SO(2n+1)$, we see $X_*({}^dT) = P_*({}^\vee G, {}^dT)$ consists of n-tuples with integral entries, and $Q_*({}^\vee G, {}^dT)$ is the subset whose entries sum to an even integer.

Let $\epsilon = (\epsilon_0, \epsilon_1, \ldots, \epsilon_n)$ denote the element

$$(\epsilon_1, \epsilon_2, \ldots, \epsilon_n) - \epsilon_0(\frac{1}{2}, \frac{1}{2}, \ldots, \frac{1}{2}) \quad (\epsilon_i \in \{0,1\}).$$

Then $\{\epsilon\}$ is a set of representatives of $P_*(G,T)/2X_*(T)$. Let $\pi(\epsilon)$ be the corresponding module; this is a representation of $Sp(2n, \mathbf{R})$ if $\epsilon_0 = 0$, and of $Sp(k, n-k)$, $k = \sum_1^n \epsilon_i$, if $\epsilon_0 = 1$.

Similarly let $\mu = (\mu_0, \mu_1, \ldots, \mu_n)$ denote the element

$$(\mu_1, \mu_2, \ldots, \mu_n) + 2\mu_0(0, 0, \ldots, 1) \quad (\mu_i \in \{0,1\}).$$

Then $\{\mu\}$ is a set of representatives of $X_*({}^dT)/2Q_*({}^\vee G, {}^dT)$. Let $\tilde{s}(\mu)$ be the element of $\tilde{\mathbf{S}}_\phi$ corresponding to μ, and let $|\mu| = \sum_1^n \mu_i$.

Under the pairing of Theorem (3-2),

$$< \pi(\epsilon), \tilde{s}(\mu) > = \exp(\pi i \sum_i \epsilon_i\mu_i - \frac{1}{2}\epsilon_0|\mu| + 2\mu_0\epsilon_n - \epsilon_0\mu_0)$$

$$= (-i)^{\epsilon_0|\mu|}(-1)^{\epsilon_0\mu_0}\prod_{i=1}^n(-1)^{\epsilon_i\mu_i}. \tag{3-17}$$

□

We now give an alternative description of these coefficents. Given $\pi \in \mathbf{\Pi}$, choose L-data S so that π is the irreducible module $J(S)$ defined by S. Let ${}^d\Lambda$ be the character of the two-fold cover of ${}^dT(\mathbf{R})$ defined by ${}^\vee\rho$ constructed in the proof of Theorem (3-2) applied to ${}^\vee G$. We consider this as a character of the inverse image ${}^dT(\mathbf{R})^{can}$ of ${}^dT(\mathbf{R})$ in ${}^\vee G^{can}$.

(3-18) Proposition ([1, Lemma 5.5]). $< I, \tilde{s} >= \frac{{}^d\Lambda}{e^{{}^d\lambda}}(\tilde{s})$.

This enables us to study how $< I, \tilde{s} >$ varies as I runs over an L-packet. Fix a strong real form x of G and a representation π of this strong real form given by L-data S as above. Let W_{im} be the Weyl group generated by the imaginary roots of T in G. Then letting $\times$ denote the cross action of W on representations (cf. [1, 3.18]), the L-packet containing π is the set

$$\{w \times \pi | w \in W_{im}\} \tag{3-19}$$

([1, 3.18]). Thus for $\tilde{s} \in \tilde{\mathbf{S}}_\phi$, let

$$\delta_{I,\tilde{s}}(w) =< w \times I, \tilde{s} > / < I, \tilde{s} > \quad (w \in W_{im}). \tag{3-20}$$

(3-21) Proposition ([1, 5.7]). *Let s be the image of $\tilde{s}$ in ${}^dT$. Then*

$$\delta_{I,s}(w) = \frac{{}^\vee w({}^\vee w^{-1} \times {}^d\Lambda)}{{}^d\Lambda}(s). \tag{3-22}$$

(As the notation indicates, the right side depends only on s, and so we let $\delta_{I,s} = \delta_{I,\tilde{s}}$.)

As a consequence, note that $\delta_{I,s}$, considered as a character of ${}^dT(\mathbf{R})$, is trivial on the identity component and hence factors to a character of the component group, which is a two-group. Parts (1) and (2) of the following Corollary follow immediately, and the third follows from properties of the cross action ([12, Lemma 8.3.17]).

(3-23) Corollary ([1, Lemma 5.8]).

(1) *For all* $w \in W_{im}$, $\delta_{I,s}(w) \in \{\pm 1\}$.
(2) *For all* $u, v \in W_{im}$, $\delta_{I,s}(uv) = \delta_{I,s}(v)\delta_{v \times I,s}(u)$.
(3) *For* α *a simple imaginary root of* T, *let* ${}^\vee\alpha = \zeta(\alpha^\vee)$ *be a corresponding real root of* ${}^dT$ *in* ${}^\vee G$. *Then*

$$\delta_{I,s}(s_\alpha) = \begin{cases} 1 & \text{if } \alpha \text{ is compact} \\ sgn({}^\vee\alpha(s)) & \text{if } \alpha \text{ is non-compact.} \end{cases}$$

Note that one may use this Proposition inductively to calculate $\delta_{I,s}(w)$ explicitly. Let $\kappa_{\zeta,s}(w)$ be defined as in [**10**], where the subscripts indicate the dependence of κ on s and an isomorphism $\zeta : X_*(T) \simeq X^*({}^dT)$ (cf. [**1**, 5.12]). Then comparing Corollary (3-23) with [**9**, Propositions 2.1 and 3.1] we obtain:

(3-24) Corollary. *Given L-data* $S = (x, T^\Gamma, B, y, {}^dT^\Gamma, {}^dB)$, *let* $\zeta = \zeta_{B,{}^dB}$ *(cf. §1). Let I be the standard module given by the L-data S.*

Then for all $w \in W_{im}$,

$$\delta_{I,s}(w) = \kappa_{\zeta,s}(w). \tag{3-25}$$

Note that we have a canonical bijection

$$\Pi_\phi = \cup_x \Pi_{\phi,x} \leftrightarrow (\tilde{\mathbf{S}}_\phi)\hat{} \tag{3-26}$$

(disjoint union). This differs from [9] and [10] in which, for all x, there is an embedding

$$\Pi_{\phi,x} \hookrightarrow (S_\phi / S_\phi^0 Z^\Gamma)\hat{}. \tag{3-27}$$

This embedding is non-canonical, depending on a choice of $\pi \in \Pi_{\phi,x}$ (going to the trivial character). Furthermore the images of $\Pi_{\phi,x}$ as x varies are not disjoint; in particular the trivial character is in the image of $\Pi_{\phi,x}$ for all x.

4. Endoscopic Groups

We turn now to the definition of endoscopic groups, and discuss their structural aspects. As usual we are given G and an inner class of real forms, with L-groups for G and ${}^\vee G$.

Recall (cf. §2) we defined the canonical cover ${}^\vee G^{can}$ of ${}^\vee G$. Let π denote the covering map ${}^\vee G^{can} \to {}^\vee G$. We say an element $\tilde{s}$ of ${}^\vee G^{can}$ is elliptic if $\pi(\tilde{s})$ is an elliptic element of ${}^\vee G$.

A set of endoscopic data for G is primarily an E-group ${}^\vee H^\Gamma$ contained in ${}^\vee G^\Gamma$ whose identity component ${}^\vee H$ is the centralizer of a semisimple element $s \in {}^\vee G$. We proceed to make this precise.

We choose to keep track not just of the group ${}^\vee H$, but also of the central element s. There is no harm in assuming that s is elliptic. Furthermore we replace s by an inverse image $\tilde{s}$ of s in ${}^\vee G^{can}$. The requirements on ${}^\vee H^\Gamma$ are spelled out in:

(4-1) Definition ([1, Definition 6.3]). *A set of* <u>*weak*</u> <u>*endoscopic*</u> <u>*data*</u> <u>*for*</u> <u>*G*</u> *is a pair* $(\tilde{s}, {}^\vee H^\Gamma)$ *where* $\tilde{s}$ *is an elliptic element of* ${}^\vee G^{can}$ *and* ${}^\vee H^\Gamma$ *is a subgroup of* ${}^\vee G^\Gamma$*. These satisfy the following conditions:*

(1) *The identity component* ${}^\vee H$ *of* ${}^\vee H^\Gamma$ *is the identity component of the centralizer of* $s = \pi(\tilde{s})$ *in* ${}^\vee G$,

(2) ${}^\vee H^\Gamma$ *has two components, the non-identity component of which is contained in* ${}^\vee G^\Gamma - {}^\vee G$,

(3) $s \in Z({}^\vee H^\Gamma)$.

The notion of equivalence of weak endoscopic data is a bit tricky. Of course we allow arbitrary conjugation by ${}^\vee G$ (${}^\vee G^{can}$ on $\tilde{s}$), and we also allow $\tilde{s}$ to be replaced by $\tilde{z}\tilde{s}$ where $\tilde{z}$ is contained in the identity component of the inverse image of $Z({}^\vee H^\Gamma) \cap {}^\vee H$ (cf. [1, Definition 6.4]).

(4-2) Definition ([1, Definition 6.9]). *The endoscopic group determined by* $(\tilde{s}, {}^\vee H^\Gamma)$ *is the quasisplit group* H *which has* ${}^\vee H^\Gamma$ *as an E-group.*

Thus H is a connected reductive quasisplit algebraic group dual to ${}^\vee H$. The inner class of ${}^\vee H$ is determined by the involution $\operatorname{int}({}^\vee\delta)|_{{}^\vee H}$. Note that H only depends (up to isomorphism) on the equivalence class of $(\tilde{s}, {}^\vee H^\Gamma)$ (i.e. not on $\tilde{s}$, and only on the conjugacy class of ${}^\vee H^\Gamma$).

We use the terminology *weak* for the same reason that Definition (1-3) is preliminary: in order for ${}^\vee H^\Gamma$ to be an E-group we need to choose an E-group structure on it, i.e. a conjugacy class ${}^\vee\mathfrak{D}_H$. We return to this in a moment.

Note that ${}^\vee H^\Gamma$ is generated by ${}^\vee H$ and any element $y \in {}^\vee H^\Gamma - {}^\vee H$. As in the proof of ([1, Lemma 6.5]) we may in fact choose $y \in {}^\vee\mathfrak{D}$, and furthermore require that $\operatorname{int}(y)|_{{}^\vee H}$ be a principal involution. Conjugating by ${}^\vee G$ we obtain the following Lemma (cf. [1,Lemma 6.7 and Corollary 6.8]) which we use to compute weak endoscopic data.

(4-3) Lemma. *Fix* ${}^\vee\delta_0 \in {}^\vee\mathfrak{D}$, *let* ${}^\vee\theta = \operatorname{int}({}^\vee\delta_0)$, *and let* ${}^\vee K = ({}^\vee G)^{{}^\vee\theta}$.

(1) *Given an elliptic element* $\tilde{s}$ *with* $\pi(\tilde{s}) \in {}^\vee K$, *let* ${}^\vee H^\Gamma$ *be the group generated by the centralizer* ${}^\vee H$ *of* $s = \pi(\tilde{s})$ *and* ${}^\vee\delta_0$. *Then* $(\tilde{s}, {}^\vee H^\Gamma)$ *is a set of weak endoscopic data, and every set of weak endoscopic data is equivalent to a set of this form.*

(2) *Furthermore we may assume that* ${}^\vee\theta\,|_{{}^\vee H}$ *is a principal involution of* ${}^\vee H$. *Two such elements* $\tilde{s}$ *and* $\tilde{s}'$ *correspond to equivalent weak endoscopic data if and only if they are conjugate via* ${}^\vee K$, *up to an element* $\tilde{z}$ *as above.*

In the following examples we fix ${}^\vee\delta_0$ and ${}^\vee K$ as in this Lemma.

(4-4) Example: H=G. The equivalence classes of endoscopic data with ${}^\vee H^\Gamma = {}^\vee G^\Gamma$ (so H is the quasisplit form of G) are described as follows. We choose $\tilde{s} \in \pi^{-1}(Z({}^\vee G)^{{}^\vee\theta})$. We can modify $\tilde{s}$ by $\tilde{z}$ contained in the identity component of $\pi^{-1}(Z({}^\vee G^\Gamma)\cap {}^\vee G) = \pi^{-1}(Z({}^\vee G)^{{}^\vee\theta})$. Thus the equivalence classes of data with H the quasisplit form of G are parametrized by

$$\pi^{-1}(Z({}^\vee G)^{{}^\vee\theta})/(\pi^{-1}(Z({}^\vee G)^{{}^\vee\theta}))^0 \tag{4-5}$$

Note that if ${}^\vee G$ is inner to a compact group then ${}^\vee\theta$ acts trivially on $Z({}^\vee G)$ and this becomes $Z({}^\vee G^{can})/Z({}^\vee G^{can})^0$. □

(4-6) **Example:** *Tori.*

(1) If G is a split one-dimensional torus, then ${}^\vee G^{can}$ is the two-fold cover of $\mathbf{C}^*$ (cf. Example (2-47)). Then any element $\tilde{s}\in {}^\vee G^{can}$ is allowed; however all such elements are equivalent so we may as well assume $\tilde{s}=1$.
(2) If G is compact, then ${}^\vee G^{can}$ is the universal cover $\mathbf{C}\to\mathbf{C}^*$ given by exp. Now ${}^\vee G^{{}^\vee\theta} = \{\pm 1\}$, so $\tilde{s}\in\{\pi i k | k\in\mathbf{Z}\}$. These elements all give inequivalent weak endoscopic data. □

(4-7) **Example:** *SL(2).* There are a total of four equivalence classes of weak endoscopic data for $G = SL(2)$ as follows. We have ${}^\vee G = PGL(2)$ and ${}^\vee G^{can} = SL(2)$ (cf. Example(2-48)). As usual choose ${}^\vee\delta$ acting by $\text{int}(t_0)$. Then ${}^\vee K^0$ is the diagonal torus ${}^dT_c$, ${}^\vee K\simeq O(2)$, and ${}^\vee\tilde{K}$ is the group generated by ${}^dT_c$ and $\begin{pmatrix}0 & 1\\ -1 & 0\end{pmatrix}$. Then ${}^\vee K\simeq O(2)$ and ${}^\vee\tilde{K}\simeq O(2)$.

(1) If s is singular then $s=1$, $\tilde{s}=\pm1$, and $H=G$.
(2) If $\tilde{s}\in SO(2)$ ($\tilde{s}\neq\pm1$) then ${}^\vee H = {}^dT_c$ and ${}^\vee\theta$ acts trivially on ${}^\vee H$. It follows that all such elements $\tilde{s}$ are equivalent to one, say $\tilde{s}_h = t_0$, and H is a split torus.
(3) If $s\in O(2)-SO(2)$ then ${}^\vee H = {}^dT_s$ so that ${}^\vee\theta$ acts by $t\to t^{-1}$ on ${}^\vee H$. Then up to equivalence we choose $\tilde{s}=\tilde{s}_e = \begin{pmatrix}0 & 1\\ -1 & 0\end{pmatrix}$. In this case H is a compact torus.

If G is $PGL(2)$, then ${}^\vee K = {}^\vee\tilde{K} = SO(2)$ is connected. As a result we lose the endoscopic group which is a compact torus. Up to equivalence the classes of weak data are given by $\tilde{s}=\pm1$ ($H=G$), and $\tilde{s}_h = t_0$ (H is a split torus). □

(4-8) **Example:** *H inner to a compact group.* In this case ${}^\vee H$ is split, ${}^\vee G$ is split, and the split Cartan subgroup of ${}^\vee G$ is contained in ${}^\vee H$. Now we may conjugate s to be in this Cartan subgroup, so ${}^\vee\theta\, s = s^{-1}$. However, ${}^\vee\theta\, s = s$, so $s^2=1$. Now ${}^\vee G^{can} = {}^\vee G^{sc}$ (cf. Example (2-48)), so there are $|Z(G)|2^n$ choices of $\tilde{s}$ ($n = rank(G)$). Two such elements yield equivalent data if they are conjugate by ${}^\vee G^{can}$; these are the endoscopic groups which come into lifting of discrete series. □

We now specify the E-group structure on ${}^\vee H^\Gamma$.

(4-9) Definition ([1, Definition 6.3]). *A set of endoscopic data for G is a triple*

$$(\tilde{s}, {}^\vee H^\Gamma, {}^\vee\mathfrak{D}_H),$$

where $(\tilde{s}, {}^\vee H^\Gamma)$ is a set of weak endoscopic data, and $({}^\vee H^\Gamma, {}^\vee\mathfrak{D}_H)$ is an E-group determined by ${}^\vee z_\rho$ (Definition (1-3)).

Suppose $(\tilde{s}, {}^\vee H^\Gamma)$ is a set of weak endoscopic data. By Lemma (4-3) there is an element $y \in {}^\vee H^\Gamma - {}^\vee H$ with $y^2 = z_\rho$ (take $y = {}^\vee\delta$). Thus the final condition of the preceding definition is a natural one, and weak endoscopic data may always be extended to endoscopic data. In fact the amount of choice is given by:

(4-10) Lemma. *Given $(\tilde{s}, {}^\vee H^\Gamma)$ the set of possible E-group structures ${}^\vee\mathfrak{D}_H$ making $(\tilde{s}, {}^\vee H^\Gamma, {}^\vee\mathfrak{D}_H)$ endoscopic data for G is parametrized by*

$$\{z \in Z({}^\vee H) | z\,{}^\vee\theta_H(z) = 1\} / \{w\,{}^\vee\theta_H(w^{-1}) | w \in Z({}^\vee H)\}. \tag{4-11}$$

(The automorphism ${}^\vee\theta_H$ of $Z({}^\vee H)$ is defined by the action of any element of ${}^\vee H^\Gamma - {}^\vee H$.)

(4-12) **Example:** *Tori.* If G is a one-dimensional split torus, then the set of (4-11) is $\{\pm 1\}$, so there are two choices of endoscopic data for G. These are given by $(1, {}^\vee G^\Gamma, {}^\vee\delta)$ and $(1, {}^\vee G^\Gamma, -{}^\vee\delta)$. If G is compact each set of weak endoscopic data extends uniquely. □

(4-13) **Example:** *SL(2).* If $G = SL(2)$ and $\tilde{s} = 1, -1$, or $\tilde{s}_e$, there is a unique way to extend this data, whereas if $\tilde{s} = \tilde{s}_h$ there are two such choices. If $G = PGL(2)$ there are always two choices. □

Recall (cf. §2) that maps of $W_{\mathbb{R}}$ into ${}^\vee H^\Gamma$ parametrize representations of certain covering groups of real forms of H which we now describe. Since the element ${}^\vee z_\rho$ defining the E-group ${}^\vee H^\Gamma$ is of order two, we are in the setting of Theorem (2-44). Choose a Borel subgroup B_H and a Cartan subgroup $T_H \subset B_H$, and let ρ_H be one-half the sum of the roots of T_H in B_H. We may define the element $\gamma = \rho - \rho_H$ as an element of of $P^*(H, T_H)$. We obtain the two-fold cover H_γ of H defined by γ (cf. (2-41)). Let θ_H be a principal involution of H, let $K_H = H^{\theta_H}$, and let $\tilde{K}_H$ be the inverse image of K_H in this covering group. Theorem (2-44) then gives:

(4-14) Lemma. *Fix a set $(\tilde{s}, {}^\vee H^\Gamma, {}^\vee\mathfrak{D}_H)$ of endoscopic data, with H the corresponding endoscopic group. Then maps of $W_{\mathbb{R}}$ into ${}^\vee H^\Gamma$ parametrize L-packets of genuine $(\mathfrak{h}, \tilde{K}_H)$-modules.*

(4-15) **Example:** *SL(2).* Suppose $G = SL(2)$. Then, for $\tilde{s} = \pm 1$, we have $\rho - \rho_H = 0$, and the corresponding cover is trivial. For H a compact or split torus, $\rho - \rho_H = \rho$ exponentiates to H so the covering of H is trivial in this case also.

If $G = PGL(2)$ and $\tilde{s} = \pm 1$, H_γ is still the trivial covering group. However, if H is a torus, $\rho - \rho_H = \rho$ does *not* factor to G. Thus H_γ is the non-trivial, connected, two-fold covering group of H, i.e., $H(\mathbf{R}) \simeq \mathbf{R}^* \cup i\mathbf{R}^*$ (cf. Example (2-42)). □

We also construct an E-group $(H^\Gamma, \mathfrak{D}_H)$ containing H.

(4-16) Definition ([1, Definition 6.11]). *Given endoscopic data*

$$S = (\tilde{s}, {}^\vee H^\Gamma, {}^\vee\mathfrak{D}_H),$$

let H be a corresponding endoscopic group and let x be a strong real form of G. The E-group defined by S and x is the E-group H^Γ containing H, determined by the element $x^2 \in Z(G) \hookrightarrow Z(H)$.

Note that H^Γ depends on x while ${}^\vee H^\Gamma$ does not. The reason for this choice will be seen in §5 (cf. the discussion preceding Lemma (5-5)).

(4-17) **Example:** *SL(2).* Let $G = SL(2)$.

(1) Let $x = \delta$, so the strong real form is $SU(1,1)$, and $x^2 = z_\rho = -1$. Recall (cf. Example (4-7)) $\tilde{s}$ equals $\pm 1, \tilde{s}_e$, or $\tilde{s}_h$, and H is $SL(2)$, a compact torus, or a split torus respectively. The corresponding E-group H^Γ is an L-group in the first case, whereas in the latter cases we obtain E-groups corresponding to the two-fold covers of ${}^\vee H(\mathbf{R}) = S^1$ and $\mathbf{R}^*$ of Example (2-42).
(2) Let $x = \pm t_o\delta$, corresponding to the compact strong real forms of G, so that $x^2 = 1$. Then the E-group containing $H = SL(2)$ is a genuine E-group as in Example (1-10), while the E-groups containing the tori become ordinary L-groups.

□

For later use we note that we obtain a several-to-one map from infinitesimal characters for $\mathfrak{h}$ to those for $\mathfrak{g}$. We say χ corresponds to χ_H if they may both be written $\chi(\lambda)$, for some $\lambda \in {}^d\mathfrak{t}$, ${}^d\mathfrak{t}$ a Cartan subalgebra of both ${}^\vee\mathfrak{h}$ and ${}^\vee\mathfrak{g}$.

We briefly describe the connection between endoscopic data defined above and as defined in [10]. Using ([10, Lemmas 2.1.6 and 2.1.7]), a set of endoscopic data may be described as a pair (s,ω), where s is a semisimple element of ${}^\vee G^\Gamma$, and ω is a component of the centralizer of s in ${}^\vee G^\Gamma - {}^\vee G$. Then we let ${}^\vee H$ be the identity component of this centralizer, and define an L-group by taking ${}^\vee H$ together with the action of int(y), for $y \in \omega$. (We require the associated element of Out(${}^\vee H$) have order two.)

Now given weak endoscopic data $(\tilde{s}, {}^\vee H^\Gamma)$ in the sense of Definition 6.1, we obtain a pair (s,ω) by taking $s = \pi(\tilde{s})$, and $\omega = {}^\vee H^\Gamma - {}^\vee H$. This defines a surjection ([1, Theorem 6.14]) from equivalence classes of weak endoscopic data as given by Definition (4-1), to equivalence classes of endoscopic data as defined in [10].

5. Method of Descent

In this section we state the results of [3] on Harish-Chandra's method of descent in the terms we need. As usual we are given G, along with an inner class of real forms and a pair of L-groups G^Γ and ${}^\vee G^\Gamma$. For the duration of this section we fix a strong real form x of G. Let $\theta = \theta_x = \text{int}(x)$ and $K = K_x = G^\theta$ as usual, and suppose s is an elliptic element of K. Choose an antiholomorphic involution γ of G fixing s corresponding to the Cartan involution θ, and let $G(\mathbf{R}) = G^\gamma$. Thus $G(\mathbf{R}) \cap K$ is a maximal compact subgroup of $G(\mathbf{R})$ (cf. §1).

Let H be the identity component of the centralizer of s in G. Then γ and θ stabilize H, and $\gamma|_H$ is a real form of H corresponding to the Cartan involution $\theta|_H$. Let $H(\mathbf{R}) = H^\gamma$; then $s \in H(\mathbf{R})$. Let G^{can} be the canonical covering of G (cf. §2).

Now suppose we are given an inverse image $\tilde{s}$ of s in G^{can}, and a Borel subgroup B of G containing s. Suppose Θ is a virtual $(\mathfrak{g}, K)$-module, with integral infinitesimal character χ. Then the descent of Θ, written

$$Des_{(\tilde{s},B,\chi)}(\Theta), \tag{5-1}$$

is defined by ([3, Definition 2.4]). It is an $H(\mathbf{R})$-invariant eigendistribution defined in a neighborhood of the identity Ω in $H(\mathbf{R})$.

Let $\tilde{H}$ be the two-fold cover of H defined by $\rho - \rho_H$, where ρ (resp. ρ_H) is given by the Borel subgroup B (resp. $B_H = B \cap H$) of G (resp. H). Then ([3, Corollary 2.13]) there exists a complex virtual character Θ_H of $\tilde{H}(\mathbf{R})$ such that

$$Des_{(\tilde{s},B,\chi)}(\Theta)|_\Omega = \Theta_H|_\Omega. \tag{5-2}$$

(By a complex virtual character we mean a finite linear combination of irreducible characters with complex coefficents.) Thus consider $Des_{(\tilde{s},B,\chi)}(\Theta)$ as a complex virtual $(\mathfrak{h}, \tilde{K}_H)$-module, modulo the space of complex virtual characters vanishing on a neighborhood of the identity. Here $\tilde{K}_H$ is the inverse image of $K_H = H^\theta$ in $\tilde{H}$.

Now if Θ is a standard module then $Des_{(\tilde{s},B,\chi)}(\Theta)$ may be computed explicitly ([3, Corollary 2.13]). We compute this in terms of L-data as follows.

Let $S = (x, T^\Gamma, B, y, {}^dT^\Gamma, {}^dB)$ be a set of L-data, and $\Theta = I(S)$ the standard $(\mathfrak{g}, K)$-module defined by S. Suppose $\tilde{s}$ and B' are specified as in (5-1), so $Des_{(\tilde{s},B',\chi)}$ is defined, and let $s = \pi(\tilde{s})$. Then $Des_{(\tilde{s},B',\chi)}(\Theta)$ is zero unless $s \in T$ and $B = B'$ (up to conjugation by K and HK respectively), so we assume this is the case.

First we define E-groups containing H and ${}^\vee H$. Let $H^\Gamma = H \cup Hx \subseteq G^\Gamma$. Let δ_H^0 be any element in Hx such that $\text{int}(\delta_H^0)$ is a principal involution of H. Let $\mathfrak{D}_H$ be the conjugacy class of (δ_H^0, B_H^0) where B_H^0 is any Borel subgroup of H which is large with respect to δ_H^0. Thus $(H^\Gamma, \mathfrak{D}_H)$ is an

E-group containing H. Let ${}^\vee H^\Gamma$ be the E-group containing ${}^\vee H$ defined by the element ${}^\vee z_\rho$.

Choose a Cartan subgroup ${}^dT_H$ of ${}^\vee H^\Gamma$ dual to T^Γ. As in ([**1**, discussion preceding Theorem 7.4]) we obtain a Borel subgroup ${}^dB_H$ of ${}^\vee H$ containing ${}^dT_H$, and carry over the element y to an element y_H of ${}^dT_H$.

(5-3) Theorem ([1,Theorem 7.4]). *Given $S = (x, T^\Gamma, B, y, {}^dT^\Gamma, {}^dB)$, with corresponding standard module Θ as above, let*

$$S_H = (x, T^\Gamma, B \cap H, y_H, {}^dT_H, {}^dB_H),$$

and let $\Theta_H = I(S_H)$ be the corresponding standard $(\mathfrak{h}, \tilde{K}_H)$-module.

Then $Des_{(\tilde{s},B,\chi)}(\Theta) = \sigma\Theta_H$ for some non-zero constant $\sigma \in \mathbf{C}$.

The dependence of $Des(\Theta)$ on the inverse image $\tilde{s}$ of s is only in the constant σ. Thus if $\tilde{s}$ and $\tilde{s}'$ have the same image in G, then $Des_{(\tilde{s},B,\chi)}(\Theta) = \xi Des_{(\tilde{s}',B,\chi)}(\Theta)$ for some $\xi \in \mathbf{C}$.

(5-4) **Example:** *SL(2).*

(1) Let $G = SL(2)$, and let $\Theta = \pi^+$ be the holomorphic discrete series representation of $SU(1,1)$. Let s be a regular element of the compact Cartan subgroup $T(\mathbf{R})$ of G (note that $G^{can} = G$), so $H(\mathbf{R}) = T(\mathbf{R})$. Let χ be the infinitesimal character of Θ. Then $Des_{(s,B^+,\chi)}(\Theta)$ is the virtual character $e^{in\theta}$ of $T(\mathbf{R})$, and $Des_{(s,B^-,\chi)}(\Theta) = 0$. In this case $\rho - \rho_H = \rho$ exponentiates to H, so $\tilde{H} = H \times \mathbf{Z}/2\mathbf{Z}$ is identified with H.

(2) Suppose $G = PGL(2)$ and Θ is a discrete series representation of $PU(1,1)$. With the remaining choices set as above in case (1), $Des_{(s,B^+,\chi)}(\Theta) = e^{i(n+\frac{1}{2})\theta}$, which is a virtual character of the two-fold cover $\tilde{T}(\mathbf{R})$ of $T(\mathbf{R})$. In this case $G^{can} = SL(2)$ is the two-fold cover of G, but the resulting freedom in the choice of $\tilde{s}$ has no effect. □

Note that Θ_H and Θ come from the same Cartan subgroup T. Thus if Θ is a discrete series representation, so is Θ_H. In fact, in this case $Des(\Theta)$ is simply a constant times the discrete series representation of $\tilde{H}$ with the same Harish-Chandra parameter as Θ.

For the applications to lifting we need to regard $Des(\Theta)$ as belonging to the block of a finite-dimensional representation, which is not necessarily true in Theorem (5-3) (cf. Example (5-6)). However we may take advantage of the fact that $Des(\Theta)$ is defined only near the identity, so we have some flexibility in the choice $Des(\Theta)$. In order to do this without changing the infinitesimal character we need to pass to the canonical cover of H. This is the reason for the choice of H^Γ in Definition (4-16).

The character Θ_H of Theorem (5-2) belongs to the block of a finite dimensional representation if and only if y_H defines a principal involution of ${}^\vee H$ (cf. (2-31)). Thus we would like to replace y_H by ${}^\vee\delta \in {}^\vee\mathfrak{D}_H$; however, this would change the infinitesimal character of Θ_H if $y_H^2 \neq {}^\vee\delta^2 = {}^\vee z_\rho$. We compensate for this by changing the E-group ${}^\vee H^\Gamma$. Thus, in the setting of Theorem (5-3), let ${}^\vee H^{\Gamma'}$ be the extended group containing ${}^\vee H$ defined by $y^2 \in Z({}^\vee G) \subset Z({}^\vee H)$. L-data for this group defines representations of the canonical cover H^{can} of H (cf. Theorem (2-50)). Choose ${}^d T_H$ and ${}^d B_H$ as in Theorem 7.4, and choose ${}^\vee\delta'_H \in {}^d T_H \cap {}^\vee\mathfrak{D}'_H$ as in ([1, Corollary 7.7]).

(5-5) Lemma ([1, Corollary 7.7]). *The conclusion of Theorem 7.4 holds with ${}^\vee H^\Gamma$ replaced by ${}^\vee H^{\Gamma'}$ and S_H replaced by*

$$S'_H = (x, T^\Gamma, B \cap H, {}^\vee\delta'_H, {}^d T_H, {}^d B_H).$$

The standard module Θ_H defined by the data S'_H is in the block of a finite dimensional representation. The module Θ_H is a standard $(\mathfrak{h}, K_H^{can})$-module, of type $\tau(y^2\,{}^\vee z_{\rho_H})$ (Definition (2-49)).

(5-6) **Example:** $H{=}G$. Suppose $\tilde{s} = 1$, and let Θ be any standard character. Let $S = (x, T^\Gamma, B, y, {}^d T^\Gamma, {}^d B)$ be L-data with $I(S) = \Theta$. Then $H = G$, and clearly $Des(\Theta) = \Theta$, which is the conclusion of Theorem (5-3), with $S_H = S$. Lemma (5-4) then says that Θ agrees near the identity with a character Θ' in the block of a finite dimensional representation.

For example (cf. Example (2-22)), let $G = SL(2)$ and let Θ be the irreducible non-spherical principal series ps_o of $SU(1,1)$ with infinitesimal character ρ. Then Θ is itself a block, but it agrees near the identity with the reducible spherical principal series ps_e, which is in the block of the trivial representation. In terms of L-data this is achieved by replacing $y = t_0\,{}^\vee\delta$ in Example (2-14(4)) with ${}^\vee\delta$, to obtain the data of (2-14(3)). In this case there is no change in E-groups.

Let $G = PGL(2)$, and let Θ be the irreducible principal series $ps_+(2\rho)$ of $PU(1,1)$. In this case $PU(1,1)$ has *no* finite dimensional representations at this infinitesimal character. Thus we need to pass to G^{can}, and obtain a character Θ' of $\widetilde{PU}(1,1)$, which has the same infinitesimal character, and is in the block of a finite dimensional representation. Thus Θ' is obtained by pulling back the reducible spherical principal series $ps_{trivial}$ of $PU(1,1)$ with infinitesimal character ρ to $\widetilde{PU}(1,1)$, and translating from infintesimal character ρ to 2ρ. This is achieved on the level of L-data by replacing $(\delta, T_s, B_s, t_0\,{}^\vee\delta, {}^d T_c, {}^d B^+)$ (for the L-group of G) with the data $(\delta, T_s, B_s, {}^\vee\delta, {}^d T_c, {}^d B^+)$ for the genuine E-group of G. □

The constant σ of Theorem (5-3) is computed as follows. Recall the length function ℓ_0 was defined in (2-34).

(5-7) Corollary ([1, Corollary 7.11]). *In the setting of Theorem (5-3), let Λ be the as in the discussion preceding (2-9), lifted to the inverse image $\tilde{T}(\mathbf{R})$ of $T(\mathbf{R})$ in $\tilde{H}(\mathbf{R})$. Then the constant σ is given by:*

$$\sigma = (-1)^{\ell_0(\Theta)+\ell_0(\Theta_H)} \frac{\Lambda}{e^{d\Lambda}}(\tilde{s}).$$

Note that Λ and $e^{d\Lambda}$ have the same differential, so the quotient factors to the component group of $\tilde{T}(\mathbf{R})$. If G^{can} is a finite covering of G (i.e. if there is no compact torus in the center of G), then this is a finite group and σ is a root of unity.

6. Stable Characters

Once again we are given G, along with L-groups ${}^{\vee}G^{\Gamma}$ and G^{Γ}. We can now define stable characters in this setting.

(6-1) Definition ([1, Definition 8.1]).

(1) *Fix an admissible pair (x,y) (Definition (2-24)), and let $\mathfrak{B}$ and ${}^{\vee}\mathfrak{B}$ be the corresponding dual blocks (Corollary (2-25)). We say a virtual $(\mathfrak{g}, K_x)$-module $\Theta \in \mathfrak{B}$ is stable if*

$$< \Theta, {}^{d}Z >= 0$$

for all ${}^{d}Z \in {}^{\vee}\mathfrak{B}$ which vanish near the identity.

(2) *A virtual $(\mathfrak{g}, K_x)$-module Θ is said to be stable if the projection of Θ on $\mathfrak{B}$ is stable for all blocks $\mathfrak{B}$.*

Recall (cf. §2) we consider an L-packet Π to be a $\mathbb{Z}$-module of virtual $(\mathfrak{g}, K_x)$-modules (spanned by the irreducible modules in an L-packet in the usual sense). We say a standard module I is contained in Π if the unique irreducible submodule J of I is contained in Π. If Π is non-tempered this does *not* imply all the constituents of I are contained in Π.

Definition (6-1) agrees with the usual one as is seen by:

(6-2) Theorem ([1, Theorem 8.2]).

(1) *Let Π be an L-packet of $(\mathfrak{g}, K_x)$-modules, and let $\Theta_\Pi = \sum_{I\in\Pi} I$. Then θ_Π is stable.*

(2) *Any stable virtual $(\mathfrak{g}, K_x)$-module is a finite sum of terms of this form.*

(6-3) **Example:** *SL(2).* Consider the blocks (a-d) of Example (2-27). The only virtual character of $PGL(2)$ ***contained in a block*** which vanishes near the identity is $\pi^{1,1}_{trivial} - \pi^{1,1}_{sgn}$ of $PU(1,1)$. Therefore $\pi_+ + \pi_-$, ps_o, and $\pi^{1,1}_{trivial}$ for $SU(1,1)$ are stable, as are $\pi^{2,0}_{trivial}$ and $\pi^{0,2}_{trivial}$ for $SU(2,0)$ and $SU(0,2)$.

On the other hand no non-zero virtual characters of $SL(2)$ which are contained in a block vanish near the identity, and all virtual characters of $PGL(2)$ are stable. □

(6-4) **Example:** *Discrete Series.* Let $\Pi = \{\pi_1, \pi_2, \ldots, \pi_n\}$ be an L-packet of discrete series representations of the quasisplit form δ of G. Then Π is contained in the block $\mathfrak{B}$, with dual block ${}^{\vee}\mathfrak{B}$, defined by the admissible pair $(\delta, {}^{\vee}\delta)$ (cf. Examples (2-20),(2-29), and (2-31)).

The form ${}^\vee\delta$ of ${}^\vee G$ is split, and the discrete series representations of G are dual to a subset $\{{}^\vee I_1, {}^\vee I_2, \dots, {}^\vee I_n\}$ of the minimal principal series (standard modules) of ${}^\vee G$. These standard modules are induced from characters of the split Cartan of ${}^\vee G$, which agree on the identity component. By the induced character formula the induced characters ${}^\vee I_i$ each agree on a neighborhood of the identity. Furthermore the differences ${}^\vee I_i - {}^\vee I_j$ span the subspace of $< \{{}^\vee I_i\} >$ consisting of the virtual characters with this property (cf. [**1**, Lemma 7.5]). It follows immediately that (up to scalar) $\theta_\Pi = \sum_i \pi_i$ is the unique stable element of Π. (The proof for general L-packets is basically the same.)

Dual to this example is the statement that a minimal principal series (standard) module itself is stable. □

Recall a semisimple element $g \in G$ is said to be strongly regular if the centralizer of g is a Cartan subgroup. The strongly regular elements are dense in G, and for such elements "stable" conjugacy is the same as conjugacy by G [**11**]. Given a strong real form x, let $G^\#$ be the set of strongly regular semisimple elements g of G satisfying the following condition: there exists a real form γ corresponding to a conjugate of x such that $g \in G^\gamma$. Note that $G^\#$ is invariant under conjugation by G.

If $\Theta \in \mathfrak{V}(\mathfrak{g}, K_x)$, we may try to define a function Φ_Θ on $G^\#$ as follows. For $g \in G^\#$ choose γ corresponding to some conjugate x' of x such that $g \in G(\mathbf{R}) = G^\gamma$. Consider Θ as an element of $\mathfrak{V}(\mathfrak{g}, K_{x'})$ and let F_Θ denote the function on the strongly regular semisimple elements of $G(\mathbf{R})$ which represents Θ [**6**]. Let $\Phi_\Theta(g) = F_\Theta(g)$. This is not necessarily well defined: it may depend on the choice of γ.

(6-5) Theorem ([1, Theorem 8.3.]). *The following conditions on a virtual* $(\mathfrak{g}, K_x)$*-module* Θ *are equivalent:*

1. *Fix* γ *associated to x, and let* $G(\mathbf{R}) = G^\gamma$*. Then* $F_\Theta(h') = F_\Theta(h)$ *for all strongly regular semisimple elements* $h, h' \in G(\mathbf{R})$ *which are conjugate via* G*;*
2. Φ_Θ *is a well defined function on* $G^\#$*;*
3. $\Theta = \sum_{i=1}^n \alpha_i \Theta_{\Pi_i}$ *where* $\{\Pi_1, \Pi_2, \dots, \Pi_n\}$ *is a set of L-packets;*
4. Θ *is stable.*

This suggests a stronger version of the notion of stability, which takes into account different real forms of G. By a virtual $(\mathfrak{g}, K)$-module Θ, we mean a formal sum $\Theta = \sum_x \Theta_x$, where Θ_x is a virtual $(\mathfrak{g}, K_x)$-module, and the sum runs over representatives of the equivalence classes of strong real forms of G. We allow formal infinite sums (which are possible only if the center of G is infinite). Note that K is merely a formal symbol, and does not denote the fixed points of an involution of G. Let $\mathfrak{V}(\mathfrak{g}, K)$ denote the vector space of virtual $(\mathfrak{g}, K)$-modules:

(6-6) $$\mathfrak{V}(\mathfrak{g}, K) = \Pi_x \mathfrak{V}(\mathfrak{g}, K_x).$$

Projection of $\mathfrak{V}(\mathfrak{g}, K)$ onto $\mathfrak{V}(\mathfrak{g}, K_x)$ is defined with respect to this decomposition.

We now define a *super-block* $\mathfrak{B}$ to be a union of blocks of different strong real forms, by analogy with the definition of super L-packets (Definition (2-15)). By Corollary (2-25) a super block is associated to the conjugacy class of a semi-simple element $y \in {}^\vee G^\Gamma - {}^\vee G$, with $y^2 \in Z({}^\vee G)$:

$$\mathfrak{B}(y) = \cup_x \mathfrak{B}(x, y). \tag{6-7}$$

Here and below, x runs over a set of representatives for the equivalence classes of strong real forms of G.

From Definition (2-38) we obtain a perfect pairing

$$\mathfrak{B}(y) \times (\oplus_x {}^\vee \mathfrak{B}(x, y)) \rightarrow \mathbf{Z}. \tag{6-8}$$

We want to introduce super-stability by analogy with Definition (6-1), using virtual characters in $\oplus_x {}^\vee \mathfrak{B}(x, y)$ that vanish near the identity. Note that the modules in ${}^\vee \mathfrak{B}(x, y)$ have infinitesimal character associated to x^2, so this may vary as x varies. We get around this by passing from ${}^\vee G$ to its canonical cover ${}^\vee G^{can}$. Thus, given an admissible pair (x, y), by Proposition (2-50) we obtain a block ${}^\vee\widetilde{\mathfrak{B}}(x, y)$ of $({}^\vee\mathfrak{g}, {}^\vee\widetilde{K}_y)$-modules. Now each block ${}^\vee\widetilde{\mathfrak{B}}(x, y)$ may be specialized to the same infinitesimal character, and we obtain a perfect pairing

$$\mathfrak{B}(y) \times (\oplus_x {}^\vee\widetilde{\mathfrak{B}}(x, y) \rightarrow \mathbf{Z}. \tag{6-9}$$

It is now necessary to renormalize our definition of length. Recall A_f is the split part of a fundamental Cartan subgroup of G.

(6-10) Definition. *Fix a strong real form x of G and define*

$$q(G, x) = \frac{1}{2}[\dim(G/K_x) - \dim A_f] \in \mathbf{Z}, \text{ and } q(G) = q(G, \delta).$$

Fix an admissible pair (x, y). For $\pi \in \mathfrak{B}(x, y)$, define $\ell(\pi)$ by (2-34) with

$$c = q({}^\vee G, y) + q(G).$$

Let $<,>$ be the pairing of Definition (2-38) with this normalization of ℓ.

(6-11) Definition.

(1) *Fix a strong real form y of ${}^\vee G$, with corresponding super-block $\mathfrak{B} = \mathfrak{B}(y)$. We say a virtual $(\mathfrak{g}, K)$-module $\Theta \in \mathfrak{B}$ is super-stable if $< \Theta, {}^d Z >= 0$ for all ${}^d Z \in \widetilde{\mathfrak{B}}(y)$ vanishing near the identity.*

(2) *A virtual (g, K)-module Θ is said to be super-stable if the projection of Θ on a super-block $\mathfrak{B}$ is super-stable for all $\mathfrak{B}$.*

If Θ is super-stable, then for all x the projection of Θ on $\mathfrak{V}(\mathfrak{g}, K_x)$ is stable. The converse is false: super-stability includes a compatibility condition on signs as the real form varies.

(6-12) **Example:** *SL(2).* Return as in Example (6-3) to the blocks of Example (2-27). We now obtain the additional virtual character $ps_+(2\rho) - ps_-(2\rho)$ of $PU(1,1)$ which vanishes near the identity (but is not contained in a block). This imposes an extra condition on super-stability for $SL(2)$. Furthermore, lifting to $\widetilde{PU}(1,1)$, we have four characters which agree near the identity of $\widetilde{PU}(1,1)$: $ps_+, ps_-, ps_+(2\rho)$, and $ps_-(2\rho)$ (lifted to $\widetilde{PGL}(2,\mathbf{R})$, and translated to infinitesimal character ρ). Taking into account the signs coming from the pairing $<,>$ we see the following virtual characters span the superstable virtual characters of $SL(2)$: $\pi_+ + \pi_- - \pi^{2,0}_{trivial} - \pi^{0,2}_{trivial}$, ps_e, and ps_o.

Similarly the super-stable virtual characters for $PGL(2)$ are $\pi_d - \pi^{2,0}_{trivial}$, ps_+, ps_-, $ps_+(2\rho)$, and $ps_-(2\rho)$. □

(6-13) **Example:** *Discrete Series.* Group together the discrete series L-packets Π_x of Example (6-4) into a super-packet $\mathbf{\Pi} = \cup_x \Pi_x$. Note that $|\mathbf{\Pi}| = |Z(G)|2^n$ (cf. Example (2-29)). Dual to these representations are *all* minimal principal series of the split form of the dual group. As a result there is only one super-stable sum in $\mathbf{\Pi}$, given by

$$\sum_x (-1)^{\dim(G/K_x)} \Theta_{\Pi_x}. \tag{6-14}$$

□

Fix a (conjugacy class of a) quasi-admissible homomorphism ϕ of the Weil group into ${}^\vee G^\Gamma$. For x a strong real form of G let Π_x be the corresponding L-packet (which may be empty), and let $\Theta_x = \Theta_{\Pi_x}$ be the stable virtual character of Theorem (6-2). Let

$$\Theta_{\mathbf{\Pi}} = \sum_x (-1)^{q(G,x)} \Theta_x. \tag{6-15}$$

We obtain a super version of Theorem (6-5) by taking all strong real forms at once. Let $G^\natural$ denote the set of strongly regular semisimple elements of G such that there exists a strong real form $x \in G^\Gamma$, γ corresponding to a conjugate of x, and $g \in G^\gamma$ (see the discussion preceding Theorem (6-5)). If Θ is a virtual $(\mathfrak{g}, K)$-module, we repeat the construction preceding Theorem (6-5), and define a function $\Phi_\Theta(g) = F_{\Theta_x}(g)$, where now we choose both x and γ.

(6-15) Theorem ([1, Theorem 8.12]). *The following conditions are equivalent:*

1. *For all strong real forms x and x' of G the following condition holds. Choose γ and γ' associated to x and x', let $G(\mathbf{R}) = G^\gamma$, and let $G(\mathbf{R})' = G^{\gamma'}$. Then $F_\Theta(h) = F_{\Theta'}(h')$ for all strongly regular semisimple elements $h \in G(\mathbf{R})$ and $h' \in G(\mathbf{R})'$ which are conjugate via G;*

(2) Φ_Θ *is well-defined;*
(3) $\Theta = \sum_i \alpha_i \Theta_{\Pi_i}$ *where* $\{\Pi_1, \Pi_2, \ldots, \Pi_n\}$ *is a set of super L-packets;*
(4) Θ *is super-stable.*

(6-16) **Example:** *Trivial Representation.* Let $\pi_{trivial,x}$ denote the trivial representation of the strong real form x. Then by (1),

$$\sum_x \pi_{trivial,x}$$

is super-stable. □

7. Lifting of Characters

We now assemble the pieces and define lifting. Once again fix G, an inner class of real forms of G, and L-groups G^Γ and ${}^\vee G^\Gamma$ (Definition (1-10)).

Let S be an equivalence class of endoscopic data, with $(\tilde{s}, {}^\vee H^\Gamma, {}^\vee\mathfrak{D}_H)$ an element of S (Definition (4-9)). The identity component ${}^\vee H$ of ${}^\vee H^\Gamma$ is the identity component of the centralizer of $s = \pi(\tilde{s})$ in ${}^\vee G$. Let H be the endoscopic group determined by S, so H is a quasisplit group with ${}^\vee H^\Gamma$ an E-group for H (Definition (4-2)). Let θ_H be a Cartan involution of H, $K_H = H^{\theta_H}$, and let $\tilde{K}_H$ denote the cover of K_H corresponding to the element $\rho - \rho_H$ (cf. §2, and [**2**, Proposition 7.12]).

Now suppose x is a strong real form of G, and write $\theta = \operatorname{int}(x)$ and $K_x = G^\theta$ as usual. Then choose an extended group $(H^\Gamma, \mathfrak{D}_H)$ containing H by Definition (4-16). Fix $\delta_H \in \mathfrak{D}_H$ with $\operatorname{int}(\delta_H) = \theta_H$; then $(\delta_H)^2 = x^2$.

Furthermore suppose we are given a stable virtual module Θ_H for the strong real form δ of H (Definition (6-1)). Thus Θ_H is contained in a block $\mathfrak{B}_H$ for H, defined by an admissible pair (δ, y) (Definition 2-24).

We define blocks for $G, {}^\vee G$ and ${}^\vee H$. Let $\mathfrak{B} = \mathfrak{B}(x, y)$ be the block of $(\mathfrak{g}, K_x)$-modules defined by (x, y), and let ${}^\vee\mathfrak{B}$ be the dual block of $({}^\vee\mathfrak{g}, {}^\vee K_y)$-modules. Here ${}^\vee\theta = {}^\vee\theta_y$ and ${}^\vee K = {}^\vee G^{{}^\vee\theta}$ are as usual. Let ${}^\vee\tilde{K}_y$ be the preimage of ${}^\vee K_y$ in ${}^\vee G^{can}$. Let ${}^\vee K_H = {}^\vee H^{{}^\vee\theta}$, and let ${}^\vee\tilde{K}_H$ be the preimage of ${}^\vee K_H$ in ${}^\vee H^{can}$. Then dual to $\mathfrak{B}_H$ is a block ${}^\vee\mathfrak{B}_H$ of $({}^\vee\mathfrak{h}, {}^\vee\tilde{K}_H)$-modules of type $\tau(x^2 z_{\rho_H})$ (cf. Proposition (2-49)).

We next choose infinitesimal characters for $G, H, {}^\vee G$ and ${}^\vee H$. Let χ_H be the infinitesimal character of Θ_H. Fix a Borel subgroup ${}^d B_H$ and a Cartan subgroup ${}^d T_H \subset {}^d B_H$ of ${}^\vee H$. Choose $\lambda \in {}^d\mathfrak{t}_H$, ${}^d B_H$-dominant, so that $\chi_H = \chi(\lambda)$. We assume λ is regular for G, and let $\chi = \chi(\lambda)$ be the associated infinitesimal character for G; thus both χ and χ_H are regular. Let ${}^d B_{des}$ be the Borel subgroup (containing ${}^d B_H$) of ${}^\vee G$ making λ dominant.

Let ${}^d\chi$ be a regular infinitesimal character for ${}^\vee G$ associated to x^2. Thus if T is a Cartan subgroup of G, then ${}^d\chi = {}^d\chi({}^d\lambda)$ for some ${}^d\lambda \in \mathfrak{t}$, with $e^{2\pi i\, {}^d\lambda} = x^2$. Using the Borel subgroups ${}^d B_H$ and ${}^d B_{des}$ we obtain an isomorphism $\zeta : T \simeq T_H$ for some Cartan subgroup T_H of H (cf. [**1**,9.3(4)]). Let ${}^d\chi_H$ be the infinitesimal character for ${}^\vee H$ defined by $\zeta({}^d\lambda)$.

Define $Des = Des_{(\tilde{s},{}^dB_{des},{}^d\chi)}$ as in (5-1).

Given these choices and definitions we define *lifting* from H to G to be dual (more precisely, adjoint) to Harish-Chandra's method of descent on ${}^\vee G$ as follows.

(7-1) Definition ([1, Definition 9.4]).

(1) *Suppose Θ_H is a stable virtual $(\mathfrak{h}, \tilde{K}_H)$ module contained in the block $\mathfrak{B}_H$. The lift of Θ_H to G (and the strong real form x), written $Lift(\Theta_H)$, is the unique complex virtual $(\mathfrak{g}, K_x)$-module contained in $\mathfrak{B}$ satisfying the following condition:*

$$< Lift(\Theta_H), {}^dZ > = < \Theta_H, Des({}^dZ) >$$

for all ${}^dZ \in {}^\vee\mathfrak{B}$. The pairing on the left is on $\mathfrak{B} \times {}^\vee\mathfrak{B}$, and that on the right is on $\mathfrak{B}_H \times {}^\vee\mathfrak{B}_H$.

(2) *If Θ_H is any stable virtual $(\mathfrak{h}, \tilde{K}_H)$-module we define $Lift(\Theta_H)$ by linearity, and projection on blocks.*

If it is necessary to specify the strong real form x to which we are lifting we will write $Lift_x(\Theta_H)$.

Recall $Des({}^dZ)$ is only defined up to virtual characters which vanish on a neighborhood of the identity in ${}^\vee H$. Precisely because Θ_H is stable, and hence orthogonal to such characters, we see that $< \Theta_H, Des({}^dZ) >$, and hence $Lift(\Theta_H)$, is independent of the choice of $Des({}^dZ)$. Thus $Lift(\Theta_H)$ is well-defined.

We now compute the lift of a stable character. By Theorem (6-2) and Definition (7-1(2)) it is enough to do this for the stable sum of the characters in an L-packet. Thus let Π_H be an L-packet of $(\mathfrak{h}, \tilde{K}_H)$-modules, and let Θ_H be the stable virtual character $\sum_{I \in \Pi_H} I$ (Theorem (6-2)). We obtain an L-packet (possibly empty) of $(\mathfrak{g}, K_x)$-modules as follows. Choose $\phi_H : W_{\mathbf{R}} \rightarrow {}^\vee H^\Gamma$ giving rise to Π_H, and let ϕ denote the map $W_{\mathbf{R}} \rightarrow {}^\vee H^\Gamma \hookrightarrow {}^\vee G^\Gamma$. If ϕ is admissible we let Π denote the L-packet for G corresponding to ϕ; otherwise it is empty. We say Π_H is *relevant* to G if Π is non-empty.

We now state the main Theorem.

(7-2) Theorem ([1, Theorem 9.7]).

(1) *$Lift(\Theta_H) = 0$ if Π_H is not relevant to G.*

(2) *Suppose Π_H is relevant to G. Then*

$$Lift(\Theta_H) = \sum_{I \in \Pi} c_I I \tag{7-3}$$

for some constants $c_I \in \mathbf{C}^*$.

(3) *Let* $<,>: \Pi \times \tilde{S}_\phi \to \mathbf{C}^*$ *be the pairing of (3-4) restricted to* Π. *Let* $\tilde{s}$ *also denote the image of* $\tilde{s}$ *in* $\tilde{S}_\phi$. *Then*

$$(7\text{-}4) \qquad c_I = (-1)^{q(G,x)+q(H)} < I, \tilde{s} > .$$

We note that $c_I/c_{I'} \in \{\pm 1\}$ for all I and I'; we will write this in a different form in Theorem (7-7). Also note that (7-4) shows that $Lift(\Theta_H)$ is independent of the choices of blocks, infinitesimal characters, and $\tilde{s}$ (within the given equivalence class of endoscopic data) ([**1**, Lemma 9.6]).

(7-5) **Example:** $H{=}G$. Assume we are in the setting of case (2) in Theorem (7-2), and let $\Theta = \Theta_\Pi$ be the stable sum of the characters in the L-packet Π. Suppose $\tilde{s} \in \pi_1({}^\vee G^{can})(\mathbf{R})$, so ${}^\vee H = {}^\vee G$ and ${}^\vee H^\Gamma = {}^\vee G^\Gamma$. Choose endoscopic data by letting ${}^\vee\mathfrak{D}_H = {}^\vee\mathfrak{D}$. Let τ_x be the character $\tau(z_\rho x^{-2})$ of $\pi_1({}^\vee G^{can})(\mathbf{R})$ (Definition (2-49)); then $Lift_x(\Theta) = (-1)^{q(G,x)+q(G)}\tau_x(\tilde{s})\Theta$, and is thus stable. In particular, $\tilde{s} = 1$ gives $Lift_x(\Theta) = (-1)^{q(G,x)+q(G)}\Theta$. Finally, if the strong form x of G is quasisplit, the coefficent is one. If we do not choose ${}^\vee\mathfrak{D}_H = {}^\vee\mathfrak{D}$ this may change by tensoring with a one-dimensional representation of G (cf. Example (7-11)). □

(7-6) **Example:** *SL(2)*. With $G = SL(2)$, we obtain the following lifts from endoscopic groups. We use the notation of Examples (2-14),(2-22), (3-14) and (4-13). For more details on this case see ([**1**, §11]). In each case the covering $\tilde{H} \to H$ splits, and we identify $\tilde{H}$ with H.

(1) $\tilde{s} = 1$, so $\tilde{H}(\mathbf{R}) = SU(1,1)$. Then from the preceding example we get
  (a) $Lift_\delta(\pi^+ + \pi^-) = \pi^+ + \pi^-$,
  (b) $Lift_{t_0\delta}(\pi^+ + \pi^-) = -\pi^{2,0}_{trivial}$, $Lift_{-t_0\delta}(\pi^+ + \pi^-) = -\pi^{0,2}_{trivial}$,
  (c) $Lift_\delta(ps_e) = ps_e$, $Lift_\delta(ps_o) = ps_o$,
  (d) $Lift_{\pm t_0\delta}(ps_e) = Lift_{\pm t_0\delta}(ps_o) = 0$.
  Writing $\pi^{1,1}_{trivial} = ps_e - (\pi^+ + \pi^-)$, we also see:
  (d) $Lift_\delta(\pi^{1,1}_{trivial}) = \pi^{1,1}_{trivial}$,
  (f) $Lift_{t_0\delta}(\pi^{1,1}_{trivial}) = \pi^{2,0}_{trivial}$, $Lift_{-t_0\delta}(\pi^{1,1}_{trivial}) = \pi^{0,2}_{trivial}$.

(2) $\tilde{s} = -1$, with $\tilde{H}(\mathbf{R}) = SU(1,1)$. Note that $\tau_x(-1) = 1$ if $x = \delta$ or -1 if $x = \pm t_0\delta$. From Example (7-5), this is the same as (1) with the following changes:
  (b) $Lift_{t_0\delta}(\pi^+ + \pi^-) = \pi^{2,0}_{trivial}$, $Lift_{-t_0\delta}(\pi^+ + \pi^-) = \pi^{0,2}_{trivial}$,
  (f) $Lift_{t_0\delta}(\pi^{1,1}_{trivial}) = -\pi^{2,0}_{trivial}$, $Lift_{-t_0\delta}(\pi^{1,1}_{trivial}) = -\pi^{0,2}_{trivial}$.

(3) $\tilde{s} = \tilde{s}_e$. Now $\tilde{H}(\mathbf{R}) = S^1$. From Example (3-14) we may read off:
  (a) $Lift_\delta(e^{i\theta}) = -\pi^+ + \pi^-$,
  (b) $Lift_{t_0\delta}(e^{i\theta}) = i\pi^{2,0}_{trivial}$, $Lift_{-t_0\delta}(e^{i\theta}) = -i\pi^{0,2}_{trivial}$.

(4) $\tilde{s} = -\tilde{s}_e$. This is similar to (3):
  (a) $Lift_\delta(e^{i\theta}) = -\pi^+ + \pi^-$,
  (b) $Lift_{t_0\delta}(e^{i\theta}) = -i\pi^{2,0}_{trivial}$, $Lift_{-t_0\delta}(e^{i\theta}) = i\pi^{0,2}_{trivial}$.

(5) $\tilde{s} = \tilde{s}_h$. In this case $\tilde{H}(\mathbf{R}) = \mathbf{R}^*$, and lifting is simply induction from the minimal parabolic (possibly twisted). Define characters $\alpha_\pm$ of $\mathbf{R}^*$ by $\alpha_+(t) = |t|$, $\alpha_- t = t$. In this case there are two choices of endoscopic data, given by ${}^\vee\delta_H = \pm {}^\vee\delta$. In the first case we have:
  (a) $Lift_\delta(\alpha_+) = -ps_e$, $Lift_\delta(\alpha_-) = -ps_o$,
  (b) $Lift_{\pm t_0\delta}(\alpha_\pm) = 0$.
  In the second case this is changed by tensoring with sgn:
  (a′) $Lift_\delta(\alpha_+) = -ps_o$, $Lift_\delta(\alpha_-) = -ps_e$,
  (b′) $Lift_{\pm t_0\delta}(\alpha_\pm) = 0$.

If $G = PGL(2)$, the situation is as follows.

(1) $\tilde{s} = 1$. In this case the covering of H is trivial, so $\tilde{H}(\mathbf{R}) = PU(1,1)$. There are two choices of strong endoscopic data. For the choice ${}^\vee\mathfrak{D}_H = {}^\vee\mathfrak{D}$ we obtain:
  (a) $Lift_\delta(\pi_d) = \pi_d$,
  (b) $Lift_{t_0\delta}(\pi_d) = -\pi^{2,0}_{trivial}$,
  (c) $Lift_\delta(ps_{trivial}) = ps_{trivial}$, $Lift_\delta(ps_{sgn}) = ps_{sgn}$,
  (d) $Lift_\delta(ps_\pm(2\rho)) = ps_\pm(2\rho)$,
  (e) $Lift_{t_0\delta}(ps_{trivial}) = Lift_{t_0\delta}(ps_{sgn}) = Lift_{t_0\delta}(ps_\pm(2\rho)) = 0$.
  Writing $\pi^{1,1}_{trivial} = ps_{trivial} - \pi_d$ and $\pi^{1,1}_{sgn} = ps_{sgn} - \pi_d$ gives:
  (f) $Lift_\delta(\pi^{1,1}_{trivial}) = \pi^{1,1}_{trivial}$,
  (g) $Lift_{t_0\delta}(\pi^{1,1}_{trivial}) = \pi^{2,0}_{trivial}$,
  (h) $Lift_\delta(\pi^{1,1}_{sgn}) = \pi^{1,1}_{sgn}$,
  (i) $Lift_{t_0\delta}(\pi^{1,1}_{sgn}) = \pi^{2,0}_{trivial}$.
  Changing ${}^\vee\mathfrak{D}_H$ to $-{}^\vee\mathfrak{D}_H$ has the effect of interchanging $ps_{trivial}$ and ps_{sgn}, and $ps_\pm(2\rho)$:
  (c′) $Lift_\delta(ps_{trivial}) = ps_{sgn}$, $Lift_\delta(ps_{sgn}) = ps_{trivial}$,
  (d′) $Lift_\delta(ps_\pm(2\rho)) = ps_\mp(2\rho)$,
  (f′) $Lift_\delta(\pi^{1,1}_{trivial}) = \pi^{1,1}_{sgn}$,
  (h′) $Lift_\delta(\pi^{1,1}_{sgn}) = \pi^{1,1}_{trivial}$.

(2) $\tilde{s} = -1$. This is the same as (1), with the following changes:
  (b) $Lift_{t_0\delta}(\pi_d) = \pi^{2,0}_{trivial}$,
  (g) $Lift_{t_0\delta}(\pi^{1,1}_{trivial}) = -\pi^{2,0}_{trivial}$,
  (i) $Lift_{t_0\delta}(\pi^{1,1}_{sgn}) = -\pi^{2,0}_{trivial}$.

(3) $\tilde{s} = \tilde{s}_h$. In this case $\tilde{H}$ is the genuine two-fold cover of H, and $\tilde{H}(\mathbf{R}) = \mathbf{R}^* \cup i\mathbf{R}^*$ (cf. Example (2-42)). Let $\tilde{\alpha}_\pm$ be the genuine character of $\tilde{H}(\mathbf{R})$ defined by $\tilde{\alpha}_\pm(t) = t$ ($t \in \mathbf{R}^*$), $\tilde{\alpha}_\pm i = \pm i$. Define $\tilde{\alpha}_\pm(2\rho)$ similarly. Again there are two choices of endoscopic data. One such choice yields the following.
  (a) $Lift_\delta(\tilde{\alpha}_+) = -ps_{trivial}$, $Lift_\delta(\tilde{\alpha}_-) = -ps_{sgn}$,
  (b) $Lift_\delta(\tilde{\alpha}_\pm(2\rho)) = -ps_\pm(2\rho)$,
  (c) $Lift_{t_0\delta}(\tilde{\alpha}_\pm) = Lift_{t_0\delta}(\tilde{\alpha}_\pm(2\rho)) = 0$.
  In the second case this is changed as follows.
  (a′) $Lift_\delta(\tilde{\alpha}_+) = -ps_{sgn}$, $Lift_\delta(\tilde{\alpha}_-) = -ps_{trivial}$,

(b′) $Lift_\delta(\tilde{\alpha}_\pm(2\rho)) = -ps_\mp(2\rho)$. □

In the setting of Theorem (7-2) fix a standard module $I \in \Pi$. The set of standard modules contained in Π is then $\{w \times I | w \in W_{im}\}$ (cf. (3-19)). Let $\delta_{I,s} : W_{im} \to \{\pm 1\}$ be as in (3-20). The following Theorem computes the coefficents in lifting.

(7-7) Theorem ([1, Theorem 9.11]). *For $w \in W_{im}$, let $I_w = w \times I$. Define $c \in \mathbf{C}$ and $\varepsilon_I(w) \in \{\pm 1\}$ $(w \in W_{im})$ by $\varepsilon_I(1) = 1$ and*

$$Lift(\Theta_H) = c \sum_{(W_{im} \cap W(K_x,T)) \backslash W_{im}} \varepsilon_I(w) I_w. \tag{7-8}$$

Then

(1) $\varepsilon_I(w) = \delta_{I,s}(w) \in \{\pm 1\}$,

(2) $c = (-1)^{q(G,x)+q(H)} \dfrac{{}^d\Lambda}{e^{d_\lambda}}(\tilde{s})$, *where ${}^d\Lambda$ is as in Proposition (3-18).*

In particular $\varepsilon_I(w)$ may be computed using Corollary (3-23)).

This definition of lifting differs from the usual in part because $\tilde{s} \in {}^\vee G^{can}$. This only plays a serious role in super-lifting (discussed below). Thus if x is given and we replace $\tilde{s}$ with $\tilde{s}'$ having the same image in ${}^\vee G$, then $Lift_x(\Theta_H)$ is replaced by $\tau Lift_x(\Theta_H)$ for some constant τ ([1, Corollary 9.14]).

In some sense only the choice of weak endoscopic data is important, and not the choice of ${}^\vee\mathfrak{D}_H$. It is natural to fix $\phi : W_\mathbf{R} \to {}^\vee G^\Gamma$ and consider endoscopic groups through which this map factors. These endoscopic groups depend only on the weak endoscopic data. The next result says that the lifted character $Lift(\Theta_H)$ depends only on the weak data, and not on the choice of ${}^\vee\mathfrak{D}_H$, even though the character Θ_H is *not* independent of this choice.

Thus suppose $(\tilde{s}, {}^\vee H^\Gamma, {}^\vee\mathfrak{D}_H)$ is a set of endoscopic data such that a map $\phi : W_\mathbf{R} \to {}^\vee G^\Gamma$ factors through ${}^\vee H^\Gamma$. Let Π_H be the corresponding L-packet, with stable virtual character Θ_H. Fix a strong real form x of G.

(7-9) Corollary.

$$Lift_x(\Theta_H) = (-1)^{q(G,x)+q(H)} \sum_{I \in \Pi} < I, \tilde{s} > I, \tag{7-10}$$

which is independent of the choice of ${}^\vee\mathfrak{D}_H$.

(7-11) **Example:** H=G. The effect of the choice of ${}^\vee\mathfrak{D}_H$ can be seen by considering the case $\tilde{s} = 1$. We may as well also assume we are lifting to the quasisplit form of G, so let $x = \delta$. Then if we choose ${}^\vee\mathfrak{D}_H = {}^\vee\mathfrak{D}$, Theorem (7-2) says $Lift(\Theta) = \Theta$ (cf. Example (7-5)). However if we take another choice of ${}^\vee\mathfrak{D}_H$ (cf. (1-12)) we see $Lift(\Theta) = \xi \otimes \Theta$ for ξ a one dimensional representation of $G(\mathbf{R})$ trivial on the identity component.

For example, let G be a split one-dimensional torus. Then (cf. Example (1-13)) there are two choices for ${}^\vee\mathfrak{D}_H$, and we obtain $Lift(trivial) = trivial$ and $Lift(trivial) = sgn$ respectively. □

(7-12) **Example:** *SL(2)*. The phenomenon of the previous example for $\mathbf{R}^*$ persists by lifting (induction) from $\mathbf{R}^*$ to $SU(1,1)$: compare Examples (7-6(4a) and (4a′)). □

Lifting as defined above differs from that of [**10**] by a constant, as follows directly from Theorem (7-7) and Corollary (3-23) (cf. [**1**, 9.17-9.18]).

We now vary the strong real form x of G, and define super-lifting.

(7-13) Definition. *Suppose we are in the setting of Definition (7-1), with x allowed to vary. Then the super-lift of Θ_H to G is defined to be the complex virtual $(\mathfrak{g}, K)$-module (cf. §6):*

$$Lift^*(\Theta_H) = \sum_x Lift_x(\Theta_H). \tag{7-14}$$

Here for any x, $Lift_x(\Theta_H)$ is the virtual $(\mathfrak{g}, K_x)$-module of Definition (7-1), and the sum runs over equivalence classes of strong real forms x of G.

(7-15) **Example:** *Discrete series for SL(2)*. From Example (7-6) we have:

(1) $\tilde{s} = 1 \quad \text{Lift}^*(\pi^+ + \pi^-) = \pi^+ + \pi^- - \pi^{2,0}_{trivial} - \pi^{0,2}_{trivial}$,
(2) $\tilde{s} = -1 \quad \text{Lift}^*(\pi^+ + \pi^-) = \pi^+ + \pi^- + \pi^{2,0}_{trivial} + \pi^{0,2}_{trivial}$,
(3) $\tilde{s} = \tilde{s}_e \quad \text{Lift}^*(\pi^+ + \pi^-) = -\pi^+ + \pi^- + i\pi^{2,0}_{trivial} - i\pi^{0,2}_{trivial}$,
(4) $\tilde{s} = -\tilde{s}_e \quad \text{Lift}^*(\pi^+ + \pi^-) = -\pi^+ + \pi^- + i\pi^{2,0}_{trivial} - i\pi^{0,2}_{trivial}$.
□

We obtain a super-version of Theorem (7-2). If Π_H is an L-packet for H, let ϕ_H and ϕ be as in the discussion preceding Theorem (7-2), and let $\mathbf{\Pi} = \mathbf{\Pi}_\phi$ be the super-packet for G defined by ϕ. For x a strong real form of G, let $\Pi_{\phi,x}$ be the L-packet for this strong real form defined by ϕ. Let Θ_H be the stable virtual $(\mathfrak{h}, \tilde{K}_H)$-module associated to the L-packet Π_H.

(7-16) Theorem.

$$\begin{aligned} Lift^*(\Theta_H) &= (-1)^{q(H)} \sum_{I \in \mathbf{\Pi}} (-1)^{q(G,x(I))} < I, \tilde{s} > I \\ &= (-1)^{q(H)} \sum_x (-1)^{q(G,x)} \sum_{I \in \Pi_{\phi,x}} < I, \tilde{s} > I. \end{aligned} \tag{7-17}$$

Here $x(I)$ is the strong real form of which I is a representation.

(7-18) **Example:** *H=G (Example (7-5))*. As in Theorem (7-16) suppose $\tilde{s} \in \pi_1({}^\vee G^{can})$, so ${}^\vee H = {}^\vee G$, and we choose endoscopic data $(\tilde{s}, {}^\vee G^\Gamma, {}^\vee\mathfrak{D})$. Given a strong real form x of G, consider Example (7-5). Let Θ_x be the stable virtual character of the strong real form of G as in that Example, and let $\tau_x = \tau(z_\rho x^{-2})$. Then, if I is a representation of the strong real form x, $< I, \tilde{s} >$ equals $\tau_x(\tilde{s})$ and depends only on x. Consequently

$$\begin{aligned} Lift^*(\Theta_H) &= (-1)^{q(G)} \sum_x (-1)^{q(G,x)} \tau_x(\tilde{s}) \sum_{I \in \Pi_x} I \\ &= (-1)^{q(G)} \sum_x (-1)^{q(G,x)} \tau_x(\tilde{s}) \Theta_x. \end{aligned} \tag{7-19}$$

In particular, if $\tilde{s} = 1$, then this equals $(-1)^{q(G)}\Theta_{\Pi}$ (the super-stable character attached to the super L-packet $\mathbf{\Pi}$. □

(7-20) **Example:** *Trivial Representation.* Let $\pi_{trivial,x}$ denote the trivial representation of the strong real form x of G. Specialize the previous example to the case $\Theta = \pi_{trivial,\delta}$. From Theorem (6-15) (cf. Example (6-16)) we obtain:

$$Lift^*(\pi_{trivial,\delta}) = \sum_x \pi_{trivial,x}. \tag{7-21}$$

□

(7-22) **Example:** *Discrete Series for Sp(2n).*

We use the notation of Example (3-16). Fix $\mu = (\mu_0, \mu_1, \dots, \mu_n)$, with corresponding element $\tilde{s}(\mu) \in \tilde{S}$. Given $\epsilon = (\epsilon_1, \epsilon_2, \dots, \epsilon_n)$, recall $\pi(\epsilon)$ is the corresponding discrete series representation of Sp; it is a representation of $Sp(2n, \mathbf{R})$ if $\epsilon_0 = 0$, and of $Sp(k, n-k)$, $k = \sum_1^n \epsilon_i$, if $\epsilon_0 = 1$.

Let $q(\mu) = q(H)$ where H is the endoscopic group defined by $\tilde{s}(\mu)$. Let $Lift^*_\mu(\Theta_\mu)$ be the super-lifting from this endoscopic group to G. We obtain:

$$\begin{aligned} Lift^*_\mu(\Theta_\mu) = (-1)^{q(\mu)}(-1)^{\frac{1}{2}n(n+1)} \sum_{\substack{\epsilon \\ \epsilon_0=0}} \left[\prod_{i=1}^n (-1)^{\epsilon_i\mu_i}\right] \pi(\epsilon) + \\ (-1)^{q(\mu)}(-i)^{|\mu|}(-1)^{\mu_0} \sum_{\substack{\epsilon \\ \epsilon_0=1}} \left[\prod_{i=1}^n (-1)^{\epsilon_i\mu_i}\right] \pi(\epsilon). \end{aligned} \tag{7-23}$$

□

Inversion now takes a simple form. Let $\phi : W_{\mathbf{R}} \to {}^\vee G^\Gamma$ be a quasi-admissible homomorphism with corresponding super L-packet $\mathbf{\Pi} = \mathbf{\Pi}_\phi$, ${}^dT$ the centralizer of $\phi(\mathbf{C}^*)$, and ${}^\vee\delta \in {}^\vee\mathfrak{D}$ chosen so that $\delta|_{{}^dT} = \text{int}(\phi(j))|_{{}^dT}$. Then S_ϕ is the group of fixed points of $\text{int}({}^\vee\delta)$ on ${}^dT$. By Lemma (4-2), to define weak endoscopic data it is enough to choose $\tilde{s} \in {}^\vee G^{can}$ such that the image of $\tilde{s}$ is elliptic and fixed by $\text{int}({}^\vee\delta)$. We change notation slightly and let $\tilde{S}_\phi$ denote such a set of representatives (which is readily seen to exist). For $\tilde{s} \in \tilde{S}_\phi$ let $H_{\tilde{s}}$ be a corresponding endoscopic group, and choose ${}^\vee\mathfrak{D}_{H_s}$, making a set $(\tilde{s}, {}^\vee H_{\tilde{s}}^\Gamma, {}^\vee\mathfrak{D}_{H_s})$ of endoscopic data. Now ϕ factors through ${}^\vee H_{\tilde{s}}^\Gamma$, and we obtain a stable virtual $(\mathfrak{h}, \tilde{K}_H)$-module Θ_{H_s}. Then by Corollary (7-9) $Lift^*(\tilde{s}, \Theta_{H_s}) \in \mathfrak{V}(\mathfrak{g}, K)$ is independent of the choice of ${}^\vee\mathfrak{D}_{H_s}$. To avoid technicalities we assume the center of G is finite, so Π and $\tilde{S}_\phi$ are finite. The following Theorem follows immediately from Theorems (5-1), (7-16), and Fourier inversion on the group $\tilde{S}_\phi$.

(7-24) Theorem. *For $I \in \Pi_\phi$ a standard $(\mathfrak{g}, K_x)$-module,*

$$I = \frac{(-1)^{q(G,x)}}{|\tilde{S}_\phi|} \sum_{\tilde{s} \in \tilde{S}_\phi} (-1)^{q(H_s)} \overline{< I, \tilde{s} >} Lift^*(\tilde{s}, \Theta_{H_s}). \tag{7-25}$$

Fix a strong real form x of G. Recalling $Lift_x()$ denotes lifting for this strong real form, and projecting both sides of (7-25) onto $(\mathfrak{g}, K_x)$-modules, we obtain:

$$\frac{(-1)^{q(G,x)}}{|\tilde{S}_\phi|} \sum_{\tilde{s} \in \tilde{S}_\phi} (-1)^{q(H_s)} \overline{< I, \tilde{s} >} Lift_x(\tilde{s}, \Theta_{H_s})$$

$$= \begin{cases} I & \text{if } I \text{ is a } (\mathfrak{g}, K_x)\text{-module} \\ 0 & \text{if } I \text{ is a } (\mathfrak{g}, K_{x'})\text{-module for } x' \text{ not conjugate to } x. \end{cases} \tag{7-26}$$

This explains the ghosts of [**10**].

REFERENCES

1. J. Adams and D. Vogan, *Lifting of Characters and Harish-Chandra's Method of Descent*, submitted to Journal of AMS (preprint).

2. ———, *L-Groups, Projective Representations, and the Langlands Classification*, to appear in Amer. Journal Math.

3. ———, *Lifting of Characters and Harish-Chandra's Method of Descent*, to appear in Amer. Journal Math.

4. J. Arthur, *On Some Problems Suggested By the Trace Formula*, in Proc. of the Special Year in Harmonic Analysis, University of Maryland, R. Herb, R. Lipsman and J. Rosenberg, eds., SLN, vol. 1024, Springer-Verlag, New York, 1983.

5. A. Borel, *Automorphic L-Functions*, Proc. Symp. Pure Math., vol. 33, American Math. Soc., Providence, Rhode Island , 1979.

6. Harish-Chandra, *Invariant Eigendistributions on a Semi-simple Lie Group*, Trans. Amer. Math. Soc. **119** (1965), 457–508.

7. S. Helgason, in Differential Geometry, Lie Groups, and Symmetric Spaces, Academic Press, New York, 1978.

8. R. Langlands, *On the Classification of Irreducible Representations of Real Algebraic Groups*, in Representation Theory and Harmonic Analysis on Semisimple Lie Groups, P. Sally, D. Vogan, eds., Mathematical Surveys and Monographs, vol. 31, AMS, Providence, 1989.

9. D. Shelstad, *Orbital Integrals and a Family of Groups attached to a Real Reductive Group*, Ann. École Nat. Sup. Méc. Nantes **12** (1979), 1–31.

10. ———, *L-Indistinguishability for Real Groups*, Math. Ann. **259** (1982), 385–430 .

11. R. Steinberg, *Regular Elements of Semisimple Algebraic Groups*, Publ. I.H.E.S. (1965).

12. D. Vogan, *Representations of Real Reductive Lie Groups*, Birkhauser, Boston, 1981.

13. ———, *Irreducible Characters of Semisimple Lie Groups IV. Character- Multiplicity Duality*, Duke Math. J. **49**, No. 4 (1982), 943–1073.

14. G. Zuckerman, *Tensor Products of Finite and Infinite Dimensional Representations of Semisimple Lie Groups*, Ann. of Math. (2) (1977).

Department of Mathematics, University of Maryland, College Park, MD 20742

E-mail: jda @ ida.umd.edu

HANDLING THE INVERSE SPHERICAL FOURIER TRANSFORM

JEAN–PHILIPPE ANKER

Princeton University and Cornell University

We use the standard notation and refer to [GV], [H] for more details. Let $X = G/K$ be a Riemannian symmetric space of the noncompact type. The basis of harmonic analysis on X was settled around the sixties, mainly by Harish–Chandra and S. Helgason. Briefly:

(1) We have a *Fourier transform*

$$\mathcal{H}f(\lambda, kM) = \int_G dx\, f(x)\, e^{-(i\lambda+\varrho)(H(x^{-1}k)}\,, \tag{1}$$

involving the *spherical principal series*

$$\{\pi_\lambda(x)\xi\}(kM) = e^{-(i\lambda+\varrho)(H(x^{-1}k))}\,\xi(k(x^{-1}k)M)\,; \tag{2}$$

(2) We know its behavior on $L^2(X)$ (the *Plancherel Theorem*), on $C_c^\infty(X)$ (the *Paley – Wiener Theorem*), and on certain more complicated objects such as the L^p *Schwartz spaces* $\mathcal{C}^p(X)$ (for the values $0 < p \leq 2$); and

(3) We have an *inversion formula*

$$f(x) = \text{const.} \int_{\mathfrak{a}^*} d\lambda\, |\,\mathrm{c}(\lambda)\,|^{-2} \int_K dk\; \mathcal{H}f(\lambda, kM)\, e^{(i\lambda-\varrho)(H(x^{-1}k))}\,. \tag{3}$$

For bi-K-invariant functions on G, (1) and (3) reduce to the expressions

$$\mathcal{H}f(\lambda) = \int_G dx\, f(x)\, \varphi_\lambda(x) \tag{4}$$

and

$$f(x) = \text{const.} \int_{\mathfrak{a}^*} d\lambda\, |\,\mathrm{c}(\lambda)\,|^{-2}\, \mathcal{H}f(\lambda)\, \varphi_{-\lambda}(x)\,, \tag{5}$$

Supported by the Swiss National Science Foundation, Grant 8220–025043 (formerly Grant 82.592.0.88).

involving the *elementary spherical functions*

$$\varphi_\lambda(x) = \int_K dk\, e^{(i\lambda-\varrho)(H(xk))} . \tag{6}$$

Although these things were known for quite a long time, handling the above formulas remained problematic. We were recently able to overcome some of the difficulties, in the bi-K-invariant setting (see [A 2, 3, 4]). Previously, comparable computations could be carried out only in some particular cases, using specific information:

— when G has *real rank* 1 (see [K] and the references cited therein),
— when G is *complex* (see, e. g., [CS], [F; § 3]),
— when G is a *normal real form* (see [AL], [F]),
— when $G = \mathbf{SU}(p,q)$ (see [A 1] and the references cited therein).

The main problem consists in estimating the global behavior of a function f on X, knowing its Fourier transform. Simple decay conditions on X are given by

$$f(x) = \mathrm{O}(\,|x|^{-N}\ \mathrm{Cosh}_{\varepsilon\varrho}(x)^{-1}\,) \qquad (\,|x| \to +\infty\,), \tag{7}$$

where $|x|$ denotes the *geodesic distance to the origin* in X and

$$\mathrm{Cosh}_{\varepsilon\varrho}(x) = \tfrac{1}{|W|} \sum_{w \in W} e^{\varepsilon(w.\varrho)(H)} ,$$

$$x = k_1(\exp H)k_2 \text{ in the } \textit{Cartan decomposition.} \tag{8}$$

(These exponentials are closely related to the growth of the Haar measure.)

Problem 1. *Find matching conditions on the Fourier transform side.*

By analogy with the Euclidean case, one would expect that multiplication by $|x|^N$ corresponds via $\mathcal{H}$ to $(-\Delta_\lambda)^{\frac{N}{2}} + \textit{error terms}$ and multiplication by $\mathrm{Cosh}_{\varepsilon\varrho}(x)$ to some λ-shifts in the tube $\mathcal{T}_\varepsilon = \mathfrak{a}^* + i\ \mathrm{co}(W.\varepsilon\varrho)$. We shall give below (Lemma 5) a sharp L^2 answer to Problem 1 in the bi-K-invariant setting. Apart from the previously mentioned cases, nothing comparable was available before. [TV] (see also [E], [GV]) does contain an implicit answer to Problem 1: the shift is right, but the other terms are not accurate. [V] also contains a remarkable formula, but the shift is too big.

Our motivation originated from the L^p *multiplier problem.*

Problem 2. *Describe those W-invariant tempered distributions h on $\mathfrak{a}^*$ for which $Tf = f * \mathcal{H}^{-1}h$ is a bounded operator on $L^p(X)$.*

For $p = 2$, the solution is simple: T is a bounded operator on $L^2(X)$ if and only if h is a bounded measurable function. For $p \neq 2$ ($1 < p < +\infty$), the problem gets very complicated. There is a necessary condition: if T is a bounded operator on $L^p(X)$, then h extends to a bounded holomorphic function inside the tube $\mathcal{T}_\varepsilon$, with $\varepsilon = 2\,|\,\frac{1}{p} - \frac{1}{2}\,|$. But what about sufficient conditions? We have obtained in [A 3] the following analogue of the *Hörmander–Michlin criterion.*

Theorem 3. *Let $0<\varepsilon<1$, let $h(\lambda)$ be a W-invariant holomorphic function inside the tube $\mathcal{T}_\varepsilon$, and let N be the smallest integer $>\frac{\varepsilon}{2}\dim X$. Assume that $h(\lambda)$ extends continuously to the closed tube $\overline{\mathcal{T}_\varepsilon}$, together with its first N derivatives, and that*

$$\sup_{\lambda\in\mathcal{T}_\varepsilon}\,(\,|\lambda|+1)^\ell\,|\nabla^\ell h(\lambda)|\quad<+\infty\qquad(\ell=0,\dots,N). \tag{9}$$

Then $Tf=f*\mathcal{H}^{-1}h$ is a bounded operator on $L^p(X)$ for $\left|\frac{1}{p}-\frac{1}{2}\right|\le\frac{\varepsilon}{2}$.

This result (actually slightly weaker statements) was established before in some particular cases:

— when G is *complex* [CS],
— when G has *real rank* 1 [ST],
— when G is a *normal real form* [AL],
— when h is *radial* [T].

Our proof was inspired by [T] (and previous works of M. E. Taylor). There, functions of the Laplacian are studied, using the representation

$$h(\sqrt{-\Delta-|\varrho|^2})=\frac{1}{2\pi}\int_{-\infty}^{+\infty}dt\;\widehat{h}(t)\,\cos\sqrt{-\Delta-|\varrho|^2} \tag{10}$$

and the *propagation at speed* ≤ 1 *principle* for solutions to the *wave equation*

$$\frac{\partial^2}{\partial t^2}f(t,x)=(\Delta_x+|\varrho|^2)\,f(t,x)\,. \tag{11}$$

This approach can be extended to our general multipliers using the hyperbolic system introduced in [Se] (see also [Sh]). On the kernel level, it amounts to replacing (10) by the abstract formula

$$\mathcal{H}^{-1}h=\text{const. }\mathcal{A}^{-1}(\mathcal{F}h) \tag{12}$$

involving the *Euclidean Fourier transform*

$$\mathcal{F}h(H)=\int_{\mathfrak{a}^*}d\lambda\;h(\lambda)\,e^{-i\lambda(H)} \tag{13}$$

and the inverse of the *Abel transform*

$$\mathcal{A}f(H)=e^{\varrho(H)}\int_N dn\;f((\exp H)n)\,, \tag{14}$$

and to consider the following *conservation of support property*.

Proposition 4. *Let* $\Lambda \in \overline{\mathfrak{a}^{*+}}$, $\mathfrak{a}_\Lambda = \{ H \in \mathfrak{a} \mid (w.\Lambda)(H) \leq 1,\ w \in W \}$, *and* $G_\Lambda = K(\exp \mathfrak{a}_\Lambda)K$. *Then* $\mathcal{A}$ *is an isomorphism between* $\{ f \in C_c^\infty(K\backslash G/K) \mid \operatorname{supp} f \subset G_\Lambda \}$ *and* $\{ g \in C_c^\infty(\mathfrak{a})^W \mid \operatorname{supp} g \subset \mathfrak{a}_\Lambda \}$.

The corresponding assertion for balls is due to S. Helgason and is a consequence of his *Paley - Wiener theorem*. Proposition 4 is proved similarly [A 2].

We use it as follows: when computing $\mathcal{H}^{-1}h(\exp H)$ for $H \in \overline{\mathfrak{a}^+}$, one can remove (smoothly) that part of $\mathcal{F}h$ which lives in

$$\mathfrak{a} \setminus \bigcup_{w \in W} (H + \overline{{}^+\mathfrak{a}})$$

(see the figure below).

Combining this cut-off argument with the Plancherel identities for $\mathcal{H}$ and $\mathcal{F}$ yields a powerful way of handling the inverse Fourier transform $\mathcal{H}^{-1}$. For instance, one gets in this way the following inequality, which is a major step in the proof of Theorem 3 and gives an L^2 answer to Problem 1.

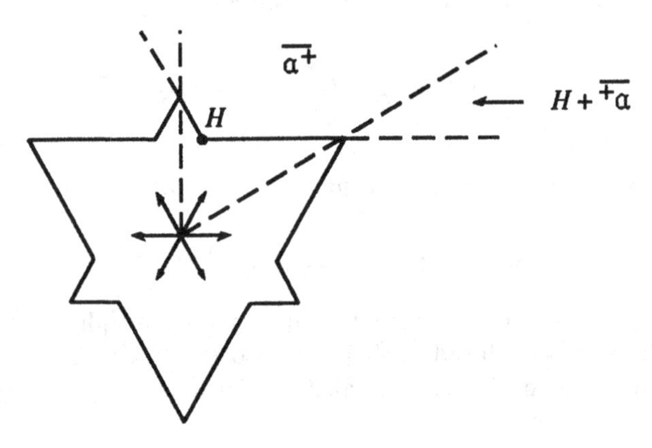

Lemma 5. *Fix an integer* $N \geq 0$ *and a constant* $\varepsilon \geq 0$. *Then*

$$r^N e^{\varepsilon |\varrho| r} \left[\int_{\Omega_r} dx\, |f(x)|^2 \right]^{\frac{1}{2}} \leq$$

$$\text{const.} \sum_{\ell=0}^{N} \left[\int_{\mathfrak{a}^*} d\lambda\, \left\{ (|\lambda|+1)^{((n-a)/2)-N+\ell}\, |\nabla^\ell h(\lambda + i\varepsilon\varrho)| \right\}^2 \right]^{\frac{1}{2}} \qquad (15)$$

for every $r \geq 1$ *and every* $f \in C_c^\infty(K\backslash G/K)$, *where*

$$\Omega_r = G_{(r+1)^{-1}|\varrho|^{-1}\varrho} \setminus G_{r^{-1}|\varrho|^{-1}\varrho},$$

$n = \dim X$, $a = \dim \mathfrak{a}$, *and* $h = \mathcal{H}f$.

Sharp pointwise estimates are harder to obtain. In [A 4] we analyze some classical operators associated to the Laplacian Δ on X such as the *heat diffusion semigroup* $e^{t\Delta}$, the *potentials* $(-\Delta)^{-s}$, and the *resolvent* $(zI+\Delta)^{-1}$. The kernel estimates we get are rather precise (although not optimal). Consequences worth mentioning are the following endpoint results.

Theorem 6. *The* heat maximal operator

$$H_* f(x) = \sup_{t>0} \left| e^{t\Delta} f(x) \right| \tag{16}$$

and the Riesz transform

$$Rf = \nabla(-\Delta)^{-\frac{1}{2}} f \tag{17}$$

are both of weak type $(1,1)$.

These facts were proven before in particular cases (see [A 1], [AL], [CG], [LZ]). The $L^p - L^p$ boundedness, for $p > 1$, was established in much wider contexts (see [B], [L 1, 2], [St]).

Finally, as an unexpected byproduct, we have found [A 2] a short proof of the main result in [TV] (see also [GV]), namely the Fourier transform characterization of the *bi-K-invariant L^p Schwartz spaces* $\mathcal{C}^p(K\backslash G/K)$ $(0<p\leq 2)$.

Theorem 7. *Let* $0<p\leq 2$ *and* $\varepsilon = 2\left(\frac{1}{p}-\frac{1}{2}\right)$. *Then* $\mathcal{H}$ *is a topological isomorphism between* $\mathcal{C}^p(K\backslash G/K)$ *and the* W-invariant Schwartz space $\mathcal{C}(\mathcal{T}_\varepsilon)^W$ *in the tube* $\mathcal{T}_\varepsilon$.

We wonder whether our approach might work for other K-types. In particular, we wonder if the full L^p *Schwartz space* $\mathcal{C}^p(X)$ on X [E] could be successfully analyzed by our techniques.

Acknowledgements. It was a pleasure and a honor for me to participate in this conference. I would like to thank the organizers — especially William Barker — for their invitation and the warm hospitality at Bowdoin College. I am grateful to the Swiss National Science Foundation for financial support.

References

[A1] J.-Ph. Anker, *Le noyau de la chaleur sur les espaces symétriques* $\mathbf{U}(p,q)$ / $\mathbf{U}(p)\times\mathbf{U}(q)$, Harmonic analysis, Luxembourg 1987, Lecture Notes Math., Vol 1359 (P. Eymard, J.-P. Pier, eds.), Springer-Verlag, 1988, pp. 60–82.

[A2] ———, *The spherical Fourier transform of rapidly decreasing functions — a simple proof of a characterization due to Harish-Chandra, Helgason, Trombi and Varadarajan*, J. Funct. Anal. **96** (1991), 331–349.

[A3] ———, *L_p Fourier multipliers on Riemannian symmetric spaces of the noncompact type*, Ann. Math. **132** (1990), 597–628.

[A4] ———, *Sharp estimates for some functions of the Laplacian on noncompact symmetric spaces*, preprint (1990).

[AL] J.-Ph. Anker and N. Lohoué, *Multiplicateurs sur certains espaces symétriques*, Amer. J. Math. **108** (1986), 1303–1354.

[B] D. Bakry, *Etude des transformations de Riesz dans les variétés Riemmanniennes à courbure de Ricci minorée*, Séminaire de probabilites XXI, Lecture Notes Math., Vol. 1247 (J. Azéma, P. A. Meyer, and M. Yor, eds.), Springer Verlag, 1987, pp. 137–172.

[CG] M. Cowling, G. Gaudry, S. Giulini, and G. Mauceri, *Weak type (1,1) estimates for heat kernel maximal functions on Lie groups*, Trans. Amer. Math. Soc. **323** (1991), 637–649.

[CS] J.-L. Clerc and E. M. Stein, *L^p-multipliers for noncompact symmetric spaces*, Proc. Nat. Acad. Sci. USA **71** (1974), 3911–3912.

[E] M. Eguchi, *Asymptotic expansions of Eisenstein integrals and Fourier transform on symmetric spaces*, J. Funct. Anal. **34** (1979), 167–216.

[F] M. Flensted-Jensen, *Spherical functions on a real semisimple Lie group; a method of reduction to the complex case*, J. Funct. Anal. **30** (1978), 106–146.

[GV] R. Gangolli and V. S. Varadarajan, *Harmonic analysis of spherical functions on real reductive groups*, Ergebnisse Math. Grenzgeb. **101**, Springer Verlag, 1988.

[H] S. Helgason, *Groups and geometric analysis — integral geometry, invariant differential operators, and spherical functions*, Academic Press, 1984.

[K] T. H. Koornwinder, *Jacobi functions and analysis on noncompact semisimple groups*, Special functions — group theoretical aspects & applications (R. A. Askey, T. H. Koornwinder, and W. Schempp, eds.), D. Reidel Publ. Co., 1984, pp. 1–85.

[L1] N. Lohoué, *Comparaison des champs de vecteurs et des puissances du Laplacien sur les variétés Riemanniennes à courbure non positive*, J. Funct. Anal. **61** (1985), 164–201.

[L2] ———, *Transformées de Riesz et fonctions de Littlewood-Paley sur les groupes non moyennables*, C. R. Acad. Sci. Paris Série **I**, **306** (1988), 327–330.

[LZ] N. Lohoué and Zhu Fu Liu, *Estimation faible de certaines fonctions maximales*, C. R. Acad. Sci. Paris Série **I**, **302** (1986), 303–305.

[Se] M. A. Semenov-Tjan-Šanskiĭ, *Harmonic analysis on Riemannian symmetric spaces of negative curvature and scattering theory*, Math. USSR Izvestija **10** (1976), 535–563.

[Sh] M. M. Shahshahani, *Invariant hyperbolic systems on symmetric spaces*, Differential Geometry, Progress in Math. **32** (R. Brooks, A. Gray, and B. L. Reinhart, eds.), Birkhäuser, 1983, pp. 203–233.

[ST] R. J. Stanton and P. A. Tomas, *Expansions for spherical functions on noncompact symmetric spaces*, Acta Math. **140** (1978), 251–276.

[St] E. M. Stein, *Topics in harmonic analysis related to the Littlewood-Paley theory*, Ann. Math. Stud. **63**, Princeton Univ. Press, 1970.

[T] M. E. Taylor, *L^p estimates on functions of the Laplace operator*, Duke Math. J. **58** (1989), 773–793.

[TV] P. C. Trombi and V. S. Varadarajan, *Spherical transforms on semisimple Lie groups*, Ann. Math. **94** (1971), 246–303.

[V] L. Vretare, *On a recurrence formula for elementary spherical functions on symmetric spaces and its applications to multipliers for the spherical Fourier transform*, Math. Scand. **41** (1977), 99–112.

Present address: Université de Nancy I, Dépt. de Mathématiques, B.P. 239, F-54506 Vandoeuvre-lès-Nancy, France

E-mail: Anker @ ciril.fr

SOME PROBLEMS IN LOCAL HARMONIC ANALYSIS

JAMES ARTHUR

University of Toronto

INTRODUCTION

The purpose of this article is to discuss some questions in the harmonic analysis of real and p-adic groups. We shall be particularly concerned with the properties of a certain family of invariant distributions. These distributions arose naturally in a global context, as the terms on the geometric side of the trace formula. However, they are purely local objects, which include the ordinary invariant orbital integrals. One of our aims is to describe how the distributions also arise in a local context. They appear as the terms on the geometric side of a new trace formula, which is simpler than the original one, and is the solution of a natural question in local harmonic analysis. The local trace formula seems to be a promising tool. It might have implications for the difficult local problems which are holding up progress in automorphic forms.

We have organized the paper loosely around three general problems. We shall describe the problems and the distributions together in §1. This section is entirely expository. In §2, which is also largely expository, we shall discuss the role of the distributions in the local trace formula. Finally, in §3, we shall see how the local trace formula can be applied to some questions in local harmonic analysis. We shall sketch an application to each of the three general problems of §1. These results are all more or less immediate consequences of the same kind of approximation argument. It remains to be seen whether a deeper study of the local trace formula will lead to further applications.

1. WEIGHTED ORBITAL INTEGRALS AND WEIGHTED CHARACTERS

Let G be a connected reductive algebraic group over a local field F. We assume that F is of characteristic 0. Then F equals the real field $\mathbb{R}$, or a p-adic field $\mathbb{Q}_p$, or a finite extension of one of these. In particular, our discussion of the group $G(F)$ of rational points applies to both real and p-adic groups.

Partially supported by NSERC Operating Grant A3483.

We fix a suitable maximal compact subgroup K of $G(F)$. We can then form the Hecke algebra

$$\mathcal{H}(G) = \mathcal{H}(G(F), K)$$

of smooth, compactly supported, K-finite functions on $G(F)$. The Hecke algebra is contained in the space $C_c^\infty(G(F))$ of smooth functions of compact support (the two spaces are in fact equal in the p-adic case), and $C_c^\infty(G(F))$ is contained in Harish-Chandra's Schwartz space $\mathcal{C}(G(F))$. There are natural topologies on the three spaces for which the embeddings

$$\mathcal{H}(G) \subset C_c^\infty(G(F)) \subset \mathcal{C}(G(F))$$

are continuous, and have dense image.

By a *distribution* on $\mathcal{H}(G)$, we shall mean a continuous linear functional I on $\mathcal{H}(G)$. This is a slight abuse of terminology, for it is only those I which extend to continuous linear functionals on $C_c^\infty(G(F))$ which are distributions on $G(F)$. The functionals which in addition extend continuously to $\mathcal{C}(G(F))$ are of course the *tempered* distributions on $G(F)$. The functional I is said to be *invariant* if

$$I(f * g) = I(g * f), \qquad f, g \in \mathcal{H}(G).$$

In case I extends to $C_c^\infty(G(F))$, this condition is easily seen to be equivalent to the more familiar property

$$I(f^y) = I(f), \qquad f \in C_c^\infty(G(F)),\ y \in G(F),$$

where $f^y(x) = f(yxy^{-1})$. The most fundamental invariant distributions are the two families of orbital integrals and tempered characters.

Recall that orbital integrals are parametrized by points γ in $G_{\text{reg}}(F)$, the set of regular semisimple elements in $G(F)$. They are defined by integrals

$$I_G(\gamma, f) = |D(\gamma)|^{\frac{1}{2}} \int_{G_\gamma(F)\backslash G(F)} f(x^{-1}\gamma x)dx, \qquad f \in \mathcal{H}(G),$$

where G_γ is the centralizer of γ in G, and

$$D(\gamma) = \det\bigl(1 - Ad(\gamma)_{\mathfrak{g}/\mathfrak{g}_\gamma}\bigr)$$

is the Weyl discriminant. Orbital integrals are invariant distributions, which have been shown by Harish-Chandra to be tempered. They are basic objects in local harmonic analysis. Orbital integrals are also very important for the theory of automorphic forms, for they are the terms on the geometric side of the Selberg trace formula for compact quotient.

The tempered characters

$$I_G(\pi, f) = \operatorname{tr}(\pi(f)), \qquad f \in \mathcal{H}(G),$$

can be regarded as dual analogues of orbital integrals. They too are invariant tempered distributions, which are parametrized by the set $\Pi_{\text{temp}}(G(F))$ of (equivalence classes of) irreducible tempered representations of $G(F)$. It is convenient to think of $I_G(\pi, f)$ as a transform as well as a distribution. We therefore define a map

$$f \longrightarrow f_G,$$

from $\mathcal{H}(G)$ to the space of complex valued functions on $\Pi_{\text{temp}}(G(F))$, by setting

$$f_G(\pi) = I_G(\pi, f) = \operatorname{tr}(\pi(f)), \qquad \pi \in \Pi_{\text{temp}}(G(F)).$$

An invariant distribution I on $\mathcal{H}(G)$ is said to be *supported on characters* if $I(f) = 0$ for every function $f \in \mathcal{H}(G)$ such that $f_G = 0$. Our first problem is a classification question, which we mention for the sake of general orientation.

Problem A. *Show that any invariant distribution I on $\mathcal{H}(G)$ is supported on characters.*

Remark 1. In the p-adic case, the problem was solved by Kazhdan [18], who used global methods (specifically, a simple form of the global trace formula) to show that orbital integrals are supported on characters. Harish-Chandra had earlier given an argument based on Shalika germs which reduced the question to the case of orbital integrals. For real groups, the problem has not been solved in its present form.

Remark 2. For tempered distributions on the Schwartz space, the analogous question can be answered for general real groups by using the characterization of $\mathcal{C}(G(F))$ [1] under the full Fourier transform. On the other hand, for p-adic groups this version of the problem has been solved only for $GL(n)$ [22].

One reason for considering the Hecke algebra is that there is a nice characterization [12], [14] of the space

$$\mathcal{I}(G) = \{f_G : f \in \mathcal{H}(G)\}$$

of functions on $\Pi_{\text{temp}}(G(F))$. This leads to a natural topology on $\mathcal{I}(G)$ in terms of the co-ordinates of the domain $\Pi_{\text{temp}}(G(F))$, for which the map $f \rightarrow f_G$ is continuous. One checks that if I is supported on characters, there is a unique distribution $\hat{I}$ on $\mathcal{I}(G)$ such that

$$\hat{I}(f_G) = I(f), \qquad f \in \mathcal{H}(G).$$

The next problem we state informally as

Problem B. *Given some natural invariant distribution I which is supported on characters, deduce information about $\hat{I}$.*

Remark 1. One does not generally expect to be able to compute $\hat{I}$ explicitly. Instead, one could try to determine the qualitative properties of $\hat{I}$. For example, if I is tempered, one could ask whether $\hat{I}$ is a function on the space $\Pi_{\mathrm{temp}}(G(F))$. Given Harish-Chandra's Plancherel formula one can construct a variety from $\Pi_{\mathrm{temp}}(G(F))$ which is a disjoint union of Euclidean spaces (F-Archimedean) or compact tori (the p-adic case). For a tempered I, one could try to determine explicitly the singular support of $\hat{I}$.

Remark 2. Suppose that

$$I = I_G(\gamma), \qquad \gamma \in G_{\mathrm{reg}}(F).$$

Using results of Shelstad on L-indistinguishability, R. Herb has computed $\hat{I}$ in the case $F = \mathbb{R}$. For p-adic groups, the problem in this case is very important, but little is known. In particular, even though I is tempered, there is to my knowledge no general result on the singular support of $\hat{I}$.

The orbital integrals $I_G(\gamma)$ are part of a larger family of invariant distributions. Suppose that M is a Levi component of some parabolic subgroup of G defined over F. The set $\mathcal{P}(M)$ of all parabolic subgroups with Levi component M is in bijective correspondence with the chambers in the real vector space

$$\mathfrak{a}_M = Hom(X(M)_F, \mathbb{R}) \xrightarrow{\sim} Hom(X(A_M), \mathbb{R}),$$

where $X(\cdot)$ stands for the module of rational characters, and A_M is the split component of the center of M. For any group $P = MN_P$ in $\mathcal{P}(M)$, there is the usual map

$$H_P : G(F) \longrightarrow \mathfrak{a}_P = \mathfrak{a}_M$$

that comes from the decomposition

$$G(F) = M(F)N_P(F)K.$$

Suppose $x \in G(F)$. Let $\Pi_M(x)$ be the convex hull in $\mathfrak{a}_M$ of the finite set

$$\{H_P(x) : P \in \mathcal{P}(M)\}.$$

We write $v_M(x) = v_M^G(x)$ for the volume in $\mathfrak{a}_M/\mathfrak{a}_G$ of the projection of $\Pi_M(x)$.

The convex polytopes $\Pi_M(x)$ are nice objects whose geometric properties are tied up with the structure of G. For example, there is a bijection

$$Q \in \mathcal{F}(M) \longrightarrow \Pi_M^Q(x)$$

between the finite set $\mathcal{F}(M)$ of parabolic subgroups which contain M, and the *facets* of the polytope $\Pi_M(x)$. The facet $\Pi_M^Q(x)$ is equal to the convex hull of the set

$$\{H_P(x) : P \in \mathcal{P}(M),\ P \subset Q\}.$$

We write $v_M^Q(x)$ for the volume of the projection of $\Pi_M^Q(x)$ onto $\mathfrak{a}_M/\mathfrak{a}_Q$. Another property of $\Pi_M(x)$ concerns the finite set $\mathcal{L}(M)$ of Levi subgroups L of G which contain M. These are in bijective correspondence with the vector subspaces $\mathfrak{a}_L$ of $\mathfrak{a}_M$ which are orthogonal complements of facets, or rather, orthogonal complements of affine spaces generated by facets. There is thus a bijection

$$L \in \mathcal{L}(M) \longrightarrow \Pi_L(x)$$

between the finite set of Levi subgroups which contain M, and convex polytopes obtained by projecting $\Pi_M(x)$ onto orthogonal complements of facets. More precisely, $\Pi_L(x)$ is the projection of $\Pi_M(x)$ onto the orthogonal complement $\mathfrak{a}_L$ of the affine space generated by any of the facets $\{\Pi_M^Q(x) : Q \in \mathcal{P}(L)\}$. The notation makes sense, for $\Pi_L(x)$ is just the convex hull of $\{H_Q(x) : Q \in \mathcal{P}(L)\}$, which is the polytope associated to L in its own right. In particular, $v_L(x)$ is the volume of the projection of $\Pi_M(x)$ onto $\mathfrak{a}_L/\mathfrak{a}_G$.

The usual diagram for $G = SL(3)$ is a useful reminder of these relationships. Taking M to be minimal, we identify $\mathfrak{a}_M$ as a Euclidean space with the plane. There are six chambers and six minimal parabolic subgroups $P \in \mathcal{P}(M)$. There are three subspaces $\mathfrak{a}_L$ of dimension 1, and for each such L there are two maximal parabolic subgroups $Q \in \mathcal{P}(L)$.

Figure 1.

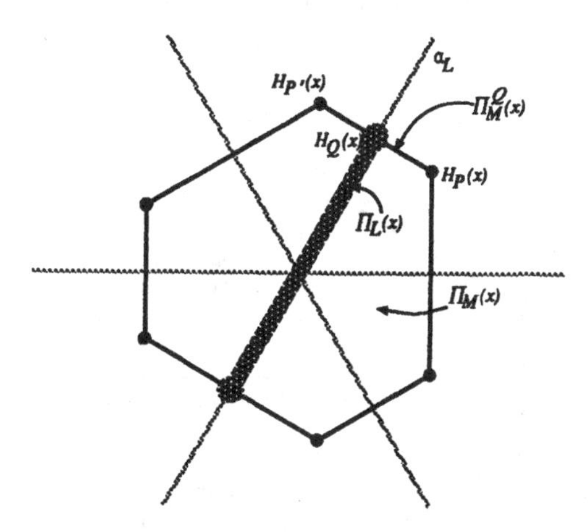

In general, the function $v_M(x)$ can be used to define a noninvariant measure on any G-regular class in $M(F)$. Observe first that if m is any point in $M(F)$, the polytope $\Pi_M(mx)$ is the translate of $\Pi_M(x)$ by a vector $H_M(m)$. Therefore, $v_M(mx)$ equals $v_M(x)$. Now, suppose that γ belongs to

$M(F) \cap G_{\text{reg}}(F)$. Then G_γ is contained in M, so that $v_M(x)$ is left invariant under $G_\gamma(F)$. One can therefore define the weighted orbital integral

$$J_M(\gamma, f) = |D(\gamma)|^{\frac{1}{2}} \int_{G_\gamma(F)\backslash G(F)} f(x^{-1}\gamma x) v_M(x) dx,$$

for any function $f \in \mathcal{H}(G)$.

Although it is a generalization of the ordinary orbital integral, the weighted orbital integral $J_M(\gamma, f)$ is not invariant. There is in fact a rather explicit formula for the lack of invariance.

It can be shown that $J_M(\gamma)$ is a tempered distribution, so to analyze its lack of invariance it suffices to look at $J_M(\gamma, f^y)$ for any element $y \in G(F)$. Changing variables in the integral over x, we first write

$$J_M(\gamma, f^y) = |D(\gamma)|^{\frac{1}{2}} \int_{G_\gamma(F)\backslash G(F)} f(x^{-1}\gamma x) v_M(xy) dx.$$

If P belongs to $\mathcal{P}(M)$, set

$$x = n_P\, m_P\, k_P, \qquad n_P \in N_P(F),\ m_P \in M(F),\ k_P \in K.$$

Then

$$H_P(xy) = H_P(m_P) + H_P(k_P y) = H_P(x) + H_P(k_P y).$$

To sketch how to evaluate $v_M(xy)$, we argue geometrically from the diagram for $SL(3)$. For simplicity, assume that for each $P \in \mathcal{P}(M)$, the point $H_P(k_P y)$ lies in the chamber of P. Then for $SL(3)$ we have Figure 2:

Figure 2.

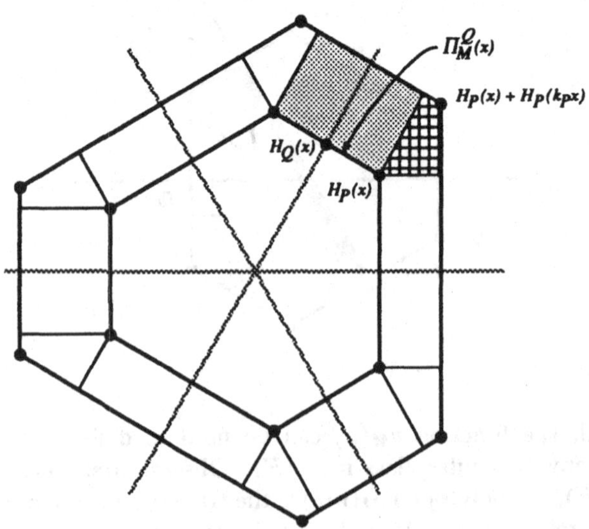

The inner hexagon is $\Pi_M(x)$, while the outer hexagon is just $\Pi_M(xy)$. Let $u_P(k_P y)$ denote the area of the hatched quadrilateral. This region is separated from the shaded rectangle by a line segment whose length equals the analogous number $u_Q(k_Q y)$ for the maximal parabolic subgroup Q. The other side of the rectangle is $\Pi_M^Q(x)$, a line segment of length $v_M^Q(x)$, so the area of the shaded rectangle is the product of $u_Q(k_Q y)$ with $v_M^Q(x)$. In this way, we can account for the area of each of the pieces that comprise $\Pi_M(xy)$. We obtain a formula

$$v_M(xy) = \sum_{Q \in \mathcal{F}(M)} v_M^Q(x) u_Q(k_Q y),$$

which expresses $v_M(xy)$ as a sum of mixed volumes. Substituting this expansion into the integral above, we obtain the formula

$$J_M(\gamma, f^y) = \sum_{Q \in \mathcal{F}(M)} J_M^{M_Q}(\gamma, f_{Q,y}), \tag{1.1}$$

where

$$f_{Q,y}(m) = \delta_Q(m)^{\frac{1}{2}} \int_K \int_{N_Q(F)} f(k^{-1} mnk) u_Q(ky) dn\, dk$$

for any $m \in M_Q(F)$ [2, Lemma 8.2]. Here M_Q is the Levi component of Q which contains M, and δ_Q is the modular function of $Q(F)$. Notice that the summand in (1.1) with $M = G$ equals $J_M(\gamma, f)$. We therefore can write (1.1) as

$$J_M(\gamma, f^y) - J_M(\gamma, f) = \sum_{Q \neq G} J_M^{M_Q}(\gamma, f_{Q,y}),$$

more clearly displaying it as an obstruction to the invariance of $J_M(\gamma)$.

Just as the tempered characters are dual to the invariant orbital integrals, there are dual analogues of weighted orbital integrals, which can be regarded as weighted (tempered) characters. If $\pi \in \Pi_{\mathrm{temp}}(M(F))$ and $\Lambda \in i\mathfrak{a}_M^*$, the representation

$$\pi_\Lambda(m) = e^{\Lambda(H_M(m))}\pi(m), \qquad m \in M(F),$$

also belongs to $\Pi_{\mathrm{temp}}(M(F))$. We can form the (parabolically) induced representations

$$\mathcal{I}_P(\pi_\Lambda), \qquad P \in \mathcal{P}(M),$$

of $G(F)$, and the normalized intertwining operators

$$R_{Q|P}(\pi_\Lambda) : \mathcal{I}_P(\pi_\Lambda) \longrightarrow \mathcal{I}_Q(\pi_\Lambda), \qquad P, Q \in \mathcal{P}(M),$$

between them [6, §1-2]. Both $\mathcal{I}_P(\pi_\Lambda)$ and $R_{Q|P}(\pi_\Lambda)$ are analytic operator valued functions of $\Lambda \in i\mathfrak{a}_M^*$. For any $P \in \mathcal{P}(M)$, set

$$\mathcal{R}_M(\pi, P) = \lim_{\Lambda \to 0} \sum_{Q \in \mathcal{P}(M)} R_{Q|P}(\pi)^{-1} R_{Q|P}(\pi_\Lambda) \theta_Q(\Lambda)^{-1},$$

where if $\Delta_Q^\vee$ denotes the set of simple "co-roots" of Q,

$$\theta_Q(\Lambda) = \operatorname{vol}\big(\mathfrak{a}_M / \mathbb{Z}(\Delta_Q^\vee) + \mathfrak{a}_G\big)^{-1} \prod_{\alpha^\vee \in \Delta_Q^\vee} \Lambda(\alpha^\vee).$$

It is a simple matter to show that the limit exists [2, Lemma 6.3]. Consequently $\mathcal{R}_M(\pi, P)$ is a well defined operator on the underlying space of $\mathcal{I}_P(\pi)$. For example, if P is a maximal parabolic with simple root $\alpha, \mathcal{R}_M(\pi, P)$ is a constant multiple of the logarithmic derivative

$$R_{\overline{P}|P}(\pi)^{-1} \cdot \lim_{z \to 0} \Big(\frac{d}{dz} R_{\overline{P}|P}(\pi_{z\alpha})\Big).$$

In general, the weighted character of a function $f \in \mathcal{H}(G)$ is defined as the trace

$$J_M(\pi, f) = \operatorname{tr}\big(\mathcal{R}_M(\pi, P)\mathcal{I}_P(\pi, f)\big).$$

As a spectral analogue of $J_M(\gamma, f)$, the distribution $J_M(\pi, f)$ is not invariant. However, the considerations that lead to the formula (1.1) can be adapted to the study of $J_M(\pi, f^y)$. Observing first that

$$v_M(x) = \lim_{\Lambda \to 0} \Big(\int_{\Pi_M(x)/\mathfrak{a}_G} e^{\Lambda(H)} dH \Big),$$

one can then rewrite the right hand limit in a general form that is similar to the expression for $\mathcal{R}_M(\pi, P)$. The mixed volume expansion for $v_M(xy)$ translates into a special case of an expansion that applies to certain functions of Λ, and in particular, to the functions of which both $v_M(x)$ and $\mathcal{R}_M(\pi, P)$ are the limits. This provides an expansion for

$$J_M(\pi, f^y) = \operatorname{tr}\big(\mathcal{I}_P(\pi, y)\mathcal{R}_M(\pi, P)\mathcal{I}_P(\pi, y)^{-1}\mathcal{I}_P(\pi, f)\big)$$

as a sum over groups $Q \in \mathcal{F}(M)$. The result is a formula

$$J_M(\pi, f^y) = \sum_{Q \in \mathcal{F}(M)} J_M^{M_Q}(\pi, f_{Q,y}) \tag{1.2}$$

which is parallel to (1.1) [2, Lemma 8.3].

Because the distributions $\{J_M(\gamma)\}$ and $\{J_M(\pi)\}$ have similar behaviour (1.1) and (1.2) under conjugation, we might suspect that they are related to each other by invariant distributions. This is indeed the case. One must first interpret $J_M(\pi)$ as a transform

$$\phi_M(f) = \phi_M^G(f) : \pi \longrightarrow J_M(\pi, f), \qquad \pi \in \Pi_{\mathrm{temp}}(M(F)),$$

that maps functions on $G(F)$ to functions on $\Pi_{\mathrm{temp}}(M(F))$. There is a technical problem that the function $\phi_M(f)$ does not belong to $\mathcal{I}(M)$. For example, in the case of maximal parabolic P, the logarithmic derivative

$$R_{\overline{P}|P}(\pi_{z\alpha})^{-1} \frac{d}{dz} R_{\overline{P}|P}(\pi_{z\alpha}), \qquad z \in i\mathbb{R},$$

will have poles when the variable z is extended to the whole complex plane. This means that $J_M(\pi, f)$ is not a Paley-Wiener function in the co-ordinates of π. However, the problem is not serious. Let $\mathcal{H}_{ac}(G)$ be the space of smooth, K-finite functions on $G(F)$ whose restrictions to the fibres of the map $H_G : G(F) \rightarrow \mathfrak{a}_G$ all have compact support. One can define a version of the map $f \rightarrow f_G$ for $f \in \mathcal{H}_{ac}(G)$, and a variant of the main theorem in [12] and [14] provides a characterization of the image

$$\mathcal{I}_{ac}(G) = \{f_G : f \in \mathcal{H}_{ac}(G)\} .$$

(See [5, Appendix].) It can then be shown that ϕ_M maps $\mathcal{H}_{ac}(G)$ continuously to $\mathcal{I}_{ac}(M)$ [6, Theorem 12.1].

Theorem. [4, §2] *There are unique invariant distributions*

$$I_M(\gamma) = I_M^G(\gamma)$$

on $\mathcal{H}_{ac}(G)$ which are supported on characters, and such that

$$J_M(\gamma, f) = \sum_{L \in \mathcal{L}(M)} \hat{I}_M^L(\gamma, \phi_L(f)) .$$

Proof sketch. We shall sketch the formal part of the proof. We assume inductively that $I_M^L(\gamma)$ has been defined, and has the required properties, for any $L \in \mathcal{L}(M)$ with $L \neq G$. The required distribution can then be defined uniquely by the formula

$$I_M(\gamma, f) = J_M(\gamma, f) - \sum_{\{L \in \mathcal{L}(M): L \neq G\}} \hat{I}_M^L(\gamma, \phi_L(f)) .$$

We shall show that $I_M(\gamma)$ is invariant.

The formulas (1.1) and (1.2) cannot actually be applied as they stand, for f^y is not a K-finite function. However, since the formulas have simple variants which pertain to K-finite functions [4, §2], we shall ignore this difficulty. In particular, we shall interpret (1.2) as as formula

$$\phi_M(f^y) = \sum_{Q \in \mathcal{F}(M)} \phi_M^{M_Q}(f_{Q,y})$$

for the map ϕ_M. Together with (1.1) this yields

$$\begin{aligned} I_M(\gamma, f^y) - I_M(\gamma, f) &= \sum_{\{Q \in \mathcal{F}(M): Q \neq G\}} J_M^{M_Q}(\gamma, f_{Q,y}) \\ &\quad - \sum_{L \in \mathcal{L}(M)} \sum_{\{Q \in \mathcal{F}(L): Q \neq G\}} \hat{I}_M^L(\gamma, \phi_L^{M_Q}(f_{Q,y})) \\ &= \sum_{\{Q \in \mathcal{F}(M): Q \neq G\}} \Big(J_M^{M_Q}(\gamma, f_{Q,y}) \\ &\quad - \sum_{\{L \in \mathcal{L}(M): L \subset M_Q\}} \hat{I}_M^L(\gamma, \phi_L^{M_Q}(f_{Q,y})) \Big) . \end{aligned}$$

Applying the induction hypothesis to M_Q, we see that the last expression vanishes. Therefore $I_M(\gamma)$ is invariant.

Since the distributions satisfy the required formula by definition, it remains only to show that $I_M(\gamma)$ is supported on characters. This was done by global means in [5, Theorem 5.1]. We shall later sketch how it can also be established by purely local means. □

We thus obtain a family $\{I_M(\gamma)\}$ of invariant distributions on $\mathcal{H}(G)$ which are parame-trized by Levi subgroups M and G-regular conjugacy classes γ in $M(F)$. These distributions should be regarded as the true generalizations of the orbital integrals $\{I_G(\gamma)\}$. They are important in the theory of automorphic forms, for they are the terms on the geometric side of the global trace formula when the quotient is assumed only to have finite volume. They are also intimately tied up with local harmonic analysis, as we shall presently see. Thus, the distributions are natural objects for which it is appropriate to ask questions as in Problem B. What is the "discrete part" of $\hat{I}_M(\gamma)$? That is, what are the values taken by $\hat{I}_M(\gamma)$ on the discrete components of $\Pi_{\text{temp}}(G(F))$? (Actually, we mean the discrete components of the variety attached to $\Pi_{\text{temp}}(G(F))$ by taking into account the reducibility of induced representations.) What is the singular support of $\hat{I}_M(\gamma)$ on the continuous components? More generally, can one compute $\hat{I}_M(\gamma)$ explicitly, modulo smooth functions on the continuous components?

We should remark that there are twisted versions of the various objects discussed above. One can account for this generalization by taking G to be a connected component of a nonconnected reductive group. The only ingredient that is lacking is a characterization of the analogous space $\mathcal{I}(G)$ when $F = \mathbb{R}$. (The p-adic twisted trace Paley-Wiener theorem has been established by Rogawski [23].)

The third problem we mention is to relate the distributions to the theory of endoscopy.

Problem C.

(a) *If G is a connected, quasi-split group, construct stable invariant distributions $SI_M^G(\gamma)$ on $\mathcal{H}(G)$ from the distributions $I_M^G(\gamma)$.*

(b) *If G is any connected group, or a component of a nonconnected group, establish identities between the distributions $\{I_M^G(\gamma)\}$ and $\{S\hat{I}_{M_H}^H(\gamma_H, f^H)\}$, where H ranges over endoscopic data for G, and $f \to f^H$ is the conjectured Langlands-Shelstad transfer mapping.*

Remark 1. The problem is very difficult. For example, when F is p-adic, the important special case that $M = G$ includes the "fundamental lemma", which is far from being solved. This problem is certainly the most important of the three for automorphic forms. A solution could be combined with the global trace formula to yield a general theory of endoscopy for automorphic forms, and in particular, many reciprocity laws between automorphic representations on different groups.

Remark 2. The endoscopic side of each identity would be a certain finite linear combination of distributions $\{S\hat{I}_{M_H}^H(\gamma_H, f^H)\}$. It should not be difficult to describe the coefficients of each such linear combination explicitly, but this has not been done. We refer the reader to the original article [21], and perhaps also [8, §3], for a discussion of the undefined terms in the statement of the problem.

Remark 3. For groups of general rank, there are only two cases of the problem that have been solved.

(i) G the multiplicative group of a central simple algebra.

(ii) G a connected component of the semi-direct product

$$\operatorname{Res}_{E/F}(GL(n)) \rtimes \operatorname{Gal}(E/F),$$

where E/F is a finite cyclic extension.

The solution in each case is contained Theorem A of [11, §II.5], a result which was proved by global methods.

2. The local trace formula

The connection of the distributions $\{I_M(\gamma)\}$ with harmonic analysis is through a local version of the trace formula. The distributions occur on the geometric side, in much the same way that they occur in the global trace formula. We shall describe the local trace formula in this section.

In the next section we shall sketch how it can be used to give information about each of the three problems. The noninvariant version of the local trace formula is proved in the preprint [10]. The details of the invariant version described here, as well as the applications, will be given elsewhere.

The formula begins with a problem suggested by Kazhdan. Consider the regular representation

$$\big(R(u,y)\phi\big)(x) = \phi(u^{-1}xy), \qquad u,y \in G(F), \quad \phi \in L^2\big(G(F)\big),$$

of $G(F) \times G(F)$ on the Hilbert space $L^2\big(G(F)\big)$. Consider also a function in $\mathcal{H}(G \times G)$ of the form

$$f(y_1, y_2) = f_1(y_1)f_2(y_2), \qquad y_i \in G(F), \quad f_i \in \mathcal{H}(G).$$

(From now on, f will denote a function of $G(F) \times G(F)$, rather than on $G(F)$ as before.) Then $R(f)$ is an operator on $L^2\big(G(F)\big)$, which maps a function ϕ to the function

$$\begin{aligned}\big(R(f)\phi\big)(x) &= \int_{G(F)}\int_{G(F)} f_1(u)f_2(y)\phi(u^{-1}xy)du\, dy \\ &= \int_{G(F)}\int_{G(F)} f_1(xu)f_2(uy)\phi(y)du\, dy \\ &= \int_{G(F)} K(x,y)\phi(y)dy,\end{aligned}$$

where

$$K(x,y) = \int_{G(F)} f_1(xu)f_2(uy)du.$$

Thus $R(f)$ is an integral operator with smooth kernel $K(x,y)$. Now by the Plancherel formula we know that

$$R = R_{\mathrm{disc}} \oplus R_{\mathrm{cont}},$$

where R_{disc} is a direct sum of square integrable representations of $G(F) \times G(F)$, and R_{cont} is a subrepresentation of R which decomposes continuously. The problem is to find an explicit formula for the trace of $R_{\mathrm{disc}}(f)$.

Suppose for a moment that G is semisimple. On the one hand,

$$\mathrm{tr}\big(R_{\mathrm{disc}}(f)\big) = \sum_{\sigma} \mathrm{tr}\big(\sigma^\vee(f_1)\big)\mathrm{tr}\big(\sigma(f_2)\big),$$

where σ is summed over the discrete series $G(F)$, and $\sigma^\vee$ denotes the contragredient of σ. Since there are only finitely many discrete series that contain a given K-type, the sum can be taken over a finite set. On the other hand, if $K_{\mathrm{cont}}(x,y)$ is the kernel of $R_{\mathrm{cont}}(f)$, $R_{\mathrm{disc}}(f)$ is an integral operator whose kernel is given by the difference of $K(x,y)$ and $K_{\mathrm{cont}}(x,y)$. In particular,

$$\mathrm{tr}\big(R_{\mathrm{disc}}(f)\big) = \int_{G(F)} \big(K(x,x) - K_{\mathrm{cont}}(x,x)\big)dx .$$

The idea is to use the formula above for $K(x,x)$, and the formula for $K_{\mathrm{cont}}(x,y)$ provided by Harish-Chandra's Plancherel theorem, to get a second expression for $\mathrm{tr}\big(R_{\mathrm{disc}}(f)\big)$.

Returning to the case of reductive G, we fix a suitable minimal Levi subgroup M_0 of G. Consider first the formula for $K(x,x)$, which after a change of variables becomes

$$K(x,x) = \int_{G(F)} f_1(u)f_2(x^{-1}ux)du .$$

The Weyl integration formula gives an expansion of this into integrals over conjugacy classes. Let $\Gamma_{ell}\big(G(F)\big)$ be the set of conjugacy classes $\{\gamma\}$ in $G(F)$ such that the centralizer of γ in $G(F)$ is compact modulo $A_G(F)$. Any G-regular conjugacy class in $G(F)$ is the image of a class $\{\gamma\}$ in $\Gamma_{ell}\big(M(F)\big)$ for some Levi subgroup M which contains M_0. The pair $(M,\{\gamma\})$ is uniquely determined only modulo the action of the Weyl group W_0^G of (G, A_{M_0}), so the number of such pairs equals $|W_0^G|\,|W_0^M|^{-1}$. The Weyl integration formula can therefore be interpreted as an expansion

$$\sum_M |W_0^M|\,|W_0^G|^{-1} \int_{\Gamma_{ell}(M(F))} |D(\gamma)| \Big(\int_{A_M(F)\backslash G(F)} f_1(x_1^{-1}\gamma x_1)f_2(x^{-1}x_1^{-1}\gamma x_1 x)dx_1\Big) d\gamma \tag{2.1}$$

for $K(x,x)$. The sum here is over the groups $M \in \mathcal{L}(M_0)$. The measure $d\gamma$ is supported on the G-regular classes in $\Gamma_{ell}\big(M(F)\big)$, and is determined in the usual way by a Haar measure on the torus that centralizes γ.

The contribution from $K_{\mathrm{cont}}(x,x)$ can be regarded as a second expansion for $K(x,x)$ in terms of spectral data. Let $\Pi_2\big(G(F)\big)$ be the set of (equivalence classes of) irreducible unitary representations of $G(F)$ which are square integrable modulo $A_G(F)$. We obtain a measure $d\sigma$ on $\Pi_2\big(G(F)\big)$ by transferring a suitable measure on $i\mathfrak{a}_G^*$ by means of the action $\sigma \to \sigma_\Lambda$. (See [16,§2].) Harish-Chandra's Plancherel theorem [15], [16] is easily seen to yield an expansion

$$\sum_M |W_0^M|\,|W_0^G|^{-1} \int_{\Pi_2(M(F))} m(\sigma) \Big(\sum_S \mathrm{tr}\big(\mathcal{I}_P(\sigma,x)S(f)\big)\overline{\mathrm{tr}\big(\mathcal{I}_P(\sigma,x)S\big)}\Big) d\sigma, \tag{2.2}$$

where $m(\sigma)$ is the Plancherel density and

$$S(f) = \mathcal{I}_P(\sigma, f_2) S \mathcal{I}_P(\sigma, f_1^\vee),$$

the function $f_1^\vee$ being defined by $f_1^\vee(x_1) = f_1(x_1^{-1})$. The outer sum is over $M \in \mathcal{L}(M_0)$ as in (2.1), and for a given M, P stands for any group $P(M)$. The inner sum is over $S \in \mathcal{B}_P(\sigma)$, a fixed K-finite, orthonormal basis of the space of Hilbert-Schmidt operators on the underlying space of $\mathcal{I}_P(\sigma)$.

Notice the formal similarity of the two expansions (2.1) and (2.2). The terms with $M = G$ are simplest in each case, and are easily seen to be integrable functions of x in $A_G(F)\backslash G(F)$. Their integrals are equal to

$$\int_{\Gamma_{ell}(G(F))} |D(\gamma)| \Big(\int_{A_G(F)\backslash G(F)} f_1(x_1^{-1}\gamma x_1)dx_1 \int_{A_G(F)\backslash G(F)} f_2(x_2^{-1}\gamma x_2)dx_2\Big)d\gamma \qquad (2.3)$$

and

$$\int_{\Pi_2(G(F))} \operatorname{tr}(\sigma^\vee(f_1))\operatorname{tr}(\sigma(f_2))d\sigma \qquad (2.4)$$

respectively. In particular, if G is semisimple, the trace of $R_{\mathrm{disc}}(f)$ equals (2.4), and can consequently be expressed as the sum of (2.3) with a "parabolic term", consisting of the remaining contributions to (2.1) and (2.2). The parabolic term is of course much harder to compute. It equals the integral over $x \in A_G(F)\backslash G(F)$ of the difference of the expressions obtained from (2.1) and (2.2) by taking the sums only over $M \neq G$. None of the terms in (2.1) and (2.2) with $M \neq G$ is integrable. How then can the difference of the expressions yield a function whose integral is computable? The answer lies in a truncation process, which in the end works out surprisingly well.

Let $P_0 \in \mathcal{P}(M_0)$ be fixed minimal parabolic subgroup. The truncation depends on a point T in the chamber $\mathfrak{a}_{P_0}^+$ which is very regular, in the sense that the number

$$d(T) = \max_{\alpha \in \Delta_{P_0}} \alpha(T)$$

is large. According to the polar decomposition, $G(F)$ equals $KM_0(F)K$. Let $u(x,T)$ be the characteristic function of the set of points

$$k_1 m k_2, \qquad m \in A_G(F)\backslash M_0(F), \;\; k_1, k_2 \in K,$$

in $A_G(F)\backslash G(F)$ such that $H_{M_0}(m)$ lies in the convex hull of

$$\{sT : \; s \in W_0^G\},$$

taken modulo $\mathfrak{a}_G$. Then $u(x,T)$ is the characteristic function of a large compact subset of $A_G(F)\backslash G(F)$. In particular, the integral

$$K^T(f) = \int_{A_G(F)\backslash G(F)} K(x,x)\,u(x,T)dx$$

converges. We shall outline the three steps by which one obtains an explicit formula from $K^T(f)$.

The first step is to study the geometric and spectral expansions of $K^T(f)$ as functions of T. These are obtained from (2.1) and (2.2) by multiplying each expression with $u(x,T)$, and then integrating over x in $A_G(F)\backslash G(F)$. If $F = \mathbb{R}$, one shows that $K^T(f)$ is asymptotic to a polynomial $p_0(T,f)$ in T as $d(T)$ approaches infinity. If F is p-adic, we take T to be in the lattice

$$\mathfrak{a}_{M_0,F} = H_{M_0}(M_0(F))$$

in $\mathfrak{a}_{M_0}$. In this case it turns out that $K^T(f)$ is asymptotic to a function

$$\sum_{k=0}^{N} p_k(T,f)e^{\zeta_k(T)} ,$$

where $\zeta_1 = 0$, $\zeta_1, \ldots , \zeta_N$ are distinct points in the the compact torus

$$i\mathfrak{a}^*_{M_0}/Hom(\mathfrak{a}_{M_0,F}, 2\pi i\mathbb{Z}),$$

and each $p_k(T,f)$ is a polynomial in T. In each case, the "constant term"

$$\tilde{J}(f) = p_0(0,f)$$

of $K^T(f)$ is well defined.

The second step is to calculate $\tilde{J}(f)$ explicitly. More precisely, one must evaluate the terms in the geometric and spectral expansions of $\tilde{J}(f)$. The calculations on the spectral side are the more difficult, and were suggested by work of Waldspurger [24], who carried out the process for p-adic spherical functions on $GL(n)$. The contributions to the final formula of the terms with $M = G$ in (2.1) and (2.2) remain as before, the expressions (2.3) and (2.4). However, the contributions from $M \neq G$ are more elaborate. Their principal ingredients are essentially the weighted orbital integrals and weighted characters discussed above. It is of course the identity of the two expansions of $\tilde{J}(f)$ that yields the noninvariant trace formula. The final result is stated and proved in [10, Theorem 12.1].

The third step is to convert the noninvariant formula into an invariant local trace formula. This is a relatively simple matter, which follows the analogous procedure used in the global trace formula [5, §2]. The final result is an identity between two expansions

$$\sum_M |W_0^M|\,|W_0^G|^{-1}\,(-1)^{\dim(A_M/A_G)} \int_{\Gamma_{ell}(M)} I_M(\gamma,f)d\gamma \qquad (2.5)$$

and

$$\sum_M |W_0^M|\,|W_0^G|^{-1}\,(-1)^{\dim(A_M/A_G)} \int_{\Pi_{disc}(M)} a^M_{disc}(\pi)\,i_M(\pi,f)d\pi\ , \qquad (2.6)$$

whose terms we describe as follows.

In the geometric expansion (2.5), $\Gamma_{ell}(M)$ stands for the set of conjugacy classes in $M(F) \times M(F)$ of the form (γ, γ), where γ is a conjugacy class in $\Gamma_{ell}(M(F))$. The integrand $I_M(\gamma, f)$ is the invariant distribution discussed above, but defined for the function f on $G(F) \times G(F)$, and with the weight function taken relative to the diagonal image of $\mathfrak{a}_M$ in $\mathfrak{a}_M \oplus \mathfrak{a}_M$, rather than the full space $\mathfrak{a}_M \oplus \mathfrak{a}_M$. This distribution can in fact be decomposed in terms of the original distributions on $G(F)$. For the splitting formula [4, Proposition 9.1] asserts that

$$I_M(\gamma, f) = \sum_{M_1, M_2 \in \mathcal{L}(M)} d_M^G(M_1, M_2) I_M^{M_1}(\gamma, f_{1,P_1}) I_M^{M_2}(\gamma, f_{2,P_2}), \quad (2.7)$$

where $d_M^G(M_1, M_2)$ is a constant, and

$$f_{P_i}(m) = \delta_{P_i}(m)^{\frac{1}{2}} \int_K \int_{N_{P_i}(F)} f_i(k^{-1}mnk) dn\, dk, \qquad m \in M_{P_i}(F),$$

for any group $P_i \in \mathcal{P}(M_i)$. We point out that this constant $d_M^G(M_1, M_2)$ also reflects the geometry of the polytopes $\Pi_M(x)$. It equals 0 unless the canonical map

$$(\mathfrak{a}_{M_1}/\mathfrak{a}_G) \oplus (\mathfrak{a}_{M_2}/\mathfrak{a}_G) \longrightarrow \mathfrak{a}_M/\mathfrak{a}_G$$

between Euclidean spaces in an isomorphism, in which case it is the Jacobian determinant of the map. In other words $d_M^G(M_1, M_2)$ is the volume of the parallelepiped determined by orthonormal bases of the complementary subspaces of $\mathfrak{a}_M/\mathfrak{a}_G$ attached to M_1 and M_2.

The constituents of the spectral expansion (2.6) are defined in terms of the decomposition of a certain distribution I_{disc} into irreducible characters. By definition, $I_{\text{disc}}(f)$ equals

$$\sum_M \sum_s \sum_\sigma |W_0^M| \, |W_0^G|^{-1} |\det(s-1)_{\mathfrak{a}_M^G}|^{-1} \varepsilon_\sigma(s) \operatorname{tr}\big(R(s, \sigma^\vee \otimes \sigma) \mathcal{I}_P(\sigma^\vee \otimes \sigma, f^1)\big),$$

where M is summed over $\mathcal{L}(M_0)$, s is summed over the regular elements

$$\{s \in W(\mathfrak{a}_M) : \det(s-1)_{\mathfrak{a}_M^G} \neq 0\}$$

in the Weyl group of (G, A_M), and σ is summed over the set

$$\{\sigma \in \Pi_2(M(F)) : s\sigma \cong \sigma\}/i\mathfrak{a}_G^*$$

of orbits of $i\mathfrak{a}_G^*$ in $\Pi_2(M(F))$. For each orbit σ, $\mathcal{I}_P(\sigma^\vee \otimes \sigma)$ is a well defined induced representation of the subgroup

$$(G(F) \times G(F))^1 = \{(y_1, y_2) : H_G(y_1) = H_G(y_2)\}$$

of $G(F) \times G(F)$. It can therefore be evaluated at the restriction f^1 of f to $\left(G(F) \times G(F)\right)^1$. The only other constituent of $I_{\mathrm{disc}}(f)$ that requires comment is the function $\varepsilon_\sigma(s)$. By definition, ε_σ is the sign character on the group

$$W_\sigma = W'_\sigma \rtimes R_\sigma = \{s \in W(\mathfrak{a}_M) : s\sigma \cong \sigma\}$$

which is 1 on the R-group R_σ and is the usual sign character on the complementary subgroup W'_σ. Having thus defined $I_{\mathrm{disc}}(f)$, we take $\Pi_{\mathrm{disc}}(G)$ to be a union of orbits of $i\mathfrak{a}_G^*$ in $\Pi_{\mathrm{temp}}\left(G(F) \times G(F)\right)$ and $\{a_{\mathrm{disc}}^G(\pi)\}$ to be a corresponding set of coefficients, such that

$$\begin{aligned} I_{\mathrm{disc}}(f) &= \sum_{\pi \in \Pi_{\mathrm{disc}}(G)/i\mathfrak{a}_G^*} a_{\mathrm{disc}}^G(\pi) \operatorname{tr}\left(\pi(f^1)\right) \\ &= \int_{\Pi_{\mathrm{disc}}(G)} a_{\mathrm{disc}}^G(\pi) \operatorname{tr}\left(\pi(f)\right) d\pi . \end{aligned}$$

This accounts for all the terms in (2.6) except for the distribution $i_M(\pi, f)$. By definition,

$$i_M(\pi, f) = r_M(\pi) \operatorname{tr}\left(\mathcal{I}_P(\pi, f)\right) ,$$

where

$$r_M(\pi) = \lim_{\Lambda \to 0} \sum_{Q \in \mathcal{P}(M)} r_{Q|\overline{Q}}(\pi_1)^{-1} r_{Q|\overline{Q}}(\pi_{1,\Lambda}) \theta_Q(\Lambda)^{-1} ,$$

for any representation $\pi = \pi_1 \otimes \pi_2$ in $\Pi_{\mathrm{disc}}(M)$. Here,

$$r_{Q|\overline{Q}}(\pi_1) = r_{Q|\overline{Q}}(\pi_2)$$

is the local normalizing factor for either of the intertwining operators $R_{Q|\overline{Q}}(\pi_1)$ or
$R_{Q|\overline{Q}}(\pi_2)$. (Since π_1 and π_2 are implicitly constituents of the same induced, tempered representation, it is immaterial whether we take π_1 or π_2.) One shows that the limit $r_M(\pi)$ exists and is a nonsingular function of π.

The local trace formula, then, is the identity of (2.5) with (2.6). To get some feeling for it, one could experiment with the simple case of $G = GL(2)$. Take f_1 and f_2 to be p-adic spherical functions, and consider (2.5) and (2.6) as bilinear forms in the Satake transforms

$$\sum_{\nu \in \mathbb{Z}^2} a_i(\nu) z^\nu, \qquad z \in (\mathbb{C}^*)^2, \quad i = 1, 2,$$

of f_1 and f_2. The spectral side (2.6) consists of two innocuous terms, but the geometric side (2.5) is more interesting. The term with $M = G$ in (2.5) is a bilinear form in the elliptic orbital integrals of f_1 and f_2, while the term with $M = M_0$ reduces to an expression involving weighted orbital integrals. In [20, §5], the orbital integrals and weighted orbital integrals on $GL(2)$ were computed explicitly in terms of Satake transforms. The identity of (2.5) and (2.6) provides a new relationship between these objects.

3. Some applications

The local trace formula is capable of yielding nontrivial information on local harmonic analysis, although it is not clear at this point how far it will lead. We shall conclude by describing three applications to the three general problems of §1. These applications are only modest advances on what is presently known, but they give some idea of how the local trace formula can be used.

We begin by sketching a local proof of the following result, which completes the induction argument of §1.

Theorem A. *The distributions*

$$I_M(\gamma_1, f_1), \qquad \gamma_1 \in M(F) \cap G_{\text{reg}}(F), \quad f_1 \in \mathcal{H}(G),$$

are supported on characters.

Proof. Let $f_1 \in \mathcal{H}(G)$ be a function such that $\text{tr}(\pi_1(f_1)) = 0$ for every $\pi_1 \in \Pi_{\text{temp}}(G(F))$. We must show that $I_M(\gamma_1, f_1)$ vanishes for every $M \in \mathcal{L}(M_0)$ and $\gamma_1 \in M(F) \cap G_{\text{reg}}(F)$. We have been carrying the induction hypothesis of §1, which asserts that the theorem holds if G is replaced by a proper Levi subgroup. Combined with a descent formula [4, Corollary 8.3], it tells us that $I_M(\gamma_1, f_1) = 0$ if γ_1 belongs to a proper Levi subgroup of M. We may therefore assume that γ_1 belongs to $\Gamma_{ell}(M(F))$.

Apply the local trace formula, with f_1 as given and f_2 an arbitrary function in $\mathcal{H}(G)$. The spectral side (2.6) vanishes, in view of the condition on f_1. On the geometric side (2.5), we use the splitting formula (2.7) for $I_M(\gamma, f)$. Our induction hypothesis insures the vanishing of all terms in (2.7) with $M_1 \neq G$. Since $d_M^G(G, M_2)$ equals 0 unless $M_2 = M$, we obtain

$$I_M(\gamma, f) = I_M(\gamma, f_1) I_M^M(\gamma, f_{2,P}), \qquad P \in \mathcal{P}(M).$$

The geometric side therefore equals

$$\sum_M |W_0^M|\, |W_0^G|^{-1} (-1)^{\dim(A_M/A_G)} \int_{\Gamma_{ell}(M(F))} I_M(\gamma, f_1) I_M^M(\gamma, f_{2,P}) d\gamma .$$

Now fix M and $\gamma_1 \in M_{ell}(F)$. Choose f_2 to be supported on the set of $G(F)$-conjugates of $\Gamma_{ell}(M(F))$, and such that, as a function of $\gamma \in \Gamma_{ell}(M(F))$, $I_M^M(\gamma, f_{2,P})$ approaches the sum of Dirac measures at the $W(\mathfrak{a}_M)$-translates of γ_1. The geometric side then approaches

$$(-1)^{\dim(A_M/A_G)} I_M(\gamma_1, f_1) .$$

It follows that $I_M(\gamma_1, f_1) = 0$, as required. □

The theorem was proved by global means in [5, Theorem 5.1]. The idea of using the global trace formula goes back to Kazhdan, who established [18] the special case that $M = G$. The local argument we have given here applies equally well to the twisted case, in which G is an arbitrary component. In particular, it could be used to prove that twisted p-adic orbital integrals are supported on (twisted) characters. This has been proved by global means in [19].

The next application concerns the "discrete part" of $\hat{I}_M(\gamma_1)$. Assume for simplicity that G is semisimple and that $\pi_1 \in \Pi_2(G(F))$ is a fixed square integrable representation. Let $f_1 \in \mathcal{H}(G)$ be a pseudo-coefficient for π_1, in the sense that

$$\operatorname{tr}(\pi_1'(f_1)) = \begin{cases} 1, & \text{if } \pi_1' \cong \pi_1^\vee \\ 0, & \text{otherwise.} \end{cases}$$

Theorem B. *Suppose that $\gamma_1 \in \Gamma_{ell}(M(F))$ is G-regular. Then*

$$I_M(\gamma_1, f_1) = (-1)^{\dim(A_M/A_G)} \, |D(\gamma_1)|^{\frac{1}{2}} \, \Theta_{\pi_1}(\gamma_1) \, ,$$

where Θ_{π_1} is the character of π_1.

Proof. This result was discussed in [9, §9], so we shall be brief. Choose a function $f_2 \in \mathcal{H}(G)$, supported on the set of $G(F)$-conjugates of $\Gamma_{ell}(M(F))$, such that, as a function of $\gamma \in \Gamma_{ell}(M(F))$, $|D(\gamma_1)|^{\frac{1}{2}} I_M^M(\gamma, f_{2,P})$ approaches the sum of Dirac measures at the $W(\mathfrak{a}_M)$-translates of γ_1. One checks that $\operatorname{tr}(\pi_1(f_2))$ approaches $\Theta_{\pi_1}(\gamma_1)$. The theorem then follows from the identity of (7.5) and (7.6), and the splitting formula (7.7). □

Theorem B is already known if $F = \mathbb{R}$ [7, Theorem 6.4], or if π_1 is supercuspidal [3]. It is new if F is p-adic and π_1 is not supercuspidal.

The final application is a small contribution to the third problem which is motivated by a comparison of trace formulas. The theory of endoscopy is likely to lead to a parallel family of distributions

$$I_M^{\mathcal{E}}(\gamma_1, f_1) = I_M^{G,\mathcal{E}}(\gamma_1, f_1), \quad M \in \mathcal{L}(M_0), \; \gamma_1 \in \Gamma_{ell}(M(F)), \; f_1 \in \mathcal{H}(G),$$

obtained from stable distributions $\{S\hat{I}_{M_H}(\gamma_{1,H}, f_1^H)\}$ on endoscopic groups. We would expect that the distributions

$$I_M^{\mathcal{E}}(\gamma, f) = \sum_{M_1, M_2 \in \mathcal{L}(M)} d_M^G(M_1, M_2) I_M^{M_1,\mathcal{E}}(\gamma, f_{1,P_1}) I_M^{M_2,\mathcal{E}}(\gamma, f_{2,P_2}), \quad (2.7)^{\mathcal{E}}$$

defined for $\gamma \in \Gamma_{ell}(M(F))$ and $f \in \mathcal{H}(G \times G)$ by the analogue of (2.7), should satisfy their own version of the local trace formula. That is, the geometric expansion

$$\sum_M |W_0^M| \, |W_0^G|^{-1} \, (-1)^{\dim(A_M/A_G)} \int_{\Gamma_{ell}(M)} I_M^{\mathcal{E}}(\gamma, f) d\gamma \qquad (2.5)^{\mathcal{E}}$$

equals some form of spectral expansion (2.6). This is what happens in the comparison of global trace formulas that is required for base change for $GL(n)$ [11, Chapter II]. In general, one would like to show that the two families are in fact the same. In the special case of $GL(n)$ (and with F replaced by a certain finite product of local fields), it is shown that the two distributions differ by a multiple of the invariant orbital integral [11, (II.17.5)]. An argument that is special to $GL(n)$ [11, p. 195-196] then establishes that this multiple is actually 0. The local trace formula can be applied to this point in the general situation.

Theorem C. *Suppose that we are given a family* $\{I_M^{\mathcal{E}}(\gamma_1, f_1)\}$ *of distributions with the property that* $(2.5)^{\mathcal{E}}$ *equals (2.6). Assume also that there are functions*

$$c_M^L(\gamma_1), \qquad L \in \mathcal{L}(M), \;\; \gamma_1 \in \Gamma_{ell}(M(F)),$$

with $c_M^M(\gamma_1) = 0$, *such that*

$$I_M^{L,\mathcal{E}}(\gamma_1, g_1) - I_M^L(\gamma_1, g_1) \;=\; c_M^L(\gamma_1) I_M^M(\gamma_1, g_{1,P})$$

for all M, L, γ_1, *and all* $g_1 \in \mathcal{H}(L)$. *Then* $c_M^L(\gamma_1) = 0$ *for all* $L \in \mathcal{L}(M)$. *In other words, the two families of distributions are the same.*

Proof. Assume inductively that $c_M^L(\gamma_1) = 0$ whenever $L \neq G$. The expressions (2.5) and $(2.5)^{\mathcal{E}}$ are both equal to (2.6), so they are equal to each other. Therefore

$$\sum_M |W_0^M| \, |W_0^G|^{-1} \, (-1)^{\dim(A_M/A_G)} \int_{\Gamma_{ell}(M(F))} (I_M^{\mathcal{E}}(\gamma, f) - I_M(\gamma, f)) d\gamma = 0 \, .$$

Combining the splitting formulas (2.7) and $(2.7)^{\mathcal{E}}$ with the induction assumption, we obtain

$$I_M^{\mathcal{E}}(\gamma f) - I_M(\gamma, f) \;=\; 2 c_M^G(\gamma) I_M^M(\gamma, f_{1,P}) I_M^M(\gamma, f_{2,P}) \, .$$

Since f_1 and f_2 are arbitrary functions in $\mathcal{H}(G)$, we see without difficulty that $c_M^G(\gamma_1) = 0$ for every $\gamma_1 \in \Gamma_{ell}(M(F))$. □

Something akin to Theorem C might be required as a replacement for the argument in [11, p. 195-196], if the techniques of [11] are to be extended to general groups. One can also use the theorem, or rather its version for G a component in the nonconnected group $\mathrm{Res}_{E/F}(GL(n)) \rtimes \mathrm{Gal}(E/F)$, to strengthen one of the peripheral results of [11]. For in this case it was possible only to determine the distributions $I_M^{\mathcal{E}}(\gamma_1, f_1)$ as a linear combination of distributions $I_M^{M_1}(\gamma_1, f_{1,P_1})$ [11, Theorem 6.1]. If one combines Theorem C with the results obtained in [11] by global methods, one can prove that the distributions $I_M^{\mathcal{E}}(\gamma_1)$ and $I_M(\gamma_1)$ in [11, Theorem 6.1] are actually equal.

The local trace formula has other possible applications. For example, it seems to be a natural tool for investigating questions posed in §1 on the singular support of $\hat{I}_M(\gamma)$. However, I think that it will be necessary to work with the Schwartz space rather than the Hecke algebra.

Problem D. *Show that identity of expansions (2.5) and (2.6) remains valid if f_1 and f_2 are Schwartz functions on $G(F)$.*

We are of course especially interested in the p-adic case. The tool for handling p-adic orbital integrals of Schwartz functions is the Howe conjecture, which has been proved by Clozel [13]. Recall that if

$$\mathcal{H}(G(F)/\!\!/K_0) = C_c^\infty(K_0\backslash G(F)/K_0)$$

is the Hecke algebra of bi-invariant functions under an open compact subgroup K_0 of $G(F)$, the Howe conjecture asserts that the vector space of linear functionals on $\mathcal{H}(G(F)/\!\!/K_0)$, obtained by restricting all invariant distributions with support on the $G(F)$-conjugates of a given compact set, is finite dimensional. In particular, if ω is a compact subset of $G(F)$, the space

$$\{I_G(\gamma_1, f_1) : \gamma_1 \in \omega \cap G_{\mathrm{reg}}(F), f \in \mathcal{H}(G(F)/\!\!/K_0)\} ,$$

of linear functionals on $\mathcal{H}(G(F)/\!\!/K_0)$, is finite dimensional. However, if $M \neq G$, the invariant distribution $I_M(\gamma_1, f_1)$ is not supported on the $G(F)$-conjugates of a compact set. One would need the following analogue of the Howe conjecture.

Problem E. *Suppose that K_0 is an open compact subgroup of $G(F)$ and that ω is a compact subset of $M(F)$. Show that the space*

$$\{I_M(\gamma_1, f_1) : \gamma_1 \in \omega \cap G_{\mathrm{reg}}(F), f \in \mathcal{H}(G(F)/\!\!/K_0)\}$$

of linear functionals on $\mathcal{H}(G(F)/\!\!/K_0)$ is finite dimensional.

REFERENCES

1. J. Arthur, *A theorem on the Schwartz space of a reductive Lie group*, Proc. Nat. Acad. Sci. U.S.A. **72** (1975), 4718–4719.

2. ———, *The trace formula in invariant form*, Ann. of Math. (2) **114** (1981), 1–74.

3. ———, *The characters of supercuspidal representations as weighted orbital integrals*, Proc. Indian Acad. Sci. **97** (1987), 3–19.

4. ———, *The invariant trace formula I. Local theory*, J. Amer. Math. Soc. **1** (1988), 323–383.

5. ———, *The invariant trace formula II. Global theory*, J. Amer. Math. Soc. **1** (1988), 501–554.

6. ———, *Intertwining operators and residues I. Weighted characters*, J. Funct. Anal. **84** (1989), 19–84.

7. ———, *Intertwining operators and residues II. Invariant distributions*, Compos. Math. **70** (1989), 51–99.

8. ———, *Unipotent automorphic representations: Conjectures*, Astérisque **171-172** (1989), 13-71.

9. ———, *Towards a local trace formula*, Amer. J. Math. (to appear).

10. ______, *A local trace formula*, preprint.

11. J. Arthur and L. Clozel, *Simple Algebras, Base Change and the Advanced Theory of the Trace Formula,* Annals of Math. Studies, Vol. 120, Princeton University Press, Princeton, 1989.

12. J. Bernstein, P. Deligne, and D. Kazhdan, *Trace Paley-Wiener theorem for reductive p-adic groups*, J. Analyse Math. **47** (1986), 180–192.

13. L. Clozel, *Orbital integrals on p-adic groups: A proof of the Howe conjecture*, Ann. of Math. **129** (1989), 237-251.

14. L. Clozel and P. Delorme, *Le théoreme de Paley-Wiener invariant pour les groupes de Lie réductifs II,* Ann. Scient. Ec. Norm. Sup. (to appear).

15. Harish-Chandra, *Harmonic analysis on real reductive groups III. The Maass-Selberg relations and the Plancherel formula*, Ann. of Math. **104** (1976), 117–201, reprinted in Collected Works, Springer-Verlag, Vol. IV, 259–343.

16. ______, *The Plancherel formula for reductive p-adic groups*, reproduced in Collected Works, Springer-Verlag, Vol. IV, pp. 353–367.

17. R. Herb, *Discrete series characters and Fourier inversion on semisimple real Lie groups*, Trans. Amer. Math. Soc. **277** (1983), 241–262.

18. D. Kazhdan, *Cuspidal geometry on p-adic groups*, J. Analyse Math. **47** (1989), 1–36.

19. R. Kottwitz and J. Rogawski, *The distributions in the invariant trace formula are supported on characters*, preprint.

20. R. Langlands, *Base Change for* $GL(2)$, Annals of Math. Studies **96** (1980), Princeton University Press, Princeton.

21. R. Langlands and D. Shelstad, *On the definition of transfer factors*, Math. Ann. **278** (1987), 219–271.

22. P. Mischenko, *Invariant Tempered Distributions on the Reductive p-Adic Group* $GL_n(\mathbb{F}_p)$, Ph.D. Thesis, University of Toronto, Toronto, 1982.

23. J. Rogawski, *The trace Paley-Wiener theorem in the twisted case*, Trans. Amer. Math. Soc. **309** (1988), 215–229.

24. J.-L. Waldspurger, *Intégrales orbitales sphériques pour* $GL(N)$, Astérisque **171–172** (1989), 270–337.

UNIVERSITY OF TORONTO, DEPARTMENT OF MATHEMATICS, TORONTO, ONTARIO, M5S 1A1, CANADA

E-mail: mars @ math.toronto.edu

ASYMPTOTIC EXPANSIONS ON SYMMETRIC SPACES

ERIK VAN DEN BAN and HENRIK SCHLICHTKRULL

University of Utrecht and
Royal Veterinary and Agricultural University, Denmark

INTRODUCTION

Let G/H be a semisimple symmetric space, where G is a connected semisimple real Lie group with an involution σ, and H is an open subgroup of the fix point group G^σ. Assume that G has finite center; then it is known that G has a σ-stable maximal compact subgroup K.

In harmonic analysis on the symmetric space G/H an important role is played by the K-finite functions f on G/H which are annihilated by a cofinite ideal of the algebra $\mathbf{D}(G/H)$ of invariant differential operators on G/H. The asymptotic behaviour of such functions is examined in [B87]. Using methods originally developed in [HC60] and [CM82] for the special case where H is compact, a converging series expansion of f at infinity is obtained. In fact, these expansions are obtained more generally for functions on G that are allowed to be H-finite on the right.

Let f be as above, but considered as a right H-invariant function on G. Then f can be written as an infinite sum of functions on G which are K-finite also on the right. Explicitly $f = \sum_{\delta \in K^\wedge} f^\delta$, where $f^\delta(x) = \dim \delta \int_K f(xk^{-1})\chi_\delta(k)\, dk$ is right K-finite of type δ, χ_δ being the character of δ.

As a function which is K-finite on both sides, each f^δ has a converging series expansion at infinity, according to the above mentioned results of [HC60] and [CM82]. The purpose of the present note is to relate the coefficients in the expansion of f to the coefficients in the expansions of f^δ.

As an application of the relation between the coefficients, we prove the following result: the function f is bounded on G/H if and only if each of the functions f^δ is bounded on G. This result was obtained earlier by different methods in [FOS88], where it was used to prove (cf. Corollary 4 below): *if f generates (on the left) a unitarizable $(\mathfrak{g}, K)$-submodule of $C^\infty(G/H)$, then f is bounded.*

It is our pleasure to thank William Barker and the Organizing Committee for hosting the conference.

1. Converging expansions of $K \times H$-finite, $\mathcal{Z}(\mathfrak{g})$-finite functions

Let $\mathfrak{g} = \mathfrak{k} + \mathfrak{p}$ be a σ-stable Cartan decomposition of the Lie algebra $\mathfrak{g}$ of G, and let $\mathfrak{g} = \mathfrak{h} + \mathfrak{q}$ be the decomposition in eigenspaces of σ. Choose a σ-stable maximal abelian subspace $\mathfrak{a}$ of $\mathfrak{p}$, and decompose it as the direct sum of $\mathfrak{a}_h = \mathfrak{a} \cap \mathfrak{h}$ and $\mathfrak{a}_q = \mathfrak{a} \cap \mathfrak{q}$.

Let $\Sigma \subset \mathfrak{a}^*$ be the restricted root system, Σ^+ a set of positive roots, Δ a set of simple roots, and $\mathfrak{a}^+$ the corresponding open chamber in $\mathfrak{a}$. We assume that $\mathfrak{a}$ and Σ^+ are chosen to be $\mathfrak{q}$-maximal and $\mathfrak{q}$-compatible, respectively (cf. [S84, p. 118-119]), and denote by Σ_q and Δ_q the sets of non-zero restrictions to $\mathfrak{a}_q$ of the elements in Σ and Δ, respectively. Let $\mathfrak{a}_q^+$ be the corresponding open chamber in $\mathfrak{a}_q$. It follows, in particular, that when P is the minimal parabolic subgroup of G associated with Σ^+, then the product PH is an open subset of G.

Let $\mathcal{Z}(\mathfrak{g})$ be the center of the universal enveloping algebra $\mathcal{U}(\mathfrak{g})$ of the complexification $\mathfrak{g}_c$, and let $f \in C^\infty(G)$ be a function which is $K \times H$-finite (that is, K-finite from the left and H-finite from the right) and $\mathcal{Z}(\mathfrak{g})$-finite (for example, f could be as in the introduction). Recall from [B87, Thm. 2.5] that there exists a finite set S of complex linear forms ν on $\mathfrak{a}_q$, a finite set of M of complex polynomials p on $\mathfrak{a}_q$, and, for each ν and p, a holomorphic function $F_{\nu,p}$ on D^{Δ_q} (where $D \subset \mathbb{C}$ is the unit disk) such that

$$f(a) = \sum_{\nu \in S, p \in M} F_{\nu,p}(\bar{\alpha}(a))\, p(\log a) a^\nu$$

for all $a \in A_q^+ = \exp \mathfrak{a}_q^+$. Here $\bar{\alpha}$ is the map from A_q^+ to $(0,1)^{\Delta_q}$ given by $\bar{\alpha}(a) = (a^{-\alpha})_{\alpha \in \Delta_q}$, where $a^{-\alpha} = e^{-\alpha(\log a)}$.

Expanding each $F_{\nu,p}$ in its power series at 0 we obtain an expansion of f with polynomial coefficients:

$$f(a) = \sum_{\nu \in S - \mathbb{N}\Delta_q} P_\nu(f, \log a) a^\nu,$$

where $P_\nu(f) \in \operatorname{Span} M$ for each

$$\nu \in S - \mathbb{N}\Delta_q = \{s - \sum_{\alpha \in \Delta_q} n_\alpha \alpha \mid s \in S, n_\alpha = 0, 1, 2 \dots\}.$$

For each $\epsilon > 0$, the sum converges absolutely and uniformly on the set $\{a \in A_q \mid \alpha(\log a) > \epsilon, \forall \alpha \in \Delta_q\}$. Each polynomial $P_\nu(f)$ is uniquely determined by f and ν. For convenience we put $P_\nu(f) = 0$ for $\nu \notin S - \mathbb{N}\Delta_q$.

In the special case where σ is the Cartan involution, so that $H = K$ and $\mathfrak{a}_q = \mathfrak{a}$, the above expansion on A^+ of a $K \times K$-finite $\mathcal{Z}(\mathfrak{g})$-finite function is the same as that given in [HC60] and [CM82].

Returning to the general case where H is noncompact, we write, as in the introduction, $f = \sum_{\delta \in K^\wedge} f^\delta$. Since the functions f^δ are $K \times K$-finite and $\mathcal{Z}(\mathfrak{g})$-finite, they admit converging expansions (according to the special case just mentioned)

$$f^\delta(a) = \sum_\xi P_\xi(f^\delta, \log a)a^\xi$$

for $a \in A^+$ with polynomials $P_\xi(f^\delta)$ on $\mathfrak{a}$, $\xi \in \mathfrak{a}_c^*$. Let T denote the set of weights μ of $\mathfrak{a}_h$ occurring in the finite dimensional representation of H generated by $R_h f, h \in H$.

Theorem 1. *Let* $\nu \in (\mathfrak{a}_q)_c^*$. *Then*

$$P_\nu(f) = \sum_{\delta \in K^\wedge} \sum_\xi P_\xi(f^\delta)|_{\mathfrak{a}_q},$$

where the inner sum extends over the finite set of $\xi \in \mathfrak{a}_c^*$ *such that* $\xi|_{\mathfrak{a}_q} = \nu$ *and* $\xi|_{\mathfrak{a}_h} \in T$. *The degrees of the polynomials* $P_\xi(f^\delta)$ *in the sum are bounded by a constant independent of* δ *and* ν, *and the sum over* δ *converges locally uniformly on* $\mathfrak{a}_q$.

The proof of the theorem will be given at the end of Section 3.

2. ASYMPTOTIC EXPANSIONS OF K-FINITE $\mathcal{Z}(\mathfrak{g})$-FINITE FUNCTIONS OF AT MOST EXPONENTIAL GROWTH

Fix an ideal $I \subset \mathcal{Z}(\mathfrak{g})$ of finite codimension and a finite set $\mathcal{T} \subset K^\wedge$ of K-types, and put

$$E(I, \mathcal{T}) = \{f \in C^\infty(G; \mathcal{T}) \mid L_u f = 0, \forall u \in I\}.$$

Here $C^\infty(G; \mathcal{T})$ denotes the space of continuous functions on G whose left translates by elements of K span a finite dimensional space in which only K-types from $\mathcal{T}$ occur, and L denotes the left regular representation of G. Then G and $\mathfrak{g}$ act on $E(I, \mathcal{T})$ via the right regular representation R.

Let $J \subset \mathcal{U}(\mathfrak{g})$ be the left ideal generated by I and by the subspace $\bigcap_{\tau \in \mathcal{T}} \ker \tau$ in $\mathcal{U}(\mathfrak{k})$, and consider the $\mathcal{U}(\mathfrak{g})$-module $\mathcal{U}(\mathfrak{g})/J$. It is easily seen that every element in this module is $\mathfrak{k}$-finite, and since it is clearly finitely generated it follows from [W88, 3.4.7] that this is an admissible $(\mathfrak{g}, \mathfrak{k})$-module.

Lemma 1. *The pairing*

$$\mathcal{U}(\mathfrak{g})/J \times E(I, \mathcal{T}) \to \mathbb{C},$$

defined by $(u, f) \to L_u f(e)$, *is* $\mathfrak{g}$*-equivariant and nondegenerate in* f.

Proof. The pairing is equivariant:

$$L_{Xu} f(e) = L_X(L_u f)(e) = -R_X(L_u f)(e) = -L_u(R_X f)(e).$$

It is nondegenerate in f because f is real analytic (cf. [HC60, p. 66]). □

Corollary 1. *The space of right K-finite functions in $E(I,\mathcal{T})$ is an admissible, finitely generated $(\mathfrak{g},K)$-module for the right action of $\mathfrak{g}$ and K.*

Proof. It follows from Lemma 1 that the $\mathfrak{g}$-module $E(I,\mathcal{T})$ embeds into the linear dual of $\mathcal{U}(\mathfrak{g})/J$. Hence the K-finite functions embed into the $\mathfrak{k}$-finite dual of $\mathcal{U}(\mathfrak{g})/J$. Now apply [W88, 4.3.2]. (That all K-types occur in $E(I,\mathcal{T})$ with finite multiplicity could also be seen from [HC60, p. 65, Cor. 2]). □

For each $r \in \mathbb{R}$ we denote $C_r(G)$ the Banach space of continuous functions on G of at most exponential growth rate r, cf. [BS87, p. 113]. Then G acts continuously on $C_r(G)$ from both sides.

Fix r and let $\mathcal{E} = E(I,\mathcal{T}) \cap C_r(G)$ be equipped with the norm inherited from $C_r(G)$. With π equal to the right action R of G on $\mathcal{E}$ we then obtain an admissible Banach representation $(\pi,\mathcal{E})$ of G. Let $\mathcal{E}^\infty$ denote the space of C^∞-vectors of this representation, i.e., the space of functions $f \in E(I,\mathcal{T})$ for which $R_u f \in C_r(G)$ for all $u \in \mathcal{U}(\mathfrak{g})$, equipped with the natural Fréchet topology. Moreover, let $(\mathcal{E}^\infty)'_K$ be the space of K-finite vectors in the topological linear dual $(\mathcal{E}^\infty)'$ of $\mathcal{E}^\infty$; by [W88, 4.3.3] it can be identified with the space $V^\sim$ of K-finite vectors in the linear dual of the $(\mathfrak{g},K)$-module V underlying $\mathcal{E}$.

Lemma 2. *Let $f \in \mathcal{E}^\infty$. There exist $v \in \mathcal{E}^\infty$ and $\sigma \in (\mathcal{E}^\infty)'_K$ such that, for all $x \in G$,*

$$f(x) = \sigma(\pi(x)v).$$

Proof. Let $v = f$ and let σ be the restriction to $\mathcal{E}^\infty$ of evaluation at the identity. Then $\sigma \in (\mathcal{E}^\infty)'$. Moreover, it follows from Lemma 1 that there is a surjection of the finite dimensional space $\mathcal{U}(\mathfrak{k})/\mathcal{U}(\mathfrak{k}) \cap J$ onto $\mathcal{U}(\mathfrak{k})\sigma$. Hence σ is K-finite (and the K-types occurring in the span of the K-translates of σ are contragredient to those in $\mathcal{T}$). □

Conversely, it follows from [W83, Lemma 5.1] that, for every admissible, finitely generated Banach representation $(\pi,\mathcal{H})$ and every $\sigma \in (\mathcal{H}^\infty)'_K$, there exists r, I, and $\mathcal{T}$ such that (1) the generalized matrix coefficient $\sigma(\pi(x)v)$ belongs to $\mathcal{E}^\infty$ for all $v \in \mathcal{H}^\infty$, and (2) the map taking $v \in \mathcal{H}^\infty$ to $\sigma(\pi(\cdot)v) \in \mathcal{E}^\infty$ is continuous.

The following theorem is now a direct consequence of [W88, Thm. 4.4.3, cf. also BS87] for the case of K-fixed functions. (Wallach only states the theorem for Hilbert representations, but it holds as well for Banach representations, cf. [W83, Thm. 5.8]).

Theorem 2 (Wallach). *Fix r, I, and $\mathcal{T}$ as above. Then there exists a finite set $E^\circ \subset \mathfrak{a}_c^*$ with the following properties. For every $f \in \mathcal{E}^\infty$ and every $\xi \in E^\circ - \mathbb{N}\Delta$ there exists a polynomial $p_\xi(f)$ on $\mathfrak{a}$ such that*

$$f(\exp tH) \underset{t\to\infty}{\sim} \sum_\xi p_\xi(f,tH)e^{t\xi(H)},$$

for every $H \in \mathfrak{a}^+$. Here the asymptotic relation $\underset{t\to\infty}{\sim}$ means that, for all $N \in \mathbb{R}$, there exist positive numbers C and ϵ such that

$$|f(\exp tH) - \sum_{\mathrm{Re}\,\xi \geq 0} p_\xi(f, tH)e^{t\xi(H)}| \leq Ce^{(N-\epsilon)t} \text{ for all } t \geq 0,$$

and that this inequality is locally uniform in $H \in \mathfrak{a}^+$.

Moreover, there exists $d \in \mathbb{N}$ such that, for each ξ, the map $f \to p_\xi(f)$ is continuous and linear from $\mathcal{E}^\infty$ to the space P_d of all polynomials on $\mathfrak{a}$ of degree $\leq d$.

Remark 1. Let $j(V^\sim)$ be the Jacquet module (cf. [W88, 4.1.5]) of the Harish-Chandra module $V^\sim = (\mathcal{E}^\infty)'_K$. It follows from [W88, Lemma 4.1.4] that $j(V^\sim)$ is generated as a $\mathcal{U}(\mathfrak{g})$-module by a finite dimensional $\mathfrak{a}$-stable subspace, say of dimension d_0. Since the adjoint action of $\mathfrak{a}$ on $\mathcal{U}(\mathfrak{g})$ is semisimple, it then follows that the representation of $\mathfrak{a}_c$ in $j(V^\sim)$ admits a simultaneous Jordan decomposition whose nilpotent part has nilpotent order at most d_0. This implies that, for each $k \in \mathbb{N}$ and $\xi \in \mathfrak{a}_c^*$, the generalized weight space $(V^\sim/\mathfrak{n}^k V^\sim)_\xi$ for the weight ξ is annihilated by $(H - \xi(H))^{d_0}$ for all $H \in \mathfrak{a}$. For the constant d in the final statement of the theorem one may take d_0.

Remark 2. The theorem stated here deals with the asymptotic behavior of f in the direction of the open chamber A^+. In fact a more general result, describing also the asymptotics 'along the walls,' is contained in [W88]. In [BS89] we study these expansions (for the K-fixed case) and prove a relation between coefficients in the expansions along the walls and coefficients p_ξ in Theorem 2 (cf. [BS89, Thm. 3.1]). However, these results are not needed here.

Let $\mathcal{S}$ be Wallach's space of rapidly decreasing functions on G (cf. [W88, 7.1.2]; $\mathcal{S} = \cap_{p>0} \mathcal{C}^p(G)$, where $\mathcal{C}^p(G)$ is Harish-Chandra's L^p-Schwartz space), and let $\mathcal{S}'$ be the dual space. Fix a Haar measure dx on G. Following [BS87, part II] (to obtain congruence with [BS87], replace $f(x)$ by $f(x^{-1})$), we obtain:

Corollary 2. *For every $f \in \mathcal{E}$ and $\xi \in E^\circ - \mathbb{N}\Delta$ there exists a polynomial $p_\xi(f)$ on $\mathfrak{a}$ with coefficients in $\mathcal{S}'$ such that*

$$f(\exp(tH)x) \sim \sum_\xi p_\xi(f, tH)(x)e^{t\xi(H)}$$

as $t \to +\infty$, for every $H \in \mathfrak{a}^+$. Here the relation $\sim$ means that the following asymptotic relation holds for all $\phi \in \mathcal{S}$ (in the sense described in Thm. 2):

$$\int_G f(\exp(tH)x)\phi(x)dx \underset{t\to\infty}{\sim} \sum_\xi p_\xi(f, tH)(\phi)e^{t\xi(H)}.$$

Moreover, for each ξ, the map $f \to p_\xi(f)$ is continuous and G-equivariant (for the right actions) from $\mathcal{E}$ to $P_d \otimes \mathcal{S}'$.

3. Application to $K \times H$-finite, $\mathcal{Z}(\mathfrak{g})$-finite functions

Here is the relation between the expansions in the previous two sections:

Theorem 3. *Let $f \in E(I, \mathcal{T})$ and assume that f is right H-finite. Then there exists $r \in \mathbb{R}$ such that $f \in C_r(G)$, and hence Corollary 2 applies to f. The restrictions to the open set PH of the distribution coefficients of $p_\xi(f)$ are, via the Haar measure dx, given by real analytic functions, and hence they can be evaluated at the identity. They satisfy the following relation with the polynomials P_ν of Section 1: for all $\nu \in (\mathfrak{a}_q)^*_c$*

$$P_\nu(f) = \sum_\xi p_\xi(f, e)|_{\mathfrak{a}_q},$$

where the sum extends over the finite set of $\xi \in \mathfrak{a}_q^*$ such that $\xi|_{\mathfrak{a}_q} = \nu$ and $\xi|_{\mathfrak{a}_h} \in \mathcal{T}$.

Proof. For the existence of r, see Remark 14.5 in [BS87]. In the special case where $\mathcal{T}$ consists only of the trivial K-type (so that f is left K-invariant), the theorem can be derived from [BS87, Sections 14-16] as follows.* That $p_\xi(f)$ is real analytic on PH is stated in Corollary 16.2. For the τ-spherical function F associated to f (cf. [BS87, p. 148]) it follows from (14.8), (16.4) and (16.9) that $p_\xi(F, e)|_{\mathfrak{a}_q}$ can be obtained from $P_\nu(F)$, where $\nu = \xi|_{\mathfrak{a}_q}$, by projecting it onto the generalized weight space in E_τ of $\mathfrak{a}_h$-weight $\mu = \xi|_{\mathfrak{a}_h}$. Hence the summation over all ξ such that $\xi|_{\mathfrak{a}_q} = \nu$ and $\xi|_{\mathfrak{a}_h} \in \mathcal{T}$ yields $\sum_\xi p_\xi(F, e)|_{\mathfrak{a}_q} = P_\nu(F)$, from which the stated result for f follows. The only difficulty in extending this proof to the general $\mathcal{T}$ is contained in the following lemma, which generalizes [BS87, Lemma 15.1]. □

Define $a_t \in A^+$ for $t \in (0,1)^\Delta$ by $(a_t)^{-\alpha} = t_\alpha$ for $\alpha \in \Delta$. Then $t \to 0$ is equivalent to $\alpha(\log a_t) \to +\infty$ for all $\alpha \in \Delta$.

Lemma 3. *There exist an open neighborhood Ω_0 of $(e, 0)$ in $G \times \mathbb{R}^\Delta$ and real analytic maps $h, a, k : \Omega_0 \to H, A, K$, respectively, such that:*

(i) *For all $(g, t) \in \Omega_0$, with $t \in (0,1)^\Delta$,*

$$g a_t = h(g,t) a(g,t) a_t k(g,t).$$

(ii) *If $(g, 0) \in \Omega_0$ and $x = man \in P$, then $(gx, 0) \in \Omega_0$, $h(gx, 0) = h(g, 0)$, $a(gx, 0) = a(g, 0)a$, and $k(gx, 0) = k(g, 0)m$.*

(iii) *For $t \in \mathbb{R}^\Delta$ near 0 we have $h(e,t) = a(e,t) = k(e,t) = e$.*

Proof. The existence of h and a is given in [BS87, Lemma 15.1]. To prove the existence of k we need the following lemma. Let $\bar{P} = MA\bar{N}$ be the minimal parabolic opposite to P.

---

*Notice that in [BS87] the sides from which K and H act are reversed.

Lemma 4. *There exist an open neighborhood U_1 of $(e,0)$ in $G \times \mathbb{R}^\Delta$ and unique real analytic maps $z_1, k_1 : U_1 \to \bar{N}A, K$, respectively, such that*

$$ga_t = z_1(g,t)a_t k_1(g,t)$$

for $(g,t) \in U_1$ with $t \in (0,1)^\Delta$. Moreover, $z_1(e,t) = k_1(e,t) = e$, and if $x = man \in P$ then $z_1(gx,0) = z_1(g,0)a$ and $k_1(gx,0) = k_1(g,0)m$.

Proof. The uniqueness is clear from the uniqueness in the Iwasawa decomposition. By the $\bar{N}AMN$ decomposition it suffices to prove the existence of such maps on a neighbourhood of $(e,0)$ in $N \times \mathbb{R}^\Delta$. Now use [BS87, Lemma 8.6] and its proof. □

Proof of Lemma 3. Let Ω_0, h, and a be as in [BS87, Lemma 15.1]. Then, for (z,t) an element of $(\bar{N}A \times (0,1)^\Delta) \cap \Omega_0$, we have

$$za_t k = h(z,t)a(z,t)a_t$$

for some $k \in K$. From the uniqueness in Lemma 4 we infer that

$$z = z_1(h(z,t)a(z,t),t) \text{ and } k = k_1(h(z,t)a(z,t),t)$$

for (z,t) near $(e,0)$ in $(\bar{N}A \times (0,1)^\Delta) \cap \Omega_0$. Hence

$$za_t = h(z,t)a(z,t)a_t k(z,t)$$

where $k(z,t) = k_1(h(z,t)a(z,t),t)^{-1}$ is defined and real analytic on a neighborhood of $(e,0)$ in $\bar{N}A \times \mathbb{R}^\Delta$.

For $(g,t) \in G \times (0,1)^\Delta$ near $(e,0)$, Lemma 4 gives

$$\begin{aligned} ga_t &= z_1(g,t)a_t k_1(g,t) \\ &= h(z_1(g,t),t)a(z_1(g,t),t)a_t k(z_1(g,t),t)k_1(g,t), \end{aligned}$$

and Lemma 3 follows. □

Proof of Theorem 1. From Theorem 3 we have

$$P_\nu(f) = \sum_{\xi|_{\mathfrak{a}_q}=\nu, \xi|_{\mathfrak{a}_h} \in T} p_\xi(f,e)|_{\mathfrak{a}_q}$$

for all $\nu \in (\mathfrak{a}_q)^*_c$, and, when applied to the case $H = K$,

$$P_\xi(f^\delta) = p_\xi(f^\delta, e)$$

for all $\xi \in \mathfrak{a}^*_c$. In particular, $\deg P_\xi(f^\delta) = \deg p_\xi(f^\delta, e) \leq d$. From the continuity and linearity of the map $f \to p_\xi(f)$ it follows that

$$p_\xi(f) = \sum_\delta p_\xi(f^\delta)$$

for all ξ. Combining these equations we get

$$P_\nu(f) = \sum_\xi p_\xi(f,e)|_{\mathfrak{a}_q} = \sum_{\xi,\delta} p_\xi(f^\delta,e)|_{\mathfrak{a}_q} = \sum_{\xi,\delta} P_\xi(f^\delta)|_{\mathfrak{a}_q},$$

and the theorem follows. □

4. Application to boundedness

For simplicity we consider in this section only functions on G that are right H-fixed. Notice that in this case the set $T \subset (\mathfrak{a}_h)^*_c$ in Theorem 1 consists only of the element 0. In [B87, Thm. 6.4] (and in [CM82, Thm. 7.5] for the case of $H = K$) a criterion for $f \in L^p(G/H)$ is given in terms of the coefficients $P_\nu(f)$, where $1 \le p < \infty$. The following theorem supplements this (at $p = \infty$), and the proof is essentially the same (it is in fact slightly easier). It is convenient to rewrite the series expansion of f as follows

$$f(a) = \sum_{\nu \in S - \mathbb{N}\Delta_q, n \in \mathbb{N}^{\Delta_q}} c_{\nu,n} (\log a)^n a^\nu \qquad (a \in A_q^+),$$

where $c_{\nu,n} \in \mathbb{C}$, and where $(\log a)^n$ is defined as $\prod_{\beta \in \Delta_q} (\beta(\log a))^{n_\beta}$. The $c_{\nu,n}$ are uniquely determined by f and by the choice of basis Δ_q defining the open chamber A_q^+.

Theorem 4. *Let f be a K-finite, $\mathbf{D}(G/H)$-finite function on G/H. Then the following three statements are equivalent:*

(i) *f is bounded on G/H.*

(ii) *For every choice Δ_q of basis for Σ_q, and for every ν, the function $a \to P_\nu(\log a)a^\nu$ is bounded on A_q^+.*

(iii) *For every choice Δ_q of basis for Σ_q, for every $\nu = \sum_{\beta \in \Delta_q} \nu_\beta \beta$ and $n \in \mathbb{N}^{\Delta_q}$ with $c_{\nu,n} \neq 0$, and for every $\beta \in \Delta_q$, we have:*

$$\operatorname{Re} \nu_\beta \le 0; \text{ in fact, if } n_\beta \neq 0, \text{ then } \operatorname{Re} \nu_\beta < 0.$$

We can now derive the following result, which was first obtained in joint work of Flensted-Jensen, Oshima and the second author ([FOS88, Lemma 3.2]).

Corollary 3. *Let f be as above, and let $f = \sum_{\delta \in K^\wedge} f^\delta$. Then f is bounded if and only if each f^δ is bounded.*

Proof. If f is bounded, then obviously each f^δ is bounded. Conversely, assume that every f^δ is bounded, and fix Δ_q and $\nu \in S - \mathbb{N}\Delta_q$.

From Theorem 4 (applied to the special case $H = K$) it follows that $P_\xi(f^\delta, \log a)a^\xi$ is bounded on A^+ for all $\xi \in \mathfrak{a}_c^*$. Hence it is also bounded on A_q^+ because A_q^+ is contained in the closure of A^+. In particular, this holds with ξ given by $\xi|_{\mathfrak{a}_q} = \nu$ and $\xi|_{\mathfrak{a}_h} = 0$. For this ξ we write

$$P_\xi(f^\delta, \log a) = \sum_{n \in \mathbb{N}^{\Delta_q}} c_n^\delta (\log a)^n$$

for $a \in A_q$. If $P_\xi(f^\delta) \neq 0$, then the boundedness of $P_\xi(f^\delta, \log a)a^\nu$ on A_q^+ implies that, for each $\beta \in \Delta_q$, we have $\operatorname{Re} \nu_\beta \le 0$. Moreover, if $n \in \mathbb{N}^{\Delta_q}$,

$c_n^\delta \neq 0$, and $n_\beta \neq 0$, then $\operatorname{Re}\nu_\beta < 0$. However, Theorem 1 gives that

$$c_{\nu,n} = \sum_\delta c_n^\delta,$$

and hence, if $c_{\nu,n} \neq 0$, then $c_n^\delta \neq 0$ for some δ. Hence $\operatorname{Re}\nu_\beta \leq 0$, and if $n_\beta \neq 0$, then $\operatorname{Re}\nu_\beta < 0$. Now Theorem 4 can be applied once more. □

Finally, we notice the following corollary, also from [FOS88]:

Corollary 4. *Let f be as above, and assume that the $(\mathfrak{g}, K)$-module V_f generated by f (on the left) is unitarizable. Then f is bounded.*

Proof. Since projection onto the space of functions of right K-type δ is a left homomorphism, V_{f^δ} is a unitarizable representation. Since f^δ is a $K \times K$-finite matrix coefficient of V_{f^δ} (cf. Lemma 2), it is thus in fact a matrix coefficient of a unitary representation, and hence it is bounded. Now apply Corollary 3. □

References

[B87] Ban, E.P. van den, *Asymptotic behaviour of matrix coefficients related to reductive symmetric spaces*, Proc. Kon. Nederl. Akad. Wet. **90** (1987), 225-249.

[BS87] Ban, E.P. van den and H. Schlichtkrull, *Asymptotic expansions and boundary values of eigenfunctions on Riemannian symmetric spaces*, J. reine angew. Math. **380** (1987), 108-165.

[BS89] ———, *Local boundary data of eigenfunctions on a Riemannian symmetric space*, Invent. math. **98** (1989), 639-657.

[CM82] Casselman, W. and D. Miličić, *Asymptotic behaviour of matrix coefficients of admissible representations*, Duke Math. J. **49** (1982), 869-930.

[FOS88] Flensted-Jensen, M., T. Oshima and H. Schlichtkrull, *Boundedness of certain unitarizable Harish-Chandra modules*, Adv. Stud. in Pure Math. **14** (1988), 651-660.

[HC60] Harish-Chandra, *Differential equations and semisimple Lie groups, Collected Papers*, vol. **3**, Springer-Verlag, 1983, pp. 57-120.

[S84] Schlichtkrull, H., *Hyperfunctions and harmonic analysis on symmetric spaces*, Birkhäuser, 1984.

[W83] Wallach, N., *Asymptotic expansions of generalized matrix entries of representations of real reductive groups*, Lecture Notes in Math. **1024** (1983), 287-369.

[W88] ———, *Real reductive groups*, vol. 1, Academic Press, 1988.

University of Utrecht, Department of Mathematics, P.O.Box 80.010, 3508 TA Utrecht, the Netherlands.

E-mail: ban @ math.ruu.nl

The Royal Veterinary and Agricultural University, Department of Mathematics and Physics, Thorvaldsensvej 40, 1871 Frederiksberg C, Denmark.

E-mail: rvamath @ vm.uni-c.dk

THE ADMISSIBLE DUAL OF GL_N VIA RESTRICTION TO COMPACT OPEN SUBGROUPS

COLIN J. BUSHNELL and PHILIP C. KUTZKO

King's College and University of Iowa

Let G be a reductive group over a p-adic field F. Then, as with reductive groups over any field, it is natural to cast the representation theory of G in terms of parabolic induction. This leads to the notion of supercuspidal representation and, in the case of GL_N, to the classification of irreducible (admissible) representations given in the work of Bernstein–Zelevinski [BZ], [Z]. On the other hand, the fact that G is a totally disconnected, locally compact group accounts for the existence of open, compact modulo center subgroups of G which in turn has a strong influence on its representation theory. In particular, one is led to consider the possibility that supercuspidal representations may be constructed by induction from such subgroups (see [Ku2] for historical background) and, more generally, to inquire into the possibility of classifying admissible representations of G by considering the subrepresentations they may have when restricted to such subgroups (the possibility of classifying the admissible dual in this fashion was first raised in [H].) In what follows, we report on recent progress in this direction in the case $G = GL_N(F)$; we begin with some general background.

1. HEREDITARY ORDERS AND COMPACT, OPEN SUBGROUPS

We fix throughout a p-adic field F with ring of integers $\mathcal{O} = \mathcal{O}_F$ and prime ideal $P = P_F$; we set $k = k_F = \mathcal{O}/P$. We fix as well a vector space V of dimension N over F and set $A = \mathrm{End}_F(V)$ and $G = A^\times$ so that G is isomorphic to $GL_N(F)$ after a choice of basis for V.

1.1. Definition.

1. By an *$\mathcal{O}_F$-lattice chain* (or simply *lattice chain*) in V we mean a doubly infinite sequence $L = \ldots L_{-1} \supset L_0 \supset L_1 \ldots$ of free, rank N, $\mathcal{O}$-submodules of V such that $L_i \neq L_{i+1}$ for all i and such that

The first author was supported in part by SERC grant GR/E 47650. The second author was supported in part by NSF Grant DMS-8704194 and by SERC grant GR/F 73366. Both authors wish to thank the Institute for Advanced Study for their hospitality during Academic Year 1988–1989. This visit was supported in part by NSF Grant DMS 8610730 .

there is an integer $e = e(L) \geq 1$ for which $L_{i+e} = PL_i$ for all i. We call e the *period* of L.

2. Let L be a lattice chain in V. Then we say that L is *uniform* if $\dim_k(L_i/L_{i+1})$ is independent of i for all integers i.
3. Let L and L' be lattice chains in V. Then we say that L and L' are *equivalent* if there is an integer k for which $L'_{i+k} = L_i$ for all integers i. We denote the equivalence class of L by [L].
4. Let L be a lattice chain in V. Then we denote by $\mathcal{A} = \mathcal{A}(L)$ the subring of A consisting of elements x for which $xL_i \subseteq L_i$ for all integers i.

1.2. *Remarks.* (See [BF] for details.)

1. $\mathcal{A}$ depends only on the equivalence class of L. It is an hereditary order in A whose radical, $\mathcal{P} = \mathcal{P}(\mathcal{A})$, is the set of x in A for which $xL_i \subseteq L_{i+1}$ for all integers i.
2. The map $L \to \mathcal{A}(L)$ induces a bijection of the set of equivalence classes of lattice chains in V with the set of hereditary orders in A.
3. Let $\mathcal{A}$ be an hereditary order in A. Then by the above remark, $\mathcal{A} = \mathcal{A}(L)$ for some lattice chain L in V, unique up to equivalence. We may thus define the *ramification index* $e = e(\mathcal{A} \mid \mathcal{O})$ by setting $e = e(L)$. The index e has the property that $P\mathcal{A} = \mathcal{P}^e$.
4. $\mathcal{A}(L)$ is a *principal order* (i.e., $\mathcal{P}$ is principal as a left ideal of $\mathcal{A}$) if and only if L is uniform.

We associate various subgroups of G to a given hereditary order $\mathcal{A}$ in A:

1.3. **Definition.**

1. We set $U(\mathcal{A}) = U^0(\mathcal{A}) = \mathcal{A}^\times$. For $n \geq 1$, we set $U^n(\mathcal{A}) = 1 + \mathcal{P}^n$.
2. We set $K(\mathcal{A}) = N_G(U(\mathcal{A}))$.

1.4. *Remarks.* (See [BF].)

1. $K(\mathcal{A}) = N_G(U^n(\mathcal{A}))$ for all positive integers n. $K(\mathcal{A})$ may also be defined as the set of x in G for which $x\mathcal{A}x^{-1} = \mathcal{A}$.
2. The map $\mathcal{A} \to K(\mathcal{A})$ is a bijection between the set of principal orders in A and the set of subgroups of G which are maximal with the property of being open and compact modulo center.

2. Complementary Sets — Duality

We are interested in describing the complex dual of the groups

$$U^{r+1}(\mathcal{A})/U^{n+1}(\mathcal{A}), \quad 0 \leq [n/2] \leq r < n,$$

where $\mathcal{A}$ is an hereditary order in A.

2.1. **Definition.** For a subset S of A, we define the complementary set, S^*, to be the set of elements x in A for which $\operatorname{tr}(xS) \subset P$.

2.2. Lemma. *If $\mathcal{A}$ is a hereditary order in A then $(\mathcal{P}^n(\mathcal{A}))^* = \mathcal{P}^{1-n}(\mathcal{A})$ for all $n \in \mathbb{Z}$.*

2.3. Corollary. *Let r and n be positive integers with $[n/2] \leq r < n$, fix a character ψ of F of conductor P (i.e., ψ is trivial on P but not on $\mathcal{O}$) and, for b in $\mathcal{P}^{-n}(\mathcal{A})$, define the function ψ_b on $U^{r+1}(\mathcal{A})$ by $\psi_b(x) = \psi(\operatorname{tr} b(x-1))$. Then the map $b \mapsto \psi_b$ induces an isomorphism of $\mathcal{P}^{-n}(\mathcal{A})/\mathcal{P}^{-r}(\mathcal{A})$ with $(U^{r+1}(\mathcal{A})/U^{n+1}(\mathcal{A}))^\wedge$.*

The above corollary serves as motivation for the following definition.

2.4. Definition. By a *stratum* in A, we mean a 4-tuple $\Omega = [\mathcal{A}, n, r, b]$ where $\mathcal{A}$ is an hereditary order in A, n and r are integers with $n > r$, and b is an element of $\mathcal{P}^{-n}(\mathcal{A})$. If $\Omega_i = [\mathcal{A}_i, n_i, r_i, b_i]$ are strata in A, $i = 1, 2$, then we say that Ω_1 is *equivalent* to Ω_2, written $\Omega_1 \sim \Omega_2$, if $b_1 + \mathcal{P}(\mathcal{A}_1) = b_2 + \mathcal{P}(\mathcal{A}_2)$.

2.5. *Remark.* If Ω_i as above are equivalent, $i = 1, 2$, then it follows that $\mathcal{A}_1 = \mathcal{A}_2$ and that $r_1 = r_2$. If, in addition, $\nu_{\mathcal{A}_i}(b_i) = -n_i$, then $n_1 = n_2$. (Here $\nu_{\mathcal{A}}$ is defined as usual by the criterion that x lies in $\mathcal{P}^{\nu_{\mathcal{A}}(x)}$ but not in $\mathcal{P}^{\nu_{\mathcal{A}}(x)+1}$ for each nonzero x in A, while $\nu_{\mathcal{A}}(0) = \infty$).

2.6. *Remark.* Note that $U(\mathcal{A})/U^1(\mathcal{A})$ is isomorphic to $\Pi \operatorname{Aut}_k(L_i/L_{i+1})$, the product running from $i = 0$ to $i = e(\mathcal{A}) - 1$. Thus we may (and will) identify the complex dual, $(U(\mathcal{A})/U^1(\mathcal{A}))^\wedge$, of $U(\mathcal{A})/U^1(\mathcal{A})$ with representations $\otimes\sigma_i$, σ_i a representation of $\operatorname{Aut}_k(L_i/L_{i+1})$.

3. Simple Strata

We will be interested in studying strata Ω of the form $[\mathcal{A}, n, r, \beta]$ where $\nu_{\mathcal{A}}(\beta) = -n$ and $E = F[\beta]$ is a sub*field* of A such that $E^\times$ lies in $K(\mathcal{A})$. Such strata will be called *pure*. By definition, we do not change the equivalence class of Ω if we replace β by another element γ of $\beta + \mathcal{P}^{-r}(\mathcal{A})$ such that $[\mathcal{A}, n, r, \gamma]$ is pure, but in doing so we may replace E by a field which is less felicitous for various computations which will concern us below. In particular, we will need that $[E : F] \leq [F[\gamma] \colon F]$ where γ runs through those elements of $\beta + \mathcal{P}^{-r}(\mathcal{A})$ for which $[\mathcal{A}, n, r, \gamma]$ is pure. We are interested in characterizing those elements β for which E has this property. To this end we fix an hereditary order $\mathcal{A}$ in A and an element β in A for which $E = F[\beta]$ is a field such that $E^\times$ lies in $K(\mathcal{A})$; we set $B = \operatorname{End}_E(V)$. It follows from the fact that $E^\times$ lies in $K(\mathcal{A})$ that the lattice chain L which defines $\mathcal{A}$ is in fact an $\mathcal{O}_E$-lattice chain. Hence $\mathcal{B} = \mathcal{A} \cap B$ is an hereditary order in B.

We introduce the following definitions.

3.1. Definition. We define the map $a_\beta \colon A \to A$ by $a_\beta(x) = \beta x - x\beta$. For each integer k we let $\mathcal{N}_k = \mathcal{N}_k(\mathcal{A}, \beta)$ be the set of elements x in $\mathcal{A}$ for which $a_\beta(x)$ lies in $\mathcal{P}^k(\mathcal{A})$.

3.2. *Remarks.*

1. For each integer k, $\mathcal{N}_k$ is an open subring of $\mathcal{A}$; in addition, $\mathcal{N}_k$ is a $(\mathcal{B},\mathcal{B})$-bilattice and $\mathcal{N}_k \cap B = \mathcal{B}$.
2. Clearly $\mathcal{N}_k = \mathcal{A}$ for $k \le \nu_{\mathcal{A}}(\beta)$. On the other hand, an easy compactness argument shows $\mathcal{N}_k \subset \mathcal{B} + \mathcal{P}$ for k sufficiently large.

In light of this last remark we may now make the following definition.

3.3. **Definition.** Suppose $E \neq F$. Then define the integer $k_0 = k_0(\mathcal{A},\beta)$ to be the minimal integer k with the property that $\mathcal{N}_{k+1} \subset \mathcal{B}+\mathcal{P}$. (If $E = F$, we set $k_0 = -\infty$.)

We may now state one of our fundamental results.

3.4. **Definition.** A pure stratum $[\mathcal{A},n,r,\beta]$ is *simple* if $r < -k_0(\mathcal{A},\beta)$.

3.5. **Theorem.**

1. *Let $[\mathcal{A},n,r,\beta]$ be a pure stratum in A. Then there exists a simple stratum $[\mathcal{A},n,r,\gamma]$ in A such that $[\mathcal{A},n,r,\beta] \sim [\mathcal{A},n,r,\gamma]$. For any simple stratum satisfying this condition, $e(F[\gamma] \mid F)$ divides $e(F[\beta] \mid F)$ and $f(F[\gamma] \mid F)$ divides $f(F[\beta] \mid F)$. In particular, among all pure strata $[\mathcal{A},n,r,\beta']$ equivalent to $[\mathcal{A},n,r,\beta]$, the simple ones are precisely those for which the field extension $F[\beta']/F$ has minimal degree.*
2. *Let $[\mathcal{A},n,r,\gamma_i]$, $i=1,2$, be equivalent simple strata in A. Then*
   a. $k_0(\gamma_1) = k_0(\gamma_2)$,
   b. $e(F[\gamma_1]|F) = e(F[\gamma_2]|F)$ *and* $f(F[\gamma_1]|F) = f(F[\gamma_2]|F)$.

Recall that if G is any group and if χ is a representation of a subgroup H of G, then an element x in G is said to *intertwine* χ if there is a non-trivial $H \cap xHx^{-1}$ homomorphism from χ to χ^x. In the case that $\chi = \psi_b$ and $H = U^{r+1}(\mathcal{A})$ for some stratum $[\mathcal{A},n,r,b]$, $r \ge [n/2]$, the condition that x intertwine χ translates itself, via Corollary 2.3, into the following property. (This property also has meaning without restriction on r.)

3.6. **Definition.** Let $\Omega = [\mathcal{A},n,r,b]$ be a stratum in A. Then we say that an element x of G *intertwines* Ω if $x^{-1}(b+\mathcal{P}^{-r}(\mathcal{A}))x \cap (b+\mathcal{P}^{-r}(\mathcal{A})) \neq \emptyset$. We denote the set of elements which intertwine Ω by $\mathcal{T}_G(\Omega)$.

Clearly $\mathcal{T}_G(\Omega)$ depends only on the equivalence class of Ω. Thus Theorem 3.5 will enable us to compute $\mathcal{T}_G(\Omega)$ for pure strata Ω once we can compute it for simple strata. Our next major result is then:

3.7. **Theorem.** *Let $\Omega = [\mathcal{A},n,r,\beta]$ be a simple stratum in A, $E = F[\beta]$, $B = \mathrm{End}_E(V)$, $\mathcal{B} = B \cap A$, and $\mathcal{Q} = \mathrm{rad}(\mathcal{B})$. Write $k = k_0(\mathcal{A},\beta)$ and $\mathcal{N} = \mathcal{N}_k(\mathcal{A},\beta)$. Then*

$$\mathcal{T}_G(\Omega) = \left(1 + \mathcal{Q}^{-(r+k)}\mathcal{N}\right) \cdot B^\times \cdot \left(1 + \mathcal{Q}^{-(r+k)}\mathcal{N}\right).$$

3.8. *Remark.* In case $r = n - 1$ above, the condition that Ω be simple is seen to be equivalent to Ω being pure and β being *E/F-minimal* in the sense of [Ku1]. Theorem 3.7 is then a generalization of Theorem 2.4 of [KuMa1].

4. Simple Characters

We fix a simple stratum $\Omega = [\mathcal{A}, n, 0, \beta]$ and, as above, set $E = F[\beta]$, $B = \mathrm{End}_E(V)$, $\mathcal{B} = B \cap \mathcal{A}$, $\mathcal{P} = \mathcal{P}(\mathcal{A})$, and $\mathcal{Q} = \mathrm{rad}(\mathcal{B}) = \mathcal{P} \cap B$. In addition, we set $r = -k_0(\mathcal{A}, \beta)$. If $r < n$, then we pick a simple stratum $\Omega' = [\mathcal{A}, n, r, \gamma]$ with $\Omega' \sim \Omega$. We will now define two subrings of $\mathcal{A}$ which in fact depend only on Ω (although this will not be clear from their definition).

4.1. **Definition.**

1. If β is E/F minimal (so that $r = n$ or ∞) we set

$$\mathfrak{h} = \mathfrak{h}(\beta) = \mathcal{B} + \mathcal{P}^{[n/2]+1} \qquad \text{and}$$
$$\mathfrak{j} = \mathfrak{j}(\beta) = \mathcal{B} + \mathcal{P}^{[(n+1)/2]}.$$

If $r < n$, we set

$$\mathfrak{h} = \mathcal{B} + \mathfrak{h}(\gamma) \cap \mathcal{P}^{[r/2]+1} \qquad \text{and}$$
$$\mathfrak{j} = \mathcal{B} + \mathfrak{j}(\gamma) \cap \mathcal{P}^{[(r+1)/2]}.$$

2. For $m \geq 0$, we set $\mathfrak{h}^m = \mathfrak{h} \cap \mathcal{P}^m$ and $\mathfrak{j}^m = \mathfrak{j} \cap \mathcal{P}^m$.

4.2. *Remark.* $\mathfrak{h}$ and $\mathfrak{j}$ are subrings of $\mathcal{A}$, $\mathfrak{h}^m$ and $\mathfrak{j}^m$ are two-sided ideals in $\mathfrak{h}$ and $\mathfrak{j}$, respectively, and $[\mathfrak{j}^m, \mathfrak{j}^{m'}] \subset \mathfrak{h}^{m+m'}$ when m and m' are at least 1.

4.3. **Definition.** Let $H = H(\beta) = \mathfrak{h}^\times$, $J = J(\beta) = \mathfrak{j}^\times$ and, for $m > 0$, $H^m = 1 + \mathfrak{h}^m$ and $J^m = 1 + \mathfrak{j}^m$.

We are now going to define a set, $\mathcal{C}(\Omega, m)$, of one-dimensional representations (characters) of the groups H^{m+1}, $0 \leq m \leq n-1$, which will extend the character ψ_B on $H^{m+1} \cap U^{[n/2]+1}(\mathcal{A})$. As before, this set will depend only on Ω and m although this will not be clear from the definition; what will also be unclear is whether these sets are non-empty. The fact that they are non-empty will be one of our major results. We begin with the case that β is E/F-minimal.

4.4. **Definition.** Suppose β is E/F-minimal and let $0 \leq m \leq n-1$. Then denote by $\mathcal{C}(\Omega, m)$ the set of characters θ on H^{m+1} satisfying

a. $\theta \mid H^{m+1} \cap U^{[n/2]+1}(\mathcal{A}) = \psi_\beta$, and
b. $\theta \mid H^{m+1} \cap B^\times$ factors through $\det_{B^\times/E}$.

4.5. *Remark.* The fact that $\psi_\beta|_{U^{[n/2]+1}(\mathcal{B})}$ factors through $\det_{B^\times/E}$ ensures that $\mathcal{C}(\Omega, \beta)$ is non-empty.

We now give the definition of in $\mathcal{C}(\Omega, \beta)$ in case β is not minimal.

4.6. **Definition.** Suppose that $r < n$. Then, for $0 \leq m \leq r-1$, let $\mathcal{C}(\Omega, m)$ be the set of characters θ of H^{m+1} such that

a. $\theta \mid H^{m+1} \cap B^\times$ factors through $\det_{B^\times/E}$,
b. θ is normalized by $K(\mathcal{B})$, and
c. for $m' = \max\{m, [r/2]\}$, $\theta \mid H^{m'+1}(\beta)$ is of the form $\theta_0 \cdot \psi_{\beta-\gamma}$ for θ_0 in $\mathcal{C}(\Omega', m')$.

For $m \geq r$, we set $\mathcal{C}(\Omega, m) = \mathcal{C}(\Omega', m)$.

The conditions defining $\mathcal{C}(\Omega, m)$ enable us to compute, for θ in $\mathcal{C}(\Omega, m)$, the set $I_G(\theta)$ of elements in G which intertwine θ. The fact that $B^\times$ lies in $I_G(\theta)$ will in turn enable us to determine $\mathcal{C}(\Omega, m)$ explicitly. This is the content of our next two results.

4.7. **Theorem.** *Let θ lie in $\mathcal{C}(\Omega, m)$ as above. Then, for $0 \leq m \leq r-1$,*

$$I_G(\theta) = \left(1 + \mathcal{Q}^{r-m}\mathcal{N} + \mathfrak{j}^{[(r+1)/2]}\right) \cdot B^\times \cdot \left(1 + \mathcal{Q}^{r-m}\mathcal{N} + \mathfrak{j}^{[(r+1)/2]}\right).$$

In particular,

$$I_G(\theta) = J^{m+1} \cdot B^\times \cdot J^{m+1}$$

whenever $m \leq [r/2]$.

4.8. **Corollary.**

1. *Suppose that $r = n$. Then for $[n/2] \leq m \leq n-1$, the set $\mathcal{C}(\Omega, m)$ consists of the single element ψ_β.*
2. *Let $[r/2] \leq m \leq r-1$ and assume that $r < n$. Then, with Ω' as above and $c = \beta - \gamma$, the map $\theta \mapsto \theta \cdot \psi_c$ is a bijection of $\mathcal{C}(\Omega', m)$ with $\mathcal{C}(\Omega, m)$.*
3. *Let $0 \leq m \leq [r/2]$. Then the restriction map from $\mathcal{C}(\Omega, m)$ to $\mathcal{C}(\Omega, [r/2])$ is surjective. The fibers of this map are of the form $\theta \cdot X$ where θ lies in $\mathcal{C}(\Omega, m)$ and X is the group of characters of $U^{m+1}(\mathcal{B})/U^{[r/2]}(\mathcal{B})$ which factor through $\det_{B^\times/E}$.*

We close this section with some results concerning the dependence of the set $\mathcal{C}(\Omega, m)$ on the order $\mathcal{A}$. To this end, we fix an element β in A which generates a subfield E of A and, as above, set $B = \mathrm{End}_E(V)$. We say that an hereditary order $\mathcal{A}$ in A is an *E-order* if $E^\times$ lies in $K(\mathcal{A})$. Our first result is

4.9. **Lemma.** *Let $\mathcal{A}$ be an E-order. Then $k_0(\mathcal{A}, \beta)/e(\mathcal{A} \cap B \mid \mathcal{O}_E)$ is independent of $\mathcal{A}$. In particular, the stratum $[\mathcal{A}, n, 0, \beta]$ is simple for one E-order $\mathcal{A}$ if and only if it is simple for all E-orders $\mathcal{A}$.*

4.10. **Proposition.** *Suppose that $\mathcal{A}_i$, $i = 1,2,3$, are E-orders and that $\Omega_i = [\mathcal{A}_i, n, 0, \beta]$ are simple strata. Then there is a unique bijection, $\iota(\Omega_1, \Omega_2)$ from $\mathcal{C}(\Omega_1, 0)$ to $\mathcal{C}(\Omega_2, 0)$, having the property that, if θ_2 in $\mathcal{C}(\Omega_2, 0)$ corresponds to θ_1 in $\mathcal{C}(\Omega_1, 0)$, then θ_1 and θ_2 have the same restriction to $H^1(\Omega_1) \cap H^1(\Omega_2)$. Furthermore, we have that $\iota(\Omega_2, \Omega_3) \circ \iota(\Omega_1, \Omega_2) = \iota(\Omega_1, \Omega_3)$.*

4.11. **Definition.** Let $\mathcal{S}$ be a set of E-orders and, for $\mathcal{A}$ in $\mathcal{S}$, define $\Omega_{\mathcal{A}}=[\mathcal{A},n,0,\beta]$. Suppose further that the strata $\Omega_{\mathcal{A}}$ are simple. For each $\mathcal{A}$ in $\mathcal{S}$ pick a character $\theta_{\mathcal{A}}$ in $\mathcal{C}(\Omega_{\mathcal{A}},0)$. Then the set of characters $\theta_{\mathcal{A}}$ is *coherent* if $\iota(\mathcal{A},\mathcal{A}')(\theta_{\mathcal{A}})=\theta_{\mathcal{A}'}$ for all orders $\mathcal{A}$ and $\mathcal{A}'$ in $\mathcal{S}$.

5. Simple Types

We fix once again a simple stratum $\Omega=[\mathcal{A},n,0,\beta]$ and, as before, set $E=F[\beta]$; all unexplained notation will be as in §4. Our first result is:

5.1. **Proposition.** *Let θ be a character in $\mathcal{C}(\Omega,0)$. Then there is a unique irreducible representation $\eta=\eta(\theta)$ of the group $J^1(\Omega)$ such that $\eta|_{H^1(\theta)}$ contains θ. Moreover, $\eta|_{H^1(\theta)}$ is a multiple of θ and $\dim(\eta)=(J^1(\Omega):H^1(\Omega))^{1/2}$. Finally, we have that $I_G(\eta)=J^1(\Omega)\cdot B^{\times}\cdot J^1(\Omega)$.*

5.2. *Remark.* Proposition 5.1 follows easily from the fact that the pairing $(1+x,1+y)\mapsto\theta([1+x,1+y])$ induces a nondegenerate alternating bilinear form on $J^1(\Omega)/H^1(\Omega)$. What is much less clear is whether the representation η has an extension to $J(\Omega)$ and, if so, whether this extension can be picked in such a way that its set of intertwining elements can be computed. (For an elaboration on this point see §2.1 of [Wa].)

Our approach to the problem of producing an appropriate extension of η to $J(\Omega)$ is to introduce two new E-orders, $\mathcal{A}_m$ and $\mathcal{A}_M$. We take $\mathcal{A}_m$ minimal over $\mathcal{O}_E$ and $\mathcal{A}_M$ maximal over $\mathcal{O}_E$ (that is, we pick $\mathcal{A}_m$ so that $e(\mathcal{A}_m\cap B\mid\mathcal{O}_E)=\dim_E(V)$ and $\mathcal{A}_M$ so that $e(\mathcal{A}_M\cap B\mid\mathcal{O}_E)=1$) such that $\mathcal{A}_m\subset\mathcal{A}\subset\mathcal{A}_M$. In addition, we pick characters θ_m in $\mathcal{C}(\Omega_m,0)$ (in the obvious notation) and θ_M in $\mathcal{C}(\Omega_M,0)$ so that the characters θ_m, θ, θ_M form a coherent set. Our result (again with the obvious notation) is then

5.3. **Theorem.**

1. *There exists an extension κ_M of η_M to J_M with the following property: the representation of $U^1(\mathcal{A}_m)$ induced by $\kappa_M|_{U^1(\mathcal{B}_m)\cdot J^1_M}$ is irreducible and equivalent to the representation of $U^1(\mathcal{A}_m)$ induced by η_m. This property determines κ_M uniquely up to tensoring with a character of the form $\chi\circ\det_{B^\times/E}$ where χ is a character of $\mathcal{O}_E^\times/1+P_E$.*
2. *Fix a representation κ_M as above. Then there exists a unique extension κ of η to J such that the representation induced by κ on $U(\mathcal{A})$ is irreducible and equivalent to the representation of $U(\mathcal{A})$ induced by $\kappa_M|_{U(\mathcal{B})J^1_M}$.*

5.4. **Definition.** We call a representation κ of J constructed as above a *β-extension* of η.

We are now able to say what we mean by a simple type. To begin with, we will now assume that our order $\mathcal{A}$ is *principal.* In that case, we have that $J/J^1\simeq U(\mathcal{B})/U^1(\mathcal{B})\simeq GL_f(k_E)\times GL_f(k_E)\times\cdots\times GL_f(k_E)(e(\mathcal{B}/\mathcal{O}_E)$

copies) where $f \cdot e(\mathcal{B}/\mathcal{O}_E) = \dim_E(V)$. Thus, given an irreducible cuspidal representation σ_0 of $GL_f(k_E)$, we may view $\sigma = \sigma_0 \otimes \sigma_0 \otimes \cdots \otimes \sigma_0$ as a representation of J which is trivial on J^1. Our definition is then

5.5. **Definition.** By a *simple type* we mean either

1. a representation $\lambda = \kappa \otimes \sigma$ of the group $J(\Omega)$ where $\Omega = [\mathcal{A}, n, 0, \beta]$ with $\mathcal{A}$ principal, κ is a β-representation of J and σ is a representation of J/J^1 constructed as above, or
2. a representation λ of $U(\mathcal{A})$ of the form $\sigma_0 \otimes \sigma_0 \otimes \cdots \otimes \sigma_0 (e(\mathcal{A}/\mathcal{O}_F)$ factors) where σ_0 is a cuspidal representation of $GL_f(k_F)$, and $f \cdot e(\mathcal{A}/\mathcal{O}_F) = N$ as above.

5.6. *Remarks.*

1. In what follows, we will only explicitly treat case 5.5.1. One obtains the results for the case 5.5.2 by substituting $U(\mathcal{A})$ for $J(\Omega)$, $K(\mathcal{A})$ for $K(\Omega)$, and putting $E = F$.
2. As a special case of 5.5.2 we have the *trivial type* consisting of the trivial character of an Iwahori subgroup of G (see below).

6. An Isomorphism of Hecke Algebras

We fix a simple stratum $\Omega = [\mathcal{A}, n, r, \beta]$ and a simple type $\lambda = \kappa \otimes \sigma$ of $J(\Omega)$ as above; all other notation will be as in §5. In addition, we pick a subfield L of A which contains E, is unramified over E of degree $f = \dim_E V/e(\mathcal{B}/\mathcal{O}_E)$, and which has the additional property that $\mathcal{A}$ is an L-order. (There are, in fact, many such fields.) We set $C = \mathrm{End}_L(V)$ and $\mathcal{C} = C \cap \mathcal{A}$. Then $\mathcal{C}$ is, in fact, a *minimal* order in C, whence $U(\mathcal{C})$ is what is commonly called an *Iwahori subgroup* of $C^\times$. We may then pick an *affine Weyl group,* $W(\mathcal{C})$; that is, $W(\mathcal{C})$ is the subgroup of $C^\times$ consisting of matrices which are monomial with prime power entries with respect to an appropriate L-basis for V and a prime element Π_E of E. $W(\mathcal{C})$ is a complete set of representatives for the set of $(U(\mathcal{C}), U(\mathcal{C}))$ double cosets in $C^\times$. Our first result is then as follows:

6.1. **Proposition.** *$I_G(\lambda) \subset J(\Omega) \cdot W(\mathcal{C}) \cdot J(\Omega)$. Furthermore, for x in G,*

$$\dim(Hom_{J \cap x J x^{-1}}(\lambda, \lambda^x)) \leq 1.$$

6.2. *Remark.* We say that the simple type λ is *maximal* if $[L : F] = N$. In this case, $W(\mathcal{C})$ is just the cyclic group generated by the prime element, Π_E, of L. Further, one shows that λ may be extended to a representation of $K(\Omega) = N_G(J) = \langle \Pi_E \rangle \cdot J$. Since any such extension is intertwined only by elements of $K(\Omega)$, we obtain the following corollary.

6.3. **Corollary.** *Let π be an irreducible (smooth) representation of G containing a maximal simple type λ on J. Then π is induced from an extension of λ to $K(\Omega)$ and hence is supercuspidal [Car].*

We now recall the definition of the Hecke algebra $\mathcal{H}_G(J,\lambda)$. If we denote by Y the complex vector space on which the representation λ acts, then $\mathcal{H}_G(J,\lambda)$ is the set of $\mathrm{End}_{\mathbb{C}}(Y)$ valued functions f on G which satisfy $f(hxk) = \lambda(h)f(x)\lambda(k)$ for x in G, h, k in J. Since $J(\Omega)$ is a subgroup of $U(\mathcal{A})$, it is easy to see that distinct elements of $W(\mathcal{C})$ lie in distinct (J,J) double coset in G. It follows that we may view $\mathcal{H}_G(J,\lambda)$ as a space of functions on $W(\mathcal{C})$. On the other hand, the Hecke algebra $\mathcal{H}_{\mathcal{C}^\times}(U(\mathcal{C}), 1_{U(\mathcal{C})})$ may also be so viewed. We may now state the main result of this paper.

6.4. **Theorem.** *There exists a canonical family of support-preserving (in the sense described above) isomorphisms $\mathcal{H}_{\mathcal{C}^\times}(U(\mathcal{C}), 1_{U(\mathcal{C})}) \simeq \mathcal{H}_G(J,\lambda)$ of $\mathbb{C}$-algebras with 1.*

Using an argument similar to that found in [Ku 3], one may now deduce the following:

6.5. **Corollary.** *Suppose π is an irreducible representation of G which contains a non-maximal simple type λ. Then π is not supercuspidal.*

7. Connections with Parabolic Induction

Let $(N_1, N_2, \ldots, N_s)$ be partition of N and let each π_i, $i = 1, \ldots, s$, be a representation of $GL_{N_i}(F)$. Then we may view $\otimes\pi_i$ as a representation of the standard parabolic in G having Levi factor $\Pi GL_{N_i}(F)$; following [BZ], we denote by $\pi_1 \times \pi_2 \times \cdots \pi_s$ the representation of G induced by $\otimes\pi_i$. Then it is well-known [Cas] that any irreducible representation π of G is a subrepresentation of some $\times\pi_i$ where the representations π_i are supercuspidal. Further, the multiset $\{\pi_i\}$ is determined by π. Following [BZ], we call this multiset the *support* of π.

7.1. **Definition.** We say that a multiset $\{\pi_i\}$ of supercuspidal representations is *simple* if each π_i has the form $\pi_0 \otimes \chi_i \circ \det$ where the characters χ_i are unramified (i.e., $1 + P$ lies in $\ker \chi_i$).

Our major result in this section is then as follows:

7.2. **Theorem.** *An irreducible representation of G contains a simple type if and only if its support is simple.*

7.3. **Corollary.** *All irreducible supercuspidal representations of G are induced.*

7.4. *Remark.* Indeed, one may be more precise with the help of the following result.

7.5. **Proposition.** *Let λ_i, $i = 1, 2$, be maximal simple types on subgroups J_i, and suppose that, for some element x in G, $\mathrm{Hom}_{J_1 \cap J_{2^{x^{-1}}}}(\lambda_1, \lambda_2) \neq (0)$. Then there is an element y in G such that $J_2 = yJ_1y^{-1}$ and $\lambda_2 = \lambda_1^y$.*

7.6. **Corollary.** *Let π be an irreducible supercuspidal representation of G. Then π contains a simple type λ on a subgroup $J(\Omega)$, the pair $(\lambda, J(\Omega))$ being unique up to conjugacy. Further, π is induced from a unique extension λ' to λ to $K(\Omega)$.*

We are now in a position to describe precisely the set of irreducible representations of G which contain a simple type. To this end, we fix a simple stratum $\Omega = [\mathcal{A}, n, 0, \beta]$ as above and continue with the corresponding notation. We will also need to consider the algebra $A(L) = \mathrm{End}_F(L)$. This algebra is endowed with a unique L-order $\mathcal{A}(L)$ ($\mathcal{A}(L)$ is just the order associated to the lattice chain $\{P^m(L)\}$) and the stratum $\Omega_0 = [\mathcal{A}(L), n, 0, \beta]$ is seen to be simple. Now let λ_0 be a simple type for $J_0 = J(\Omega_0)$. Then λ_0 is maximal by construction. Let λ_0' be some extension of λ_0 to $K(\Omega_0)$ and set $\pi_0 = \mathrm{Ind}\, \lambda_0'$. Then

7.7. **Theorem.** *Let λ_0 be a maximal simple type in $A^\times(L)$ as above. Then there is a unique simple type λ on $J(\Omega)$ with the property that an irreducible representation π of G contains λ if and only if π is supported on a multiset of the form $\{\pi_0 \otimes \chi_i \circ \det\}$ where the characters χ_i are unramified.*

7.8. **Corollary.** *Every discrete series representation of G contains a simple type.*

7.9. *Remarks.*

1. One should note that Theorem 7.7 may be viewed as a generalization of [Bo]. Indeed, this result corresponds to the case that λ is the trivial type.
2. Corollary 7.8 follows immediately from the Zelevinski classification and Theorem 7.7. However, one may avoid using the Zelevinski classification with a little more work. Indeed, one may deduce most of the results of [Z] from results contained here.

References

[BZ] I. N. Bernstein and A. V. Zelevinsky, *Induced representations of reductive p-adic groups*, Ann. Scient. Ec. Norm. Sup. (4) **10** (1977), 441–472.

[Bo] A. Borel, *Admissible representations of a semisimple group over a local field with vectors fixed under an Iwahori subgroup*, Invent. Math. **35** (1976), 233–259.

[Bu] C. Bushnell, *Hereditary orders, Gauss sums, and supercuspidal representations of GL_N*, J. Reine Angew. Math. **375/376** (1987), 184–210.

[BF] C. Bushnell and A. Fröhlich, *Non-abelian congruence Gauss sums and p-adic simple algebras*, Proc. London Math. Soc. (3), **50** (1985), 207–264.

[Car] H. Carayol, *Représentations cuspidales du group linéaire*, Ann. Sci. Ecole Norm. Sup. (4) **17** (1984), 191–225.

[Cas] W. Casselman, *Introduction to the theory of admissible representations of p-adic reductive groups*, preprint.

[H] R. E. Howe, *Some qualitative results on the representation theory of GL_n over a p-adic field*, Pac. J. Math. **73** (1977), 497–538.

[Ku1] P. Kutzko, *Towards a classification of the supercuspidal representations of GL_N*, J. London Math. Soc. (2) **37** (1988), 265–274.

[Ku2] P. Kutzko, *On the supercuspidal representations of GL_N and other reductive groups*, Proc. Int. Cong. Math, Berkeley 1986 (AMS 1987), 853–861.

[Ku3] P. Kutzko, *On the restriction of supercuspidal representations to compact, open subgroups*, Duke Math. J. **52** (1985), 753–764.

[KuMa1] P. Kutzko and D. Manderscheid, *On intertwining operators for $GL_N(F)$, F a non-archimedean local field*, Duke Math J. **57** (1988), 275–293.

[KuMa2] P. Kutzko and D. Manderscheid, *On the supercuspidal representations of GL_N, N the product of two primes*, Ann. Sci. Ec. Norm. Sup. (4) **23** (1990), 39–88.

[Wa] J.-L. Waldspurger, *Algebres de Hecke et induites de representations cuspidales, pour $GL(N)$*, J. Reine Angew. Math. **370** (1986), 127–191.

[Z] A. V. Zelevinsky, *Induced representations of reductive p-adic groups II; On irreducible representations of $GL(n)$*, Ann. Scient. Norm. Sup. (4) **13** (1980), 165–210.

Department of Mathematics, King's College, Strand, London WC2R 2LS, Great Britain

Department of Mathematics, University of Iowa, Iowa City, IA 52242
E-mail: pkutzko @ umaxc.weeg.uiowa.edu

INVARIANT HARMONIC ANALYSIS ON THE SCHWARTZ SPACE OF A REDUCTIVE p-ADIC GROUP

LAURENT CLOZEL

Université Paris-Sud

1. INTRODUCTION

At the Williamstown conference on harmonic analysis, R. Howe publicized two conjectures concerning the orbital integrals of functions on a reductive p-adic group and on its Lie algebra. He proved the Lie algebra case of the conjecture himself; recently we have proved the conjecture on the group, at least in zero characteristic [4c,4d].

At about that same time, Harish-Chandra had developed a rather complete theory of harmonic analysis on the Schwartz space of a p-adic reductive group; he announced his results in [7a], and a fairly complete exposition was later given by Silberger [15]. (A notable omission is the Plancherel formula: a sketch of Harish-Chandra's proof can be found in [7d], but the detailed proof has yet to appear). However, when he wanted to imitate in the p-adic case the *invariant* harmonic analysis on the Schwartz space that is the cornerstone of this work in the discrete series and the Plancherel formula in the *real* case, Harish-Chandra found that the Howe conjecture for the group was an indispensible ingredient. Consequently, this theory has remained incomplete. Two notable statements that depended (in Harish-Chandra's approach) on the Howe conjecture were the convergence of *arbitrary* orbital integrals on the Schwartz space, and the proof of the orthogonality relations between characters of discrete series representations. In a 1982 letter to Vignéras, Harish-Chandra sketched the proof of the orthogonality relations.

Since then, Kazhdan has found another proof of the orthogonality relations (indeed, more general than for the discrete series) that does not rely on the Schwartz space. In fact, the analysis in $C_c^\infty(G)$—rather than the Schwartz space $\mathcal{C}(G)$—developed by Bernstein and Kazhdan can be used to prove many results that would have been attacked earlier using $\mathcal{C}(G)$. However, the last word, in our opinion, has not been said. For example, in our proof [4d] of the fundamental lemma for stable base change, the fact that regular orbital integrals are tempered plays a fundamental role, and spaces of functions other than $C_c^\infty(G)$ will often be used in the comparison of trace formulae (see e.g. [1, Ch. II]). The reader should also consult

Arthur's paper in this volume, and his forthcoming paper on the (invariant) local trace formula, for further problems on the Schwartz space.

In this paper, we have given proofs of the following results :

(1) Temperedness of orbital integrals.
(2) Orthogonality relations for the discrete series.
(3) Existence of pseudo-coefficients for discrete series and limits of discrete series. In the case of limits of discrete series, the best possible result is proved only (in § 4) for SL_p. A weaker result (Theorem 5) is proved in § 5 for all groups.

In § 5, we also state some open problems.

Of these results, (2) and (3) (the latter for discrete series) were already known, thanks to the work of Bernstein, Deligne and Kazhdan [3,9a]. The proofs we give rely on Harish-Chandra's methods; in the case of (2) we have simply reproduced his proof [7e]. The proof of (3) should illustrate the fundamental flexibility of $\mathcal{C}(G)$ as opposed to $C_c^\infty(G)$: technically, this is due to the fact that $\mathcal{C}(G)$ is sent to a *fine* sheaf of functions by the matrix-valued Fourier transform. Here Harish-Chandra's Plancherel theorem, and its refinement by Mischenko, play a fundamental role.

Granted the Howe conjecture, most of the results in this paper (Theorem 2, and its Corollaries 2, 3; Proposition 2, Proposition 3 and its Corollary, Proposition 4 and its Corollary, Theorem 3) were known to Harish-Chandra.

The account we give is expository: we often eschewed slicker proofs (e.g. for Theorem 2) in order to recall known facts and to point out the important phenomena.

I would like to thank D. Keys and M.-F. Vignéras for discussions concerning the material in § 5.

2. The Howe conjecture and orbital integrals

For the rest of the paper, $\underline{G}$ will denote a reductive, connected group over a p-adic field F of characteristic 0, and $G = \underline{G}(F)$ its set of points over F. We define $C_c^\infty(G)$ to be the space of smooth compactly supported functions on G, and $\mathcal{C}(G)$ the Schwartz space.

Recall that $\mathcal{C}(G) = \varinjlim_K \mathcal{C}_K(G)$, where K runs over compact-open subgroups,[1] and

$$\mathcal{C}_K(G) = \{f \in C(K \setminus G/K) : \sup_{x\in G} |f(x)|\Xi(x)^{-1}(1+\|x\|)^N < \infty \ \forall N > 0\}.$$

Here Ξ is the usual spherical function [15] and $\|x\|$ the distance on G, denoted by $\sigma(x)$ in [15]. The topology on $\mathcal{C}_K$ is defined by the semi-norms that appear in its definition, and the topology on $\mathcal{C}(G)$ is the inductive limit topology.

[1] When "K" is used without comment, it will always denote an arbitrary compact open subgroup.

Other usual notations will be used without comment, cf. [7a], [15].

We recall the statement of the Howe conjecture. Let $\Omega \subset G$ be closed, invariant by conjugation, and compact modulo conjugation [4c]. Let $\mathcal{I}(\Omega)$ be the space of invariant distributions on G supported on Ω, and define $\mathcal{H}_K = C_c^\infty(G//K)$ to be the Hecke algebra of K.

Theorem 1 ([4c], see also [4d]). *The restriction to $\mathcal{H}_K$ of the distributions in $\mathcal{I}(\Omega)$ is finite-dimensional.*

Recall that a distribution on G is *tempered* if it extends continuously to the Schwartz space. The most important invariant distributions on G are the irreducible characters (which are tempered if the representation is tempered [15]) and the *orbital integrals*: for $\gamma \in G$, $f \in C_c^\infty(G)$, we set

$$O_\gamma(f) = \int_{G_\gamma(F)\backslash G(F)} f(g^{-1}\gamma g)\frac{dg}{dg_\gamma}\,, \tag{2.1}$$

where dg and dg_γ are Haar measures on $G(F)$ and the centralizer $G_\gamma(F)$ of γ, respectively. (Thus O_γ depends on the choice of measures). The integral is convergent by a result of Rao.

The following theorem was announced in [4a, 4c].

Theorem 2. *The linear form O_γ extends continuously to $\mathcal{C}(G)$. In fact, for $f \in \mathcal{C}(G)$, the integral defining $O_\gamma(f)$ is absolutely convergent, and defines a continuous linear form on $\mathcal{C}(G)$.*

We first note that if O_γ extends continuously to $\mathcal{C}(G)$, the integral will be absolutely convergent on $\mathcal{C}(G)$. This follows from the next lemma :

Lemma 1. *O_γ extends continuously to $\mathcal{C}(G)$ if and only if*

$$\int_{O_\gamma} \Xi(1+\|\ \|)^{-N} < \infty$$

for $N \gg 0$, O_γ being the orbit of γ furnished with the invariant measure.

Proof. By definition of the topology of $\mathcal{C}(G)$, the integral will extend continuously if it converges for large N. Conversely, assume O_γ continuous. Fix K. Then there exists $N > 0$ such that, for $f \in C_c^\infty(G//K)$,

$$\left|\int_{O_\gamma} f\right| \leq \sup_{x\in G} |f(x)|\Xi(x)^{-1}(1+\|x\|)^N. \tag{2.2}$$

Consider the function $\varphi = \Xi(1+\|\ \|)^{-N-1}$. It is bi-invariant by a Bruhat-Tits subgroup K_0, and we may assume $K \subset K_0$. For $n \geq 0$, let $\varphi_n = \xi_n\varphi$, where ξ_n is the characteristic function of $\{x \in G : \|x\| \leq n\}$. Then φ_n belongs to $C_c^\infty(G//K)$, φ belongs to the completion of $C_c^\infty(G//K)$ for the semi-norm on the right-hand side of (2.2), and φ_n converges to φ in this completion. Therefore, $\int_{O_\gamma} \varphi_n$ converges to $\int_{O_\gamma} \varphi$, where the right-hand side denotes the continuous extension of $\int_{O_\gamma}$ to the completion. In particular, $\int_{O_\gamma} \varphi_n$ remains bounded as n increases without bound, and this clearly implies that $\int_{O_\gamma} \varphi$ is (absolutely) convergent. This establishes the Lemma (with N replaced by $N+1$). □

Proof of Theorem 2. We first recall that Theorem 2 is known (and due to Howe and Harish-Chandra) for γ *semi-simple* and *regular*. Let us first consider this case. An easy argument, based on the descent properties of orbital integrals, reduces to the case of elliptic elements. (Use the fact that $f \longmapsto \bar{f}^{(P)}$ is, for P a parabolic subgroup, a continuous map on the Schwartz space [15, Thm. 4.4.3], and that for $\gamma \in M(F) \subset G(F)$, M a Levi subgroup, $O_\gamma(f)$ is equal, up to a Jacobian factor, to $O_\gamma^M(\bar{f}^{(P)})$, where O_γ^M is taken in M [15, § 4.7].)

For γ *elliptic regular*, the argument of Harish-Chandra [15, § 4.7] first shows that, for N large, we have $\int_{O_\gamma} \Xi(1+\|\ \|)^{-N} < \infty$ for *almost all* γ. At this point the Howe conjecture is not needed to extend this to all regular elliptic γ: there are two arguments, one due to Howe (cf. [15]) and the other to Harish-Chandra [7c]. We will give a third, based on the Howe conjecture, which treats in the same fashion all remaining elements, semi-simple (regular) or not. At this point we may assume that Theorem 2 has been proved for all regular semi-simple γ except those in a set of measure zero.

Now take $\gamma \in G$, and let $\gamma = \sigma\nu$ be its Jordan decomposition, with σ semisimple and ν unipotent. Let $L = G_\sigma$ be the centralizer of σ, and U the (finite) set of unipotent orbits in L. We will use the theory of Shalika germs about σ, summarized in two assertions :

(i) If T is a Cartan subgroup of G containing σ (whence $T \subset L$), and $f \in C_c^\infty(G)$, then there exist functions $\Gamma_u^T(t)$ defined near σ on T_{reg}, the set of regular elements of T, such that

$$O_t(f) = \sum_{u \in U} \Gamma_u^T(t)\ O_{\sigma u}(f) \tag{2.3}$$

for all $t \in T_{reg}$ close enough to σ. Of course some (fixed) normalizations of measures are assumed. In order for O_t to be smooth in t, we define it by the following formula (dg, dt being fixed measures):

$$O_t(f) = \int_{T\backslash G} f(g^{-1}tg)\frac{dg}{dt}\ ,$$

which may differ from the definition (2.1) if $G_t \neq T$, i.e., if t is not strongly regular. The two integrals are, of course, proportional.

(ii) Linear independence of germs. The germs $\Gamma_u = (\Gamma_u^T)$, seen as functions on the collection of Cartan subgroups T, are linearly independent, i.e., if a_u are complex constants such that $\sum_u a_u\Gamma_u^T(t) = 0$ for all t in $V_T \cap T_{reg}$, where V_T is a fixed neighborhood of σ in T, and T runs over representatives of the Cartan subgroups through σ modulo conjugation, then $a_u = 0$ for all u.

These results are due to Shalika, Harish-Chandra and Rogawski [14 a]; see also Vignéras [16 a].

Now fix a compact-open subgroup K. A basic consequence of the Howe conjecture is the following:

Proposition 1. *There exists a neighborhood V_T of σ in T such that, for $f \in \mathcal{H}_K$, the expansion (2.3) is valid in $V_T \cap T_{reg}$.*

Proof. The Howe conjecture implies that, if a neighborhood V'_T of σ in T is compact, then the linear forms $t \longmapsto O_t(f)$ $(t \in T_{reg} \cap V'_T)$ span a finite-dimensional space for $f \in \mathcal{H}_K$. Thus we can find $f_1, \cdots, f_R \in \mathcal{H}_K$ such that, for all $f \in \mathcal{H}_K$, $O_t(f) = \sum \lambda_i \; O_t(f_i)$, with the λ_i uniquely determined by f. We may then take V_T equal to the intersection of the V'_T associated to the f_i (... and the domain of definition of the Γ_u^T). □

Fix a family $\mathcal{T}$ (finite) of representatives for the Cartan subgroups through σ, let V_T $(T \in \mathcal{T})$ be fixed as in Proposition 1, and define the set $T_{conv} \subset T_{reg}$ to be the subset (of comeasure zero) for which the orbital integrals are known to converge on $\mathcal{C}(G)$. Since the germs Γ_u^T are locally constant and linearly independent, then the relation $\sum_u a_u \Gamma_u^T(t) = 0$ $(t \in \mathcal{T},\; t \in V_T \cap T_{conv})$ implies $a_u = 0$. In particular, by (2.3), the relation $O_t(f) = 0$ $(t \in \mathcal{T},\; t \in V_T \cap T_{conv})$ implies $O_{\sigma u}(f) = 0$ for all u.

Since the linear forms $f \longmapsto O_t(f)$ span a finite-dimensional space, we see that the orbital integrals $O_{\sigma u}$ are in the span of a finite number of convergent orbital integrals $O_{t_i}(f)$ on $\mathcal{H}_K$. Therefore they are continuous on $\mathcal{H}_K$ with the Schwartz topology. Taking into account Lemma 1, this proves Theorem 2. □

As a corollary, we obtain :

Proposition 2. *The germ expansion* (2.3) *is true for f in $\mathcal{C}_K(G)$, in a neighborhood determined by K.*

In fact, for fixed t, the two sides are continuous linear forms on $\mathcal{C}_K(G)$, and the identity is true on the dense subspace $\mathcal{H}_K$.

As an example of its usefulness, this germ expansion can be applied to the coefficients of a discrete series representation (cf. Rogawski [14 b, Prop. 5.5]).

We will now prove two other corollaries to Theorem 2. Recall that, for regular γ in a Cartan subgroup $T \subset G$, $D_G(\gamma)$ is the determinant of $Ad(\gamma) - 1$ acting on $\mathfrak{g}/\mathfrak{t}$, where $\mathfrak{g}, \mathfrak{t}$ denote the Lie algebras of G and T respectively. We write D for D_G if G is understood.

Corollary 2. *Assume $T \subset G$ is a Cartan subgroup. Then there exists $N \geq 0$ such that*

$$\sup_{\gamma \in T_{reg}} |D(\gamma)|^{1/2} \int_{T\backslash G} \Xi(g^{-1}\gamma g)(1 + \|g^{-1}\gamma g\|)^{-N} \frac{dg}{dt} < \infty.$$

Corollary 3. *There exists $N \geq 0$ such that*

$$\int_G |D(g)|^{-1/2} \Xi(g)(1 + \|g\|)^{-N} < \infty.$$

Proof. Clearly Corollary 3 follows from Corollary 2 by the Weyl integration formula. We now prove Corollary 2.

First note that, for $N \gg 0$ and $\varphi = \Xi(1 + \|\ \|)^{-N}$, the orbital integral $O_\gamma(\varphi)$ will be absolutely convergent for any $\gamma \in G$. This follows from the arguments used to prove Theorem 2: first, by Harish-Chandra's argument recalled after Lemma 1, this is true for almost all regular γ; then, since φ is fixed by a Bruhat-Tits subgroup K_0, the orbital integral $O_\gamma(\varphi)$ can be obtained as the extension to the completion of $\mathcal{H}_{K_0}$ of a linear combination of convergent regular orbital integrals. This verifies the claimed absolute convergence.

We notice next that we may assume $T \subset G$ elliptic. Indeed, assume T elliptic in a Levi subgroup M. Then, with standard notations [15], we have $|D(\gamma)|^{1/2} O_\gamma^M(\varphi^{(P)})$ where $\gamma \in T_{reg}$ and $\varphi^{(P)}$ is the constant term of φ along a parabolic subgroup $P = MU$. O_γ^M is taken in M. Since the map $\varphi \longmapsto \varphi^{(P)}$ is continuous for the topologies defining the Schwartz spaces [15, Thm. 4.4.3], it is clear that Corollary 2 for $T \subset M$ implies it for $T \subset G$ (with a different N).

We may therefore assume T elliptic. Clearly the center plays no role, so we may assume G has compact center and T is compact. It is then enough to prove that the supremum in Corollary 2 is finite in a neighborhood of a fixed $\gamma_0 \in T$. We then have, for $f \in \mathcal{H}_{K_0}$ in a *fixed* neighborhood of γ_0 (Proposition 1), a germ expansion

$$O_\gamma(f) = \sum_u O_{\gamma_0 u}(f)\Gamma_u(\gamma);$$

it is well-known [7b, 14a, 16a] that $|D(\gamma)|^{1/2}\Gamma_u(\gamma)$ is bounded near γ_0. By the argument given for Proposition 2, this germ expansion is still true for the orbital integrals of φ (... which is not in the Schwartz space, but these arguments necessitate only the semi-norm associated to N). Clearly this implies Corollary 2. $\square$

Finally, we note that the previous arguments easily imply:

Proposition 3. *If $\Omega \subset G$ is closed and compact modulo conjugation, then any invariant distribution Θ with support in Ω is tempered.*

In the real case, this is an old result of Harish-Chandra.

3. Tempered characters

We can now prove some results about the tempered characters of G.

Let π be an irreducible admissible representations of G, and Θ_π its character. In particular, Θ_π is a locally integrable function on G, smooth on the set G_{reg} of regular elements.

Proposition 4. *The representation π is tempered if and only if we can choose, given any Cartan subgroup $T \subset G$, positive constants $c = c(T)$ and $r = r(T)$ such that*

$$|D_G(\gamma)|^{1/2}|\Theta_\pi(\gamma)| \leq c(1 + \|\gamma\|)^r$$

for all $\gamma \in T_{reg} = G_{reg} \cap T$.

Proof. If the growth conditions hold, the distribution Θ_π is tempered by Corollary 3 to Theorem 2.

Conversely, assume Θ_π tempered. We will use the Casselman-Deligne stratification [4a,4c]. Recall that we can write G as a finite union of open and closed subsets,

$$G = \coprod_{P=MN} G(M_c^+),$$

where P ranges over a set $\mathcal{P}$ of representatives for the conjugacy classes of parabolic subgroups (say, standard parabolic subgroups containing a fixed parabolic subgroup P_0). For each $P \in \mathcal{P}$, with $P = MN$, we let M_c denote the compact part of M, composed of elements $m \in M$ such that the eigenvalues of Ad m have absolute value 1, and

$$M_c^+ = \{m \in M_c : \text{Ad } m \text{ strictly contracting on } N\}.$$

Then $G(M_c^+) = \{g \in G : g = xmx^{-1},\ x \in G,\ m \in M_c^+\}$.

In particular, we have $T = \coprod T(M_c^+)$ by intersection—of course $T(M_c^+)$ is nonempty only if $T \subset M$ modulo conjugation. Since it is enough to prove the desired estimate on $T(M_c^+)$, we will assume $T \subset M$.

By Casselman's theorem [19], $\Theta_\pi(\gamma) = \Theta_{\pi_N}(\gamma)$ for regular $\gamma \in T(M_c^+)$, where π_N is the (unormalized) Jacquet module with respect to N. Writing, in the usual fashion, $D_G(\gamma) = D_{G/M}(\gamma)\ D_M(\gamma)$, we have

$$|D_G(\gamma)|^{1/2}\Theta_\pi(\gamma) = |D_{G/M}(\gamma)|^{1/2}\ |D_M(\gamma)|^{1/2}\Theta_{\pi_N}(\gamma).$$

The usual computation (cf. [4c]) shows $|D_{G/M}(\gamma)| = \delta_P^{-1}(\gamma)$ if $\gamma \in M_c^+$, hence

$$|D_G(\gamma)|^{1/2}|\Theta_\pi(\gamma)| = |D_M(\gamma)|^{1/2}|\Theta_{\bar{\pi}_N}(\gamma)|,$$

where $\bar{\pi}_N = \delta_P^{-1/2}\pi_N$ is the normalized Jacquet module, a representation of finite length of M.

The representation $\bar{\pi}_N$ has a Jordan-Hölder sequence composed of irreducible representations τ of M, and the theory of exponents of Harish-Chandra and Casselman implies the following. Let $A_M \subset M$ be the split component of M, ω the restriction to A_M of the central character of τ, and $|\omega|$ its absolute value ($|\omega|$ can be considered an element of $\mathfrak{a}_M^*$, where $\mathfrak{a}_M^* = X^*(A) \underset{\mathbb{Z}}{\otimes} \mathbb{R}$ is the "real Lie algebra" of A_M). Then $|\omega(a)| = \prod\limits_{\alpha} |a^\alpha|^{\lambda_\alpha}$, with $\lambda_\alpha \geq 0$ and α ranging over the simple roots of A in N. Now $T \cap M_c^+$ can be written as a finite union of sets of the form $A^+T_c t_i$, where T_c is the maximal compact subgroup of T, and $t_i \in T$ (cf. [4a]). Write $\gamma = act_i$ accordingly. Then

$$|D_M(\gamma)|^{1/2}|\Theta_{\bar{\pi}_N}(\gamma)| = |D_M(ct_i)|^{1/2}|\omega(a)|\Theta_{\bar{\pi}_N}(ct_i)|.$$

Since ct_i remains in a compact subset, the theory of germs of characters [7b,4b] shows that $|D_M(ct_i)|^{1/2}|\Theta_{\bar{\pi}_N}(ct_i)|$ is bounded by a fixed constant. Moreover, since $|a^\alpha| < 1$ for all the roots α of N, $|\omega(a)|$ remains bounded, which gives the desired result. (Note that, in fact, we can take $r = 0$.) □

Remark. In [7e], Harish-Chandra asserts that Proposition 4 is true for any admissible distribution Θ in lieu of Θ_π. Clearly, an admissible distribution satisfying the estimates is tempered, but I do not know how to prove the converse.

Corollary 4. *If π is tempered, and $f \in \mathcal{C}(G)$, then*

$$< \Theta_\pi, f > = \int_G \Theta_\pi(g) f(g) dg.$$

This follows easily from Proposition 4 and Corollary 3 of Theorem 2.

We now consider the characters of square-integrable representations of G. If Z is the center of G, these are irreducible unitary representations whose matrix coefficients are square-integrable mod Z. We first recall a well-known property of the coefficients of square-integrable representations which does not depend on the Howe conjecture.

Proposition 5. *Assume π is square-integrable mod Z, and let f be a K-finite matrix coefficient of π. Then, if γ is a regular element in a Cartan subgroup $T \subset G$,*

(i) $\displaystyle\int_{T\backslash G} f(g^{-1}\gamma g)\frac{dg}{dt} = 0$ *(T non-elliptic),*

(ii) $\displaystyle\int_{T\backslash G} f(g^{-1}\gamma g)\frac{dg}{dt} = \frac{1}{d(\pi)} f(1)\ \Theta_\pi(\gamma)$ *(T elliptic).*

In (ii) the normalization of measures is as follows. A measure $\frac{dg}{dz}$ on $Z\backslash G$ is used to compute the formal degree $d(\pi)$ [7a, 15] of π; the same measure dg is used to compute the orbital integral, with a measure dt such that the volume of $Z\backslash T$ is equal to 1.

Remark. Part (i) is the "Selberg principle." Note that, contrary to rather widespread misconceptions, the "Selberg principle" is an easy consequence of standard facts from Harish-Chandra's Williamstown lecture, and was known to him [7e]—as well, of course, as (ii). Also note that the proof of (ii) does not require the Howe conjecture.

Proof of Proposition 5. We recall the proof[2] (due to Jacquet-Langlands, cf. [8, Lemma 7.4.1], for supercuspidal representations). In order to prove (i), note that by a theorem of Harish-Chandra ([7a, Thm. 2.9] ; [15, Cor. 4.4.7]) the constant term $f^{(P)}$ of f vanishes for any proper parabolic subgroup $P \subset G$. This is still true for $\bar{f}^{(P)}$, where $\bar{f}$ is the conjugation average for f under a Bruhat-Tits subgroup. Let M be the Levi component of P. Since orbital integrals of f at regular elements of $T \subset M \subset G$ are orbital integrals of $\bar{f}^{(P)}$, this implies (i).

For (ii) we need a lemma.

Lemma 2. *Suppose π is square-integrable and S is an endomorphism of $\mathcal{H}_\pi$ supported on a finite subspace of K-finite vectors of $\mathcal{H}_\pi$. Then*

$$\int_{Z\backslash G} (u, \pi(x)S\pi(x^{-1})v)\frac{dx}{dz} = \frac{(u,v)}{d(\pi)} \text{ trace } S.$$

This is a direct consequence of the Schur orthogonality relations (cf., e.g., [8], loc. cit.). If $\varphi \in C_c^\infty(G)$, applying Lemma 2 to $S = \pi(\varphi)$ yields

$$\text{trace } \pi(\varphi) = d(\pi) \int_{Z\backslash G} \int_G f(xhx^{-1})\varphi(h)dh \;\; dx/dz$$

(cf. [8]), where $f \in \mathcal{C}(G)$ denotes the coefficient $g \longmapsto (u, gv)$. We assume—as we may—that $(u,v) = 1$.

Assume now that γ is elliptic regular. We choose a compact neighborhood ω of γ, contained in the set of elliptic regular elements, and apply this identity to the characteristic function φ of ω. We want to show that we may interchange the integrations, which will be the case if

$$\int_{Z\backslash G} \int_G |f(xhx^{-1})| \;\; |\varphi(h)| dh \frac{dx}{dz} < \infty.$$

This integral equals

$$\int_\omega \int_{Z\backslash G} |f(xhx^{-1})| \frac{dx}{dz} dh.$$

[2] See [8], [7e], [20 : Prop. A.3.e].

However, $|f|$ belongs to the Schwartz space. The arguments of Howe and Harish-Chandra mentioned after Lemma 1 show that the inner integral (in essence the orbital integral of $|f|$ at $h \in \omega$, since h is elliptic and its centralizer T is compact modulo Z) is absolutely convergent. Moreover, it behaves as the orbital integral of a compactly supported function (by the arguments given there) and is therefore locally constant in h. The convergence of the double integral is, therefore, clear. (Of course, now the original argument of Harish-Chandra may also be replaced by the use of the Howe conjecture as in § 2).

Interchanging the integrations, we obtain

$$\text{trace } \pi(\varphi) = d(\pi) \int_G \varphi(h) \Big\{ \int_{Z\backslash G} f(xhx^{-1})dx \Big\} dh.$$

Letting φ tend to the Dirac distribution at γ, this yields (ii). The proof of Proposition 5 is complete. □

We can now prove the following theorem, first established by Kazhdan [9a]. The proof here is Harish-Chandra's [7e][3].

Theorem 3. *Let π_1, π_2 two square-integrable representations of G with the same central character. Then*

$$\sum_{\substack{T \subset G \\ T \text{ elliptic}}} |W(G,T)|^{-1} \int_{Z\backslash T} |D(\gamma)| \bar{\Theta}_{\pi_1}(\gamma) \Theta_{\pi_2}(\gamma) d\bar{\gamma} = \begin{cases} 1 & \text{if } \pi_1 \simeq \pi_2, \\ 0 & \text{otherwise.} \end{cases}$$

Remark. The sum runs over elliptic Cartan subgroups modulo conjugation, $W(G,T)$ is the Weyl group, and the Haar measure $d\bar{\gamma}$ gives mass 1 to $Z\backslash T$.

Proof. Assume first, for simplicity, that π_1 and π_2 have a trivial central character. Assume $f \in \mathcal{C}(G/Z)$. Then f gives, by integration, a linear operator $\pi_1(f)$ on the space $\mathcal{H}_{\pi_1}$. It is K-finite on both sides. By the Corollary to Proposition 4, we have

$$\text{(3.1)} \qquad \text{trace } \pi_1(f) = \int_{Z\backslash G} \Theta_{\pi_1}(x) f(x) \overline{dx},$$

the integral being absolutely convergent.

We now take $f_i = \bar{h}_i$, where h_i is a K-finite matrix coefficient of π_i such that $h_i(1) = 1$. Then trace $\pi_1(f_i) = \frac{1}{d(\pi_1)}\delta_{1,i}$ by the Schur orthogonality relations. On the other hand, since (3.1) is absolutely convergent, we may rewrite it using the Weyl integration formula. Theorem 3 now follows from Proposition 5. □

[3]See also [20, § A.3].

In general, if the central characters are not trivial, we consider instead $f_i \in \mathcal{C}(G,\xi^{-1})$, where $\mathcal{C}(G,\xi^{-1})$ is the space of Schwartz functions transforming according to ξ^{-1} under Z. The rest of the proof is the same.

As a consequence of Proposition 5 and the Howe conjecture, we obtain a different proof of the following result of Bernstein, Deligne, and Kazhdan [3].

Proposition 6 (Existence of pseudo-coefficients). *Assume π is square-integrable. Then there exists a function $f_\pi \in C_c^\infty(G,\xi^{-1})$, where ξ is the central character of π, such that*

$$\text{trace } \delta(f_\pi) = \begin{cases} 0 & (\delta \text{ tempered}, \quad \delta \not\cong \pi) \\ 1 & (\delta = \pi). \end{cases}$$

Proof ($Z = 1$). Let $h_\pi \in \mathcal{C}(G)$ be a coefficient of π, with $h_\pi(1) = 1$. Then h_π satisfies the desired trace conditions, up to a scalar. Moreover, by Proposition 5, the non-elliptic orbital integrals of h_π vanish.

Consider the elliptic set G_{ell} in G, and let Ω be an open, closed, subset of G, compact modulo conjugation, containing G_{ell}. Let ξ_Ω denote the characteristic function of Ω. The Howe conjecture easily implies (cf. the arguments in § 2) that there is a function $f'_\pi \in C_c^\infty(G)$ having the same orbital integrals as h_π on Ω. The function $\xi_\Omega f'_\pi = f_\pi$ then has the same orbital integrals as h_π on all of G; by the Corollary to Proposition 4, f_π and h_π have the same traces when integrated against a tempered representation. □

4. The case of limits of discrete series

For arithmetic (and other) applications, it would be interesting to obtain analogues of Theorem 3 and Proposition 6 in the case of limits of discrete series (rather than discrete series) representations. We do not know how to prove the natural extension of Theorem 3, but the natural conjecture (which was first formulated by Harish-Chandra) will be stated in the next paragraph. We will show here how to prove the analogue of Proposition 6 in the special case of $G = SL(p, F)$ for p *prime*. We do not claim that the results here are original: we only want to illustrate a method.

First some general notions. Let G be any connected reductive F-group. Recall that a representation π of G is *elliptic* if it is tempered (irreducible), and if its character is not identically zero on the elliptic set. As a tempered representation, π is a submodule of an induced representation

$$\rho = \operatorname{ind}_{MN}^G(\delta \otimes 1)$$

(unitary induction), δ being a *unitary* discrete series representation of the Levi subgroup M. In fact, we have $\rho = \pi_1 \oplus \cdots \oplus \pi_r$ (direct sum) with π_1 (say) equal to π. Following Kazhdan [9a], the components π_i are called *relatives* of π. Of course, π will have elliptic summands only for very special representations δ (cf. [4a, § 4]). Elliptic representations are the analogues, in the p-adic case, of the non-degenerate limits of discrete series of real groups [11].

Now let $G = SL(p, F)$. Then the following facts are known.

(i) Elliptic summands may occur only for $P = MN$ equal to G or a Borel subgroup.

(ii) Let $P_0 = HN_0$ be the Borel subgroup, where H is the diagonal (split) torus in G. Consider $\rho = \mathrm{ind}_{P_0}^G(\varphi \otimes 1)$, with φ being a unitary character of H. Then ρ is reducible if and only if φ is the restriction to H of a character of the diagonal torus $(F^\times)^p$ of $GL(p)$ of the form $(\xi, \xi\omega, \cdots, \xi\omega^{p-1})$ with ω of exact order p.

For these facts see [6, 10, 17]. We recall the argument. It is known that a tempered representation of $GL(p)$ is of the form

$$\rho = \mathrm{ind}_{G_{p_1}\times\cdots\times G_{p_r}\times N}(\delta_1 \otimes \cdots \otimes \delta_r \otimes 1)$$

for some partition $p = p_1 + \cdots + p_r$, where each δ_i is a discrete series representations of $GL(p_i)$. Such an induced representation is always irreducible [2], and two such represenations are isomorphic only if their data (δ_i) differ by a permutation of the set $(1, \cdots, r)$.

By Mackey theory [6], the restriction of ρ to $SL(p)$ will be reducible only if $\rho \simeq \rho \otimes \omega$ for some non-trivial character $\omega : F^\times \longrightarrow \mathbb{C}^\times$ vanishing on $(F^\times)^p$. Let R be the Abelian group of characters satisfying $\rho \simeq \rho \otimes \omega$. It is clearly a p-group. If $\omega \subset R$, we must have $\delta_{s \cdot i} \simeq \delta_i \otimes \omega$ for a permutation s of $\{1, \cdots, r\}$. Moreover, since δ_i is a representation of $GL(p_i)$, it cannot be fixed by a twist of order p for $p > p_i$. Thus, considering the orbits of s, the only possibilities for ω to be non-trivial are (i) $r = 1$ (ρ discrete) or (ii) $r = p$, $p_i = 1$, and $\rho = \mathrm{ind}(\xi \otimes \xi\omega \otimes \cdots \otimes \xi\omega^{p-1} \otimes 1)$. Note that, in case (ii), R must be equal to $\mathbb{Z}/p\mathbb{Z}$ since, by the same arguments, the relation $\rho \simeq \rho \otimes \omega'$ implies that $\omega' \simeq \omega^i$ for some $i \in \{0, \cdots, p-1\}$. In this case it is known [6, 10, 17] that R is isomorphic to the R-group (see the cited papers for a definition of the R-group) and that ρ restricted to $SL(p)$ has p inequivalent summands. In particular, *the reducibility points* in case (ii) are *isolated* in the orbit

$$\mathcal{O}_\varphi = \{\varphi_1 |\ |^{s_1}, \cdots, \varphi_n |\ |^{s_n} : s_i \in i(\mathbb{R}/2\frac{\pi}{\log q}\mathbb{Z})\}$$

parametrizing the continuous family of tempered principal series representations through ρ. Here we have $\varphi = (\varphi_1, \cdots, \varphi_n)$, and a character $(\varphi_1, \cdots, \varphi_n)$ is considered modulo the "diagonal" characters $(\varphi_0, \cdots, \varphi_0)$ which restrict trivially to $H = (F^\times)^p \cap SL(p)$.

From this discussion we see that the only elliptic representations of $SL(p)$ are the discrete series and, possibly, the representations occurring as summands in (the reducible) case (ii).

Theorem 4. *Suppose φ is a character of H obtained by restriction of a character*

$$(\xi, \xi\omega, \cdots, \xi\omega^{p-1})$$

of $(F^\times)^n$ with ω of exact order p. Let $\rho = \bigoplus_{i=1}^{p} \pi_i$ be the reduction of ρ, and $(a_i)_{i=1,\cdots,p}$ a set of complex numbers such that $\sum a_i = 0$. Then there exists $f \in C_c^\infty(G)$ such that

(i) trace $\pi_i(f) = a_i$, and

(ii) trace $\pi(f) = 0$ for any tempered representation of f non-isomorphic to π_i.

Corollary 5 (Kazhdan [9b]). *For each $i = 1, \cdots, p$, the representation π_i is elliptic.*

Proof of Corollary. Fix (a_i), and choose f as in Theorem 4. By condition (ii) and the usual descent argument, one easily sees that the non-elliptic orbital integrals of f vanish (cf. Proposition 5). Thus, choosing $a_{i_0} \neq 0$, condition (i) obviously implies that π_{i_0} is elliptic. □

In order to prove Theorem 4 we will use Harish-Chandra's Plancherel theorem for the Schwartz space. We first, rather informally, recall its content. Let $P = MN$ be a parabolic subgroup of G, and δ a discrete series representation of M. Let $\mathfrak{a}$ be the usual "real Lie algebra" of the split component of M and $H : M \longrightarrow \mathfrak{a}$ the Harish-Chandra map. For $\lambda \in i\mathfrak{a}^*$, $\delta(m)e^{2\pi < H(m),\lambda >}$ is still a discrete series representation of M (modulo the center). We denote it by δ_λ; here λ may be replaced by its image in $i\mathfrak{a}^*/iL^*$ where L is the lattice $H(M) \subset \mathfrak{a}$, and L^* the dual lattice. For $s \in i\mathfrak{a}^*/iL^* = \mathcal{O}$, we write δ_s for the twisted representation.

The representations $\rho_s = \mathrm{ind}_p^G(\delta_s)$ can be realized on a fixed Hilbert space $\mathcal{H}_\delta$, containing a well-defined subspace V_δ of K-finite vectors. If $f \in \mathcal{C}(G) = \varinjlim \mathcal{C}_K(G)$, we consider the operators $\rho_{\delta,s}(f) : \mathcal{H}_\delta \longrightarrow \mathcal{H}_\delta$ for all δ and s. They have the following properties:

(i) $\rho_{\delta,s}(f)$ is a matrix with a finite number of non-zero entries in a basis of V_δ ("$\rho_{\delta,s}(f)$ is K-finite").

(ii) If we consider the orbits $\mathcal{O}$ up to isomorphism, only a finite number of orbits are such that $\rho_{\delta,s} \neq 0$ for $\delta \in \mathcal{O}$. When it is non-zero, $\rho_{\delta,s}$ is a smooth function of $s \in \mathcal{O}$ (with values in a finite-dimensional space, by (i)).

(iii) $\rho_{w\delta,ws}(f)A(w,\delta,s) = A(w,\delta,s)\delta_{\delta,s}(f)$ where $A(w,\delta,s) : \mathcal{H}_s \longrightarrow \mathcal{H}_{w\delta}$ is a standard intertaining operator between $\rho_{\delta,s}$ and $\rho_{w\delta,ws}$ (cf. e.g. [15]).

It is known that the $A(w,\delta,s)$ can be normalized so as to be holomorphic and unitary for unitary s ($s \in i\mathfrak{a}^*/iL^*$) (cf. [15]). Moreover, up to a scalar, they obey the obvious composition rule.

Conversely, an important theorem of Mischenko [18 a,b], which strengthens Harish-Chandra's Plancherel theorem [7d], asserts that any family of operators $(\delta, s) \longmapsto T(\delta, s)$ satisfying conditions (i), (ii), and (iii), is of the form $\rho_{\delta,s}(f)$ (for varying δ, s) for some fixed function $f \in \mathcal{C}(G)$.

Proof of Theorem 4. We now prove Theorem 4, first for a Schwartz function f. Consider the reducibility point of the principal series associated to ρ. It is isolated, as we pointed out above. Consider the reduction $\mathcal{H} = \oplus \mathcal{H}_i$ associated to the reduction of ρ, $\mathcal{H}$ being the (fixed) Hilbert space of the corresponding principal series. Let s_0 be associated to ρ (e.g., $s_0 = 0$). We can choose a smooth function $s \longmapsto T(s) \in \mathcal{L}(\mathcal{H}, H)$ of $s \in \mathcal{O} = i\mathfrak{a}_0^*/iL^*$ such that (1) $T(s) = \oplus T_i(s)$ with $T_i : \mathcal{H}_i \longrightarrow \mathcal{H}_i$, (2) $T(s)$ is K-finite, (3) T has support in a neighborhood of 0 containing 0 as its only reducibility point, (4) trace $T(s) = 0$ (all s), and (5) trace $T_i(s) = a_i$. We extend T by 0 on the other orbits of generalized principal series. We obtain a family of operators verifying conditions (i), (ii) above. In order to obtain (iii), note that the Weyl group (only the Weyl group $W \simeq \mathfrak{S}_n$ for the minimal parabolic subgroup intervenes) acts on our space of functions by $w \longmapsto A_w$,

$$A_w : T_{\delta,s} \longmapsto (A_w T)_{\delta,s} = A(w, \delta, s)^{-1} T_{w\delta, ws} A(w, \delta, s).$$

It is indeed a linear group action because of the cocycle relation (composition up to a scalar) for the $A(w, \delta, s)$.

Condition (iii) is equivalent to the fact that $(T_{\delta,s})$ is W-invariant for this action, and we insure this by replacing T by its average under W. (Note that T will then be non-zero on some other orbits, i.e., those conjugate by W to the orbit of φ). We now obtain an operator $T = (T_{\delta,s})$ which is, by Harish-Chandra's theorem, the scalar-valued Fourier transform of a Schwartz function f on G. By construction, trace $\pi(f) = 0$ if π is not a submodule of a unitary principal series, or if π is a full principal series. So we only have to check the reducibility points of principal series. Consider first the point $s_0 \in \mathcal{O}$. Only elements $w \in W$ that fix the original character $(\xi, \xi\omega, \cdots, \xi\omega^{p-1})$ will give transforms of T that contribute to the trace. The corresponding operators must preserve the decomposition $\mathcal{H} = \oplus \mathcal{H}_i$ (since they commute with ρ). It follows that the trace of the averaged operator in $\mathcal{H}_i$ is a scalar multiple (independant of i) of the trace of the original operator. Finally, by our condition on the support of T, we see that the averaged operator has non-zero values only at reducibility points that are W-conjugate to ρ. The summands of the corresponding representations are again $(\pi_1, \cdots, \pi_p)$, and this completes the proof. □

5. Problems on tempered characters, pseudo-coefficients, and a theorem

In this section we first state a few problems. We start with Harish-Chandra's formulation of the (conjectural) orthogonality and density properties of elliptic characters.

Assume $P = MN \subset G$ is a parabolic subgroup, and δ belongs to the unitary discrete series of M. Let $G_{ell} \subset G$ be the elliptic set. Let Φ_δ be the (finite-dimensional) space of functions on G_{ell} generated by the (elliptic) summands of $\mathrm{ind}_p^G(\delta \otimes 1)$. Thus $\Phi_\delta = \{0\}$ if the induced representation has no elliptic summands.

If π is an irreducible admissible representation of G, then $|D|^{1/2}\Theta_\pi$, where D is the usual discriminant, is locally bounded on G [7b, 4b]. In particular, we see that $|D|^{1/2}\Theta_\pi$ is bounded on G_{ell} if the central character of π is unitary.

This implies that Θ_π belongs to the space

$$L^2(\mathbb{T})^{\mathbf{W}} = \bigoplus_{T \in \mathcal{T}} L^2(T,\omega)^{W_T},$$

where $\mathcal{T}$ is a set of representatives for the *elliptic* tori, ω is the central character, W_T is the Weyl group of T, and $L^2(T,\omega)$ denotes the space of functions on T which transform under Z by ω and are square-integrable for the scalar product

$$\int_{Z\backslash T} \Theta(t)\bar{\Theta}'(t)|D(t)|\frac{dt}{dz}.$$

The Haar measure $\frac{dt}{dz}$ gives mass 1 to $Z \setminus T$ (cf. Theorem 3).

We endow the space $L^2(\mathbb{T})^{\mathbf{W}}$ with the scalar product which occurs in Theorem 3.

Fix a unitary central character ω on Z. We consider only representations with central character ω.

Conjecture 1 (Harish-Chandra).

1. **Orthogonality**. *Assume M_1, M_2 are Levi subgroups, and δ_1, δ_2 are discrete series representations of M_1, M_2 respectively (unitary, with central characters restricting to ω). If the data (M_1, δ_1) and (M_2, δ_2) are not conjugate in G, then Φ_{δ_1} and Φ_{δ_2} are orthogonal in $L^2(\mathbb{T})^{\mathbf{W}}$.*
2. **Density**. *The spaces Φ_δ (for all M, δ) generate $L^2(\mathbb{T})^{\mathbf{W}}$.*

In fact, part 1 of the Conjecture has been proved by Kazhdan [9a, Cor. to Prop. 5.5]. Part 2 was proved by Harish-Chandra in the real case [7f], and remains an open problem in the p-adic case. For another formulation of the density problem, see the Conjecture after Theorem K of the Introduction to Kazhdan's paper [9a].

A preliminary question connected with this problem is the following[4]. Recall that π is *elliptic* if it is tempered and its character does not vanish on G_{ell}.

[4]This was pointed out to me by D. Keys. The analogous statement seems to follow, for *connected* real reductive groups, from Knapp-Zuckerman [11]

Conjecture 2. *π is elliptic if and only if π cannot be realized as a full induced representation from a proper parabolic subgroup.*

Clearly an induced representation cannot be elliptic, by the formulas for induced characters. The problem is, then, the converse. As Keys explained to me at the conference, Kazhdan's results in [9a] imply that a representation that is not elliptic is a *linear combination* (in the Grothendieck group of representations of G) of properly induced representations. However, this does not suffice to prove that π is induced.

As regards Conjecture 1, one would like to have a more precise formulation (cf. Theorem 3) giving the L^2-norms of selected elements of Φ_δ. This is intimately connected with problems of L-indistinguishability [9b, 13], and perhaps L-indistinguishability will, indeed, be the way to attack the density problem.

Finally, we state the natural formulation of the problem of pseudo-coefficients for elliptic representations[5]. It is, in fact, a special case of the scalar-valued Paley-Wiener problem for the Schwartz space. This has been solved in [5] for *real* groups, but only for smooth compactly supported functions rather than Schwartz functions. However, the proof, and in particular the combinatorics of Appendix C, immediately extends to Schwartz functions. For the Schwartz space, the solution of the scalar-valued "Paley-Wiener" problem had, in fact, been announced long ago by Arthur.

We call a representation π of G a *limit of discrete series* if it is tempered and not properly induced. (This is the analogue of the "non-degenerate limits of discrete series" of [11]; according to Conjecture 2, these representations should be exactly the elliptic representations). Let $MN = P \subset G$ and $\mathfrak{a} = \mathfrak{a}_M$; L will denote the image of the Harish-Chandra map, and L^* the dual lattice. If δ is a limit of discrete series for M, we can consider, as in §4, δ_s for $s \in i\mathfrak{a}^*/iL^*$. Write $I(\delta_s)$ for $\mathrm{ind}_P^G(\delta_s \otimes 1)$. We call the set of all δ_s for varying s an *orbit*.

There are *relations* between such representations. Assume $P_1 \subset P_2 \subset G$, which implies $M_1 \subset M_2$ and $\mathfrak{a}_2 \subset \mathfrak{a}_1$. If δ_1 is a limit of discrete series for M_1, let $\mathrm{ind}_{P_1 \cap M_2}^{M_2}(\delta_{1,t} \otimes 1) = I^{M_2}(\delta_{1,t})$. For certain values of δ_1, it may happen that

$$I^{M_2}(\delta_1) = \bigoplus_{i=1}^{S} \delta_2^i, \tag{5.1}$$

where the δ_2^i are limits of discrete series for M_2.

Conjecture 3. *Let $(F(\delta))_{\delta \in \Delta}$ be a set of scalars, indexed by the collection Δ of all limits of discrete series representations of all Levi subgroups M of G. Suppose the $F(\delta)$ have the following properties.*

(i) K**-finiteness:** *$F(\delta) = 0$ unless δ belongs to a finite number of orbits; equivalently, $F(\delta) = 0$ unless $I^G(\delta)$ has non-zero vectors fixed by a compact-open subgroup of G determined by F.*

[5] This section is the result of conversations with M.-F. Vigneras, cf. [16b].

(ii) **Invariance**: *If (M,δ) and (M',δ') are conjugate by G, then the scalars $F(\delta')$ and $F(\delta)$ are equal.*

(iii) **Compatibility**: *If P_1 and P_2 are given, and δ_1 and $(\delta_2^i)_{i=1,\dots,S}$ verify relation (5.1), then*

$$F(\delta_1)=\sum_{i=1}^{S} F(\delta_2^i).$$

(iv) **Smoothness**: *For δ a representation of M, and $s\in i\mathfrak{a}^*/iL^*$, the function defined by $s\longmapsto F(\delta_s)$ is smooth in s.*

Under these conditions there exists $f\in \mathcal{C}(G)$ such that, for all δ,

$$F(\delta)=\operatorname{trace}(I(\delta)(f)).$$

A scalar-valued Paley-Wiener theorem (for functions in $C_c^\infty(G)$) has been proved by Bernstein, Deligne, and Kazhdan [3], but, as Vignéras pointed out, it seems difficult to extract from it the existence of pseudo-coefficients for limits of discrete series. For $G=GL(n,F)$, Conjecture 3 has been proved by Mischenko [18b] (in this case condition (iii) is empty).

Conjecture 3 would have the following consequence. Assume δ_1 is a limit of discrete series for G. Then there exists a Levi subgroup $M\subset MN\subset G$ and a discrete series representation δ of M such that

$$\operatorname{ind}_{MN}^{G}(\delta\otimes 1)=\delta_1\oplus\cdots\oplus\delta_S, \tag{5.2}$$

$\delta_2,\cdots,\delta_S$ being other tempered representations. The "Langlands disjointness theorem" (due to Langlands in the real case) states that (M,δ) is then unique up to conjugation in G. Note that it follows readily from Harish-Chandra's Plancherel theorem and Mischenko's theorem (§ 4): since the conditions imposed on the matrix-valued Fourier transforms in this theorem only involve intertwining operators between conjugate orbits, this implies immediately that the representations $\operatorname{ind}_P^G(\delta)$ are disjoint for non-conjugate δ.

On the other hand, in general there will be Levi subgroups $M_1\supset M$ and limits of discrete series δ_{M_1} of M_1 such that

$$\operatorname{ind}_{M_1N_1}^{G}(\delta_{M_1})\simeq\delta_1\oplus\delta_2'\oplus\cdots\oplus\delta_R'. \tag{5.3}$$

In this case, by the disjointness theorem, δ_{M_1} is contained in $\operatorname{ind}_{M_1\cap(MN)}\delta$, and the δ_j' form a subset of the δ_i. We now have the following Corollary.

Corollary 6 (to Conjecture 3). *Assume $(c_i)_{i=1,\dots,S}$ is a set of complex numbers, indexed by the δ_i occurring in the decomposition* (5.2), *such that $\sum_{i\in\Delta} c_i=0$ for each subset Δ of $\{1,\cdots,S\}$ corresponding to a decomposition* (5.3). *Assume, moreover, that $c_i=c_j$ if $\delta_i\simeq\delta_j$, and $c_i=0$ if δ_i is not a limit of discrete series. Then there exists $f\in C_c^\infty(G)$ such that*

$$\operatorname{trace}\delta_i(f)=c_i \text{ if } i=1,\cdots,S,$$

$$\operatorname{trace}\pi(f)=0 \text{ if } \pi \text{ is not one of the } \delta_i.$$

This follows readily from the Conjecture.

Consider the function $\delta \longmapsto F(\delta)$ given by $\delta_i \longmapsto c_i$, and $\delta \longmapsto 0$ for $\delta \notin \{\delta_i\}$. We contend that it satisfies the assumptions of Conjecture 3. Properties (i), (ii), (iv) are obvious, and the only non-trivial cases of (iii) are assumed by definition. By the Conjecture, we obtain $f \in \mathcal{C}(G)$; the arguments used in the proof of Proposition 6 replace it by an $f \in C_c^\infty(G)$.

We note that two technical questions appear naturally here. We state them as conjectures for simplicity, although we do not have strong evidence for their validity in the p-adic case; in the real (connected) case, they are true [11].

Conjecture 4. *Assume δ_1 is a limit of discrete series for G, and*

$$\operatorname{ind}_{MN}^G \delta = \delta_1 \oplus \cdots \oplus \delta_S$$

with δ discrete. Then all the δ_i are limits of discrete series.

Conjecture 5. *Assume δ is a tempered representation of G, and*

$$\delta = \operatorname{ind}_{MN}^G \delta_M,$$

with δ_M a limit of discrete series. (Note that, by definition, any δ can be so obtained). Then (M, δ_M) is unique modulo conjugation.

We will now prove, without restrictive conditions, a result weaker than Corollary 6. Assume δ_1 is *elliptic*, and consider the decomposition (5.2) associated to δ_1. Let $\Theta_1^e, \cdots, \Theta_S^e$ be the elliptic characters of $\delta_1, \cdots, \delta_S$ (as functions on G_{ell}).

Theorem 5. *Assume $a_1, \cdots, a_S$ are complex numbers subject to the following condition: if the linear relation $\lambda_1\Theta_1^e + \cdots + \lambda_s\Theta_S^e = 0$ is valid, then $a_1\lambda_1 + \cdots + a_S\lambda_S = 0$. Then there exists $f \in C_c^\infty(G)$ such that*

$$\text{trace } \delta_i(f) = a_i \text{ if } i = 1, \cdots, S,$$
$$\text{trace } \pi(f) = 0 \text{ if } \pi \text{ is not one of the } \delta_i.$$

Note that, in particular, trace $\delta_i(f) = a_i = 0$ if δ_i is not elliptic; also, a_i must be equal to a_j if $\Theta_i^e = \Theta_j^e$.

Corollary 7. *If δ_1 is elliptic, then there exists $f \in C_c^\infty(G)$ such that trace $\delta_1(f) = 1$ and trace $\pi(f) = 0$ for $\pi \notin \{\delta_i\}$.*

Proof of Corollary. We merely start with a compactly supported function h on G_{ell} such that trace $\delta_1(h) = 1$, and set $a_i =$ trace $\delta_i(h)$. □

Proof of Theorem 5. As in the proof of the Corollary, we start with a function $h \in C_c^\infty(G_{ell})$ such that trace $\delta_i(h) = a_i$; such a function exists by condition (c). We consider the matrix-valued Fourier transform $\delta \longmapsto \rho_\delta(h)$ for all (M, δ) with M a Levi subgroup of G (§4). Let (M_0, δ_0) be the Levi subgroup, and its discrete representation, occurring in the left-hand side of (5.2) for δ_1. We may truncate the family of operators $\rho_\delta(h)$ by replacing all those corresponding to orbits non-conjugate (in G) to the orbit of δ_0 by zero. The defining equations of the Fourier transform of the Schwartz space are still satisfied. Therefore we obtain a function $f' \in \mathcal{C}(G)$ that has the same (matrix) Fourier transform as h on orbits conjugate to that of δ_0, and vanishes elsewhere. (In Harish-Chandra's terminology, we have kept only the wave-packet associated to δ_0).

What can we say about f'? By construction, it has the same trace as h in all the representations contained in the induced representations of the orbit of δ_0, and other traces vanish. Consider the orbits conjugate to that of δ_0. They contain a finite number of points containing elliptic summands [4a]. Thus, localizing (by multiplying the operators $\rho_\delta(f')$ by a scalar-valued, smooth function of δ supported near δ_0) we may replace f' by a function f'' having the same traces in the components of δ_0, and trace zero in all other elliptic representations in this orbit. Since we have multiplied by a scalar, we have not changed the traces in non-elliptic representations, which are zero as h had elliptic support. Finally, we see that the new function f'' has all requested properties. It belongs to $\mathcal{C}(G)$, but we replace it by a function $f \in C_c^\infty(G)$ with the same traces by the sempiternal argument based on the Howe conjecture. □

The following consequence of Corollary 7 may be useful :

Corollary 8. [6] *Assume $G = \underline{G}(F)$, with $\underline{G}$ a connected F-group. Then there exists a number field k, a finite prime v of k such that $k_v \simeq F$, and a connected k-group $\underline{\underline{G}}$ with $\underline{\underline{G}} \times_{\mathrm{Spec}\ k} \mathrm{Spec}\ k_v \simeq \underline{G}$, having the following properties. If δ_1 is an elliptic representation of $G = \underline{G}(F)$, there exists an irreducible, admissible representation π of $\underline{\underline{G}}\,(\mathbb{A}_k)$, occurring in the space of cusp forms on $\underline{\underline{G}}\,(k)\backslash\,\underline{\underline{G}}\,(\mathbb{A}_k)$, whose local component at v is a representation δ_i for $i \in [1, \cdots, s]$ (a relative of δ_1 in Kazhdan's terminology).*

Proof Sketch. (For simplicity we consider just the case where the δ_i are the only unitary representations of G whose expansions as linear combinations of standard representations involve one of the δ_i.) Use the fact that the orbital integrals of the function f of Theorem 5 and its Corollary do not vanish on the elliptic set, along with the methods of Kazhdan in the Appendix to [9a]—see also [12]. □

We should emphasize the difference between the proofs of Theorem 5—where we have used the ellipticity of δ_1—and of Theorem 4—where we have not used it, having obtained it as a corollary. The method of proof of Theorem 4 seems to show, more generally, that a limit of discrete series δ_1 that corresponds to an *isolated* reducibility point in a generalized principal

[6] *Added in Proof:* We have to make an extra assumption in the proof sketch.

series (i.e., in equation (5.2), ind($\delta_s \otimes 1$) is irreducible for any $s \in i\mathfrak{a}^*$ that is non-zero and small enough) admits a pseudo-coefficient and is, therefore, elliptic. I have not checked the details. Note that in the non-isolated case, the extension of this argument would require an analog of the delicate analysis of "reducibility along strata" contained in [5] (in particular its Appendix C). In the p-adic case, where R-groups are not abelian, this would be complicated.

Finally, M.-F. Vignéras has pointed out to me that some results on the existence of pseudo-coefficients of limits of discrete series should follow from her work on projective modules in the category of tempered representations [16 b], and recent unpublished results of Bernstein and Kazhdan.

References

[1] J. Arthur and L. Clozel, *Simple Algebras, Base Change, and the Advanced Theory of the Trace Formula,* Annals of Math. Studies, Vol. 120, Princeton University Press, Princeton, New Jersey, 1989.

[2] J. Bernstein, *P-invariant distributions on GL(N) and the classification of unitary representations of GL(N) (non archimedian case),* Lie Group Representations II, Lecture Notes in Math., Vol. 1041 (R. Herb, ed.), Springer-Verlag, Berlin-New York, 1984, pp. 50–102.

[3] J. Bernstein, P. Deligne, and D. Kazhdan, *Trace Paley-Wiener theorem for reductive, p-adic groups,* J. Analyse Math. **47** (1986), 180–192.

[4a] L. Clozel, *Sur une conjecture de Howe-I,* Compositio Math. **56** (1985), 87–110.

[4b] ———, *Characters of non-connected, p-adic reductive groups,* Can. J. Math. **39** (1987), 149–167.

[4c] ———, *Orbital integrals on p-adic groups : a proof of the Howe conjecture,* Ann. of Math. **129** (1989), 237–251.

[4d] ———, *The fundamental lemma for stable base change,* Duke Math. J. **61** (1990), 255–302.

[5] L. Clozel and P. Delorme, *Le théorème de Paley-Wiener invariant pour les groupes de Lie réductifs II,* Ann. Sc. E.N.S. (to appear).

[6] S. Gelbart and A. Knapp, *L-indistinguishability and R-groups for the special linear group,* Adv. in Math. **43** (1982), 101–121.

[7a] Harish-Chandra, *Harmonic analysis on reductive p-adic groups,* Proc. Symp. Pure Math. **XXVI** (1974), 167–192.

[7b] ———, *Admissible invariant distributions on reductive p-adic groups,* Queen's Papers in Pure and Applied Math. **48** (1978), 281–347.

[7c] ———, *A submersion principle and its applications,* Proc. Indian Acad. Sci. (Math. Sci.) **90** (1981), 95–102.

[7d] ———, *The Plancherel formula for reductive p-adic groups,* Collected Papers, Vol. IV, Springer-Verlag, Berlin-Heidelberg-New York, 1984.

[7e] ———, Letter to M.-F. Vignéras, 1982.

[7f] ———, *Supertempered distributions on real reductive groups,* Stud. App. Math. Adv. Math. **8** (1983), 139–153.

[8] H. Jacquet and R. P. Langlands, *Automorphic forms on GL(2), Lecture Notes in Math., Vol 114,* Springer-Verlag, Berlin-New York, 1970.

[9a] D. Kazhdan, *Cuspidal geometry of p-adic groups*, J. Analyse Math. **47** (1986), 1–36.

[9b] ———, *On lifting*, Lie groups representations II, Lecture Notes in Math., Vol. 1041 (R. Herb, ed.), Springer-Verlag, Berlin-Heidelberg-New York, 1984.

[10] D. Keys, *On the decomposition of reducible principal series representations of p-adic Chevalley groups*, Pacific J. Math. **101** (1982), 351–388.

[11] A. Knapp and G. Zuckerman, *Classification of irreducible tempered representations of semi-simple groups*, Ann. Math. **116** (1982), 389–501.

[12] R. Kottwitz and J. Rogawski, *The distributions in the invariant trace formula are supported on characters*, preprint.

[13] R. P. Langlands, *Les débuts d'une formule des traces stable*, Publ. Math. Univ., Paris 7, Paris, undated.

[14a] J. Rogawski, *Applications of the building to orbital integral*, Ph.D. Thesis, Princeton University, Princeton, 1980.

[14b] ———, *Representations of $GL(n)$ and division algebras over a p-adic field*, Duke Math. J. **50** (1983), 161–169.

[15] A. Silberger, *Introduction to Harmonic Analysis on Reductive p-adic Groups*, Princeton Univ. Press, Princeton, New Jersey, 1979.

[16a] M.-F. Vignéras, *Caractérisation des intégrales orbitales sur un groupe réductif p-adique*, J. Fac. Sc. Univ. Tokyo, Sec. IA **29** (1981), 945–962.

[16b] ———, *On formal degrees of projective modules for reductive p-adic groups*, preprint.

[17] N. Winarsky, *Reducibility of principal series representations of p-adic Chevalley groups*, Amer. J. Math. **100** (1978), 941–956.

[18a] P.A. Mischenko, *Invariant tempered distributions on the reductive p-adic group $GL_n(F_p)$*, C. R. Math. Acad. Sc., Soc. Roy. du Canada **4** (1982), 123–127.

[18b] ———, *Invariant tempered distributions on the reductive group $GL_n(F_p)$*, Ph.D. Thesis, University of Toronto, Toronto, 1982.

[19] W. Casselman, *Characters and Jacquet modules*, Math. Ann. **230** (1977), 101-105.

[20] P. Deligne, D. Kazhdan, M.-F. Vignéras, *Représentations. des algèbres centrales simples p-adiques*, Représentations des Groupes Réductifs sur un Corps Local, Hermann, Paris, 1984, pp. 33–117.

URA D 0752 du CNRS, Université Paris-Sud, Département de Mathématique, 91405 Orsay

CONSTRUCTING THE SUPERCUSPIDAL REPRESENTATIONS OF $GL_n(F)$, F p-ADIC

LAWRENCE CORWIN

Rutgers University

INTRODUCTION

In this paper I give a description of a construction of the supercuspidal representations of $GL_n(F)$. This construction, the result of work done in late 1988 and early 1989, was first given in [6], and the description here follows the same lines.

The title of this paper is incomplete. The construction of the supercuspidals was in large part based on a construction of the unitary dual of the multiplicative groups of central division algebras over F, worked out in the summer of 1988 and given in [5]. I include an account of that construction here as well. The present account is essentially that in [5], but I have added some details that (I hope) make it easier to read. (The reader will still need [5] for some details.) I would like to thank Allen Moy, Paul Sally, and Marie-France Vigneras for their suggestions; my thanks also go to William Barker and Ann Kostant for their careful editorial work. Any errors remaining in the paper are solely the responsibility of the author.

The construction of these representations is "elementary" in the sense that one does not need to know an advanced mathematical machine to understand it. (What is required is some elementary ramification theory and a bit about the structure of division algebras and matrix algebras.) However, the construction is also fairly complicated. I have tried to give some motivation for the procedure used here.

The paper divides naturally into three parts. Part I (Sections 1–5) is concerned with division algebras. In Part II (Sections 6–10) a collection of supercuspidals for $GL_n(F)$ is constructed. Finally, in Part III (Section 11) I describe the Matching Theorem and give an indication of why the construction of Part II in fact gives all supercuspidals of $GL_n(F)$.

Supported by NSF Grant DMS 89–02993.

I. Representations of Division Algebras

1. The Theorem on Representations of Division Algebras .

We begin by giving some information on the structure of the division algebra D_n. (All the facts we need are found in [21].) Let F_f be the unramified extension of F of residue class degree f, and let k be the residue class field of F (k has q elements, q a power of p). Then $D = D_n$ is generated by F_n and an element ϖ such that

(1.1)
(1) $\varpi^n = \varpi_F$, a uniformizing (= prime) element of F;
(2) ϖ normalizes F_n, and $\varpi a \varpi^{-1} = a^\sigma$ for $a \in F_n$ (where σ generates $Gal(F_n/F)$.

In fact, σ determines D up to isomorphism.

Let $\bar{k}_n$ be the set of coset representatives for k_n in F_n given by the roots of $X^{q^n} - X = 0$. Every element in $D^\times$ is uniquely of the form

$$x = \sum_{j=j_0}^{\infty} \alpha_j \varpi^j, \quad \alpha_j \in \bar{k}_n, \ \alpha_{j_0} \neq 0. \tag{1.2}$$

We often do not distinguish between k_n and $\bar{k}_n$ (writing, e.g., $Tr_{k_n/k}\alpha$ for $\alpha \in \bar{k}_n$); this does not cause difficulty and should not cause confusion.

Let $\mathcal{O} = \{0\} \cup \{x : j_0 \geq 0\}$, $P = \{0\} \cup \{x : j_0 > 0\}$; $\mathcal{O}$ is the maximal compact subring of D, and P its unique maximal (right, left, or 2 sided) ideal. Set $K^0 = K = \{x : j_0 = 0\}$ = units of $\mathcal{O}$, $K^m = 1 + P^m$ for $m \geq 0$. The K^m form a neighborhood basis of 1 in $D^\times$, and the K^m are all normal in $D^\times$; $D^\times/K \cong \langle\varpi\rangle \cong \mathbb{Z}$.

Let ψ be a character of F nontrivial on the integers $\mathcal{O}_F$ of F but trivial on the prime ideal P_F. (Thus ψ is also a nontrivial character of $\mathcal{O}_F/P_F = k$.) The (additive) characters χ_x of D are given by $\chi_x(y) = \psi \circ Tr_{D/K}(xy)$, where Tr equals the reduced trace. Under this identification of D with $D\hat{}$, $(P^m)^\perp = P^{1-m}$. For $m < r \leq 2m$, the map $1 + y \mapsto y$ gives an isomorphism, $K^m/K^r \cong P^m/P^r$, and this and the above identification give $(K^m/K^r)\hat{} \cong P^{1-r}/P^{1-m}$. Thus we can define $\chi_x(y)$ for $y \in K^m$, $x \in P^{1-r}$; x is specified only mod P^{1-m}.

In constructing $(D^\times)\hat{}$ there is no loss of generality in restricting attention to those irreducible representations π with $\pi(\varpi^n) = I$, since $\pi(\varpi^n)$ is always scalar and we can make the scalar 1 by tensoring with a character trivial on K. This assumption means that we are essentially dealing with the compact group $D^\times/\langle\varpi^n\rangle$. Furthermore, π is trivial on some K^m. [Let $\pi = D^\times \to U(n_0, \mathbb{C})$, $n_0 = \dim \pi$; let V be a neighborhood of 1 in $U(n_0, \mathbb{C})$ containing no subgroup of $U(n_0, \mathbb{C})$ except $\{1\}$. Then $\pi^{-1}(V)$ is open in $D^\times$ and so contains some K^m. Since $\pi(K^m)$ is a subgroup of V, $\pi(K^m) = \{1\}$.] Let m be as small as possible with $K^m \subseteq \operatorname{Ker} \pi$. We call

m the *conductor* of π. If $m = 0$, then it is simple to describe π, and the case $m = 1$ is also easy. So assume $m > 1$, and set $m = s_1 + 1$. Since $K^{s_1}/K^{s_1+1} \cong P^{s_1}/P^{s_1+1}$ is Abelian, $\pi|_{K^{s_1}}$ reduces to a sum of characters χ_x, $x \in P^{-s_1}$ (mod P^{1-s_1}), and the different x are conjugate. If $x \in F^\times$, then $\pi|_{K^{s_1}}$ is scalar, there is a character χ_0 of $D^\times$ agreeing with χ_x on K^{s_1}, and $\pi \otimes \chi_0^{-1}$ is trivial on K^{s_1}. Thus we may assume $x \notin F$. (We call such π *generic*.) We want to construct all generic π containing χ_x.

The result when $p \nmid n$ (the "tamely ramified" case) has been known for some time (see [4], [11], or [7]; an account of the result is in [17]). Let $D_1 =$ algebra of elements commuting with x (hence with $E_1 = F[x]$), $D_1^\perp = \{y \in D : Tr(yw) = 0, \text{ for all } w \in D_1\}$. Define $s_1' = [s_1/2] + 1$, $s_1'' = s_1 + 1 - s_1'$. (If s_1 is odd, then $s_1' = s_1'' = (s_1+1)/2$; if s_1 is even, then $s_1' = (s_1/2) + 1$, $s_1'' = s_1/2$.) Assume for the moment that s_1 is odd. One can then verify fairly straightforwardly that if we set $H_1 = D_1^\times K^{s_1'}$, then:

(1.3)
(1) $H_1 = D_1^\times N_1$, where $N_1 = K^{s_1'} \cap (1 + D_1^\perp)$;
(2) H_1/K^{s_1+1} is a semidirect product of $N_1/N_1 \cap K^{s_1+1}$ (normal) with $D_1^\times/D_1^\times \cap K^{s_1+1}$;
(3) $\chi|_{K^{s_1}}$ is trivial on $N \cap K^{s_1}$.

We get all representations of π containing χ as follows: let τ be any irreducible representation of $D_1^\times$ whose restriction to $D_1^\times \cap K^{s_1}$ agrees with χ. Extend τ to H_1 by making it trivial on N_1. Then induce to $D^\times$ to get π. Furthermore, distinct τ yield inequivalent π.

Notice that $\chi|_{D_1^\times \cap K^{s_1}}$ extends to a character of $D_1^\times$. Therefore we can write $\tau = \tau_0 \otimes \chi$, where τ_0 is trivial on $D_1^\times \cap K^{s_1}$. We have thus reduced our problem to one involving a smaller division algebra and a smaller conductor. On $D_1^\times$, we can extend χ as $\chi_{E_1} \circ N_{D_1/E_1}$ for some character χ_{E_1} of E_1, where N_{D_1/E_1} is the reduced norm.

If s_1 is odd, it is no longer true that the representation τ on H_1 induces to an irreducible of $D^\times$. We first need to create a representation on $J_1 = H_1 K^{s_1''}$. A set of coset representatives for J_1/H_1 is given by the elements $y_\delta = 1 + \delta\varpi^{s_1''}$, $\delta \in \bar{k}_n \cap D_1^\perp$. Let $\tilde{\tau_1}$ be given on J_1 by $\tilde{\tau_1}\,(y_\delta w) = \tau(w)$ for $w \in H_1$. Then $\tilde{\tau_1}$ is a projective representation of J_1, and

$$(\delta, \varepsilon) \mapsto \chi(1 + (\delta\varepsilon^{\sigma_1^{s_1''}} - \varepsilon\delta^{\sigma_1^{s_1''}})\varpi^{s_1''} = \chi(y_\delta y_\varepsilon y_\delta^{-1} y_\varepsilon^{-1})$$

is a nondegenerate bilinear form on $\bar{k}_n \cap D_1^\perp$ (or, more properly, on the image of $\bar{k}_n \cap D_1^\perp$ in k_n). There is a canonical projective representation W of J_1, trivial on $H_1 \cap K^1$, such that $\tilde{\tau_1} \otimes W$ is an (ordinary) irreducible representation on J_1. Induce to get π.

We will prove later that the ramification index $e(E_1/F)$ is given by $e_1 = n/(n, s_1)$, where $(\ , \ )$ is the greatest common denominator. Write $f_1 = f(E_1/F)$; $f_1 \mid s_1$.

Since $p \nmid [D_1 : E_1]$, the representations of $D_1^\times$ are analyzed similarly. Thus we have an inductive method for constructing $(D_n^\times)\hat{}$. Howe showed (for $GL_n(F)$, in [12]) how to combine the inductive information. Given π, there are triples $(e_1, f_1, s_1), \ldots, (e_r, f_r, s_r)$, fields $F = E_0 \subsetneqq E_1 \subsetneqq \cdots \subsetneqq E_r$, associated division algebras D_i (= commutant of E_i), and a character χ' of $E_r^\times$, such that:

(1.4) (1) $s_1 > \cdots > s_r \geq 0$;

(2) $1 = e_0 | e_1 | \cdots | e_r$, $1 = f_0 | f_1 | \cdots | f_r$, and $e_r f_r \mid n$;

(3) $e_i f_i > e_{i-1} f_{i-1}$ for $1 \leq i \leq r$, and $e_r = e_{r-1}$ if $s_r = 0$;

(4) $e_i = n/(n/e_{i-1}, s_i)$ and $f_r \mid s_i$, for all i;

(5) $e(E_i/F) = e_i$, and $f(E_i/F) = f_i$;

(6) $\chi'|_{E_r \cap K^{s_i+1}}$ factors through $N_{E_r/E_{i-1}}$, but the smallest field $E \supseteq F$ such that $\chi'|_{E_r \cap K^{s_1}}$ factors through $N_{E_r/E}$ is E_i.

If $r = 0$ (so that the sequence is empty), then χ' is a character of F and the representation of $D^\times$ can be taken to $\chi' \circ N_{D/F}$.

To construct π from the data, one lets $s_i' = [s_i/2] + 1$, $s_i'' = s_{i+1} - s_i'$ (as before, $s_i' = s_i'' = (s_i + 1)/2$ if s_i is odd, $s_i' = (s_i/2) + 1$, $s_i'' = s_i/2$ if s_i is even). Set $H = K^{s_1'}(K^{s_2'} \cap D_1^\times) \cdots (K^{s_r'} \cap D_{r-1}^\times) D_r^\times$. Construct a character χ on H as follows: $\bmod K^{s_1+1}$, H is the semidirect product of a normal subgroup N with $\tilde{H} = (K^{s_2'+1} \cap D_1^\times) \cdots (K^{s_r'+1} \cap D_{r-1}^\times) D_r^\times$. On $(K^{s_i'+1} \cap D_{i-1}^\times)$, $\chi = (\chi' \circ N_{D_{i-1}/E_{i-1}})^{n/e_{i-1}}$ is a character, and these characters agree on their common domain. This gives a character on $\tilde{H}$; make it 1 on N to get χ on H. If the $s_i > 0$ are all odd, $\mathrm{Ind}_H^{D^\times} \chi$ is irreducible; if some of the s_i are even and positive, we may need to tensor with Weil-type representations as above. The representation π determines the triples (e_i, f_i, s_i), and if $\tilde{E}_1, \ldots, \tilde{E}_r$, $\tilde{\chi}'$ also give π, then $E_r \cong \tilde{E}_r$ by an isomorphism taking each E_i to $\tilde{E}_i$ and χ' to $\tilde{\chi}'$.

This description of the data is highly redundant, since E_r and χ determine the intermediate E_i and the triples (s_i, e_i, f_i), Howe's description thus involved only E and appropriate ("admissible") χ. (Moy [17] gives a good description for the division algebra case.)

The description of π and of $(D^\times)\hat{}$ is similar in the general case, but is more complicated. There are three basic problems:

(a) We don't have a decomposition like (1.3).

(b) There are often too many fields. In the tame case, the element $\alpha\varpi^{-m}$ that was used to start the construction has the following property: if $x \equiv x' \bmod P^{1-m}$, then a conjugate of x' commutes with x. (That's why tame ramification is tame.) In the general case, however, x and x' may generate different extensions of the same degree.

(c) There are often too few fields. In the tame case, if $[E\colon F]$ is composite, then there exists a proper intermediate field. This is false in general.

As a consequence, it seems difficult to associate π with a character on some $E_r^\times$. Instead, we get the following:

1.5. Theorem. *Let π be an irreducible representation of $D^\times$. Then there are triples $(e_i, f_i, s_i), \cdots, (e_r, f_r, s_r)$, fields E_i $(1 \le i \le r)$, associated division algebras D_i (= commutant of E_i), $1 \le i \le r$, and a character χ on a subgroup with the following properties (where s_i', s_i'' are defined as before):*

(1) $s_1 > \cdots > s_r \ge 0$.
(2) $1 = e_0 | e_1 | \cdots | e_r$, $1 = f_0 | f_1 | \cdots | f_r$, *and* $e_r f_r \mid n$.
(3) $e_i f_i > e_{i-1} f_{i-1}$ *for* $1 \le i \le r$, *and* $e_r = e_{r-1}$ *if* $s_r = 0$.
(4) $e_i = n/(n/e_{i-1}, s_i)$ *and* $f_r \mid s_i$, *all* i.
(5) $e(E_i/F) = e_i$, *and* $f(E_i/F) = f_i$.
(6) *Define* s_i', s_i'' *as before. Then*

$$H = K^{s_1'}(K^{s_2'} \cap D_1^\times) \ldots (K^{s_r'} \cap D_{r-1}^\times) D_r^\times,$$

and, for each i, $K^{s_1'}(K^{s_2'} \cap D_1^\times) \cdots K^{s_1'}(K^{s_i'} \cap D_{i-1}^\times)$ is a normal subgroup.

(7) *Write $\chi^x(y) = \chi(xyx^{-1})$ on $x^{-1}Hx$. Then $\chi^x = \chi$ on their common domain $\Leftrightarrow x \in J = K^{s_1''}(K^{s_2''} \cap D_1^\times) \cdots (K^{s_r''} \cap D_{r-1}^\times) D_r^\times$. There is a canonical way of getting a representation σ on J such that $\sigma\big|_{H \cap K^1}$ is a multiple of $\chi\big|_{H \cap K^1}$: one extends to a projective representation on J and tensors with a Weil-type representation.*

(8) $\mathrm{Ind}_J^{D^\times} \sigma = \pi$.

The number of generic representations π trivial on ϖ^n with a given sequence of triples $(s_1, e_1, f_1), \ldots, (s_r, e_r, f_r)$ is $(n/f_r) \prod_{j=0}^{s_1} C_j$, where

$$C_j = \frac{f_{i-1}}{f_i} \sum_{d | f_i / f_{i-1}} (q^{f_{i-1} d} - 1) \mu(f_i / f_{i-1} d) \text{ if } j = s_i \ (\mu = \text{Möbius function})$$

$$C_j = 1 \text{ if } s_{i-1} < j < s_i \text{ and } \frac{n}{e_i} \nmid j, \text{ or if } 0 < j < s_r \text{ and } \frac{n}{e_r} \nmid j;$$

$$C_j = q^{f_i} \text{ if } s_{i-1} < j < s_i \text{ and } \frac{n}{e_i} \mid j, \text{ or if } 0 < j < s_r \text{ and } \frac{n}{e_r} \mid j;$$

$$C_j = q^{f_r} - 1 \text{ if } j = 0 < s_r.$$

This description is incomplete in some respects. For example, the E_i need to fit together well, so that the set H in (6) is indeed a group. One way to make this precise is to recall that D is generated by F_n and ϖ; similarly, D_i will be generated by F_{n/e_i} and an element η_i such that

(i) η_i generates P^{f_i};
(ii) η_i normalizes F_{n/e_i}, and conjugation by η_i is σ^{f_i} on F_{n/e_i};
(iii) $E_i = F_{f_i}[\eta_i^{n/e_i f_i}]$, and η_i is uniformizing for E.

The property of fitting together well means that η_i is congruent to an element of $D_{i-1} \bmod P^{f_i+s_{i-1}-s_i}$.

The E_i are generally not determined by π, but for any other possible choice E_i', we have $e(E_i'/F) = e_i$ and $f(E_i'/F) = f_i$. Furthermore, E_i and E_i' are "close" in the sense, for instance, that if $0 \neq x' \in E_i'$, then there exists $x \in E$ with $x^{-1}x' \in K^{s_i - s_{i+1}}$.

Thus we end up with a list of representations, but not a very good parametrization of the list.

2. General Remarks on the Proof.

The construction of the irreducibles of $\hat{G} = D^\times$ is long and complicated, and it is easy to get lost in the details. In this section I will try to give an idea of the strategy of the proof.

Except for certain easy special cases (e.g., representations trivial on K), the construction works as follows: we construct a group H_0 and a character χ on H_0 such that $J_0 = \{x \in G : \chi^x = x \text{ on } H_0 \cap x^{-1}H_0x\}$ is a group and constructing all irreducibles of J_0 whose restriction to H_0 is a multiple of χ is easy. In particular, J_0 will normalize H_0. One especially nice case is $J_0 = H_0D_0^\times$, where D_0 is a sub-division algebra of D, χ extends to a character of J_0, and $H_0 \cap D_0^\times = K^m \cap D_0^\times$ for some m. In that case, all the representations of J extending χ are of the form $\chi^\sim \otimes \rho$, where $\chi^\sim$ denotes an extension of χ to J and ρ is an irreducible of $D_0^\times$ trivial on $D_0^\times \cap K^m$. Since D_0 will have smaller degree than D, we may assume inductively that the ρ are known. It is then easy to see that the $\chi^\sim \otimes \rho$ induce to distinct irreducibles of $D^\times$. In general J_0 is not quite so nice, and there is a further step involving the Heisenberg-type representations mentioned in the tame case. Of course, there will be lots of pairs (H_0, χ) to construct.

The construction of (H_0, χ) begins as follows: we assume that χ is trivial on K^{s_1+1} but not on K^{s_1}, and that $\chi|_{K^{s_1}}$ is not equal to any character of $D^\times$ restricted to s_1. Then χ is given on K^{s_1} by

$$\chi(1+\gamma\varpi^{s_1}) = \psi \circ Tr_{k_n/k}(\alpha_{s_1}\gamma^{\sigma^{-s_1}}) = \psi \circ Tr_{k_n/k}(\alpha_{s_1}^{\sigma^{s_1}}\gamma)$$

for some unique $\alpha_1 \in k_n$. Because χ does not extend to a character of $D^\times$, $\alpha_1\varpi^{-s_1}$ is not central.

It is not hard to prove the following fact (a more precise version is given in Prop. 3.12):

2.1. Suppose we extend χ to K^j for each $j < s_1$, and assume that

$$x \equiv 1 + \delta\varpi^{s_1-j} (\operatorname{mod} K^{s_1-j+1})$$

satisfies $\chi^x = \chi$ on K^j. Then $[\delta\varpi^{s_1-j}, \alpha_1\varpi^{s_1}] = 0$.

This means that we cannot extend χ to all of K^j if $j \leq s_1'$. For then $x = 1 + \delta\varpi^{s_1-j} \in K^j$, for all $\delta \in \bar{k}_n$. If $x \in \operatorname{Dom}\chi$, we must have $\chi^x = \chi$; but $\chi^x = \chi$ implies that x and $\alpha_1\varpi^{s_1}$ commute. This gives a restriction on

H_0. On the other hand, there is no obstacle to extending χ to K^j if $2j > s_1$. This suggests that $H_0 \supseteq K^{s'_1}$. However, $\{x : \chi^x = \chi \text{ on } K^{s'_1}\}$ will usually be a complicated group whose representations are not easy to describe. So generally $H_0 \not\supseteq K^{s'_1}$.

As we go along, H_0 will depend to some extent on χ. In order to avoid circularities, we proceed as follows: let $H_0^j = H_0 \cap K^j$. From time to time, we define H_0^j for certain j; we then say that H_0 is defined "up to" level j. This is certainly so when $\chi|_{H_0^{\tilde{j}}}$ is defined for some $\tilde{j} \leq j$. If it ever happens that the definition of χ "catches up" to that of H_0, in that H_0 is defined up to level j but not further, χ is defined on H_0^j, and we do not immediately define H_0 up to some $j^{\#} < j$, then the construction stops, $H_0 = H_0^j$, and we have defined (H_0, χ).

We begin by defining $H_0^{s'_1} = K_0^{s'_1}$. (Recall: s'_1 = smallest integer with $2s'_1 > s_1$.) Unless $s_1 \leq 2$, $s'_1 < s_1$. Thus χ, which is defined only on K^{s_1}, is not yet defined on $H_0^{s'_1}$ if $s_1 > 2$. As we go on, we extend the definition of H by setting $H_0^{s'_2} = H_0^{s'_1}(K_0^{s'_2} \cap D_1^\times)$, $H_0^{s'_3} = H_0^{s'_2}(K_0^{s'_3} \cap D_2^\times), \ldots$, where s'_2, s'_3 are defined in the construction and the D_i are sub-division algebras of D. Of course, we must verify that we do not change H_0^j when we extend H_0 beyond level j.

We also extend χ to H^{s_1-1}, then to H^{s_1-2}, and so on. Suppose that χ is defined on H_0^{j+1} and that H_0 is defined at least up to level j. We want to extend χ to H_0^j. The first question to answer is whether χ has any extension to H_0^j. The answer will be yes; in fact, we will know of one particularly nice extension χ_0 to H^j, and this extension will have useful properties that help us determine properties of the extension χ that is eventually chosen. We also need to know exactly what other extensions there are. This will amount to knowing H^j/H^{j+1} (which will be Abelian).

We now need to do the following:

(a) describe which extensions of χ we want to use;
(b) determine whether or not we want to set j = one of the s_i, and, if we do, determine e_i and f_i;
(c) determine whether we want to extend the definition of H_0 beyond the level to which it was defined;
(d) compute $J_0^j = \{x \in G : \chi^x = \chi \text{ on } H^j \cap x^{-1}H^jx\}$.

Then we are ready to continue the induction.

As a result, we need to carry some information along inductively. In addition, other technical problems in the proof means that we need additional information:

1. Suppose that when we defined χ on H_0^{j+1}, we had defined $s_1, \cdots, s_i$ (hence $e_1, \cdots e_i$, $f_1, \cdots f_i$). This will turn out to mean that the largest division algebra $D_{(j+1)}$ such that

$$x \in D_{(j+1)}^\times \text{ implies } \chi^x = \chi$$

has dimension n^2/e_if_i over F, and its center $E_{(j+1)}$ has ramification index e_i and residue class degree f_i over F. ("Largest" here means only that $[D_{(j+1)}\colon F]$ is maximal; there may be many choices for $D_{(j+1)}$, but we will have chosen one.) We need to know what $D_{(j+1)}$ is.

2. When we extend χ to H_0^j it may be that the largest division algebra $D_{(j)}$ such that $x \in D_{(j)}^\times$ implies $\chi^x = \chi$ is of the same dimension as $D_{(j)}$, or it may be smaller. We need to know which is the case, and we need to know $D_{(j)}$.

3. It turns out that we define $j = s_{i+1}$ iff $[D_{(j)}\colon F] < [D_{(j+1)}\colon F]$. Then we need to define e_i, f_i (their definitions are determined by what we said in 2). This means checking the arithmetic facts (Theorem 1.5, (1)–(4)) about (s_i, e_i, f_i).

4. It also turns out that H_0 has been defined up to level s_i' (= smallest integer with $2s_i' > s_i$). If $j = s_{i+1}$, we define H up to level S_{i+1}' by

$$H_0^{s'_{i+1}} = H_0^{s'_i}(D_{(j+1)}^\times \cap K^{s'_{i+1}}).$$

We need to check that this is consistent with what went before (in essence, that $D_{(j+1)}^\times \cap K^{s'_i} \subseteq H_0^{s'_i}$). Therefore we need information about the elements of $D_{(j+1)}^\times$.

5. In computing J_0^j, we will of course need to know J_0^{j+1}. We will need some information about χ at the levels $H^{s_1}, H^{s_2}, \cdots$. Suppose for definiteness that $j = s_{i+1}$. We have χ and the reference character χ_0 on H_0^j; the information we need is about $\chi \cdot \chi_0^{-1}$, a character on H^j/H^{j+1}.

This may help to explain why the main theorem in constructing (H_0, χ) is long and complex. Much of its proof consists of small inductions involving calculations of similar types, and I have tried to isolate most of these as separate lemmas.

As the above outline shows, it is vital to have information about the division algebras $D_{(j)}$. Two other points deserve special attention. First, we need to approximate elements in one of these algebras by elements in another. (For example, we need to know that for every $x \in D_{(j)}^\times$, there exists $x \in D_{(j+1)}$ with $xx_1^{-1} \in K^1$.) Second, we need an efficient way of proving that, for example, all elements of $D_{(j)}^\times$ commute with χ on H_0^j. $D_{(j)}$ may not be large as a division algebra, but any description of $D_{(j)}^\times$ as a group is fairly complicated.

Thus the next section is devoted to an analysis of sub-division algebras of D.

3. Sub-division Algebras of Division Algebras.

Let $D = D_n$ be a division algebra of dimension n^2 over its center F, where F is p-adic. There are a few facts about the structure of D that we shall use (all are proved in [21]).

3.1. If E is any extension of F with $[E: F] = n$, then there is an embedding of E in D. In particular, F_n (the unramified extension $/F$ of degree n) embeds in D. Fix an embedding. There exists an element $\varpi \in D^\times$ such that (i) ϖ^n is a prime element of F; (ii) ϖ normalizes F_n; (iii) the map $x \mapsto \varpi x \varpi^{-1} = x^\sigma$ is an automorphism of F_n generating $Gal(F_n/F)$. (As noted earlier, σ determines D up to isomorphism.) In fact, for every prime element ϖ_F of F we can choose ϖ so that (i)–(iii) hold with $\varpi^n = \varpi_F$.

3.2. F_n and ϖ generate D. In fact, if $\{a_1, \cdots a_n\}$ is an F-basis for F_n, then $\{a_i \varpi^j : 1 \leq i,\ j \leq n\}$ is an F-basis for D.

3.3. Let $\bar{k}_n = \{a \in F_n : \alpha^{q^n} = \alpha)$, where q is the cardinality of the residue class field k of F. It follows from (3.2) that every nonzero element of E has a unique "power series" expansion as in (1.2). Let $\nu(x) = j_0$, $\nu(0) = \infty$. Then ν is a valuation on D, and $\nu|_F$ is n times the "usual" valuation. As in section 1, set $\mathcal{O} = \mathcal{O}_D = \{x \in D : \nu(x) \geq 0\}$, and $P^j = P^j_D = \{x \in D : \nu(x) \geq j\}$, for $j \in \mathbb{Z}$; then $\varpi\mathcal{O} = \mathcal{O}\varpi = P = P_D = P^1_D$. Furthermore, $\mathcal{O}_D/P_D \cong k_n$ (the finite field with $[k_n : k] = n$), and this map takes $\bar{k}_n$ bijectively to k_n.

3.4. If $x, y \in D$ satisfy the same minimal equation $/F$, then there is an element $w \in D$ with $wxw^{-1} = y$. (This follows from the Skolem–Noether Theorem.)

The following result is not new, but I don't know a convenient reference.

3.5. Proposition. *Let E_0 be any extension of F contained in D, and let D_0 = commutant of F_0 in D. Suppose $e(E_0/F) = e_0$, $f(E_0/F) = f_0$. Then*

(i) $e_0 f_0 \mid n$.

(ii) $[D_0 : E_0] = (n/e_0 f_0)^2$ *and* $[D_0 : F] = n^2/e_0 f_0$.

(iii) *The maximal unramified extension of F in D_0 is of degree n/e_0 over F, and $\nu(D_0^\times) = f_0\mathbb{Z}$.*

Proof. Let E_1 be the maximal unramified extension of F in E_0, so that $[E_1 : F] = f_0$. The valuation ν restricts to a multiple of the standard valuation on E_1, and therefore $E_1 \cap \mathcal{O}_D / E_1 \cap P_D \cong$ residue class field of $E_1 \cong k_{f_0}$ (the extension of k of degree f_0). Since $\mathcal{O}_D/P_D \cong k_n$, k_{f_0} is a subfield of k_n; therefore $f_0 \mid n$. From (3.4), we may assume that $E_1 = F_{f_0}$, the subfield of F_n with $[F_{f_0} : F] = f_0$. Let ξ_0 be a prime element of E_0, and suppose that $\nu(\xi_0) = r$. Then $\xi_0 = \sum_{j \geq r_0} \alpha_j \varpi^j$, with $\alpha \neq 0$. Let $\gamma \in \bar{k}_{f_0}$ (= pre-image of k_{f_0} in $\bar{k}_n$). Then $\gamma\xi \equiv \alpha_r \gamma \varpi^r \bmod P^{r+1}$, while $\xi\gamma \equiv \alpha_r \gamma^{\sigma^r} \varpi^r \bmod P^{r+1}$. Therefore $\gamma^{\sigma^r} = \gamma$ for all $\gamma \in \bar{k}_{f_0}$ because $\gamma\xi = \xi\gamma$. So σ^r fixes F_{f_0} (since $\bar{k}_{f_0}$ generates F_{f_0}), so that $f_0 \mid r$. Since $\nu(E_0^\times) = r\mathbb{Z}$, $e_0 = e((E_0^\times)F) = n/r$ divides n/f_0. That proves (i).

Now let E_0' be a maximal unramified field in D_0. Since E_0' is a field in D containing F and satisfying $e(E_0'/F) = e(E_0'/E_0)e(E_0/F) = e_0$, (i) implies that $f(E_0'/E) \leq n/e_0$. For the other inequality, there exists an unramified extension $E_0^{\sim}$ with $[E_0^{\sim}: E_0] = n/e_0 f_0$. Then $[E_0^{\sim}: F] = n$, so that $E_0^{\sim}$ has an embedding in D_0, and $f(E_0^{\sim}) = n/e_0$. Therefore $f(E_0'/F) = n/e_0$. From (3.4) we may assume $F_{n/e_0} \subseteq E_0'$, where $F_{n/e_0} \subseteq F_n$. This is half of (iii).

Since any element of $D_0^{\times}$ commutes with F_{f_0}, the same argument used in (i) shows that for $x \in D_0^{\times}$, $\nu(x)$ is a multiple of f_0. On the other hand, $E_0[\xi^{e_0 f_0/n}]$ is a totally ramified extension of E_0 whose degree over F is n. Therefore this field embeds in D. From (3.4), we may assume that the embedding contains E_0, so that it is in D_0. Let $\eta = \xi^{e_0 f_0/n}$. Then

$$\nu(\eta) = \nu(\xi)\frac{e_0 f_0}{n} = \frac{n}{e_0} \cdot \frac{e_0 f_0}{n} = f_0.$$

To prove (ii) and the second part of (iii), it now suffices to show that the elements $\eta, \dots, \eta^{n/e_0 - 1}$ form an F_{n/f_0}-basis for D_0. From valuation considerations, these elements are linearly independent. If they fail to span, then we can find $y \in D_0$ such that (a) $y \notin F_{n/f_0}\eta^0 + \cdots F_{n/f_0}\eta^{n/e_0-1}$; (b) $y \in \mathcal{O}_D \cap D_0$; (c) for $w \in F_{n/f_0}\eta^0 + \cdots + F_{n/f_0}\eta^{n/e_0-1}$, $\nu(y-w) = \nu(y)$. Let $\nu(y) = jf_0 + h$, where $j \in \mathbb{Z}$, $j \geq 0$, and $0 \leq h < f_0$. Then $\nu(y\eta^{-j}) = h$. If $h \neq 0$, then $[E[y]: E_1] \geq n/(n,h) > n/f_0$ and $[E[y]: F] > n$, which contradicts (3.1). Therefore $h = 0$ and $y\eta^{-1} \equiv \alpha \in \bar{k}_n \bmod P_0$ for some nonzero α. If $\alpha \in \bar{k}_{n/e_0}$, $w = \alpha\eta^j$ violates (c). If $\alpha \notin \bar{k}_{n/e_0}$, then a Hensel's lemma argument shows that D_0 contains an element $\alpha^{\sim}$ generating a field $\cong F[\alpha]$. This violates the first part of (iii). □

Remark. Once we have conjugated so that $F_{n/e_0} \subseteq D_0$, an argument like the one for part (i) shows that in the expansion of ξ, $\xi = \sum_{j=n/e_0}^{\infty} \alpha_j \varpi^j$, the only nonzero α_j have $(n/e_0) \mid j$. From (3.2), we can then choose η so that conjugation by η induces σ^{f_0} on F_{n/e_0}. (Notice that σ^{f_0} generates $Gal(F_{n/e_0}/F_{n/f_0})$.) We have $\eta = \sum_{j=f_0}^{\infty} \beta_j \varpi^j = \beta_{f_0}\varpi^{f_0}(1 + \sum_{j=1}^{\infty} \gamma_j \varpi^j)$, say; since $\beta_{f_0}\varpi^{f_0}$ induces σ^{f_0} on K_{n/e_0}, another argument like that for (i) shows that $\gamma_j = 0$ unless $(n/n_0) \mid j$ (hence $\beta_j = 0$ unless $j \equiv f_0 \bmod n/e_0$).

In the previous proposition we began with D_0 and constructed the generators η, F_{n/k_0}. The following proposition reverses that procedure.

3.6. Proposition. *Let $e_0 f_0 \mid n$. Suppose that $\eta_0 \in D$ satisfies*

(i) $\nu(\eta_0) = f_0$;

(ii) *η_0 normalizes F_{n/e_0}, and conjugation by η_0 is σ^{f_0} on F_{n/e_0}.*

Let D_0 be the division algebra generated by η_0 and F_{n/e_0}, and let E_0 be its center. Then

(a) $E_0 = F_{f_0}[\eta_0^{n/e_0 f_0}]$;

(b) $e(E_0/F) = e_0$ *and* $f(E_0/F) = f_0$;

(c) $[D_0: E_0] = (n/e_0 f_0)^2$ *and* $[D_0: F] = n^2/e_0 f_0$.

Proof. Regard D_0 as an F_{n/e_0}-vector space for now. Since $1, \eta_0, \dots, \eta_0^{n/f_0-1}$ are linearly independent over F_{n/e_0} (because $\nu(\eta_0^j) = f_0 j$ and the

$f_0 j$, $0 \le j < n/f_0$ are incongruent mod n), its dimension is $\ge n/f_0$. Hence $[D_0 : F] \ge n^2/e_0 f_0$. Next, consider $F_{f_0}[\eta_0^{n/e_0 f_0}]$. F_{f_0} commutes with F_{n/e_0} and with η_0, since σ^{f_0} is trivial on F_{f_0}; $\eta_0^{n/e_0 f_0}$ commutes with η_0 and with F_{n/e_0}, since $(\sigma^{f_0})^{n/e_0 f_0} = \sigma^{n/e_0}$ is trivial on F_{n/e_0}. Set $E_0' = F_{f_0}[\eta_0^{n/e_0 f_0}]$; then $E_0' \subseteq E_0$. Clearly $e(E_0'/F) \ge e_0$, since $\nu(\eta_0^{n/e_0 f_0}) = n/e_0$; also, $f(E_0'/F) \ge f_0$. Hence $e(E_0/F) \ge e_0$ and $f(E_0/F) = f_0$. From Proposition 3.5, $[D_0 : F] \le n^2/e_0 f_0$. Thus $[D_0 : F] = n^2/e_0 f_0$ and Proposition 3.5 gives $[E_0 : F] = e_0 f_0$. Now the rest of the proposition follows easily. □

We next turn to the problem of proving that all nonzero elements of a sub-division algebra $D_0^\times$ commute with a character χ. Suppose that χ is defined on a subgroup H of K^1 and that $D_0^\times$ normalizes H. Let D_0 be given as in Prop. 3.6, so that F_{n/e_0} and η_0 generate $D_0^\times$, and suppose that $\chi^x = \chi$ when $x \in \bar{k}_{n/e_0}$ $(x \neq 0)$ and when $x = \eta_0$. (Recall: $\chi^x(y) = \chi(xyx^{-1})$.) From the power series expansion of x and the fact that $\chi \equiv 1$ on a neighborhood of 1, we see that to show that $\chi^x = \chi$ for all $x \in D^\times$, it suffices to prove this for $x = 1 + \beta\eta_0^j$, $\beta \in \bar{k}_{n/e_0}$. We know that $x_0 = \beta\eta_0^j$ commutes with χ. Thus it suffices to show that for $y \in H$, $xyx^{-1}y^{-1}$ can be written as a product of commutators involving the x_0's. The following result answers our needs.

3.7. Lemma. *Let $\mathcal{A} = \mathbb{Z}[[a,b]]$, the algebra of formal power series in two noncommutating variables a, b, and let $B = [a^{-1}, b^{-1}]$. For integers $r, s > 0$, assign a weight to each word in $\mathcal{A}$ by giving a weight r and b weight s, and summing the weights of the letters in a word to get the weight of the word (thus $abab^2$ has weight $2r + 3s$). Let $\mathcal{A}_n$ =* ideal *in $\mathcal{A}$ generated by all words of length (not weight) n. Consider:*

$$\begin{aligned} x &= (1+a)(1+b)(1+a)^{-1}(1+b)^{-1} \\ &= (1+a)(1+b)(1-a+a^2-\cdots)(1-b+b^2-\cdots). \end{aligned}$$

For each n, there exist k_n and $c_1, \ldots, c_{k_n}, d_1, \ldots, d_{k_n} \in \mathcal{A}$ such that

(i) *x is congruent mod $(1 + \mathcal{A}_n)$ to the product of commutators*

$$(c_1, d_1) \cdots (c_{k_n}, d_{k_n});$$

(ii) *$d_j - 1$ is a word, c_j is one of a, a^{-1}, b, b^{-1}, and $c_j(d_j - 1)c_j^{-1} \in \mathcal{A}$,*

(iii) *if $d_j - 1$ has weight $\le 2s$, then $c_j = a$ or a^{-1}.*

(One can replace (iii) by (iii′): if $d_j - 1$ has weight $\le 2r$, then $c_j = b$ or b^{-1}.)

Proof. This is essentially what is proved in Lemma 2.2 and Remark 2.3 of [5]. The idea is this: assume the result for n, and consider the words of length $n+1$ in $x\left(\prod_{j=1}^{k_n}(c_i, d_i)\right)^{-1}$ $(= x_n$, say):

$$x_n \equiv 1 + \sum_{j=1}^{r} m_j w_j \bmod \mathcal{A}_{n+2}.$$

Suppose that w_j begins with an "a", i.e., $w_j = av_j$. Then

$$a(1+v_ja)a^{-1}(1+v_ja)^{-1} \equiv 1+(v_ja-w_j) \bmod \mathcal{A}_{n+1}.$$

Similar calculations (using the last letter also) show that multiplying by the permitted commutators can cyclically permute the w_j. (If w_j has weight $\leq 2s$, then either $w_j = b^2$ or the word for w_j has at most one b; in either case, we can get all cyclic permutations of w_j by moving only a's. That explains (iii).) So if we can show that after cyclically permuting some of the w_j we have $x_n \equiv 1 \bmod \mathcal{A}_{n+2}$, we are done. We prove this property of x by mapping $\mathcal{A}$ into an appropriate division algebra D_1 and showing that x cannot be a commutator unless it has this property. The details are found in §2 of [5]. □

We typically use Lemma 3.7 in one of three ways:

3.8. Corollary. *Let H be an open subgroup of K^r and χ a character on H; let $x = 1+x_0 \in K^\ell$ normalize H. Assume that conjugation by x_0 normalizes H and fixes χ, and suppose that for all $y = 1+y_0 \in H$ with $y_0 \neq 0$, conjugation by y_0 normalizes H and fixes χ. Then conjugation by x fixes χ.*

Proof. For $y = 1+y_0 \in H$, the commutator (x,y) is a product of commutators (x_0,u), (x_0^{-1},u), (y_0,u) and (y_0^{-1},u); all of these are mapped to 1 by χ. □

3.9. Corollary. *Let H be an open subgroup of K^r and let χ be a character on H. Let $x = 1+x_0 \in K^\ell$, $y = 1+y_0 \in K^h$ satisfy:*

(i) *x_0 commutes with $\chi\big|_{K^{\ell+h}}$ and y_0 commutes with $\chi\big|_{K^{\ell+2h}}$;*
(ii) *If w is any sum of words in x_0 and y_0 of length ≥ 2 such that no word is a power of x_0 or of y_0, then $1+w \in H$.*

Then $\chi((x,y)) = \chi((y,x)) = 1$.

3.10. Corollary. *Let H be an open subgroup of K^r. Suppose that for all $y = 1+y_0 \in H \cap K^s$, y_0 normalizes $H \cap K^{2s}$. Let D_0 be a division algebra generated by F_{n/e_0} and η_0, as in Proposition 3.7; assume $D_0^\times$ normalizes H. Let χ be a character on H. If conjugation by η_0 and by every element of $k_{n/e_0}^\times$ fixes χ, then conjugation by any element of $D_0^\times$ fixes χ.*

The following result about commutators is also useful.

3.11. Lemma. *Let $x \in (D,D) \cap K^j$, $j \geq 1$. Then there are elements u_j $(1 \leq j \leq r)$, all in $\bar{k}_n^\times \cup \{\eta\}$, and elements v_j $(1 \leq j \leq r)$, all in K^j, such that $x \equiv \prod_{j=1}^r (u_jv_j) \bmod K^{j+1}$.*

This is proved as Lemma 2.1 of [5].

The next lemma, on characters of $D^\times$, is used repeatedly in the proof of Theorem 1.5. Let ψ be a character of F trivial on the prime ideal P_F

but not on the ring of integers $\mathcal{O}_F$. As noted earlier, every character of the additive group D can be written as $x \mapsto \psi \circ Tr_{D/F}(xy)$ for a unique $y \in D$ (here, Tr is the reduced trace on D). Under this identification of $D\hat{}$ and D, $(P_D^j)^\perp = P_D^{1-j}$. Recall also that ψ also defines a character on $\mathcal{O}_F/P_F \cong k$, also called ψ. Similarly, $\psi \circ Tr_{D/F}$ gives a character $\psi \circ Tr_{k_n/k}$ on $\mathcal{O}_D/P_D \cong k_n$. As stated earlier, we often identify $\bar{k}_n$ and k_n.

Let χ be a character of K^s, $s \geq 1$, that is trivial on K^{s+1}. Since $K^s/K^{s+1} \cong P_D^s/P_D^{s+1}$ under the map $1 + y \mapsto y$, we may regard χ as a character on P_D^s. Then there is a unique $\alpha \in k_n$ such that

$$\chi(y) = \psi \circ Tr_{k_n/k}\, \alpha\gamma^{\sigma^{-s}} = \psi \circ Tr_{k_n/k}\, \alpha^{\sigma^s}\gamma$$

for all $y = 1 + \gamma\varpi^s \in P_D^s$. That, of course, determines χ for all $y \in K^s$.

3.12. Proposition. *Let χ be as above, and let h satisfy $1 \leq h < s$. For $\delta \in k_n$, define a character χ_δ on K^h by*

$$\chi_\delta(y) = \chi(wyw^{-1}y^{-1}), \quad w = 1 + \delta\varpi^{s-h}.$$

Set $y = 1 + \gamma\varpi^h \in K^h$. Then:

(a) *χ_δ is trivial on K^{h+1}.*

(b) *χ_δ is trivial iff $\delta\varpi^{s-h}$ and $\alpha\varpi^{-s}$ commute.*

(c) *$\chi_\delta(y) = 1$ for all $\delta \in k_n$ iff y and $\alpha\varpi^{-s}$ commute.*

(d) *The map $\delta \mapsto \chi_\delta$ is a homomorphism of k_n onto the characters of K^h trivial on K^{h+1} and on the elements $y = 1 + \gamma\varpi^h$ commuting with $\alpha\varpi^{-s}$. In particular, if $[\delta\varpi^{s-h}, \alpha\varpi^{-s}] = 0$ only for $\delta = 0$, then the map $\delta \mapsto \chi_\delta$ is an isomorphism of k_n onto the characters of K^h trivial on K^{h+1}.*

Remarks. 1. We often apply this proposition not to the character χ on D, but on the restriction of χ to a sub-division algebra D_0 generated by k_{n/e_0} and an element η_0 generating P^{f_0} as in Prop. 3.7. Then χ is defined on $K^s \cap D_0$ by $\chi(1 + \gamma\eta_0^{s/f_0}) = \psi \circ Tr_{k_{n/e_0}/k}(\alpha\gamma^{\sigma^{-s}})$.

2. The character χ_δ is defined on $K^{s+1}(K^s \cap D_0^\times)$ and is trivial on K^{s+1}. Sometimes χ will also be defined on a larger subgroup of K^s. When this happens, the proposition of course gives information only for χ on $K^{s+1}(K^s \cap D_0^\times)$.

Proof of Proposition 3.12. Since $(K^{h+1}, K^{s-h}) \subseteq K^{s+1}$, (a) is clear. We compute ξ_δ on K^h, using the fact that

$$\begin{aligned}(1 + \gamma\varpi^h)(1 + \delta\varpi^{s-h})&(1 + \gamma\varpi^h)^{-1}(1 + \delta\varpi^{s-h})^{-1} \\ &\equiv 1 + (\gamma\delta^{\sigma^h} - \delta\gamma^{\sigma^{s-h}})\varpi^s \bmod K^{s+1} :\end{aligned}$$

$$\begin{aligned}\chi_\delta(1 + \gamma\varpi^h) &= \psi \circ Tr_{k_n/k}\alpha^{\sigma^s}(\gamma\delta^{\sigma^h} - \delta\gamma^{\sigma^{s-h}}) \\ &= \psi \circ Tr_{k_n/k}\gamma(\delta\alpha^{\sigma^{s-h}} - \alpha\delta^{\sigma^{-s}})^{\sigma^h}.\end{aligned}$$

This is 1 for all γ iff $\delta\alpha^{\sigma^{s-h}} = \alpha\delta^{\sigma^{-s}}$, or iff $[\delta\varpi^{s-h}, \alpha\varpi^{-s}] = 0$. That proves (b). We also have

$$\chi_\delta(1+\gamma\varpi^h) = \psi \circ Tr_{k_n/k}\, \delta(\alpha\gamma^{\sigma^{-s}} - \gamma\alpha^{\sigma^h})^{\sigma^{s-h}}.$$

This is 1 for all δ iff $\alpha\gamma^{\sigma^{-s}} = \gamma\alpha^{\sigma^h}$, or iff $[\gamma\varpi^h, \alpha\varpi^{-s}] = 0$, and (c) follows.

Finally, let D_0 be the division algebra of elements commuting with $\alpha\varpi^{-s}$. Suppose that $e(F[\alpha\varpi^{-s}]/F) = e_0$ and $f(F[\alpha\varpi^{-s}]/F = f_0$. Then for $j \geq 1$, $[K^j \cap D_0^\times : K^{j+1} \cap D_0^\times] = 1$ if $f_0 \nmid j$ and $= q^{n/e_0}$ (where $q = \text{card}\, k$) if $f_0 \mid j$, from the remark after Proposition 3.5. But $f_0 \mid h$ iff $f_0 \mid s-h$ (because $f_0 \mid s$), so that Card $(\text{Ker}(\delta \mapsto \chi_\delta)) = \text{Card}\,\{\gamma \in k_n\colon 1+\gamma\varpi^h \in D_0^\times\}$. That proves (d). □

Note: We sometimes need a result related to (b) of Prop. 3.12. Let χ be defined as above, and let $y_0 = \delta\varpi^h$ with $\delta \in k_n^\times$; define

$$\chi_0(x) = \chi(y_0 x y_0^{-1}).$$

Then $\chi_0 = \chi$ iff y and $\alpha\varpi^{-s}$ commute. The calculation is quite similar to the one for (b): if $x = 1 + \gamma\varpi^s$, then

$$y_0 x_0 y_0^{-1} x_0^{-1} \equiv 1 + (\delta\gamma^{\sigma^h}(\delta^{-1})^{\sigma^s} - \gamma)\varpi^s,$$

so that

$$\begin{aligned}\chi_0(x)\chi(x)^{-1} &= \psi \circ Tr_{k_n/k}\, \alpha^{\sigma^s}[\delta\gamma^{\sigma^h}(\delta^{-1})^{\sigma^s} - \gamma] \\ &= \psi \circ Tr_{k_n/k}\, \gamma[(\delta^{-1})\alpha\delta^{\sigma^{-s}} - \alpha^{\sigma^h}]^{\sigma^{s-h}}.\end{aligned}$$

This is 1 for all γ iff the term in brackets is 0, and we conclude as in (b). We shall use this (regarding it as a part of Prop. 3.12) in what follows. It is also applied to sub-division algebras (see the remark above).

The next sort of result we need might be described as an approximation lemma involving different division algebras. We assume we have division algebras $D = D_0, D_1, \cdots, D_i$, with respective centers $F = E_0, E_1, \cdots, E_i$, that $e(E_h/F) = e_h$ and $f(E_h/F) = f_h$, and:

(a) $1 = e_0 | e_1 | \cdots | e_i$ and $1 = f_0 | f_1 | \cdots | f_i$;
(b) E_h is generated by F_{f_h} and an element η_h generating P^{f_h} such that conjugation by η_h induces σ^h on F_{f_h}.

3.13. Lemma. *Let the D_h, E_h, and η_h be as above. Suppose, in addition, there are numbers $s_1 > \cdots > s_i > 0$ such that, for $h \geq 2$,*

$$\eta_h = \delta_h \eta_{h-1}^{f'_h}(1+\zeta_{(h,h-2)})\cdots(1+\zeta_{(h,1)})(1+\zeta_{(h,0)}),$$

where $f'_h = f_h/f_{h-1}$ and $\zeta_{h,j} \in P^{s_{j+1}-s_h}$. Then:

(i) *For every $x_h \in D_h$, there are elements $x_{h-1} \in D_{h-1}, \ldots, x_0 \in D_0$ such that $x_h = \sum_{j=0}^{h-1} x_j$ and $\nu(x_j) = \nu(x_h) + s_{j+1} - s_h$, $0 \leq j \leq h-1$.*

(ii) *For every $x_h \in K^r$, there are elements $y_{h-1} \in D_{h-1}, \dots, y_0 \in D_0$ such that $x_h = y_{h-1} \cdots y_0$ and $y_j \in K^{s_h - s_{j+1} + r}$ for $j > h$.*

(iii) *If $x_h \in D^\times$ and $\nu(x_h) \neq 0$, then there are elements $z_{h-1} \in D_{h-1}$, $\dots$, $z_0 \in D_0$ with $\nu(z_{h-1}) = \nu(x_h)$, $z_j \in K^{s_{j+1} - s_h}$ for $0 \leq j \leq h-2$, and $x_h = z_{h-1} \cdots z_0$.*

(iv) *Suppose that $r_1, r_2 \in \mathbb{Z}$ and $h \leq j \leq i$. Then*

$$\begin{aligned}(P^{r_1} \cap D_h)(P^{r_2} \cap D_j) \subseteq &(P^{r_1+r_2} \cap D_{h-1}) \\ &+ (P^{r_1+r_2+s_{h-1}-s_h} \cap D_{h-2}) \\ &+ \dots + (P^{r_1+r_2+s_2-s_h} \cap D_1) \\ &+ P^{r_1+r_2+s_1-s_h}\end{aligned}$$

and similarly for $(P^{r_2} \cap D_j)(P^{r_1} \cap D_h)$.

(v) *If $r_1, r_2 \in \mathbb{Z}$ and $h < j \leq i$, then*

$$\begin{aligned}(P^{r_1} \cap D_h)(P^{r_2} \cap D_j) \subseteq &(P^{r_1+r_2} \cap D_h) \\ &+ (P^{r_1+r_2+s_h-s_{h+1}} \cap D_{h-1}) \\ &+ \cdots + P^{r_1+r_2+s_1-s_{h+1}}\end{aligned}$$

and similarly for $(P^{r_2} \cap D_j)(P^{r_1} \cap D_h)$.

(vi) *For each $r > 0$ and each $h \leq i$,*

$$\begin{aligned}1 + (P^r \cap D_{h-1}) &+ (P^{r+s_{h-1}-s_h} \cap D_{h-2}) + \cdots + P^{r+s_1-s_h} \\ &= (K^r \cap D_{h-1}^\times) \cdots (K^{r+s_2-s_h} \cap D_1^\times)(K^{r+s_1-s_h})\end{aligned}$$

is a group normalized by $D_h^\times$; furthermore,

$$\begin{aligned}(K^{s_1-s_h+r})(K^{s_2-s_h+r} \cap D_1^\times) \cdots (K^r \cap D_{h-1}^\times) \cap D_h^\times \\ = K^r \cap D_h^\times.\end{aligned}$$

Proof. We argue by induction. For $i = 1$, the only nonvacuous part is in (v), that $D_h^\times$ normalizes K^r, and this is obvious. For $i = 2$, the hypothesis says that we can write $\eta_2 = u_1 + u_0$, with $u_1 \in D_1^\times$, $\nu(v_1) = f_2 = \eta(\nu_2)$, and $u_2 \in P^{f_2+s_1-s_2}$. It suffices to prove (i) when $\nu(x_2) > 0$ (multiply by a power of ϖ^n if necessary), and then it suffices to assume that x_2 is a power of η_2. Now (i) is clear, because $\eta_2^r \equiv u_1^r \bmod P^{f_2 r + s_1 - s_2}$. For (ii), write $x_2 = 1 + y_2$, with $\nu(y_2) = r > 0$; set $y_2 = y_1 + y_0$ as in (i). Then

$$x_2 = (1 + y_1)(1 + (1 + y_1)^{-1} y_0).$$

Part (iii) is similar: we have $x = z_1 + z_0$, as in (i), then $x = z_1(1 + z_1^{-1} z_0)$. Part (iv) is trivial unless $h = j = 2$. We then need to prove, in essence, that $\eta_1^{r_2} \cdot \eta_2^{r_2} \subseteq (P^{r_1+r_2} \cap D_1) + P^{r_1+r_2+s_1-s_2}$, and this is clear from (i).

Part (v) is only interesting for $h=1, j=2$; use (ii) to write $P^{r_2}\cap D_2 \subseteq (P^{r_2}\cap D_1)+P^{r_2+s_1-s_2}$, and apply (iv). As for (vi), we need to show first that $K^{r+s_1-s_2}(K^r\cap D_1^\times)=1+(P^r\cap D_1^\times)+P^{r+s_1-s_2}$. If $v_0\in P^{r+s_1-s_2}$ and $u_1\in P^r\cap D_1$, then $(1+u_0)(1+u_1)=1+u_0+(u_1+u_0u_1)$ and $1+u_0+u_1=(1+u_0)1+(1+u_0)^{-1}u_1)$, where $(1+u_0)u_1$, $(1+u_0)^{-1}u_1\in P^{r+s_1-s_2}$. The set is clearly a group. That $D_2^\times$ intersects it in $D_2^\times\cap K^r$ is a consequence of (i). If $x_2\in D_2^\times$, write $x_2=z_0z_1$ as in (ii) or (iii). Then z_0 normalizes $K^{s_1-s_2+r}(D_2^\times\cap K^r)$ because all commutators with elements of that group lie in $K^{s_1-s_2+r}$; z_1 obviously normalizes $D_1^\times\cap K^r$, and $K^{s_1-s_2+r}$ is normal. That finishes the proof of (vi).

Now assume the result for $i-1$; we prove it for i. For property (i), it suffices, as in the case $i=2$, to prove the result for η_i^r with $r\geq 1$. We show first that η_i^r can be written as $y_{i-1,r}\cdots y_{0,r}$, where $y_{i-1,r}\in P^{f_ir}\cap D_{i-1}$ and $y_{j,r}\in K^{s_{j+1}-s_i}\cap D_j^\times$ for $0\leq r\leq j-2$. For $r=1$, this is true by hypothesis. If the result holds for r, then

$$\eta_i^{r+1}=y_{i-1,r}\cdots y_{0,r}\cdot y_{i-1,1}\cdots y_{0,1};$$

but $y_{i-1,1}^{-1}(y_{i-2,r}\cdots y_{0,r})y_{i-1,1}=v_{i-2}\cdots v_0$, with $v_j\in K^{s_{j+1}-s_i}$, from (vi) for $i-2$. Thus $\eta_i^{r+1}=y_{i-1,r}y_{i-1,1}v_{i-2}\cdots v_0y_{i-2,1}\cdots y_{0,1}$. We now use (vi) again to write this as $y_{i-1,r}y_{i-1,1}y_{i-2,r+1}\cdots y_{0,r+1}$, with $y_{j,r+1}\in K^{s_{j+1}-s_i}\cap D_j^\times$; then set $y_{i-1,r+1}=y_{i-1,r}y_{i-1,1}$ to complete the induction.

We next want to show that we can write $\eta_i^r=y_{i-1}\cdots y_0$ (say) in the form $x_0+\cdots+x_r$ required for (i). From (vi) and (v) respectively we obtain

$$\begin{aligned}y_{i-2}\cdots y_0\in 1+P^{s_1-s_i}+(P^{s_2-s_i}\cap D_1)+\cdots+(P^{s_{i-1}-s_i}\cap D_{i-2}),\ \text{and}\\ y_{i-1}(P^{s_{j+1}-s_i}\cap D_j)\subseteq(P^{s_{j+1}-s_i+rf_i}\cap D_j)+(P^{s_j-s_i+rf_i}\cap D_{j-1})\\ +\cdots+(P^{s_1-s_i+rf_i}\cap D_0).\end{aligned}$$

Summing, we get

$$\eta_i^r\subseteq P^{f_ir}\cap D_{i-1}+(P^{f_ir+s_{i-1}-s_i}\cap D_{i-2})+\cdots+(P^{f_ir+s_1-s_i}\cap D_0),$$

and (i) follows. For (iii), write $x_i=x_0+\cdots+x_{i-1}$; factoring out x_{i-1} gives $x_i=x_{i-1}(1+x_{i-1}^{-1}x_{i-2}+\cdots+x_{i-1}^{-1}x_0)$. We now apply (v) to see that $x_{i-1}^{-1}(x_{i-2}+\cdots+x_0)\in(P^{s_{i-1}-s_i}\cap D_{i-2})+\cdots+(P^{s_2-s_i}\cap D_1)+P^{s_1-s_i}$. From (vi), $x_{i-1}^{-1}x_i\in(K^{s_{i-1}-s_i}\cap D_{i-2}^\times)\cdots(K^{s_2-s_i}\cap D_1^\times)K^{s_1-s_i}$, and (iii) follows. The argument for (ii) is similar, but we use $x_i=1+x_0+\cdots+x_{i-1}=(1+x_{i-1})(1+(1+x_{i-1})^{-1}(x_0+\cdots+x_{i-2}))$. Part (iv) follows from the inductive hypothesis unless either h or $j=i$. If, say, $h=i>j$, then $P^{r_1}\cap D_i\subseteq(P^{r_1}\cap D_{i-1})+(P^{r_1+s_{i-1}-s_i}\cap D_{i-2})+\cdots+P^{r_1+s_1-s_i}$, from (i), and we again get the result from the inductive hypothesis. So the only case to check is where $h=j=i$. Then $(P^{r_1}\cap D_i)(P^{r_2}\cap D_i)=P^{r_1+r_2}\cap D_i$;

and the result again follows from (i). For (v), we may also assume that $h < j = i$; we may also prove the result for $\eta_h^{r_1}\eta_j^{r_2}$. Use (i) and induction.

Only (vi) is left. That $(K^r \cap D_{i-1}) \cdots (K^{s_2-s_i+r_i} \cap D_1^\times)K^{s_1-s_i+r} = H_r$ (say) is a group is clear from (v) applied to $i-1$, since we also have $H_r = \{D_{i-1}^\times(K^{s_{i-1}-s_i+r} \cap D_{i-2}^\times) \cdots (K^{s_2-s_i+r_i} \cap D_1^\times)K^{s_1-s_i+r}\} \cap K^r$. That $H_r = 1 + P^{s_1-s_i+r} + (P^{s_2-s_i+r} \cap D_1) + \cdots + (P^r \cap D_{i-1})$ follows from (v). To see that $D_i^\times$ normalizes H_r, it suffices to see that η_i and $D_i^\times \cap K$ normalize it. Verifying that η_i normalizes H_r is an exercise in the use of (v). Since $D_i^\times \cap K/D_i^\times \cap K^1$ has coset representatives that are also in every D_j with $j < i$, we need only prove that $D_i^\times \cap K^1$ normalizes H_r. But $(1+a)y(1+a)^{-1} = y + (ay - ya) - (aya - ya^2) + \cdots$, so that this, too, becomes an application of (v). □

3.14. Corollary. *Let $H_r = (K^{s_1-s_i+r})(K^{s_2-s_i+r} \cap D_1^\times) \cdots (K^r \cap D_{i-1}^\times)$, $r \geq 1$, and let $j \geq r$; let h be such that $s_{h+1} - s_i + r \leq j < s_h - s_i + r$. Then*

$$H_r \cap K^j / H_r \cap K^{j+1} \cong D_h \cap P^j / D_h \cap P^{j+1}, \text{ and}$$
$$H_r \cap K^j = K^{s_1-s_i+r}(K^{s_2-s_i+r} \cap D_1^\times) \cdots (K^{s_h-s_i+r} \cap D_{h-1}^\times)(K^j \cap D_h^\times).$$

Proof. Any element of H_r is of the form $x = 1 + y_0 + y_1 + \cdots + y_{i-1}$ with $y_\ell \in P^{s_{\ell+1}-s_i+r} \cap D_\ell$, from (vi). Also suppose $x \in K^j$ and $j \geq s_{i-1} - s_i + r$. Then y_{i-1} is the only term in the sum not immediately in $P^{s_{i-1}-s_i+r}$; since $x - 1 \in P^{s_{i-1}-s_i+r}$, we must have $y_{i-1} \in P^{s_{i-1}-s_i+r} \cap D_{i-1}$. From (i), $y_{i-1} \in (P^{s_{i-1}-s_i+r} \cap D_{i-2}) + \cdots + (P^{s_1-s_i+r} \cap D_0)$, so that we may delete y_{i-1} from the sum (perhaps changing the other y_ℓ). Proceeding inductively, we see that $x = 1 + y_0 + y_1 + \cdots + y_h$, $y_\ell \in D_\ell \cap P^{s_{\ell+1}-s_i+r}$ and $y_h \in D_h \cap P^h$. So mod K^{j+1}, $x \equiv 1 + y_h$, $y_h \in D_h \cap P^j$ (and y_h determined mod $D_h \cap P^{j+1}$). The last part now is a consequence of (v) and induction. □

3.15. Corollary. *$K^{s_1-s_i}$, $K^{s_2-s_i} \cap D_1^\times$, $\cdots$, $K^{s_{i-1}-s_i} \cap D_{i-2}^\times$, $K \cap D_{i-1}^\times$ all normalize the subgroup H_r above.*

Proof. For $y \in H_r$ and $w \in K^{s_h-s_i+r} \cap D_{h-1}^\times$, write $w = 1 + w_0$. Then $wyw^{-1} = y + (w_0y - yw_0) - (w_0yw_0 - yw_0^2) + \cdots$; now use (iv)–(vi) of Lemma 3.13. □

4. Construction of H_0 and χ.

In this section, we construct a subgroup H_0 and a character χ of H_0 with the properties described at the start of §2. We assume that $H_0 \supseteq K^{s_1}$ (for an $s_1 \geq 1$), that χ is trivial on K^{s_1+1}, and that χ is given on K^{s_1} by

$$\chi(1 + \gamma\varpi^{s_1}) = \psi \circ Tr_{k_n/k}\alpha_1\gamma^{\sigma^{-s_1}} = \psi \circ Tr_{k_n/k}\alpha_1^{\sigma^{s_1}}\gamma,$$

where $F[\alpha_1\varpi^{s_1}] = E_{(s_1)}$ satisfies $e(E_{s_1}/F) = e_1$, $f(F_{s_1}/F) = f_1$, and $e_1f_1 > 1$. We also set $e_0 = f_0 = 1$ and $s_0 = s_0' = \infty$, for notational ease later. The triple (s_1, e_1, f_1) is the first in the set of triples (s_i, e_i, f_i), $1 \leq i \leq r_0$, that will be associated with H_0; we will also have division algebras D_i, $1 \leq i \leq r_0$. We set $s'_i = [s_i/2]+1$, $s''_i = s_i + 1 - s'_i$. As before, we write $H_0^j = H_0 \cap K^j$. We will have $H_0 = H_0^{s'_{r_0}}$ (i.e., $H_0 \subseteq K^{s'_{r_0}}$); r_0 will be defined as we proceed.

We construct H_0 and χ inductively. Thus we need to describe H_0^j and $\chi|_{H_0^j}$ as well as the (s_i, e_i, f_i). They have the following properties:

A. Arithmetic properties of the s_i, e_i, f_i (these were listed in §1):

(1) $s_1 > s_2 > \cdots > s_{r_0}$, and $s_{i+1} \geq s'_i$ for $1 \leq i < r_0$.
(2) $1 = e_0 | e_1 | \cdots | e_{r_0}$, $1 = f_0 | f_1 | \cdots f_{r_0}$, and $e_{r_0} f_{r_0} \mid n$.
(3) $1 < e_1 f_1 < \cdots < e_{r_0} f_{r_0}$.
(4) $e_i = n/(n/e_{i-1}, s_i)$ for $1 \leq i \leq r_0$. (Hence $f_{r_0} \mid s_i$ for $1 \leq i \leq r_0$: $f_{r_0} e_i \mid n$, so that $f_{r_0} \mid (n/e_i) = (n/e_{i-1}, s_i)$.)

B. The division algebras $D_{(j)}$ (in what follows, i is the largest index with $s_i \geq j$): for each j with $s_1 \geq j \geq s'_{r_0}$, there exist an element $\eta_{(j)} \in D$, a division algebra $D_{(j)}$, and a field $E_{(j)}$ such that:

(5) $\eta_{(j)}$ generates $P_D^{f_i}$ and conjugation by $\eta_{(j)}$ induces σ^{f_i} on F_{n/e_i}.
(6) $D_{(j)}$ is generated by $\eta_{(j)}$ and F_{n/e_i}.
(7) $E_{(j)}$ is the center of $D_{(j)}$; $e(E_{(j)}/F) = e_i$, and $f(E_{(j)}/F) = f_i$.
(8) $E_{(j)}$ is generated over F by $\eta_{(j)}^{n/e_i f_i}$ and F_{f_i}.

We set $D_i = D_{(s_{i+1}+1)}$, $E_i = E_{(s_{i+1}+1)}$, $\eta_i = \eta_{(s_{i+1}+1)}$ for $1 \leq i < r_0$; $D_{r_0} = D_{(s'_{r_0})}$, $E_{r_0} = E_{(s'_{r_0})}$, $\eta_{r_0} = \eta_{(s'_{r_0})}$. We sometimes write η_0 for ϖ.

C. Relations among the $\eta_{(j)}$ and among the $D_{(j)}$:

(9) If $S_{i+1} < j < S_1$, then $\eta_{(j)}$ is of the form

$$\eta_{(j)} = \eta_{(j+1)}(1+y_{i-1})(1+y_{i-2})\cdots(1+y_0),$$

where $y_h \in P^{s_{h+1}-j} \cap D_h$; if $j = s_i$, then $\eta_{(j)}$ is of the form

$$\eta_{(j)} = \delta\eta_{(j+1)}^{f_i/f_{i-1}}(1+y_{i-2})\cdots(1+y_0),$$

where $\delta \in k_{n/e_{i-1}}$ and $y_h \in P^{s_{h+1}-j} \cap D_h$. (Here, $D_0 = D$; for $s'_{r_0} \leq j < s_{r_0}$, we take $i = r_0$.)

(10) If $s_{i+1} < j < s_i$ and $\ell > 0$, then

$$D_{(j)}^\times \cap K^\ell \subseteq (D_{(j+1)}^\times \cap K^\ell)(D_{i-1}^\times \cap K^{s_i-j+\ell})\cdots(D_0^\times \cap K^{s_1-j+\ell});$$

if $j = s_i$, then

$$D_{(j)}^\times \cap K^\ell \subseteq (D_{(j+1)}^\times \cap K^\ell)(D_{i-2}^\times \cap K^{s_{i-1}-s_i+\ell})\cdots(D_0^\times \cap K^{s_1-s_i+\ell}).$$

We define $H_0 = K^{s'_1}(K^{s'_2} \cap D_1^\times)\cdots(K^{s'_{r_0}} \cap D_{r_0-1}^\times)$ and $H_0^j = H_0 \cap K^j$.

(11) If $s'_{h+1} \leq j < s'_h$, then

$$H_0^j = K^{s'_1}(K^{s'_2} \cap D_1^\times)\cdots(K^{s'_h} \cap D_{h-1}^\times)(K^j \cap D_h^\times),$$

$$\text{and } H_0^j/H_0^{j+1} \cong D_h^\times \cap K^j/D_h^\times \cap K^{j+1}.$$

D. Properties of χ (the character defined on H_0).

(12) Given $s'_i < j \leq s_i$, any extension of $\chi|_{H_0^{s_i}}$ to H_0^j extends as a character to $H_0^{s'_i}$.

(13) Say that x *commutes* with $\chi|_{H_0^j}$ if $\chi^x = \chi$ on $H_0^j \cap x^{-1}H_0^j x$. Assume that, as above, i is the largest index with $s_i \geq j$; assume that $s'_{h+1} \leq j < s'_h$. Then

$$\{x \in D_n^\times : x \text{ commutes with } \chi|_{H_0^j}\} = K^{s''_1}(D_1^\times \cap K_0^{s''_2})\ldots$$
$$\ldots(D_{h-1}^\times \cap K^{s''_h})(D_h^\times \cap K^{s_{h+1}-j+1})\ldots$$
$$\ldots(D_1^\times \cap K^{s_i-j+1})D_{(j)}^\times;$$

this is a group, J_0^j (say), and it normalizes H_0^j. (Notice that J_0^j *decreases* with j.)

(14) Let $s'_{h+1} \leq s_i < s'_h$, so that $D_{i-1}^\times$ commutes with $\chi|_{H_0^{s_{i+1}}}$. Then one can write $\chi|_{H_0^{s_i}}$ as $\chi_{0,i}\chi_{1,i}$, where

(i) $D_{i-1}^\times$ commutes with $\chi_{0,i}$;

(ii) $\chi_{1,i}$ is trivial on $H_0^{s_i+1}$;

(iii) On $H_0^s \cap D_{i-1}^\times$ we have

$$\chi_{1,i}(1+\gamma\eta_{i-1}^{s_i/f_i}) = \psi \circ Tr_{k_n/e_{i-1}/k}(\alpha_i^{\sigma^{s_i}}\gamma), \quad \gamma \in k_{n/e_i},$$

where $E_{i-1}[\alpha_i\eta_{i-1}^{-s_i/f_i}] = E_{(s_i)}^-$ satisfies $e(E_{(s_i)}^-/F) = e_i$ and $f(E_{(s_i)}^-/F) = f_i$.

(iv) Let $D_{(s_i)}^-$ be the algebra of elements commuting with $E_{(s_i)}^-$. Then $D_{(s_i)}^-$ has a prime element $\tilde{\eta}_{(s_i)}$ satisfying the relation $\tilde{\eta}_{(s_i)} \equiv \eta_{(s_i)} \bmod P^{f_i+1}$, and such that $\tilde{\eta}_{(s_i)}$, k_{n/e_i} generate $D_{(s_i)}$ as in Proposition 3.7.

The number of (H_0, χ) associated with a given set of triples (s_1, e_1, f_1), $\cdots$, $(s_{r_0}, e_{r_0}, f_{r_0})$ will be $\prod_{j=s'_{r_0}}^{s_1} C_j$, where:

$C_j = \dfrac{f_{i-1}}{f_i} \displaystyle\sum_{d|f_i/f_{i-1}} (q^{f_{i-1}d} - 1)\mu(f_i/f_{i-1}d)$ if $j = s_i$ (μ = Mobius function);

$C_j = 1$ if $s_{i+1} < j < s_i$ and $f_i \nmid j$, or if $j < s_{r_0}$ and $f_{r_0} \nmid j$;

$C_j = q^{n/e_i}$ if $s_{i+1} < j < s_i$ and $f_i \mid j$, or if $j < s_{r_0}$ and $f_{r_0} \mid j$ (then $i = r_0$).

The construction is of course inductive. Before we begin, here are some remarks.

1. In view of the results in §3, there are considerable redundancies in this list of properties. For example, (6), (7), and (8) follow from (5) because of Prop. 3.6, and (9) implies (10) and (11) because of Lemma 3.13. To prove property (12), we have to show that $\chi \equiv 1$ on $(H_0^{s_i}, H_0^{s_i}) \cap H_0^j$. Since $(H_0^{s_i}, H_0^{s_i}) \subseteq H_0^{s_i+1}$ this result is independent of j. Assume (12) for $i-1$. From (11), we see that we need to prove that $\chi((u,v)) = 1$ if either (i) both $u, v \in K^{s'_i} \cap D_i^\times$ or (ii) $u \in K^{s'_i} \cap D_i^\times$ and $v \in H_0 \cap K^{s'_{i-1}}$ (or the roles of u, v are reversed). In each case the result is a consequence of (13) and Corollary 3.9. Thus we need only verify (1)–(5), (9), (13), and (14).

2. There is another, subtler redundancy that will occur. In proving (13), it will not be hard to show that J_0^j commutes with $\chi|_{H_0^j}$ and normalizes H_0^j. For the converse, we will prove that if x normalizes H_0^j, the x can be written as $y_0(1 + y_1 + y_2 + \cdots)$, with $y_0 \in D_{(j)}^\times$, $y_\ell \in P^\ell$, $y_\ell \notin P^{\ell+1}$ unless $y_\ell = 0$, and, e.g., $y_\ell \in D_g$ if $s_{g+1} - j + 1 \leq \ell < s_g - j + 1$ (for $g \geq h$; as usual, $s'_{h+1} \leq j < s'_h$). This not only proves (13), it shows that elements of J_0^j can be put in this special form; and this fact can easily be used to prove (11).

We need to verify everything for $\chi|_{K^{s_1}} = \chi|_{H^{s_1}}$; the verification in this case is different from that for the general step, since in this case we are handed $E_{(s_1)}$. Recall that

$$\chi(1 + \gamma\varpi^{s_1}) = \psi \circ Tr_{k_n/k}(\alpha_1^{\sigma^{s_1}}\gamma),$$

and $E_{(s_1)} = F[\alpha_1\varpi^{-s_1}] \neq F$; $e(E_{(s_1)}/F) = e_1$, $f(E_{(s_1)}/F) = f_1$. Set $E^{\sim}_{(s_1)} = E_{(s_1)}$, $\chi_{1,1} = \chi$, $\chi_{0,1} \equiv 1$ in (14). Then (14) holds, and (1)–(3) are obvious. For (4), notice first that any power of $\alpha_1\varpi^{-s_1}$ is of the form $\beta_j\varpi^{-s_1 j}$, $\beta \in k_n^\times$. Let $\tilde{e}_1 = n/(n, s_1)$. Then $\alpha_1\varpi^{-s_1}, \ldots, (\alpha_1\varpi^{-s_1})^{\tilde{e}_1}$ are linearly independent $/F_n$ (because their valuations, $-s_1, \cdots, -\tilde{e}_1 s_1$, are distinct modulo n), and hence $e(E_{(s_1)}/F) \geq \tilde{e}_1$. But $(\alpha_1\varpi^{-s_1})^{\tilde{e}_1} = \beta_n\varpi^{cn}$ for some $c \in \mathbb{Z}$, so that $E_{(s_1)}$ is totally ramified of degree $\tilde{e}_1$ over the unramified extension $F[\beta]$. Therefore $\tilde{e}_1 = e_1$, and (4) follows.

Choose $a, b \in \mathbb{Z}$ with $an - bs_1 = n/e_1$ and let $\zeta = \varpi^{an}(\alpha_1\varpi^{-s_1})^b \in E_{(s_1)}$. Valuation considerations show that ζ is a prime element of $E_{(s_1)}$. From Prop. 3.6 (and Prop. 3.1 applied to the commutant $D_{(s_1)}$ of $E_{(s_1)}$), there is an element η with $\nu(\eta) = f_1$ and $\eta^{n/e_1 f_1} = \zeta$. Write $\eta \equiv \delta\varpi^{f_1} \bmod P^{f_1+1}$. Then $\zeta = \eta^{n/e_1 f_1} \equiv (\delta\varpi^{f_1})^{n/e_1 f_1} \bmod P^{f_1+1}$; it follows that $\eta_{(s_1)} = \delta\varpi^{f_1}$ also satisfies $\eta_{(s_1)}^{n/e_1 f_1} = \zeta$. Hence $\eta_{(s_1)} \in D_{(s_1)}$. Because s is a multiple of n/e_1, $\bar{k}_{n/e_1}$ commutes with $\alpha_1\varpi^{-s_1}$ and is therefore $\subset D_{(s_1)}$. It is easy to see that conjugation by $\eta_{(s_1)}$ induces σ^{f_1} on k_{n/e_1}. That takes care of (5) and (9).

Only (13) is left. Since $(K^1, K^{s_1}) \subseteq K^{s_1+1} \subseteq \operatorname{Ker}\chi$, K^1 commutes with χ, while $D^{\times}_{(s_1)}$ commutes with χ because of Prop. 3.12 (and the note following it) and Corollary 3.10. For the converse, suppose that x commutes with χ. Write $x = \delta\varpi^{j_0}(1+\delta_1\varpi+\cdots)$. Since $1+\delta_1\varpi+\cdots$ is known to commute with χ, we may assume $x = \delta\varpi^{j_0}$. From Proposition 3.12 (and the note following it), x must commute with $\alpha_1\varpi^{-s_1}$. This proves (13).

How many choices are there for χ? Equivalently, we need to know how many nonconjugate $\alpha_1\varpi^{-s_1}$ there are that generate fields with ramification index e_1 and residue class degree f_1. We have seen that s_1 determines e_1. If $(\alpha_1\varpi^{-s_1})^{e_1} = \beta\varpi^{-cn}$, then $F[\beta]$ is the maximal unramified extension in $E_{(s_i)}$. Therefore β must be a primitive element in $F^{\times}_{f_1}$. There are $\sum_{d|f_1}(q^d-1)\mu(f_1/d)$ such primitive elements, and, since $\operatorname{Gal}(F_{f_1}/F)$ has f_1 elements, there are $(1/f_1)\sum_{d|f_1}(q^d-1)\mu(f_1/d)$ conjugacy classes of primitive elements. That gives C_{s_1}.

The inductive step assumes that we have χ defined on H_0^{j+1}, that $s_1,\ldots,s_i$ have been defined, that $j \geq s_i'$ (so that H_0^j is also defined), and that (1)–(14) hold for $\chi|_{H_0^{j+1}}$. Let h be the integer with $s'_{h+1} \leq j < s'_h$. We want to extend the construction to H_0^j. If $f_h \nmid j$, then $H_0^j = H_0^{j+1}$, and there is nothing to extend; we set $E_{(j)} = E_{(j+1)}$, $D_{(j)} = D_{(j+1)}$, $\eta_{(j)} = \eta_{(j+1)}$, and we verify that $J_0^j = J_0^{j+1}$. So the inductive step holds in that case.

That leaves the interesting case, where $f_h \mid j$. In what follows, it may help in understanding the proof to keep two guiding principles in mind:

(1) Since we more or less understand χ on H_0^{j+1}, we can concentrate attention on H_0^j/H_0^{j+1}.

(2) What *really* matters is the behavior of the character χ on $H_0^j \cap D^{\times}_{(j)}$ (mod $H_0^{j+1}\cap D^{\times}_{(j)}$); most of the analysis concentrates on this.

The following lemma is the key to the construction.

4.1. Lemma. *Use notation as above. Then χ has an extension χ_0 to H_0^j such that $D^{\times}_{(j+1)}$ commutes with χ_0.*

Proof. Recall that $D_{(j+1)}$ is generated by $\eta_{(j+1)}$ and $\bar{k}^{\times}_{n/e_i}$.

We first find an extension χ_1 such that $\bar{k}^{\times}_{n/e_i}$ commutes with χ_1. The following argument, suggested by Roger Howe, is simpler than the one in [5]. Let B be the group of characters (= homomorphisms into the circle) of H_0^j that are trivial on K^{s_1+1}, A the subgroup of characters trivial on H_0^{j+1}, and C the group of restrictions of elements of B to H_0^{j+1}. Then $0 \to A \to B \to C \to 0$ is clearly exact. For the moment, set $G = \bar{k}^{\times}_{n/e_i}$. Standard cohomology theory gives the long exact sequence

$$0 \to A^G \to B^G \to C^G \to H^1(A,G) \to \cdots.$$

Since A is a p-group and G has order prime to p, $H^1(A,G)=\{0\}$. Therefore B^G maps surjectively onto C^G, which is what we need.

Observe next that if χ_1 is fixed by $k^\times_{n/e_i}$, then so is $\chi_1^{\eta_{(j+1)}}$, since $\eta_{(j+1)}$ normalizes $k^\times_{n/e_i}$. If $(n/e_i)\nmid j$, then only one extension of $k^\times_{n/e_i}$ of χ can commute with $k^\times_{n/e_i}$. (Proof: suppose χ_1, χ_2 commute with $k^\times_{n/e_i}$. Then the character $\chi_1\chi_2^{-1}$ also commutes, and is trivial on H^{j+1}. So it must be of the form $(1+\gamma\eta_h^{j/f_h})\mapsto\psi\circ Tr_{k_{n/e_h}/k}(\delta^{\sigma^j}\gamma)$ for some $\delta\in k_{n/e_h}$. Since $k^\times_{n/e_i}$ commutes with this character, Prop. 3.12 says that $k^\times_{n/e_i}$ commutes with $\delta\eta_h^{-j/f_h}$. Conjugation by η_h^{-j/f_h} induces σ^{-j} on k_{n/e_i}. Therefore either $(n/e_i)\mid j$ or $\delta=0$.) So if $(n/e_i)\nmid j$, then we may take $\chi_0=\chi_1$.

When $(n/e_i)\mid j$, the action of $k^\times_{n/e_i}$ is in fact trivial, since $k^\times_{n/e_i}$ commutes with the coset representatives $1+\gamma\eta_{(j+1)}^{j/f_i})$, $\gamma\in k_{n/e_h}$. So let χ_1 be any extension of χ to H_0^j. For $\gamma\in k_{f_i}$, $[\gamma,\eta_{(j+1)}]=0$. Therefore the character $y\mapsto\chi_1(\eta_{(j+1)}y\eta_{(j+1)}^{-1}y^{-1})$ is trivial on H_0^{j+1} and on elements $1+\gamma\eta_{(j+1)}^{j/f_i})$, $\gamma\in k_{f_i}$. Now let $\delta\in k_{n/e_i}$, and let $\tilde{\chi_\delta}(1+\gamma\eta_{(j+1)}^{j/f_i})=\psi\circ Tr_{k_{n/e_h}/k}(\delta^{\sigma_j}\gamma)$ ($\tilde{\chi_\delta}\equiv 1$ on H_0^{j+1}), $\chi_\delta(y)=\tilde{\chi_\delta}(\eta_{(j+1)}y\eta_{(j+1)}^{-1}y^{-1})$. Then $\chi_\delta\equiv 1$ on H_0^{j+1}, and a calculation gives

$$\begin{aligned}\chi_\delta(1+\gamma\eta_{(j+1)}^{j/f_i})&=\psi\circ Tr_{k_{n/e_h}/k}\delta^{\sigma^j}(\gamma^{\sigma^{f_i}}-\gamma)\\&=\psi\circ Tr_{k_{n/e_h}/k}\gamma^{\sigma^{-j+f_i}}(\delta-\delta^{\sigma^{f_i}}).\end{aligned}$$

The first of these shows that $\chi_\delta(1+\gamma\eta_{(j+1)}^{j/f_i})=1$ if $\gamma\in k_{f_i}$; the second, that χ_δ is trivial iff $\delta\in k_{f_i}$. A counting argument (like that in Prop. 3.12) now shows that $\delta\mapsto\chi_\delta$ maps onto the characters on H_0^j/H_0^{j+1} trivial on $H_0^j\cap E^\times_{(j+1)}$. In particular, there exists δ with $\chi_1(\eta_{(j+1)}y\eta_{(j+1)}^{-1}y^{-1})=\chi_\delta(y)$ on H_0^j. But then $\chi_0(y)=\chi_1(y)\tilde{\chi_\delta}(y)^{-1}$ is fixed by $\eta_{(j+1)}$, and we are done. □

Any extension of χ to H_0^j is of the form $\chi_0\chi_1$, where χ_0 is the extension of Lemma 4.1 and χ_1 is trivial on H_0^{j+1}. We examine these extensions; there are a number of cases.

Case 1. If $f_i\nmid j$, all these extensions $\chi=\chi_0\chi_1$ are conjugate to χ_0. Here's a proof. By hypothesis, $f_h\mid j$. Let g be as small as possible with $f_g\nmid j$. We first find $w_{g-1}=1+\delta_{g-1}\eta_{g-1}^{(s_g-j)/f_{g-1}}$, $\delta\in k_{n/e_{g-1}}$, such that $\chi_0^{w_{g-1}}=\chi_0\chi_1$ on $H_0^{j+1}(H_0^j\cap D^\times_{g-1})$. From (13), $\chi_0^{w_{g-1}}=\chi_0$ on H_0^{j+1} (because $w_{g-1}\in J_0^{j+1}$), so we may restrict attention to $H_0^j\cap D^\times_{g-1}$. Since $w_{g-1}yw_{g-1}^{-1}y^{-1}\in H_0^{s_g}$ for $y\in H_0^j$, we consider χ on $H_0^{s_g}$. From (13), $\chi_0|_{H_0^{s_g}}$ equals $\chi_{0,g}\chi_{1,g}$, where $\chi_{0,g}$ commutes with $D^\times_{g-1}$ and hence with w_{g-1}; therefore $\chi_0(w_{g-1}yw_{g-1}^{-1}y^{-1})$ equals $\chi_{1,g}(w_{g-1}yw_{g-1}^{-1}y^{-1})$ for all

$y \in H_0^{s_g}$. Now (14) and Lemma 3.12(d) say that we can choose w_g such that $\chi_{1,g}(w_g y w_g^{-1} y^{-1}) = \chi_1(y)$ for all $y \in (H_0^j \cap D_{g-1}^\times)(H_0^{j+1})$. (Since $f_g \nmid j$, the part of 3.12(d) about isomorphisms applies.) Therefore $\chi_0(w_g y w_g^{-1}) = \chi_0(y)\chi_1(y)$ for these elements.

We next find $w_{g-2} = 1 + \delta_{g-2}\eta_{g-2}^{(s_{g-1}-j)/f_{g-2}}$; $\delta_{g-2} \in k_{n/e_{g-2}}$, such that $\chi_0^{w_{g-1}w_{g-2}}$ and $\chi_0\chi_1$ agree on $H_0^{j+1}(H_0^j \cap D_{g-2}^\times)$. The reasoning is almost the same: $w_{g-2} y w_{g-2}^{-1} y^{-1} \in H_0^{s_{g-1}}$ for $y \in H_0^j$, and on $H_0^{s_{g-1}}$ we have $\chi_0^{w_{g-1}} = \chi_{0,g-1}\chi_{1,g-1}$, where $\chi_{0,g-1}$ commutes with $D_{g-2}^\times$ and hence with w_{g-2}. Thus $\chi_0^{w_{g-1}}(w_{g-2} y w_{g-2}^{-1} y^{-1}) = \chi_{1,g-1}(w_{g-2} y w_{g-2}^{-1} y^{-1})$. But $\chi_{1,g-1}(w_{g-2} y w_{g-2}^{-1} y^{-1}) = 1$ if $y \in H_0^{j+1}$ (since then $w_{g-2} y w_{g-2}^{-1} y^{-1} \in H_0^{s_{g-1}+1}$) or if $y \in H_0^j \cap D_{g-1}^\times$ (by Lemma 3.12(c)). Therefore we have $\chi_{1,g-1}(w_{g-2} y w_{g-2}^{-1} y^{-1}) = 1$ if y is congruent mod K^{j+1} to an element commuting with $\alpha_{g-2}\eta_{g-2}^{-s_{g-2}/f_{g-2}}$; by (14), (9), and Lemma 3.13, this holds if $y \in H^j \cap D_{g-1}^\times (\operatorname{mod} H^{j+1})$.) Therefore Lemma 3.12(d) implies that for some w_{g-2}, $\chi_{1,g-1}(w_{g-2} y w_{g-2}^{-1}) = \chi_0(y)\chi_1(y)(\chi_0^{w_{g-1}}(y))^{-1}$, and this is the desired w_{g-2}. We proceed inductively to prove the claim.

So we lose nothing by requiring that $\chi = \chi_0$. We then declare that $j \neq s_{i+1}$, which means that (1)–(4) continue to hold. We also declare that $\eta_{(j)} = \eta_{(j+1)}$, and every other property but (13) then holds for j by induction. The proof of (13) is implicit in the calculation given above, since the elements fixing χ are the elements fixing $\chi|_{H_0^{j+1}}$ that don't conjugate χ into a different character, and we just saw what conjugations are possible. Here is a more detailed account. It is easy to check by using 3.8 that elements of $K^{s_1''}(D_1^\times \cap K^{s_2''}) \cdots (D_{h-1}^\times \cap K^{s_h''})$ fix χ and normalize H_0^j. Corollary 3.15 shows that elements of $D_h^\times \cap K^{s_{h+1}-j+1}$ normalize H_0^j; elements of this group also fix χ, by Corollary 3.9. (Let $w \in D_h^\times \cap K^{s_{h+1}-j+1}$. If $y \in D_{\ell-1} \cap K^{s_\ell'}$ for $\ell \leq h$, then y commutes with χ on $H^{2s_\ell'}$, and w commutes with χ on $K^{s_{h+1}-j+1+s_\ell'}$ because $s_\ell' \geq j$ and $D_h^\times$ commutes with χ on $K^{s_{h-1}+1}$. If $y \in D_h^\times \cap K^j$, then y commutes with χ on $H_0^{2s_h'}$ and w commutes with χ on $K^{s_{h+1}+1}$.) Similarly, Corollary 3.15 shows that elements of $D_{h+1}^\times \cap K^{s_{h+2}-j+1}$ normalize H_0^j, and Corollary 3.9 shows that these elements fix χ. An obvious induction shows that J_0^j normalizes H_0^j and commutes with χ. Suppose conversely that x commutes with χ. Then $x \in J_0^{j+1}$. Dividing by an element of J_0^j, we may assume that $x \equiv 1 + \delta_{g-1}\eta_{g-1}^{(s_g-j)/f_{g-1}} \operatorname{mod} K^{s_g-j+1}$, $\delta_{g-1} \in k_{n/e_{g-1}}$. We consider $\chi^x(y)\chi(y)^{-1} = \chi(xyx^{-1}y^{-1})$ for $y = 1 + \gamma\eta_{g-1}^{j/f_{g-1}}$, $\gamma \in k_{n/e_{g-1}}$. Since $xyx^{-1}y^{-1} \in H_0^{s_g}$, we consider $\chi|_{H_0^{s_g}}$; there, $\chi = \chi_{0;g}\chi_{1;g}$, as in (14), and $\chi_{0;g}(xyx^{-1}y^{-1}) = 1$ if $y \in D_{g-1}^\times \cap H_0^j$. For these y, therefore, $\chi(xyx^{-1}y^{-1}) = \chi_{1;g}(xyx^{-1}y^{-1})$. From Lemma 3.12(d), $\chi_{1;g}(xyx^{-1}y^{-1}) = 1$ for all $y \in D_{g-1}^\times \cap H_0^j$ iff $\delta_{g-1} \in k_{n/e_g}$. Therefore x is congruent mod K^{s_g-j+1} to an element in J_0^j, and therefore congruent to an element in J_0^j mod $K^{s_{g-1}-j}$ (to see this, compare J_0^j with J_0^{j+1}). Dividing by such an element, we have $x \equiv 1 + \delta_{g-2}\eta_{g-2}^{(s_{g-1}-j)/f_{g-2}} \bmod K^{s_{g-1}-j+1}$, $\delta_{g-2} \in k_{n/e_{g-2}}$. Now consider $\chi^x(y)\chi(y)^{-1}$ for $y = 1 + \gamma\eta_{g-2}^{j/f_{g-2}}$, $\gamma \in k_{n/e_{g-2}}$. An analysis like the one above says that $\chi^x(y)\chi(y)^{-1} = \chi_{1;g-2}(xyx^{-1}y^{-1})$, and this is 1 for all y iff $\delta \in k_{n/e_{g-1}}$. This means that x is congruent (mod $K^{s_{g-1}-j+1}$) to an element of $D_{(s_{g-1})}$. By (14) and (9), x is congruent (mod $K^{s_{g-1}-j+1}$) to an element of D_{g-1}. This element is in J_0^j. Dividing by it, we see that we may assume that $x \in K_0^{s_{g-1}-j+1} \cap J_0^{j+1}$. Hence x is congruent mod $K^{s_{g-2}-j}$ to an element $x_j \in J_0^j$. Dividing by x_j, we get $x \equiv 1 + \delta_{g-3}\eta_{g-3}^{(s_{g-2}-j)/f_{g-3}} \bmod K^{s_{g-2}-j+1}$. Continuing inductively, we get (13).

Notice that we have only one extension to j; also, $C_j = 1$.

In the remaining cases, $f_i \mid j$. Therefore we can define χ_1 by

$$\chi_1(1 + \gamma\eta_{(j+1)}^{j/f_i}) = \psi \circ Tr_{k_{n/e_h}/k}(\alpha'\gamma^{\sigma^{-j}})$$

for some $\alpha' \in k_{n/e_h}$; for $\gamma \in k_{n/e_i}$, this last is

$$\psi \circ Tr_{k_{n/e_i}/k}\gamma^{\sigma^{-j}}\alpha, \quad \alpha = \psi \circ Tr_{k_{n/e_h}/k_{n/e_i}}\alpha'.$$

Case 2. $\alpha\eta_{(k+1)}^{-j/f_i}) \in E_{(j)}$. Here, too, we have $j \neq s_{i+1}$, so that (1)–(4) are automatic. Notice that $n/e_i \mid j$, from (7) and (8).

We may as well remark now that if $Tr_{k_{n/e_h}/k_{n/e_i}}\alpha' = 0$, then $\chi_0\chi_1$ and χ_0 are conjugate (equivalently, the conjugacy class of $\chi_0\chi_1$ depends only on α rather than on α'). The proof is like that in Case 1. Let $w_{i-1} = 1+\delta_{i-1}\eta_{i-1}^{(s_i-j)/f_i}$, $\delta_i \in k_{n/e_{i-1}}$, and consider $\chi_{w_{i-1}}(y) = \chi(w_{i-1}yw_{i-1}y^{-1})$. Then $\chi_{w_i(y)} \equiv 1$ on H_0^{j+1}, and one sees (using (14)) that on H_0^j we have

$$\chi_{w_{i-1}}(y) = \chi_{1;i}(w_{i-1}yw_{i-1}y^{-1}).$$

Since $Tr_{k_{n/e_h}/k_{n/e_i}}\alpha' = 0$, $\chi_1 \equiv 1$ on $H_0^j \cap D_{(j+1)}^\times$. Prop. 3.12(d) now says that we can choose w_i so that $\chi_{w_{i-1}} = \chi_1$ on $H_0^j \cap D_{i-1}^\times$, or so that $\chi_0^{w_{i-1}} = \chi_0\chi_1$ on $H_0^{j-1}(H_0^j \cap D_{i-1}^\times)$. We next let $w_{i-2} = 1+\delta_{i-2}\eta_{i-2}^{(s_{i-1}-j)/f_i}$; similar reasoning shows that for appropriate δ_{i-2}, $\chi_0^{w_{i-1}w_{i-2}} = \chi_0\chi_1$ on $H_0^{j-1}(H_0^j \cap D_{i-2}^\times)$. Continue inductively.

Choose α' for α as follows: let $\beta \in k_{n/e_h}$ satisfy $Tr_{k_{n/e_h}/k_{n/e_i}}\beta = 1$, and let $\alpha' = \alpha\beta$. That has the advantage of forcing $\alpha' = 0$ when $\alpha = 0$. If e_i/e_h is prime to p, then β can be in the prime field $\mathbb{F}_p$; that often makes things simpler. In particular, it makes this construction conform to the one given in [4] or [11] when $p \nmid n$.

Notice that $k_{n/e_i}^\times$ commutes with χ_0 (by construction) and with χ_1 (see the note after Prop. 3.12; $k_{n/e_i}^\times$ commutes with $\alpha'\eta_{(j+1)}^{-j/f_i}$ because $\eta_{(j+1)}^{-j/f_i}$ induces σ^{-j} on $k_{n/e_i}^\times$, and $n/e_i \mid j$). We now construct

$$\eta_{(j)} = \eta_{(j+1)}(1 + \delta_{i-1}\eta_{i-1}^{(s_i-j)/f_{i-1}}) \cdots (1 + \delta_h \eta_h^{(s_{h-1}-j)/f_h}), \ \delta_\ell \in k_{n/e_\ell},$$

so that $\eta_{(j)}$ commutes with χ. Once $\eta_{(j)}$ has this form, (5)–(11) will be clear. Since (12) and (14) follow from the inductive hypothesis, we'll need only (13).

We now construct $\eta_{(j)}$. Since $\eta_{(j+1)}$ commutes with the character χ_0 and with $\alpha\eta_{(j+1)}^{j/f_i})$, $\chi^{\eta_{(j+1)}}\chi^{-1} \equiv 1$ on $H_0^{j+1}(H_0^j \cap D_{(j+1)}^\times)$. Proposition 3.12(d) shows that there exists $w_{i-1} = 1 + \delta_{i-1}\eta_{i-1}^{(s_i-j-j)/f_{i-1}}$ such that $\chi^{\eta_{(j+1)}w_{i-1}}\chi^{-1} \equiv 1$ on $H_0^{j+1}(H_0^j \cap D_{i-1}^\times)$. Another application of Proposition 3.12(d) shows that there exists $w_{i-2} = 1 + \delta_{i-2}\eta_{i-2}^{(s_{i-1}-j)/f_{i-1}}$ such that $\chi^{\eta_{(j+1)}w_{i-1}w_{i-2}}\chi^{-1} \equiv 1$ on $H_0^{j+1}(H_0^j \cap D_{i-2}^\times)$; we continue inductively to get $\eta_{(j)} = \eta_{(j+1)}w_{i-1} \cdots w_h$.

That leaves (13). We know from the above construction, (6), and Corollary 3.10, that $D_{(j)}^\times$ commutes with χ. (Lemma 3.13(vi) shows that $D_{(j)}^\times$ normalizes H_0^j.) We prove that $D_{i-1}^\times \cap K^{s_i-j+1}, \ldots, D_h^\times \cap K^{s_{h+1}-j+1}, \ldots, K^{s_1''}$ commute with χ just as in Case 1. Thus J_0^j commutes with χ. For the converse, suppose that x commutes with χ. Since x commutes with $\chi|_{H_0^{j+1}}$, $x \in J_0^{j+1}$ and we can write

$$x = \delta_0\eta_{(j+1)}^a(1 + y_1' + y_2' + \cdots)$$

where $y_\ell' \in K^\ell$ and $y_\ell' \in D_{(j+1)}$ for $\ell < s_i - j$, $y_{s_i-j}' \in D_{i-1}$. (Since $\chi^x = \chi$ on $H_o^j \cap D_{(j)}$, different χ are nonconjugate.) By (9) and Lemma 3.13(i), we may also write

$$x = \delta_0\eta_{(j)}^a(1 + y_1 + y_2 + \cdots),$$

where $y_\ell \in K^\ell$ and $y_\ell \in D_{(j)}$ for $\ell < s_i - j$, $y_{s_i-1} \in D_{i-1}$. Dividing by an element known to commute with χ, we consider

$$x = 1 + y_{s_i-j} + \cdots = 1 + \delta_{i-1}\eta_{i-1}^{(s_i-j)/f_{i-1}} + \cdots.$$

Just as in Case 1, a calculation using (14) and Lemma 3.12(d) shows that $\delta_{i-1} \in k_{n/e_j}$. This means that x is congruent mod K^{s_i-j+1} to an element in $D_{(j)}^\times$, and hence to an element in J_0^j. Dividing by this element and then an element in $J_0^{j+1} \cap J_0^j$, we consider

$$x = 1 + y_{s_{i-1}-j} = 1 + \delta_{i-2}\eta_{i-2}^{(s_{i-1}-j)/f_{i-2}} + \cdots,$$

and an inductive argument like the one in Case 1 proves that $x \in J_0^j$.

We have $|k_{f_i}| = q^{f_i}$ choices for α; that explains why $C_j = q^{f_i}$.

Case 3. $\alpha\eta_{(j+1)}^{-j/f_i} \notin E_{(j)}$. As in Case 2, if $Tr_{k_n/k_{n/e_i}}\alpha' = Tr_{k_n/k_{n/e_i}}\alpha'' = \alpha$, then the corresponding χ's are conjugate. So we begin by fixing α' corresponding to α as in Case 2. There remain some extensions that are conjugate; we'll return to that issue later.

In this case, $j = s_{i+1}$. We thus define $D_{(j+1)} = D_i$, $\eta_{(j+1)} = \eta_i$, etc., and set $\alpha_i = \alpha$, so that in (14) we will have $E_{(j)}^{\sim} = E_{(s_{i+1})}^{\sim} = E_i[\alpha_i\eta_i^{-s_i/f_i}]$. We then define $e_{i+1} = e(E_{(j)}^{\sim}/F)$, $f_{i+1} = f(E_{(j)}^{\sim}/F)$, so that $e_{i+1}/e_i = e(E_{(j)}^{\sim}/E_i)$, $f_{i+1}/f_i = f(E_{(j)}^{\sim}/E_i)$. The proof that

$$e_{i+1} = \frac{n}{(n/e_i, s_{i+1})}$$

is like the proof of the corresponding fact for e_1. Let

$$e^{\sim} = \frac{n}{(n/e_i, s_{i+1})},$$

and write η for $\alpha\eta_i^{-js_i/f_i}$. Then $\eta, \eta^2, \ldots, \eta^{e^{\sim}}$ are linearly independent over E_i because their valuations are in different residue classes modulo n/e_i. (If $n/e_i \mid \nu(\eta^r) = jr = -s_{i+1}r$, then $n/(e_i(n/e_i, s_{i+1})) = e^{\sim}|r$.) Hence $e_{i+1}/e_i \leq e^{\sim}$. But $\eta^{e^{\sim}} = \beta\eta_i^{-s_{i+1}e^{\sim}/f_i}$, $\beta \in k_n^{\times}$, and $s_{i+1}e_i^{\sim}$ is a multiple of n/e_i. Since $\eta_i^{-n_i/e_if_i} \in E_i$, we see that $e_{i+1}/e_i = e^{\sim}$. Now (4) follows; (1)–(3) (for $i+1$) are also clear.

Next, suppose that x commutes with χ, so that $x \in J_0^{j+1}$ and therefore $x = (1 + y_1 + y_2 + \cdots)\delta_0 n_i^{\ell} = x_1 x_0$, where $y_\ell \in P^\ell$ and $\delta_0 \in k_{n/e_i}$. We first get information on x_0. Suppose that $y = 1 + \gamma\eta_i^{j/f_i}$, $\gamma \in k_{n/e_i}$. Then $y \in K^j \cap D_i^{\times}$, and Corollary 3.9 says that $\chi^{x_1}(y) = \chi(y)$. Write $\chi = \chi_0\chi_1$; $\chi_0^x = \chi_0$ by construction. Thus we must have $\chi_1^{x_0}(y) = \chi_1(y)$. We calculate:

$$\begin{aligned}
\chi_1^{x_0}(y) &= \chi_1\big(1 + \delta_0\gamma^{\sigma^{f_i\ell}}(\delta_0^{-1})^{\sigma^j}\eta_i^{j/f_i}\big) \\
&= \psi \circ Tr_{k_{n/e_i}/k}\alpha^{\sigma^j}\delta_0\gamma^{\sigma^{f_i\ell}}(\delta_0^{-1})^{\sigma^j} \\
&= \psi \circ Tr_{k_{n/e_i}/k}\big[\alpha^{\sigma^{-f_i\ell}}\delta_0^{\sigma^{-f_i\ell-j}}(\delta_0^{-f_i\ell})^{-1}\big]^{\sigma^j}\gamma.
\end{aligned}$$

Thus $\alpha = \alpha^{\sigma^{f_i\ell}}\delta_0^{\sigma^{-f_i\ell-j}}(\delta_0^{\sigma^{-f_i\ell}})^{-1}$ or $(\delta_0\alpha^{\sigma^{f_i\ell}})^{\sigma^{-f_i\ell}} = (\alpha\delta_0^{\sigma^{-j}})^{\sigma^{-f_i\ell}}$. That is, $\alpha\eta_i^{-j/f_i}$ and $\delta\eta_i^{\ell}$ commute. (Conversely, if they commute, then $\delta\eta_i^{\ell}$ commutes with $\chi|_{H^j \cap D_i^{\times}}$.) Proposition 3.5 implies that ℓ is a multiple of f_{i+1}/f_i. For $\ell = 0$, we see (again by Proposition 3.5) that $\delta \in k_{n/e_{i+1}}$. This implies (by a Hensel's Lemma argument) that if $D_{(j)}$ is any division algebra commuting with χ, then the maximal unramified extension in $D_{(j)}$ has residue class degree n/e_{i+1} over F. By Proposition 3.5, the center $E_{(j)}$ of $D_{(j)}$ has $e(E_j/F) \leq e_{i+1}$ and $f(E_j/F) \leq f_{i+1}$.

On the other hand, $k^{\times}_{n/e_{i+1}}$ (or $F^{\times}_{n/e_{i+1}}$) commutes with χ_0 (because $F^{\times}_{n/e_{i+1}} \subseteq D_i$) and with χ_1 (because $\nu(\eta_i^{-j/f_i}) = \delta_{i+1}$ is a multiple of n/e_{i+1}). So if we can find

$$\eta_{(j)} = \delta_i \eta_i^{f_{i+1}/f_i}(1 + \delta_{i-1}\eta_{i-1}^{(s_i - s_{i+1})/f_{i-1}}) \cdots (1 + \delta_h \eta_h^{(s_{h+1} - s_{i+1})/f_h})$$

commuting with χ, then properties (5)–(8) will hold, as will (9)–(11). Because $\delta_i \eta_i^{f_{i+1}/f_i}$ must commute with $\alpha\eta_i^{-j/f_i}$, (14) will also hold (with $\alpha_{i+1} = \alpha$). We have seen earlier that (12) also holds. So once we find $\eta_{(j)}$, we are done except for (13).

The construction of $\eta_{(j)}$ follows a familiar pattern. Let $\delta_i \eta_i^{f_{i+1}/f_i} = \eta^{\#}$ be a prime for the division algebra of elements of D_i commuting with $\alpha\eta_i^{-j/f_i}$. Then $\chi^{\eta^{\#}} = \chi$ on $H_0^{j+1}(K^j \cap D_i^{\times})$. Just as in Cases 1 and 2, we now find $w_{i-1} = 1 + \delta_{i-1}\eta_{i-1}^{(s_i - s_{i-1})/f_{i-1}}$ such that $\chi^{\eta^{\#} w_{i-1}}$ equals χ on $H_0^{j+1}(K^j \cap D_{i-1}^{\times})$, $w_{i-2} = 1 + \delta_{i-2}\eta_{i-2}^{(s_{i-1} - s_{i+1})/f_{i-2}}$ such that $\chi^{\eta^{\#} w_{i-1} w_{i-2}}$ equals χ on $H_0^{j+1}(K^j \cap D_{i-2}^{\times})$ and so on. When we reach w_h, we get $\eta_{(j)} = \eta^{\#} w_{i-1} \cdots w_h$.

The proof of (13) is also familiar. That J_0^j commutes with χ is proved essentially as in Case 2. If x commutes with χ, then $x \in J_0^{j+1}$, so that $x = x_0(1 + y_1 + \cdots)$, $x_0 = \delta_0 \eta_i^{\ell}$, $y_i \in K^i$. We have seen that x_0 commutes with $\alpha\eta_i^{-j/f_i}$; by (14) and Lemma 3.13(i), we may assume that $x_0 \in D_{(j)}^{\times}$ (perhaps at the cost of changing the y_i). Since $D_{(j)}^{\times}$ commutes with χ, we may divide x_0 out; that is, we may assume that $x_0 = 1$. From now on the argument is exactly as in Case 2.

Which different α give conjugate extensions? The only elements that fix $\chi|_{H_0^{j+1}}$ are those in J_0^{j+1}/J_0^j. We saw in Case 2 that conjugating by elements in $J_0^{j+1} \cap K^1/J_0^j \cap K^1$ changes α' but not α. The remaining elements we can use for conjugating are of the form $\delta\eta_i^{\ell}$, $\delta \in k^{\times}_{n/e_i}$. Conjugating $\alpha\eta_i^{-j/f_i}$ by such an element gives $\beta\eta_i^{-j/f_i}$, where $\alpha\eta_i^{-j/f_i}$, $\beta\eta_i^{-j/f_i}$ have the same minimal equation; it is also not hard to show (using (3.4)) that if $\alpha\eta_i^{-j/f_i}$, $\beta\eta_i^{-j/f_i}$ have the same minimal equation, then they are conjugate by an element $\delta\eta_i^{\ell}$. So we need the number of classes of such elements. Since $(\alpha\eta_i^{-j/f_i})^{e_{i+1}/e_i} = \zeta\eta_i^{-r}$, where ζ generates $F_{f_{i+1}}/F_{f_i}$, ζ is a primitive element of $k_{f_{i+1}}/k_{f_i}$. There are $\sum_{d|f_{i+1}/f_i}(q^{df_i} - 1)\mu(f_{i+1}/df_i)$ such elements, and f_i/f_{i+1} times as many conjugacy classes (since elements conjugate under Gal $(k_{f_{i+1}}/k_{f_i})$ are conjugate.) That gives C_j.

We define s_{i+1} only in Case 3. What defines r_0 is that Case 3 does not arise for $s_{r_0}^1 \leq j < s_{r_0}$.

5. Final Details of the Construction.

The major part of the construction is now complete, but certain details still need to be dealt with.

We need to create representations of $J = J^{s'_{r_0}}$ that restrict to a multiple of χ on $H_0 = H_0^{s'_{r_0}}$. This is done in two steps. (A different procedure is described in the Remark at the end of this section.) Let $J_0 = H_0 D_{r_0}^\times$. Then χ extends to a character of J_0, which we also call χ. To prove that χ extends, we show that $\chi \equiv 1$ on $(J_0, J_0) \cap H_0$. Because J commutes with χ on H_0, $\chi(u, v) = 1$ if at least one of u, v is in H_0. Thus we need only show that $\chi \equiv 1$ on $(D_{r_0}^\times, D_{r_0}^\times) \cap H_0$. This is an immediate consequence of Lemma 3.11.

We now get all representations of J_0 containing $\chi|_{H_0}$ by tensoring χ with irreducible representations τ of $D_{r_0}^\times$ trivial on $D_{r_0}^\times \cap K^{s'_{r_0}}$. Since $[D_{r_0}: E_{r_0}] < [D: F]$, we may assume inductively that we know these representations. (They are the ones constructed as above, tensored with characters of $D_{r_0}^\times$ trivial on $D_{r_0}^\times \cap K^{s'_{r_0}}$.) These representations have associated triples $(\tilde{s_1}, \tilde{e_1}, \tilde{f_1}, \ldots, (\tilde{s_{\tilde r}}, \tilde{e_{\tilde r}}, \tilde{f_{\tilde r}})$. We associate corresponding triples $(s_{r_0+1}, e_{r_0+1}, f_{r_0+1}), \ldots, (s_r, e_r, f_r)$, $r = r_0 + \tilde{r}$, by setting $s_{r_0+1} = f_{r_0} \tilde{s_i}$ (to account for the difference in valuations between D and D_{r_0}), $e_{r_0+i} = e_{r_0} \tilde{e_i}$ (because $e(E_{r_0}/F) = e_{r_0}$), and $f_{r_0+i} = f_{r_0} \tilde{f_i}$ (because $f(E_{r_0}/F) = f_{r_0}$).

We need to say something about the case $s_r = 0$. Because of the inductive procedure, we may assume that $r = 1$. The representations considered here are representations of $D^\times$ trivial on K^1. On $K/K^1 \cong k_n^\times$, we let σ_0 be any character that does not factor through $N_{k_n/k}$, and let f_1 be the smallest divisor of n such that σ_0 factors through $N_{k_n/k_{f_1}}$. Set $e_1 = 1$. Then σ_0 (lifted to K) extends to $K \cdot \langle \varpi^{f_1} \rangle$, and the extension induces to G irreducibly. This procedure means that we satisfy (3) of Theorem 1.5.

The construction has produced a representation σ_1 on J_0; we need a representation σ on J. This, too, is done inductively. J_0 and J differ because s'_i and s''_i can differ. Assume that $s''_1 = s'_1 - 1$. Then $(J \cap K^{s''_1})J_0/J_0$ has as coset representatives the elements $y_\beta = 1 + \beta \varpi^{s''_1}$ with β running through a complementary subspace to k_{n/e_1} in k_n. If we define σ'_1 on $(J \cap K^{s''_1})J_0$ by $\sigma'_1(y_\beta w) = \sigma_1(w)$ for $w \in J_0$, then a calculation shows that σ'_1 is a projective, or multiplier, representation. To get an ordinary representation σ_1, we tensor σ'_1 with a representation that cancels out the multiplier and is trivial on $J_0 \cap K^1$. It turns out that this representation is (essentially) a Weil representation. Now suppose that $s''_2 < s'_2$; we repeat this procedure with σ_1 and $(J \cap K^{s''_2})J_0/(J \cap K^{s''_1})J_0$, and so on. Further details are provided in the Appendix. It is important (and true) that $\sigma_1 = \tilde{\sigma_1} \Leftrightarrow \sigma = \tilde{\sigma}$, where $\tilde{\sigma}$ is constructed from $\tilde{\sigma_1}$ by tensoring with Weil representations. One can prove this by comparing characters.

We now have an irreducible representation σ on J. That σ induces irreducibly to a representation π of $D^\times$ is easy: $\sigma \mid H_0$ is a multiple of

$\chi|_{H_0}$, and therefore if $\sigma^x = \sigma$, then $\chi^x = \chi$ on $H_0 \cap x^{-1}H_0x$. This means that any x intertwining σ with itself is easily seen to be in J. That different σ induce different π is similar. Let σ_1, σ_2 be representations of J_1, J_2 respectively. Let $\sigma_j|_{H_{0,j}}$ be a multiple of χ_j, and let $\{(s_i(j), e_i(j), f_i(j)) : 1 \leq i \leq r(j)\}$ be the triples associated with σ_j, $j = 1,2$. Assume that σ_1^x and σ_2 intertwine on their common domain; then χ_1^x and χ_2 agree on their common domain. Since χ_1^x is trivial on $K^{s_1(1)+1}$ but not on $K^{s_1(1)}$, we must have $s_1(1) = s_2(1)$. Hence $\chi_1^x = \chi_2$ on $K^{s_1'(1)}$. Because we chose nonconjugate χ in the construction, $\chi_1 = \chi_2$ there. Thus $x \in J_{0,1}^{s_1'(1)}$ and $(s_i(1), e_i(1), f_i(1)) = (s_i(2), e_i(2), f_i(2))$ if $s_i(1) \geq s_1'(1)$; furthermore, the division algebras $D_{(j)}(1)$, $D_{(j)}(2)$ in the proof are equal if $j \geq s_1'(1)$. If $s_2(1) \geq s_1'(1)$, we can repeat this argument with χ_1^x, χ_2 on $H_{0,1}^{s_2'(1)} = H_{0,2}^{s_2'(2)}$. We thus prove inductively that if $r_0(j)$ is the first index with $s^1_{r_o(j)} > s_{r_0(j)+1}(j)$, then $(s_i(1), e_i(1), f_i(1)) = (s_i(2), e_i(2), f_i(2))$ for $i \leq r_0(1)$, $r_0(1) = r_0(2)$, and $D_{(j)}(1) = D_{(j)}(2)$ for $j \geq s'_{r_o}(1)$; furthermore, $\chi_1 = \chi_2$ on $H_{0,1} = H_{0,2}$. Therefore $x \in J_1 = J_2$. Since σ_1^x intertwines σ_2, $\sigma_1 = \sigma_2$.

We still need to see that we have constructed all the irreducibles of $D^\times$. One procedure is to count all the representations trivial on K^m that are constructed by tensoring the above irreducible π with characters of $D^\times$ trivial on ϖ. Since Koch [15] has computed the number of conjugacy classes of $D^\times / K^m \langle \varpi \rangle$, we need only verify that the numbers are equal. This was the procedure used in [5]; the interested reader can look there for details. A second procedure, alluded to in [5], is to analyze the construction of the π and show that we have accounted for every possible representation in the construction. (To verify this in the step from J_0 to J, one uses Frobenius reciprocity and counts dimensions). A third possibility is to compute the dimensions of the π's and show that

$$\sum_{\pi \text{ trivial on } \langle \varpi \rangle K^m} (\dim \pi)^2 = [D^\times : \langle \varpi \rangle K^m] = n(q^n - 1)q^{n(m-1)}.$$

The dimensions are computed in [5], but I've never tried to sum them.

Remark. Instead of producing representations on J_0 and J, one could first construct a character χ' on a subgroup

$$H = K^{s_1'}(K^{s_2'} \cap D_1^\times) \cdots (K^{s_r'} \cap D_{r-1}^\times) D_r^\times$$

such that $\chi'^x = \chi'$ iff $x \in H'' = K^{s_1''}(K^{s_2''} \cap D_1^\times) \cdots (K^{s_r''} \cap D_{r-1}^\times) D_r^\times$. Then one extends χ' to σ on H'' via the Weil-type construction as above and induces to get π on $D^\times$. To construct H and χ', set $H = H_0 H_1$, where $H_1 = (K^{s'_{r_0+1}} \cap D_{r_0}^\times) \cdots (K^{s_r'} \cap D_{r-1}^\times) D_r^\times \subseteq D_{r_0}$. We construct H_1 and χ' by the same inductive procedure used to find H_0 and χ.

II. SUPERCUSPIDAL REPRESENTATIONS OF $GL_n(F)$

6. The Structure of $GL_n(F)$.

The construction of supercuspidal representations of $GL_n(F)$ parallels in many ways the construction of irreducibles of $D^\times$. As in the division algebra case, the supercuspidals are "parametrized" by triples (s_i, e_i, f_i) satisfying the conditions (1)–(8) of Theorem 1.5, but with the additional requirement that $e_r f_r = n$. (There are square integrable representations corresponding to $e_r f_r < n$, the "generalized special" representations. See [10] for a construction of the special or Steinberg, representation, and [22] for the generalized specials.) We give the theorem in the next section, since some of the notation of this section is needed to describe it.

In this section, we assume that the triples (s_i, e_i, f_i) satisfying (1)–(4) of Theorem 1.5 are given and that $e_r f_r = n$; we set $e = e_r$, $f = f_r$.

We introduce more structure into $GL_n(F)$. Think of $GL_n(F)$ and the matrix algebra $M_n(F)$ as acting on F^n. Given $e|n$, we define a lattice chain $\mathcal{L} = \{\dots, L_{-1}, L_0, L_1, \dots\}$ in F^n as follows: let $\mathcal{O}_F$ = ring of integers of F, P_F = prime ideal of $\mathcal{O}_F$, and set

$$\begin{aligned}
L_0 &= \mathcal{O}_F \oplus \cdots \oplus \mathcal{O}_F && (n \text{ copies of } \mathcal{O}_F)\\
L_k &= \mathcal{O}_F \oplus \cdots \oplus \mathcal{O}_F \oplus P_F \oplus \cdots \oplus P_F && (kf \text{ copies of } P_F, \text{ for } 1 \le k < e)\\
L_e &= P_F \oplus \cdots \oplus P_F && (n \text{ copies of } P_F)\\
L_{ae+b} &= P_F^a L_b, && 0 \le b < e \text{ and } a \in \mathbb{Z}.
\end{aligned}$$

Now set

$$\begin{aligned}
\mathcal{A}_e &= \{x \in M_n(F) : xL_i \subseteq L_i, \forall_i\},\\
\mathcal{A}_e^m &= \{x \in M_n(F) : xL_i \subseteq L_{i+m}, \forall_i\},\\
K_e &= \text{group of invertible elements in } \mathcal{A}_e,\\
K_e^m &= 1 + \mathcal{A}_e^m \ (\text{for } m \ge 1).
\end{aligned}$$

Let ϖ_F be a prime element in F (i.e., $\varpi_F \mathcal{O}_F = P_F$), and set

$$\varpi_n = \begin{bmatrix} 0 & 1 & 0 & & 0 \\ \vdots & 0 & 1 & \ddots & \\ \vdots & & \ddots & \ddots & \ddots \\ 0 & & & \ddots & 1 \\ \varpi_F & 0 & \cdots & \cdots & 0 \end{bmatrix}, \quad \varpi_e = \varpi_n^f,$$

(ϖ_e has f ϖ_F's); $\varpi_n^n = \varpi_e^e = \varpi_F I$. Let Z_e = group generated by ϖ_e. Then:

$Z_e K_e$ is a group, as are the K_e^m;

$Z_e K_e$ normalizes each K_e^m;

$Z_e K_e = \{x \in G : \exists \text{ some } j \in \mathbb{Z} \text{ with } xL_i = L_{i+j}, \text{ all } i\}$.

(The introduction of the $\mathcal{A}_e^m$ into this subject, and the corresponding description of these groups, is due to C. Bushnell.) Note that $\varpi_e L_j = L_{j+1}$ for all j.

It may be useful to see what these groups look like in some cases. Let $n = 6$. For $e = 1$ and $f = 6$,

$$\mathcal{A}_1 = \begin{bmatrix} \mathcal{O} & \mathcal{O} & \mathcal{O} & \mathcal{O} & \mathcal{O} & \mathcal{O} \\ \mathcal{O} & \mathcal{O} & \mathcal{O} & \mathcal{O} & \mathcal{O} & \mathcal{O} \\ \mathcal{O} & \mathcal{O} & \mathcal{O} & \mathcal{O} & \mathcal{O} & \mathcal{O} \\ \mathcal{O} & \mathcal{O} & \mathcal{O} & \mathcal{O} & \mathcal{O} & \mathcal{O} \\ \mathcal{O} & \mathcal{O} & \mathcal{O} & \mathcal{O} & \mathcal{O} & \mathcal{O} \\ \mathcal{O} & \mathcal{O} & \mathcal{O} & \mathcal{O} & \mathcal{O} & \mathcal{O} \end{bmatrix},$$

$$\mathcal{A}_1^j = \begin{bmatrix} P^j & P^j & P^j & P^j & P^j & P^j \\ P^j & P^j & P^j & P^j & P^j & P^j \\ P^j & P^j & P^j & P^j & P^j & P^j \\ P^j & P^j & P^j & P^j & P^j & P^j \\ P^j & P^j & P^j & P^j & P^j & P^j \\ P^j & P^j & P^j & P^j & P^j & P^j \end{bmatrix} \quad (j \in \mathbb{Z})$$

(where $\mathcal{O} = \mathcal{O}_F$, $P^j = P_F^j$; an entry of P^j means that the entry must be in P^j, etc.) For $e = 3$,

$$\mathcal{A}_e = \begin{bmatrix} \mathcal{O} & \mathcal{O} & \mathcal{O} & \mathcal{O} & \mathcal{O} & \mathcal{O} \\ \mathcal{O} & \mathcal{O} & \mathcal{O} & \mathcal{O} & \mathcal{O} & \mathcal{O} \\ P & P & \mathcal{O} & \mathcal{O} & \mathcal{O} & \mathcal{O} \\ P & P & \mathcal{O} & \mathcal{O} & \mathcal{O} & \mathcal{O} \\ P & P & P & P & \mathcal{O} & \mathcal{O} \\ P & P & P & P & \mathcal{O} & \mathcal{O} \end{bmatrix}, \quad \mathcal{A}_3^1 = \begin{bmatrix} P & P & \mathcal{O} & \mathcal{O} & \mathcal{O} & \mathcal{O} \\ P & P & \mathcal{O} & \mathcal{O} & \mathcal{O} & \mathcal{O} \\ P & P & P & P & \mathcal{O} & \mathcal{O} \\ P & P & P & P & \mathcal{O} & \mathcal{O} \\ P & P & P & P & P & P \\ P & P & P & P & P & P \end{bmatrix},$$

$$\mathcal{A}_3^2 = \begin{bmatrix} P & P & P & P & \mathcal{O} & \mathcal{O} \\ P & P & P & P & \mathcal{O} & \mathcal{O} \\ P & P & P & P & P & P \\ P & P & P & P & P & P \\ P^2 & P^2 & P & P & P & P \\ P^2 & P^2 & P & P & P & P \end{bmatrix}, \quad \mathcal{A}_3^3 = P\mathcal{A}_3, \quad \mathcal{A}_e^4 = P\mathcal{A}_3^1, \quad \text{etc.}$$

And for $e = 6$,

$$\mathcal{A}_6 = \begin{bmatrix} \mathcal{O} & \mathcal{O} & \mathcal{O} & \mathcal{O} & \mathcal{O} & \mathcal{O} \\ P & \mathcal{O} & \mathcal{O} & \mathcal{O} & \mathcal{O} & \mathcal{O} \\ P & P & \mathcal{O} & \mathcal{O} & \mathcal{O} & \mathcal{O} \\ P & P & P & \mathcal{O} & \mathcal{O} & \mathcal{O} \\ P & P & P & P & \mathcal{O} & \mathcal{O} \\ P & P & P & P & P & \mathcal{O} \end{bmatrix}, \quad \mathcal{A}_6^1 = \begin{bmatrix} P & \mathcal{O} & \mathcal{O} & \mathcal{O} & \mathcal{O} & \mathcal{O} \\ P & P & \mathcal{O} & \mathcal{O} & \mathcal{O} & \mathcal{O} \\ P & P & P & \mathcal{O} & \mathcal{O} & \mathcal{O} \\ P & P & P & P & \mathcal{O} & \mathcal{O} \\ P & P & P & P & P & \mathcal{O} \\ P & P & P & P & P & P \end{bmatrix},$$

$$\mathcal{A}_6^2 = \begin{bmatrix} P & P & \mathcal{O} & \mathcal{O} & \mathcal{O} & \mathcal{O} \\ P & P & P & \mathcal{O} & \mathcal{O} & \mathcal{O} \\ P & P & P & P & \mathcal{O} & \mathcal{O} \\ P & P & P & P & P & \mathcal{O} \\ P & P & P & P & P & P \\ P^2 & P & P & P & P & P \end{bmatrix}, \quad \mathcal{A}_6^3 = \begin{bmatrix} P & P & P & \mathcal{O} & \mathcal{O} & \mathcal{O} \\ P & P & P & P & \mathcal{O} & \mathcal{O} \\ P & P & P & P & P & \mathcal{O} \\ P & P & P & P & P & P \\ P^2 & P & P & P & P & P \\ P^2 & P^2 & P & P & P & P \end{bmatrix},$$

$$\mathcal{A}_6^4 = \begin{bmatrix} P & P & P & P & \mathcal{O} & \mathcal{O} \\ P & P & P & P & P & \mathcal{O} \\ P & P & P & P & P & P \\ P^2 & P & P & P & P & P \\ P^2 & P^2 & P & P & P & P \\ P^2 & P^2 & P^2 & P & P & P \end{bmatrix}, \mathcal{A}_6^5 = \begin{bmatrix} P & P & P & P & P & \mathcal{O} \\ P & P & P & P & P & P \\ P^2 & P & P & P & P & P \\ P^2 & P^2 & P & P & P & P \\ P^2 & P^2 & P^2 & P & P & P \\ P^2 & P^2 & P^2 & P^2 & P & P \end{bmatrix},$$

$\mathcal{A}_6^6 = P\mathcal{A}_6$, $\mathcal{A}_6^7 = P\mathcal{A}_6^1$, etc. There's also a sequence for $\mathcal{A}_e^2$.

Let $[E:F] = n$, with $e(E/F) = e$ and $f(E/F) = f$. We choose a basis for E/F as follows: let F_f be the maximal unramified extension of F in E, and $a_1, \dots, a_f$ to be a basis for F_f/F composed of roots of unity, ϖ_E to be a prime for E. The basis is then

$$\varpi_E^{e-1}a_1, \dots, \varpi_E^{e-1}a_f, \varpi_E^{e-2}a_1, \dots, \varpi_E^{e-2}a_f, \dots, a_f.$$

Define

$$L_0 = \varpi_E^{e-1}a_1\mathcal{O}_F \oplus \varpi_E^{e-1}a_f\mathcal{O}_F \oplus \varpi_E^{e-2}a_1\mathcal{O}_F \oplus \cdots \oplus a_f\mathcal{O}_F,$$

$$L_1 = \varpi_E^{e-1}a_1\mathcal{O}_F \oplus \cdots \oplus \varpi_E^1 a_f\mathcal{O}_F + a_1P_F + \cdots a_fP_F$$

(the number of P_F terms is f),

$$L_2 = \varpi_E^{e-1}a_1\mathcal{O}_F \oplus \cdots \oplus \varpi_E^1 a_f\mathcal{O}_F + \varpi_E^1 a_1P_F + \cdots + a_fP_F$$

(the number of P_F terms is $2f$),

and so on. Then ϖ_E maps L_0 to L_1, L_1 to L_2, etc., so that $\varpi_E \in Z_eK_e$. Since $F_{f_i}^\times \in Z_eK_e$ also, we get an embedding of $E^\times$ in Z_eK_e.

For $m \geq 1$, $K_e^m/K_e^{m+1} \cong \mathcal{A}_e^m/\mathcal{A}_e^{m+1} \cong M_f(k)^e$; $K_e/K_e^1 \cong GL_f(k)^e$. (As before, k = residue class field of F; k has q elements.) Choose coset representatives for $M_f(k)$ in $\mathcal{A}_e$ (mod $\mathcal{A}_e^1$), replicate them to get coset representatives for $M_f(k)^e$, and restrict to get coset representatives for $GL_f(k)^e$. A coset representative for $M_f(k)^e$ is thus an element of the form

$$\alpha = (\alpha_1, \dots, \alpha_e) = \begin{bmatrix} \alpha_1 & 0 & \cdots & 0 \\ 0 & \alpha_2 & \cdots & 0 \\ \vdots & \vdots & \ddots & \vdots \\ 0 & 0 & \cdots & \alpha_e \end{bmatrix},$$

where each entry represents an $f \times f$ matrix. A calculation gives

$$\varpi_e\alpha\varpi_e^{-1} = \alpha^{\sigma_e} = (\alpha_2, \dots, \alpha_e, \alpha_1). \tag{6.1}$$

Let m_e be the set of coset representatives for $M_e(k)^e$. Any nonzero element of $M_n(F)$ has a unique expression of the form

$$(6.2) \qquad x = \sum_{j=j_0}^{\infty} \alpha_j \varpi_e^j, \quad \alpha_j \in m_e \text{ for all } j, \text{ and } \alpha_{j_0} \neq 0.$$

The multiplication rule is given by (6.1) (and, of course, the multiplication law for the elements of m_e). This is quite similar to the situation for division algebras. One critical difference is that it is generally impossible to tell from (6.2) whether $x \in GL_n(F)$. But if α_{j_0} is invertible mod $\mathcal{A}_e^1$ (i.e., if it represents an element of $GL_f(k)^e$), then $x \in Z_e K_e$.

We choose the elements of m_e with some care. Recall that we are dealing with a given sequence of triples (s_i, e_i, f_i). We first embed F_f (the unramified extension of F with $[F_f : F] = f$) in $M_n(F)$ as follows: F_{f_1} is generated over F by a $(q^{f_1} - 1)^{th}$ root of unity, and we can embed this root of unity in $M_{f_1}(F)$ so that all entries are integers. Extend this embedding "diagonally" to $M_n(F)$, so that a typical element of F_{f_1} looks like $\begin{bmatrix} a & \cdots & 0 \\ \vdots & \ddots & \vdots \\ 0 & \cdots & a \end{bmatrix}$, where $a \in F_{f_1} \subseteq M_{f_1}(F)$ and the matrix is an $n/f_1 \times n/f_1$ matrix of $f_1 \times f_1$ blocks. The elements in $M_n(F)$ commuting with F_{f_1} are then matrices of the form $\begin{bmatrix} a_{11} & \cdots & a_{1d} \\ \vdots & & \vdots \\ a_{d1} & \cdots & a_{dd} \end{bmatrix}$, $d = n/f_1$, where each $f_1 \times f_1$ block $a_{ij} \in F_{f_1}$. This algebra can therefore be regarded as $M_d(F_{f_1})$. Now embed F_{f_2} in $M_d(F_{f_1})$ in the same way, and so on. The result is that F_{f_i} is embedded in $M_n(F)$ diagonally as repeated $f_i \times f_i$ matrices.

Here is a simple example, with, $n = f = 6$, $F = \mathbb{Q}_7$, and $f_1 = 2$. Then F_2 is generated over F by $\sqrt{-1}$, which we embed in $M_2(F)$ as $\begin{bmatrix} 0 & -1 \\ 1 & 0 \end{bmatrix}$. We thus embed F_2 in $M_6(F)$ as elements

$$\begin{bmatrix} a & -b & 0 & 0 & 0 & 0 \\ b & a & 0 & 0 & 0 & 0 \\ 0 & 0 & a & -b & 0 & 0 \\ 0 & 0 & b & a & 0 & 0 \\ 0 & 0 & 0 & 0 & a & -b \\ 0 & 0 & 0 & 0 & b & a \end{bmatrix}.$$

F_6 is generated over F_2 by $\sqrt[3]{2}$, which we embed as $\begin{bmatrix} 0 & I & 0 \\ 0 & 0 & I \\ 2I & 0 & 0 \end{bmatrix}$, $I = \begin{bmatrix} 1 & 0 \\ 0 & 1 \end{bmatrix}$. Then $\alpha + 1 + \alpha_2 \sqrt[3]{2} + \alpha_3 \sqrt[3]{4}$ $(\alpha_j = a_j + b_j \sqrt{-1})$ becomes

$$\begin{bmatrix} a_1 & -b_1 & a_2 & -b_2 & a_3 & -b_3 \\ b_1 & a_1 & b_2 & a_2 & b_3 & a_3 \\ 2a_3 & -2b_3 & a_1 & -b_1 & a_2 & -b_2 \\ 2b_3 & 2a_3 & b_1 & a_1 & b_2 & a_2 \\ 2a_2 & -2b_2 & 2a_3 & -2b_3 & a_1 & -b_1 \\ 2b_2 & 2a_2 & 2b_3 & 2a_3 & b_1 & a_1 \end{bmatrix}.$$

We now choose coset representatives as follows: we choose roots of unity in F_f when possible. These map under the quotient map $A_e \to A_e^1$ to elements of k_f in $M_f(k)^e$. Since k_f is its own commutant in $M_f(k)$, the commutant of k_f in $M_f(k)^e$ is k_f^e; we choose the coset representatives for k_f^e so that each component is a coset representative for k_f. Next, the coset representatives for F_{f-1} map to $k_{f_{r-1}} \subseteq M_f(k)^e$ ($k_{f_{r-1}}$ is embedded diagonally, and in each component as diagonal blocks of $f_{r-1} \times f_{r-1}$ matrices). The commutant of $k_{f_{r-1}}$ in $M_f(k)^e$ is $\simeq M_{f_r/f_{r-1}}(k_{f_{r-1}})^e$; pick coset representatives for the elements that do not already have representatives so that the representatives for each factor are the same and so that the representatives for $M_{f_r/f_{r-1}}(k_{f_{r-1}})^e$ all commute with $F_{f_{r-1}}$ (i.e., with the coset representatives for $k_{f_{r-1}}$). Furthermore, make the representatives stable under multiplication by the group $(k_f^\times)^e$ and invertible iff the corresponding elements of $M_f(k)^e$ are. Continue in the same way. The result is that if an element $\alpha \in M_f(k)^e$ commutes with k_{f_i} (so that $\alpha \in M_{f/f_i}(k_{f_i})$), then its coset representative commutes with F_{f_i}. Furthermore, the coset representatives are stable under σ and under $(k_f^\times)^e$.

We generally identify $M_f(k)^e$ with the coset representatives m_e. In m_e, k_{f_i} is the set of coset representatives for k_{f_i}, and $k_{f_i}^e$ is the set of coset representatives of the form $(\alpha_1, \ldots, \alpha_e)$, where each α_j is a coset representative for k_{f_i} (but the α_j need not be equal). We write $m_e^{f_i}$ for the set of coset representatives of elements commuting with k_{f_i} (hence with $k_{f_i}^e$). The representatives of invertible elements in these sets are denoted by $k_{f_i}^\times$, $(k_{f_i}^\times)^e$, $m_e^\times$, etc. The next lemma, which we use repeatedly, explains some of the above choices.

6.3. Lemma. *Let $t \in \mathbb{Z}$, and suppose that $e_i = e/(e,t)$ (where e_i is in one of the triples (s_i, e_i, f_i). Suppose that $\beta\varpi^t$ satisfies the following:*

(i) $\beta = (\beta_1, \ldots, \beta_e) \in (k_{f_i}^\times)^e$;

(ii) $(\beta\varpi_e^t)^{e_i} = \gamma\varpi_e^{e_i t}$, *where $\gamma \in k_{f_i}^\times$ generates F_{f_i}/F.*

Suppose that $\alpha \in m_e$ and that $\beta\varpi_e^t$ and $\alpha\varpi_e^r$ commute $\operatorname{mod} \mathcal{A}_e^{t+r+1}$ (i.e., $[\beta\varpi_e^t, \alpha\varpi_e^r] \in \mathcal{A}_e^{t+r+1}$). Then $\beta\varpi_e^t$ and $\alpha\varpi_e^r$ commute.

Proof. By induction, either $[(\beta\varpi_e^t)^{e_i}, \alpha\varpi_e^r]$ or $[\gamma\varpi_e^{e_i t}, \alpha\varpi_e^r]$ is an element of $\mathcal{A}_e^{e_i t+r+1}$. But $\varpi_e^{e_i t}$ is a power of ϖ_F, which is central; ϖ_e commutes with $k_{f_i}^\times$ and γ commutes with ϖ_e, so that $[\gamma, \alpha] \in \mathcal{A}_e^1$. That is, the images of α and γ in $M_f(k)^e$ commute. By our choice of coset representatives, α commutes with F_{f_i}. This implies that the components of α commute with those of β. Therefore β and $\alpha\beta^{\sigma^r}(\alpha^{\sigma^t})^{-1} = \beta^{\sigma^r}\alpha(\alpha^{\sigma^t})^{-1}$ are congruent $\operatorname{mod} \mathcal{A}_e^1$. Since $\alpha(\alpha^{\sigma^t})^{-1} \in (k_f^\times)^e$, our choice of coset representatives makes $\beta = \beta^{\sigma^r}\alpha(\alpha^{\sigma^t})^{-1}$, and this proves the lemma. □

Because of this lemma, it will not be important to distinguish between $M_f(k)^e$ and the set of coset representatives m_e, and we generally will not do so.

We will need to consider embeddings of other fields E_0 in $M_n(F)$, where $e(E_0/F) = e_i$ and $f(E_0/F) = f_i$ (and (s_i, e_i, f_i) is one of our triples). One part of the embedding is straightforward: $F_{f_i} \subseteq E_0$, and we have seen how to embed F_{f_i} in $M_n(F)$. The commutant of F_{f_i} is $\cong M_{n/f_i}(F_{f_i})$, which is embedded in a very specific way, and clearly we must embed E_i in $M_{n/f_i}(F_{f_i})$. So for most of the description there is no real loss of generality in assuming that $f_i = 1$, and we make this assumption.

The easiest way to visualize embedding E_0, as diagonal $e_i \times e_i$ matrices, turns out not to be useful. The example that shows what we want is the case $e = n$ and $E_0 = F[\varpi_n^{n/e_i}]$. Consider, for instance, $n = 6$ and $e_i = 2$; write $\varpi = \varpi_F$. Then

$$\varpi_6^3 = \varpi_2 = \begin{bmatrix} 0 & 0 & 0 & 1 & 0 & 0 \\ 0 & 0 & 0 & 0 & 1 & 0 \\ 0 & 0 & 0 & 0 & 0 & 1 \\ \varpi_F & 0 & 0 & 0 & 0 & 0 \\ 0 & \varpi_F & 0 & 0 & 0 & 0 \\ 0 & 0 & \varpi_F & 0 & 0 & 0 \end{bmatrix},$$

and a typical element of E_0 is

$$\begin{bmatrix} a & 0 & 0 & b & 0 & 0 \\ 0 & a & 0 & 0 & b & 0 \\ 0 & 0 & a & 0 & 0 & b \\ \varpi b & 0 & 0 & a & 0 & 0 \\ 0 & \varpi b & 0 & 0 & a & 0 \\ 0 & 0 & \varpi b & 0 & 0 & a \end{bmatrix}, \quad \varpi = \varpi_F. \tag{6.4}$$

The key properties of these matrices are:

(1) The only nonzero entries are at indices (h, j) with $h \equiv j \bmod e/e_i$ (in our example, $h \equiv j \bmod 3$).

(2) The matrices formed by looking only at rows and columns with indices in a fixed conjugacy class are all the same. (In our example, they're all $\begin{bmatrix} a & b \\ \varpi b & a \end{bmatrix}$.)

Another way of saying this is as follows: let P be the permutation matrix such that conjugation by P, $x \mapsto P^{-1}xP$, rearranges the rows and columns of an $n \times n$ matrix in the order

$$1, e/e_i + 1, 2e/e_i + 1, \ldots, (e_i - 1)e/e_i + 1, 2, e/e_i + 2, \ldots, e/e_i, 2e/e_i, \ldots, e.$$

Then conjugation by P turns the elements of E_0 into block diagonal matrices. For instance, the element in (6.4) turns into

$$\begin{bmatrix} a & b & 0 & 0 & 0 & 0 \\ \varpi b & a & 0 & 0 & 0 & 0 \\ 0 & 0 & a & b & 0 & 0 \\ 0 & 0 & \varpi b & a & 0 & 0 \\ 0 & 0 & 0 & 0 & a & b \\ 0 & 0 & 0 & 0 & \varpi b & a \end{bmatrix}.$$

Any totally ramified extension E_0 of F with $[E_0:F] = e_i$ embeds in $M_n(F)$ similarly. We then say that E_0 is "nicely embedded" in $M_n(F)$.

What about the elements commuting with E_0 in $M_n(F)$? Let ξ_0 be a prime element of E. If P is as above, then we may write

$$(6.5) \qquad P^{-1}\xi_0 P = \begin{bmatrix} \xi & 0 & \dots & 0 \\ 0 & \xi & \dots & 0 \\ \vdots & \vdots & \ddots & \vdots \\ 0 & 0 & \dots & \xi \end{bmatrix}, \quad \xi \in M_{e_i}(F).$$

Define η_0 by

$$P^{-1}\eta_0 P = \begin{bmatrix} 0 & 1 & 0 & \dots & 0 \\ 0 & 0 & 1 & & 0 \\ \vdots & \vdots & & \ddots & \\ 0 & 0 & \dots & 0 & 1 \\ \xi & 0 & \dots & 0 & 0 \end{bmatrix}.$$

Then $\eta_0^{n/e_i} = \xi_0$. Furthermore, the form of η_0 shows that

$$\eta_0 = \sum_{j=1}^{\infty} \delta_j \varpi_n^j, \ \delta_j \in k^n, \ \delta_j = 0 \text{ unless } j \equiv 1 \bmod n/e_i, \ \delta_1 \text{ invertible.}$$

Similarly, $\eta_0^{n/e} = \eta_1$ (say) satisfies $\eta_1^{e/e_i} = \xi_0$,

$$\eta_1 = \sum_{j=1}^{\infty} \varepsilon_j \varpi_e^j, \ \varepsilon_j \in k^n \cap m_e, \ \varepsilon_1 \text{ invertible,}$$
$$\varepsilon_j = 0 \text{ unless } j \equiv 1 \text{ mod } e/e_i.$$

So if we let $m_e(e_i) = \{(\alpha_1, \dots \alpha_e) \in m_e : \alpha_h = \alpha_j \text{ if } h \equiv j \bmod e/e_i\}$, then conjugation by η_1 induces σ on $m_e(e_i)$. (Note: $\varpi_e^{e/e_i} \alpha \varpi_e^{-e/e_i} = \alpha$ if α is an element of $m_e(e_i)$.) Also, $m_e(e_i)$ gives a set of coset representatives for $(M_{n/e_i}(E_0) \cap \mathcal{A}_e)/(M_{n/e_i}(E_0) \cap \mathcal{A}_e^1)$, and $\eta_1 \in M_{n/e_i}(E_0)$ generates the ideal $\mathcal{A}_e^1$ of $\mathcal{A}_e$. Dimension-counting shows η_1 and $m_e(e_i)$ generate $M_{n/e_i}(E_0)$.

We can also describe $GL_{n/e_i}(E_0)$. Recall first that from [13] or [1], $GL_n(F) = K_1^1 W K_1^1$, where W is the group of permutation matrices times diagonal matrices of the form

$$(6.6) \qquad \begin{bmatrix} \alpha_1 \varpi^{h_1} & & 0 \\ & \ddots & \\ 0 & & \alpha_n \varpi^{h_n} \end{bmatrix}, \quad j_j \in \mathbb{Z} \text{ and } \alpha_j \in k^\times.$$

(W is essentially the affine Weyl group.) Similarly,

$$GL_{n/e_i}(E_0) = K_{1,0}^1 W_0 K_{1,0}^1,$$

where $K_{1,0} = K_1 \cap GL_{n/e_i}(E_0)$ and W_0 is generated by the analogue of the permutation matrices and of the diagonal matrices. The latter are matrices x such that

$$(6.7) \qquad P^{-1}xP = \begin{bmatrix} \alpha_1\xi^{h_1} & & 0 \\ & \ddots & \\ 0 & & \alpha_d\xi^{h_d} \end{bmatrix}, \quad d = n/e_i, \quad \alpha_j \in k^\times,$$

and ξ as in (6.5). The permutation matrices permute the first n/e_i rows and columns of the matrix, the next n/e_i rows and columns "in the same way," and so on. For instance, with $n = 6$ and $e_i = 2$, the permitted permutations are the trivial one and

$$(123)(456), (132)(465), (12)(45), (13)(46), (23)(56).$$

Now reintroduce f_i. Everything said above remains true with minor changes, the basic one being that we have to regard all matrix entries as $f_i \times f_i$ matrices representing elements of F_{f_i}. For example, the permutation matrices described above are now $n/f_i \times n/f_i$ block permutation matrices in which each block is 0 or I. If, for example, $n = 12$ and $e_i = f_i = 2$, then the matrix corresponding to (12)(45) in the previous description permutes the first and second blocks, then the fourth and fifth blocks; it is therefore (13)(24)(79)(8,10).

Let $e = e_r$, $f = f_r$. The algebra M_0 of elements of $M_n(F)$ commuting with E_0 is isomorphic to $M_d(E_0)$; it is generated by $\eta^{f/f_0} = \eta_0$ and $m_e^{f_0}(e_0) = \{\alpha = (\alpha_1, \dots, \alpha_e) \in m_e\colon$ each α_i commutes with k_{f_0}, and $\alpha_i = \alpha_j$ if $i \equiv j \bmod e/e_i\}$. (Notice that ξ_0 commutes with $m_e^{f_0}(e_0)$. For in the expansion of ξ_0 as a "power series" in ϖ_e with coefficients in m_e, all coefficients are in fact in $(k_{f_0})^e$ and thus commute with these α_i; also, only powers of ϖ_e^{e/e_0} appear, and these induce σ^{e/e_0}, which is trivial on $m_e^{f_0}(e_0)$. Similarly, η_0 induces σ_e on $m_e^{f_0}(e_0)$; observe that $\eta_0^{e/e_0} = \xi_0$.) Let $G_0 =$ group of invertible elements of M_0. We have $G_0 = K_{1,0}^1 W_0 K_{1,0}^1$, where $K_{1,0}^1 = K_{e_0}^1 \cap G_0$ and W_0 is the group of "diagonal" matrices

$$\begin{bmatrix} \alpha_1\xi^{h_1} & & 0 \\ & \ddots & \\ 0 & & \alpha_d\xi^{h_d} \end{bmatrix}$$

where $d = n/e_0f_0$, each entry stands for an $e_0f_0 \times e_0f_0$ matrix (best regarded as an $e_0 \times e_0$ matrix of $f_0 \times f_0$ blocks, the $e_0 \times e_0$ matrix embedded as in (6.7), and the α_j are in k_{f_0}.

We will need two additional pieces of notation later. Let $e_0 \mid e$. For $\alpha \in m_e$, define $Tr_{e_0}\alpha = \sum_{j=1}^{e_0} \alpha^{\sigma^{je/e_0}}$. Then $Tr_{e_0}\alpha$ has periodic entries with period e_0; that is, $Tr_{e_0}\alpha \in m_e^1(e_0)$. For $\beta \in m_e^1(e_0)$, define $Tr^{(e_0)}\beta = \sum_{j=1}^{e/e_0} \beta^{\sigma^j}$. Notice that if $\alpha = (\alpha_1, \dots, \alpha_n)$ and we set $\gamma = \sum_{j=1}^n \alpha_j$, then $Tr^{(e_0)}Tr_{e_0}\alpha = (\gamma, \gamma, \dots, \gamma)$. If $e = n$, then $\gamma = Tr\,\alpha$, and we often regard $Tr^{(e_0)}Tr_{e_0}\alpha$ as γ.

7. The Theorem for $GL_n(F)$.

The theorem on supercuspidals for $GL_n(F)$ is similar to the theorem on irreducible representations of $D_n^\times$. We fix a prime ϖ_F of F in advance, and we regard $F^\times$ as embedded in $G = GL_n(F)$ in the obvious way.

7.1. Theorem. *Let $(s_1, e_1, f_1), \ldots, (s_r, e_r, f_r)$ satisfy (1)–(4) of Theorem 1.5, with (2) modified to say that $e_r f_r = n$. Let $t_i = s_i/f$; let $t_i' = [t_i/2]+1$, $t_i'' = t_i + 1 - t_i'$ (but $t_r' = t_r'' = 0$ if $t_r = 0$). Define C_j, $0 \le j \le t_1$, by*

$$C_j = 1 \text{ if } t_{i+1} < j < t_i \text{ and } e + j_{e_i};$$

$$C_{t_i} = \frac{f_{i-1}}{f_i} \sum_{d f_{i-1} | f_i} \mu(f_i/f_{i-1}d)(q^{f_{i-1}} - 1) \quad (\mu = \text{Mobius function});$$

$$C_j = q^f \text{ if } 0 < j < t_r;$$

$$C_j = q^f - 1 \text{ if } j = 0 < t_r.$$

Then there are $e\prod_{j=0}^{t_1} C_j$ choices of fields $E_1, \ldots, E_r$, subgroups H, and representations ρ of H such that:

1. *$e(E_i/F) = e_i$ and $f(E_i/F) = f_i$, for all i.*
2. *$H = K^{t_1'}(K_e^{t_2'} \cap G_1) \cdots (K_e^{t_r'} \cap G_{r-1})G_r$, where G_i = subgroup of elements of $GL_n(F)$ commuting with E_i (hence $G_r = E_r^\times$).*
3. *$H \cap K_e^{t_i'} = K_e^{t_1'}(K_e^{t_2'} \cap G_1) \cdots (K_e^{t_i'} \cap G_{i-1})$, and for all i this is a normal subgroup of H.*
4. *On $H^1 = H \cap K_e^1$, ρ is a multiple of a character χ_i; $K_e^{t_1+1} \subset \ker \rho$, and $\varpi_F \in \operatorname{Ker} \rho$.*
5. *For $x \in G$, ρ^x and ρ intertwine on their common domain if and only if $x \in J = K_e^{t_1''}(K_e^{t_2''} \cap G_1) \cdots (K_e^{t_r''} \cap G_{1-1})G_r$. There is then a canonical way of getting a representation ρ_1 on J such that $\sigma\big|_{H^1}$ is a multiple of χ: one extends to a projective representation on J and tensors with a Weil-type representation.*
6. *$\operatorname{Ind}_J^G \rho_1 = \pi$ is an irreducible supercuspidal representation of G, and distinct (H, ρ) gives distinct π.*

As in the case of $D_n^\times$, the statement does not give full information about the E_i; for example, they need to fit together so that (2) and (3) hold. We arrange that in part by making sure that the E_i are nicely embedded, but there is more to the story.

The proof is similar in form to that for division algebras. Assume for now that $t_1 > 0$. We start with χ trivial on $K_e^{t_1+1}$ and define it appropriately on $K_e^{t_1}$. We get a corresponding field $E_{(t_1)}$, with $e(E_{(t_1)}/F) = e_1$ and $f(E_{(t_1)}/F) = f_1$, such that $\chi^x = \chi$ iff $x \in K_e^1 G_{(t_1)} K_e^1$, where $G_{(t_1)} \cong GL_{n/e_1 f_1}(E_{(t_1)})$ is the group of elements in G commuting with $E_{(t_1)}$. We also decree that $H \cap K_e^{t_1'} = K_e^{t_1'}$. We next extend χ to $K_e^{t_1-1}$, then to $K_e^{t_1-2}$, and so on. At each state we define a new field $E_{(j)}$ and recompute

$\{x \in G : \chi^x = \chi\}$; when we get to t_2, we define $E_1 = E_{(t_2+1)}$, and so on. We require that $e(E_{t_2}/F) = e_2$ and $f(E_{t_2}/F) = f_2$. Then we set $H \cap K_e^{t_2'} = K_e^{t_1'}(G_1 \cap K_e^{t_2'})$, where $G_1 = G_{(t_1+1)}$. This construction continues inductively. At each stage we consider only certain extensions of χ, since the others do not yield supercuspidals. As in the division algebra case, there are a large number of details to check at each individual step. The next section gives some lemmas needed for this purpose.

We simplify somewhat by assuming from now on that $q > 2$. The difference between $k = \mathbb{F}_2$ and other finite fields k is that the invertible elements of k^n span k^n unless $k = \mathbb{F}_2$. This necessitates further work in some lemmas; see [6] for details.

8. Preliminary Lemmas.

The results here roughly parallel those in Section 3. There are differences for two sorts of reasons. Some facts about matrix algebras are better known than the corresponding facts for division algebras and don't need repeating here. On the other hand, some technical difficulties arise in the GL_n case that were absent in the $D_n^\times$ case, notably because K_e is not a normal subgroup in $GL_n(F)$. Because the proofs are given in full in [6], we will not generally give all details.

The analogue of Proposition 3.5 has been treated in §6. For Proposition 3.6, we need to work with a weaker result than would be ideal. Recall that a nicely embedded field E_0 (of ramification index f_0 and residue class e_0 over F, where (e_0, f_0) is one of the (e_i, f_i)) satisfies:

(i) E_0 is embedded as $n/f_0 \times n/f_0$ block matrices, each block consisting of elements of F_{f_0};

(ii) under the permutation P, which changes the rows and columns of blocks from $(1, 2, \ldots, n/f_0)$ to

$$\begin{aligned}(1,\ &1+n_0,\ 1+2n_0, \ldots, 1+(e_0-1)n_0,\\ &2,\ 2+n_0,\ 2+2n_0, \ldots, 2+(e_0-1)n_0,\\ &\ldots,\\ &n_0,\ 2n_0,\ 3n_0, \ldots, e_0 n_0\ (=n/f_0)),\end{aligned}$$

with $n_0 = n/(e_0 f_0)$, E_0 is embedded as block matrices

$$\lambda = \begin{bmatrix} \lambda_1 & & 0 \\ & \ddots & \\ 0 & & \lambda_{n_0} \end{bmatrix},$$

and $E_0 \subseteq Z_e K_e$;

(iii) $\lambda_1 = \lambda_2 = \cdots = \lambda_{n_0}$.

If we replace (iii) by

(iii′) the λ_h are all conjugate in $GL_{n_0}(F_{f_0})$; if λ is a prime in E_0, then the λ_h all generate $A_{e_0}^1$ (in $GL_{n_0}(F_{f_0})$),

then we simply say that E_0 is *embedded*.

8.1. Proposition. *Let (e_0, f_0) be one of the (e_i, f_i) and let $n_0 = n/e_0f_0$. Suppose that $\eta_0 \in GL_n(F)$ satisfies:*

(i) *η_0 generates A_e^1;*

(ii) *η_0 normalizes $m_e^{f_0}(e_0)$, and conjugation by η_0 is σ there.*

Let M_0 be the subalgebra of $M_n(F)$ generated by η_0 and $m_e^{f_0}(e_0)$, and let E_0 be its center. Then:

(a) *$E_0 = F_{f_0}(\eta_0^{e/e_0})$, and E_0 is an embedded field;*

(b) *$e(E_0/F) = e_0$ and $f(E_0/F) = f_0$.*

(c) *$[M_0 \colon E_0] = n_0^2$.*

Proof. This is Proposition 2.1 of [6]. Let $\xi = \eta_0^{e/e_0}$. It is easy to see that F_{f_0} and ξ commute with $m_e^{f_0}(e_0)$. They also commute with η_0. Hence $F_{f_0}[\xi]$ is central. Most of the rest is now a matter of counting dimensions. □

The analogues of Corollaries 3.8–3.10 hold, but there is one extra step involved in proving them.

8.2. Lemma. *Let H be an open subgroup of K_e^r, let χ be a character on H, and let $x = 1 + x_0 \in K_e^1$ normalize H. Suppose that x_0 is a sum of invertible elements, $x_0 = x_1 + x_2 + \cdots + x_h$, all of which are in $Z_eK_e \cap A_e^1$, normalize H, and fix χ. Then conjugation by x fixes χ.*

8.3. Lemma. *Let H, χ be as above, let E_0 be nicely embedded in $M_n(F)$, and set $M_0 = M_{n_0}(E_0)$. Let η_0 be the element of M_0 that generates A_e^1, normalizes $m_e^{e_0}(f_0) = m_0$, and acts on m_0 as σ. Suppose that for all $\alpha \in m_0$ and some $j > 0$,*

(i) *$1 + \alpha\eta_0^j$ normalizes H and $\alpha\eta_0^j$ normalizes H when invertible;*

(ii) *$\alpha\eta_0^j$ commutes with $\chi|_{H\cap K_e^{r+j}}$ when invertible.*

Assume also:

(iii) *For $y = 1 + y_0 \in H$ with y_0 invertible, y_0 commutes with $\chi|_{K_e^{2r+j}}$.*

Then $1 + \alpha\eta_0^j$ commutes with χ, for all $\alpha \in m_0$.

In both cases we use Lemma 3.7; the added problem is that elements of m_e need not be invertible. We get around this as follows: in Lemma 8.2, say that $x_0 = x_1 + x_2$, with x_1, x_2 both invertible. Then $1 + x_0 \equiv (1 + x_1)(1 + x_2)(1 - x_1x_2)\cdots$, where for given m this holds mod K_e^m after finitely many terms. Lemma 3.7 now applies to each term in the product. Details are given in Section 4 of [6].

The analogue of Lemma 3.11 is:

8.4. Lemma. *If $x \in K_e^{r_0} \cap (Z_eK_e, Z_eK_e)$, $r \geq 1$, then there are elements u_j, v_j $(1 \leq j \leq h)$ such that each $v_j \in K_e^r$, each u_j is either $\alpha_j\varpi$ or α_j for some $\alpha_j \in m_e^{\times}$, and $x \equiv (u_1, v_1)\cdots(u_h, v_h) \bmod K_e^{r+1}$.*

This is Lemma 4.1 of [6].

We need a few results in place of Proposition 3.12, but they are all proved in essentially the same way. In all cases, χ is a character of K_e^m trivial on K_e^{m+1}; on K_e^m, $\chi(1+\gamma\varpi^m) = \psi \circ Tr(\alpha^{\sigma^m}\gamma) = \psi \circ Tr(\gamma^{\sigma^{-m}})$ for all $\gamma \in m_e$, where $\alpha \in k_{f_0}^e$ and $\alpha\varpi^{-m}$ generates a nicely embedded field E_0 with $e(E_0/F) = e_0$ and $f(E_0/F) = f_0$, and (e_0, f_0) is one of the (e_i, f_i). Let $G_0 =$ subgroup of G commuting with E_0.

8.5. Lemma. *Use the above notation, and let $1 \le j < m$. For $\delta \in m_e$, define χ_δ on K_e^j by*

$$\chi_\delta(y) = \chi(wyw^{-1}y^{-1}),\ y = 1 + \gamma\varpi^j \text{ and } w = 1 + \delta\varpi^{m-j}$$

(χ_δ trivial on K_e^{j+1}). Then:

(i) *$\chi_\delta(y) = 1$ for all δ iff $y \in K_e^j \cap G_0$.*
(ii) *$\chi_\delta \equiv 1$ iff $w \in G_0$ (i.e., if $\delta\varpi^{m-j}$ commutes with $\alpha\varpi^{-m}$).*

8.6. Lemma. *In the situation of Lemma 8.5, let e', f' satisfy $e_0 \mid e'$, $e \mid je'$, $f_0 \mid f'$ (where (e', f') is one of the (e_i, f_i)) and restrict δ to be in $k_{e'}^f$. Then the χ_δ are exactly the characters $\chi^\#$ of K_e^j trivial on $K_e^{j+1}(K_e^j \cap G_0)$ and of the form $\chi^\#(1 + \gamma\varpi^j) = \psi \circ T_r(\varepsilon\gamma^{\sigma^{-j_0}})$ for some $\varepsilon \in k_{e'}^f$.*

8.7. Lemma. *Let χ, E_0, and G_0 be as in Lemma 8.5; for $\delta \in m_e^\times$, define χ_δ by $\chi_\delta(y) = \chi(wyw^{-1}y^{-1})$, $w = \delta\varpi^j$. Then:*

(i) *$\chi_\delta(y) = 1$ for all δ iff $y \in K_e^j \cap G_0$;*
(ii) *$\chi_\delta \equiv 1$ if $w \in G_0$.*

These are Lemmas 4.6–4.8 of [6].

We next need results like Lemma 3.13. We need two approximation lemmas here: one for Z_eK_e and one for the product of permutation matrices with matrices like those in (6.6). We call such matrices "power-permutation matrices"; for G, these are the matrices in W.

The notation we use in the next two lemmas is as follows: we have fields $E_0 = F$, E_1, $E_2, \dots, E_i$ with $e(E_h/F) = e_h$, $f(E_h/F) = f_h$, and $e_0 = 1|e_1|\cdots|e_i$, $f_0 = 1|f_1|\cdots|f_i$, but we do not need $e_hf_h > e_{h-1}f_{h-1}$. The F_h are all assumed to be embedded. We let $n_h = n/e_hf_h$, $M_h = M_{n_h}(E_h) =$ subalgebra of $M_n(F)$ commuting with E_h, and $G_h =$ invertible elements of M_h. Assume that M_h is generated by $M_e^{f_h}(e_h)$ and an element η_h such that η_h generates A_e^1, normalizes $m_e^{f_h}(e_h)$, and acts as σ there; we let $\varpi_e = \eta_0$. Note that $G_0 = G$.

8.8. Lemma. *Suppose that in the above situation there exist integers $t_1 > t_2 > \cdots > t_i > 0$ such that, for $h \ge 1$,*

$$\eta_h = \delta_h\eta_{h-1} + \zeta_{h,h-2} + \cdots,$$

$\zeta_{h,j} \in A_e^{t_j - t_h + 1} \cap M_h$ and $\delta_h \in (k_h^\times)^e$. Then:

(i) *For every $x_h \in A_e^r \cap M_h$, there exists $x_{h-1} \in M_{h-1}, \dots, x_0 \in M_0$ with $x_j \in A_e^{r+t_{j+1}-t_h}$, $0 \le j \le h-1$, and $x = \sum_{j=0}^{h-1} x_j$.*

(ii) *For every* $x_h \in K_e^r \cap G_h$ *there exists* $y_{h-1} \in G_{h-1}, \dots, y_0 \in G_0$ *with* $y_j \in K_e^{r+t_{j+1}-t_h}$, $0 \le j \le h-1$, *and* $x_h = y_{h-2} \cdots y_0$.

(iii) *If* $x_h \in Z_e K_e \cap G_h$ *and* $x_h \notin K_e^1$, *assume* $x_h \in A_e^r$ *but* $\notin A_e^{r+1}$. *Then there exists* $z_{h-1} \in G_{h-1}, \dots, z_0 \in G_0$ *such that* $z_{h-1} \in A_e^r \cap G_{h-1}$, $z_j \in K_e^{t_h - t_{j+1}}$ *for* $0 \le j \le h-2$, *and* $x_h = z_{h-1} z_{h-2} \cdots z_0$.

(iv) *Suppose that* r_1, r_2 *are integers and that* $h \le j \le i$. *Then*

$$\begin{aligned}(A_e^{r_1} \cap M_h)(A_e^{r_2} \cap M_j) \subseteq A_e^{r_1+r_2} &\cap M_{h-1} \\ &+ A_e^{r_1+r_2+t_{h-1}-t_h} \cap M_{h-2} \\ &+ \cdots + A_e^{r_1+r_2+t_1-t_h},\end{aligned}$$

and similarly for $(A_e^{r_2} \cap M_j)(A_e^{r_1} \cap M_h)$.

(v) *If* r_1, r_2 *are integers and* $h < j \le i$, *then*

$$\begin{aligned}(A_e^{r_1} \cap M_h)(A_e^{r_2} \cap M_j) &\subset A_e^{r_1+r_2} \cap M_h \\ &+ A_e^{r_1+r_2+t_h-t_{h+1}} \cap M_{h-1} + \cdots + A_e^{r_1+r_2+t_1-t_{h+1}}\end{aligned}$$

and similarly for $(A_e^{r_2} \cap M_j)(A_e^{r_1} \cap M_h)$.

(vi) *For each* $r > 0$ *and each* $h \le i$,

$$\begin{aligned}1 + A_e^r \cap M_{h-1} + A_e^{r+t_{h-1}-t_h} \cap M_{h-2} + \cdots + A_e^{r+t_2-t_h} \cap M_1 + A_e^{r+t_1-t_h} \\ = (K_e^r \cap G_{h-1})(K_e^{r+t_{h-1}-t_h} \cap G_{h-2}) \cdots K_e^{r+t_1-t_h} = H_{h,r} \quad \text{(say)}\end{aligned}$$

is a group normalized by $G_h \cap Z_e K_e$; $H_{h,r} \cap G_h = K_e^r \cap G_h$.

8.9. Corollary. *Let*

$$H_r = (K_e^r \cap G_{i-1})(K_e^{r+t_{i-1}-t_i} \cap G_{i-2}) \cdots (K_e^{r+t_1-t_i}), \quad j \ge r;$$

let h *be the index such that* $t_{h+1} - t_i + r \le j < t_h - t_i + r$. *(For* $j \ge t_1 - t_i + r$, $h = 0$.*) Then*

$$\begin{aligned}H_r \cap K_e^j / H_r \cap K_e^{j+1} &\cong G_h \cap K_e^j / G_h \cap K_e^{j+1}, \quad \text{and} \\ H_r \cap K_e^j &= (G_h \cap K_e^j)(K_e^{r+t_h-t_i} \cap G_{h-1}) \cdots K_e^{r+t_1-t_i}.\end{aligned}$$

Furthermore, $K_e^{t_1-t_i}$, $K_e^{t_2-t_i} \cap G_1, \dots, K_e^{t_{i-1}-t_i} \cap G_{i-2}$, $K_e \cap G_{i-1}$, *and* η_i *all normalize* H_r.

8.10. Lemma. *In the above situation, also assume the* E_h *are all nicely embedded,* $f_i = f_{i-1}$, *and* $\eta_i \equiv \eta_{i-1} \bmod A_e^2$. *If* $x \in W_i$ *is any power-permutation matrix, then there exists* $u \in (K_e^{t_i - t_{i-1}} \cap G_{i-2}) \cdots (K_e^{t_i - t_1} \cap G_0)$ *such that* $ux \in W_{i-1}$.

These are proved in §3 of [6]. The hypothesis that $f_i = f_{i-1}$ is convenient, and in any case we can satisfy it by using the compositum of E_{i-1} and F_{f_i}.

Finally, we need some results that have no good analogue for $D_n^\times$; they concern elements commuting with a character on some K_e^t. (We also use these results on $GL_{n/e_i f_i}(E_i)$, where E_i is nicely embedded.)

8.11. Lemma. *Let χ, $\tilde{\chi}$ be characters on K_e^t given by*

$$\chi(1+y) = \psi \circ Tr(xy), \quad \tilde{\chi}(1+y) = \psi \circ Tr(\tilde{x}y),$$

with $x = \alpha\varpi^{-t}$, $\tilde{x} = \tilde{\alpha}\varpi^{-t}$. (Then χ, $\tilde{\chi}$ are trivial on K_e^{t+1}.) Suppose that x, $\tilde{x}$ generate fields over F and that for some $w \in G$, $\chi(wuw^{-1}) = \tilde{\chi}(u)$ on their common domain. Then x, $\tilde{x}$ are conjugate in $Z_e K_e$.

8.12. Lemma. *Let χ be defined on K_e^j and be trivial on K_e^{t+1}, with $2j \geq t+1$, and suppose that $\chi = \chi_x$ on K_e^j, where $F[x]$ is a field. Let $G_{(x)} = \{w \in GL_n(F) : wx = xw\}$, and let $e(F[x]/F) \mid e_0$, $f(F[x]/F) \mid f_0$. Let E_0 be an extension of $F[x]$ that is nicely embedded, with $e(E_0/F) = e_0$, $f(F_0/F) = f_0$, and let b be a power-permutation matrix in $GL_{n_0}(E_0) = G_0$, $n_0 = n/e_0 f_0$ (i.e., $b \in W_0$). Let $k_1, k \in K_e^{t-j}$. If $\chi^{k_1 bk} = \chi$ on their common domain, then $k_1 bk = k_1' bk'$, with $k_1', k' \in K_e^{t-j+1}(K_e^{t-j} \cap G_{(x)})$.*

8.13. Lemma. *Let χ be as in Lemma 8.11, and let $y \in G$ satisfy $\chi^y = \chi$ on $K_e^t \cap y^{-1}K_e^t y$. Then $y \in K_e^1 G_{(x)} K_e^1$.*

Lemmas 8.11 and 8.13 are like Lemmas 8 and 13 of [12], and Lemma 8.12 is related to Lemma 2.18 of [16]. They are proved in §5 of [6]. Note that in Lemma 8.12, the hypothesis says that $\chi^b = \chi$ on their common domain.

9. Construction of H_0 and χ.

As in the division algebra case, the heart of the construction involves finding subgroups H_0 and characters χ on H_0 with many of the properties we need for the proof of Theorem 7.1. If $s_1 = 0$, then $r = 1$, $e_1 = 1$, and $f_1 = 1$. We let ρ be an irreducible cuspidal representation of $GL_n(k) \cong K_e/K_e^1$ lifted to K_e, and we extend ρ to $Z_e K_e = \langle \varpi_F \rangle K_e$ by making it trivial on ϖ_F. From [20], there are C_0 distinct ρ; from Theorem 4.1 of [3], these ρ induce to distinct irreducible supercuspidal representations of G.

Assume that $s_1 > 0$. Then there is a smallest r_0 with $s_{r_0} \geq 2s_{r_0+1}$, or, equivalently, $t_{r_0} \geq 2t_{r_0+1}$ (define $s_{r+1} = t_{r+1} = 0$, so that possibly $r_0 = r$). We now construct nicely embedded fields $E_{(j)}$, matrix algebras $M_{(j)}$ (with $G_{(j)}$ = group of invertible elements of $M_{(j)}$), elements $\eta_{(j)}$, a subgroup H_0 (we write $H_0^j = H_0 \cap K_e^j$), and a character χ on H_0 with the following properties (analogous to those in §4, except that we need not restate those under A there; our convention is that i is the largest index with $t_i \geq j$):

B. The matrix algebras $M_{(j)}$ and M_i.

(1) $\eta_{(j)} A_e = A_e^1$, and conjugation by $\eta_{(j)}$ induces σ on $m_e^{f_i}(e_i)$.

(2) $M_{(j)}$ is generated by $\eta_{(j)}$ and $m_e^{f_i}(e_i)$.

(3) $E_{(j)}$ is the center of $M_{(j)}$; $e(E_{(j)}/F) = e_i$, and $f(E_{(j)}/F) = f_i$.

(4) $E_{(j)}$ is generated by $\eta_{(j)}^{e/e_i}$ and F_{f_i}.

We set $E_i = E_{(t_{i+1}+1)}$, $M_i = M_{(t_{i+1}+1)}$, $\eta_i = \eta_{(t_{i+1}+1)}$ for $1 \leq i < r_0$; $E_{r_0} = E_{(s'_{r_0})}$, $M_{r_0} = M_{(s'_{r_0})}$, etc. We sometimes write $\eta_0 = \varpi_e$, $E_0 = F$, $M_0 = M = M_n(F)$, and $G_0 = G$.

C. Relations among the $\eta_{(j)}$ and $M_{(j)}$:

(5) If $t_{i+1} < j < t_i$, then $\eta_{(j)}$ is of the form

$$\eta_{(j)} = \eta_{(j+1)} + y_{i-1} + \cdots + y_0, \text{ with } y_h \in M_h \cap A_e^{t_{h+1}-j+1}.$$

If $j = t_i$, then $\eta_{(j)}$ is of the form $\zeta_i \eta_{i-1} + y_{i-2} + \cdots + y_0$, where the y_h are as above and $\zeta_i \in (k_{f_i}^{\times})^e$. (For $s_{r'_0} \leq j < t_{r_0}$, take $i = r_0$.)

(6) If $t_{i+1} < j < t_i$ and $\ell > 0$, then

$$G_{(j)} \cap K_e^{\ell} \subseteq (G_{(j+1)} \cap K_e^{\ell})(G_{i-1} \cap K_e^{t_i - j + \ell}) \cdots (G_0 \cap K_e^{t_1 - j + \ell});$$

for $j = t_i$,

$$G_{(j)} \cap K_e^{\ell} \subseteq (G_{i-1} \cap K_e^{\ell})(G_{i-2} \cap K_e^{t_{i-1} - t_i + \ell}) \cdots (G_0 \cap K_e^{t_1 - t_i + \ell}).$$

(7) If $t_{i+1} < j < t_i$ and $w \in W_{(j)}$, then $w = w_0 w_1$ with $w_0 \in W_{(j+1)}$ and $w_1 \in (G_{i-1} \cap K_e^{t_i - j}) \cdots (G_0 \cap K_e^{t_1 - j})$; if $j = t_i$ and $w \in W_{(j)}$, then $w = w_0 w_1$, with $w_1 \in (G_{i-1} \cap K_e^{t_i - j}) \cdots (G_0 \cap K_e^{t_1 - j})$ and w_0 a power-permutation matrix for the group of elements commuting with E_{i-1} and F_{f_i}.

We define $H_0 = K_e^{t'_1}(K_e^{t'_2} \cap G_1) \cdots (K_e^{t'_{r_0}} \cap G_{r_0 - 1})$, $H_0^j = H_0 \cap K_e^j$.

(8) If $t'_{h+1} \leq j < t'_h$, then

$$H_0^j = K_e^{t'_1}(K_e^{t'_2} \cap G_1) \cdots (K_e^{t'_h} \cap G_{h-1})(K_e^j \cap G_h)$$
$$\text{and } H_0^j / H_0^{j+1} \cong K_e^j \cap G_h / K_e^{j+1} \cap G_h.$$

D. Properties of the character χ (on H_0):

(9) Given $t'_i < j \leq t_i$, any extension of $\chi|_{H_0^{t_i}}$ to H_0^j extends as a character to $H_0^{t'_i}$.

(10) Say that x commutes with $\chi|_{H_0^j}$ if χ^x equals χ on $H_0^j \cap x^{-1} H_0^j x$. Let $t'_{h+1} \leq j < t'_h$. Then

$$\{x \in Z_e K_e : x \text{ commutes with } \chi|_{H_0^j}\}$$
$$= K_e^{t''_1}(K_e^{t''_2} \cap G_1) \ldots (K_e^{t''_h} \cap G_{h-1})(K_e^{t_{h+1}-j+1} \cap G_h) \ldots$$
$$\ldots (K_e^{t_i - j + 1} \cap G_{i-1})(Z_e K_e \cap G_{(j)}) = J_0^j \text{ (say)}.$$

This is a group, and it normalizes H_0^j. (J_0^j decreases with j.) Let $J_1^j = J_0^j \cap K_e^1$.

(11) $\{x \in G : x \text{ commutes with } \chi|_{H_0^j}\} = J_1^j G_{(j)} J_1^j = J_1^j W_{(j)} J_1^j$.

(12) Let $t'_{h+1} \leq t_i < t'_h$, so that G_{i-1} commutes with χ on $H_0^{t_i+1}$ and $H_0^{t_i}/H_0^{t_i+1} \cong G_h \cap K_e^{t_i}/G_h \cap K_e^{t_i+1}$. Then one can write $\chi|_{H_0^{t_i}}$ as $\chi_{0,i}\chi_{1,i}$, where:

(i) G_{i-1} commutes with $\chi_{0,i}$;

(ii) $\chi_{1,i}$ is trivial on $H_0^{t_i+1}$;

(iii) on $H_0^{s_i} \cap G_{i-1}$, we have

$$\chi_{1,i}(1+\eta_{i-1}^{t_i}) = \psi \circ Tr^{(e_{i-1})}(\alpha_i^{\sigma^{t_i}}\gamma), \text{ for all } \gamma \in m_e^{f_{i-1}}(e_{i-1})$$

for some $\alpha_i \in (k_{f_i})^\times \cap m_e^{f_i}(e_i)$ such that $E_{i-1}[\alpha_i\eta_{i-1}^{-t_i}] = E^{\check{}}_{(t_i)}$ is a field satisfying $e(E^{\check{}}_{(t_i)}/F) = e_i$, $f(E^{\check{}}_{(t_i)}/F) = f_i$;

(iv) the matrix algebra $M^{\check{}}_{(t_i)}$ of elements commuting with $E^{\check{}}_{(t_i)}$ has an element $\tilde{\eta}_{(t_i)}$ generating A_e^1 such that $\tilde{\eta}_{(t_i)}$, $m_e^{f_i}(e_i)$ generate $M^{\check{}}_{(t_i)}$ as in Prop. 2.1, and $\tilde{\eta}_{(t_i)} \equiv \eta_{(t_i)} \bmod A_e^2$.

There will be $\prod_{j=t'_{r_0}}^{t_1} C_j$ nonconjugate (H_0, χ) constructed with these properties.

There is some redundancy in these conditions. For example, (5) implies (6), (7), and (8), because of 8.8–8.10, and (9) holds because of Lemma 8.2 (all elements of $(H_0^{t'_i}, H_0^{t'_i})$ are in $H_0^{t_i}$, and Lemma 8.2 implies χ is trivial on them). Properties (1)–(4) are closely related because of Lemma 8.1.

As in the case with division algebras, the proof is by a backwards induction on j. The outline of the proof is similar to that for division algebras, but there are some extra complications. Here is a sketch of the major part of in the proof; details are given in Sections 6–8 of [6]. For simplicity of exposition, assume in what follows that $e = n$ and $f = 1$; write $\varpi = \varpi_n$.

1. We need the result for $j = t_1$. Any character on $K_e^{t_1}$ trivial on $K_e^{t_i+1}$ is of the form $\chi(1+\gamma\varpi^{t_1}) = \psi \circ Tr(\alpha^{\sigma^{t_1}}\gamma)$ for some $\alpha \in m_e$. We pick α so that $F[\alpha\varpi^{t_1}] = E_{(t_1)}$ is a nicely embedded field with $e(E_{(t_1)}/F) = e_1$ and $f(E_{(t_1)}/F) = f_1$ ($= 1$, by hypothesis). From Lemma 8.11, there are C_{t_1} distinct ways of doing this (since $e = 1$, $C_{t_1} = q-1$). Then $e(E_{(t_1)}/F) = e_1$ because $e_1 = (n, s_1)$, as in the case of division algebras. For (12), we also set $\alpha = \alpha_1$, $\chi = \chi_{1,1}$, $\chi_{0,1} =$ trivial character. Now we form $\eta_{(j)}$ as in §6. All the properties are reasonably straightforward; the hardest is (11), and it follows from Lemma 8.13.

2. We need a result like Lemma 4.1 (stated here for all f, not just $f = 1$):

9.1. Lemma. *Assume that χ is defined on H_0^j, $j > t'_{r_0}$, and that E is a well-embedded field with $e(E/F) = e_i$ and $f(E/F) = f_i$ (where $i =$ largest index with $t_i \geq j$). Let $G_0 = \{x \in G : x$ commutes with $E\}$ and suppose $\chi^x = \chi$ on their common domain if $x \in G_0$. Then χ has an extension χ_0 to H_0^{j-1} such that $\chi_0^x = \chi$ on their common domain if $x \in G_0$.*

Sketch of proof (for $f = 1$, $e = n$). The proof is in 3 stages.

(a) We get an extension χ_1 such that $\chi_1^x = \chi_1$ for $x \in m_n^1(e_i)^\times$.

(b) If $e \nmid (j-1)e_i$, then χ_1 is unique with this property, and $\chi_1^x = \chi_1$ for $x \in W_0 =$ group of permutation-power matrices for G_0. Hence $\chi_1^x = \chi_1$ for $x \in G_0$.

(c) If $e \mid (j-1)e_i$, then we can find χ_0 such that $\chi_0^x = \chi_0$ when $x = \eta$ (the generator of A_e^1 in G_0 as described following (6.5).) For this χ_0, $\chi_0^x = \chi_0$ if $x \in W_0$, so $\chi_0^x = \chi_0$ for $x \in G$.

We do (a) as in the case of division algebras. The set $m_n^1(e_i)$ has $(q-1)^{e/e_i}$ elements and the number of extensions of χ to H_0^j is q^{e/e_h} (where $t'_h \leq j-1 < t'_{h+1}$). So the same argument as for Lemma 4.1 applies. For (b), notice that if χ_1, χ_2 are fixed under $m_e^{f_i}(e_i)^\times$, then so is $\chi^\# = \chi_1\chi_2^{-1}$; moreover, $\chi^\#$ is trivial on K_e^j. So $\chi^\#(1+\gamma\eta^{j-1}) = \psi \circ Tr^{(e_h)}(\delta^{\sigma^{j-1}}\gamma)$, for all $\gamma \in m_n^1(e_h)$ for some $\delta \in m_n^1(e_h)$. Since $m_n^1(e_i)^\times$ commutes with $\chi^\#$, Lemma 8.7 says that $m_n^1(e_i)^\times$ commutes with $\delta\eta^{1-j}$. If $e \nmid (j-1)e_i$, this is impossible unless $\delta = 0$, as a calculation shows. [Notice that m_n is Abelian.) So the issue is whether $m_n^1(e_i)^\times$ commutes with η^{1-j}; since η^{1-j} induces σ^{1-j}, we must have $(e/e_i) \mid (1-j)$.] That's the uniqueness.

If $\chi_1^x = \chi_1$ for $x \in W_0$, then $\chi_1^\eta = \chi_1$ in particular. Lemma 8.2 gives $\chi^x = \chi$ for $x \in G_0 \cap K_e^1$; then (b) holds since $G_0 = (G_0 \cap K_e^1)W_0(G_0 \cap K_e^1)$. We need only check that $\chi_1^x = \chi_1$ for $x \in W_0$. The idea is this: consider $\chi_1^x\chi_1^{-1}$ on $H_0^{j-1} \cap x^{-1}H_0^{j-1}x$. We know $\chi_1^x\chi_1^{-1} = 1$ on $H_0^j \cap x^{-1}H_0^j x$; we must check only for elements y such that $y \in H_0^{j-1}/H_0^j$ or $xyx^{-1} \in H_0^{j-1}/H_0^j$. We can deal with coset representatives, and it suffices to work with elements y such that y or xyx^{-1} is of the form $1+\delta\eta^{j-1}$ with $\delta = (\delta_1, \dots, \delta_n)$ and only one of $\delta_1, \dots, \delta_{n/e_h}$ is nonzero (these y, xyx^{-1} generate $H_0^{j-1} \cap x^{-1}H_0^{j-1}x$ modulo $H_0^j \cap x^{-1}H_0^j x$). Assume for definiteness that y (rather than xyx^{-1}) is of the given form. We can find $\varepsilon \in m_n^1(e_i)^\times$ such that $\varepsilon y \varepsilon^{-1} \equiv y^2 \bmod H_0^j \cap x^{-1}J_0^j x$. Since ε fixes $\chi_1^x\chi_1^{-1}$, we have $\chi_1^x\chi_1^{-1}(y) = \chi_1^x\chi_1^{-1}(\varepsilon y \varepsilon^{-1} y^{-1}) = 1$. That does the trick. (We need $p \neq 2$ here; a variant works for $p = 2$.)

For (c), let χ_α be trivial on H_0^j and given the formula by $\chi_\alpha(1+\gamma\eta^{j-1}) = \psi \circ Tr(\alpha^{\sigma^{j-1}}\gamma)$ for $\gamma \in m_n^1(e_h)$, where $\alpha \in m_n^1(e_h)$. Then χ_α commutes with $m_n^1(f_i)^\times$. We want to choose α such that $\chi_0 = \chi_1\chi_\alpha$ satisfies $\chi_0^\eta = \chi$, or $\chi_1^\eta\chi_1^{-1} = (\chi_\alpha^\eta\chi_\alpha^{-1})^{-1}$. This is a counting argument like Lemma 8.5 (we used the same argument in Lemma 4.1). $\chi_1^{\eta_0}\chi_1^{-1}$ is trivial on $H_0^j(H_0^{j-1} \cap G_0)$, as is $(\chi_\alpha^{\eta_0}\chi_\alpha^{-1})^{-1}$ (for any α). Furthermore, $\chi_\alpha^\eta\chi_\alpha^{-1}$ is trivial on H_0^{j-1} iff α commutes with η. This means that the $\chi_\alpha^\eta\chi_\alpha^{-1}$ exhaust the characters on H_0^{j-1} trivial on H_0^j $(H_0^{j-1} \cap G_0)$, and we can therefore choose α as required.

Because $(e/e_0) \mid j-1$, we can pick a set of coset representatives $\{y_h\}$ for H_0^{j-1}/H_0^j such that $g y_h g^{-1} \in H^{j-1}$ for all $g \in W_0$. For this reason, it suffices to prove that $\chi_0^g = \chi_0$ when g is a diagonal matrix or a permutation matrix. It's not hard to check that if g is diagonal and $y = 1+\delta\eta^{j-1}$ with δ as above (only one of $\delta_1, \dots, \delta_{n/e_h}$ nonzero), then $gyg^{-1} = \eta^a y \eta^{-a}$ for some

a. Therefore $\chi_0^g = \chi_0$ when g is diagonal. The argument for permutation matrices is more difficult. If g is a permutation matrix, then one checks that χ_0^g is stable under $m_n^1(e_i)^\times \cong (k^\times)^d$, $d = n/e_i$. Therefore $\chi_0^g = \chi_0\chi_\alpha$ for some $\alpha \in m_n^1(e_i) = k_n^d$. The group of permutations in W_0 is $\cong S_d$ (the symmetric group on d letters) and acts on k_n^d in the obvious way (permuting the first d terms, the next d in the same way, etc.) Define $\varphi : S_d \to k_n^d$ by $\chi_0^g = \chi_0\chi_{\varphi(g)}$. Then φ is easily seen to satisfy:

(i) $\varphi(g_1g_2) = \varphi(g_1) + \varphi(g_2)^{g_1}$ (φ is a 1-cocycle);

(ii) if g_1 fixes j (g_1 permutes $\{1, 2, \cdots, d\}$) and $\varphi(g_1) = (\alpha_1, \cdots, \alpha_n)$, then $\alpha_j = 0$;

(iii) for $g_0 = (1, 2 \cdots d)$ (the cyclic permutation of order d), $\varphi(g_0) = 0$. (That's because g_0 is the permutation associated with η.)

Properties (i)—(iii) imply that $\varphi \equiv 0$. The idea is this: there is an involution b normalizing the group generated by g_0. Then $\chi_{\varphi(g)}$ is fixed by η and $(k_n^\times)^d$, and therefore $\varphi(b) = (\alpha_0, \cdots, \alpha_0)$, with all entries the same. Suppose that $\alpha_0 = 0$. Write b as a product of disjoint cyclic permutation of order 2, and let g be one of these permutations. From (ii), $\varphi(gb) = (\alpha_1, \ldots, \alpha_n)$, where $\alpha_i = 0$ if g moves i (i.e., gb fixes i). From (i), $\alpha_{g\cdot i} = \alpha_0 + (i^{th}$ entry in $\varphi(g))$. So if $\alpha_0 = 0$, then the ith entry in $\varphi(g)$ is 0 if g moves i, and is 0 (by (ii)) if g fixes i. Thus $\varphi(g) = 0$. Since these permutations and their conjugates under η generate S_n, (i) makes $\varphi \equiv 0$.

Getting $\alpha_0 = 0$ is easy if d is odd; then b fixes some letter, and (ii) applies. It's also easy if p is odd; (i), applied to $g_1 = g_2 = b$, gives $2\varphi(b) = 0$. When d is even and $p = 2$, we use the above analysis to show that

$$\varphi((h, h+1)) = (0, 0, \cdots, \alpha_0, \alpha_0, 0, \cdots, 0) \quad (\alpha_0 \text{ in the } h,\ h+1 \text{ places}),$$

and then write g_0 as a product of involutions $(h, h+1)$; the result is, once again, that $\alpha_0 = 0$. Further details are in Lemma 6.2 of [6]. □

3. The inductive step divides into three cases. Let $t_{i+1} < j \le t_i$; the cases depend on whether $j - 1 = t_{i+1}$ or not and on whether $(e/e_i) \mid j - 1$ or not. Again, we'll assume for simplicity that $f = 1$ and $e = n$.

Case 3a: $j - 1 \ne t_{i+1}$, $(e/e_i) \nmid j - 1$. (Recall that $n = e$.) We extend χ to H_0^{j-1} as χ_0 and set $E_{(j-1)} = E_{(j)}$, $\eta_{(j-1)} = \eta_{(j)}$ etc. Verifying that (1)–(12) extend inductively is easy except for (10) and (11), and the proof for these parts is exactly as in 3b (below).

Case 3b: $j - 1 \ne t_{i+1}$ and $(e/e_i) \mid j - 1$. Again, $n = e$. Any extension of χ to H_0^{j-1} is of the form $\chi_0\chi_{\alpha'}$, where $\alpha' = Tr_{e_h}(\alpha'')$ for some $\alpha'' \in m_n$ and $\chi_{\alpha'}(1 + \gamma\eta_{(j)}^{j-1}) = \psi \circ Tr^{(e_h)}(\alpha''^{\sigma^{j-1}}\gamma)$, all $\gamma \in m_n^1(e_h)$. (See the end of §6 for notation. α'' is not uniquely determined, though $\chi_{\alpha'}$ does determine α' uniquely. Note that $\alpha' \in m_n^1(e_h)$.) Let $\alpha = Tr_{e_i}\alpha''$; then α' determines α uniquely, and $\chi'(1 + \gamma\eta_{(j)}^{j-1}) = \psi \circ Tr^{(e_i)}(\alpha^{\sigma^{j-1}}\gamma)$ for all $m_n^1(e_i)$. We require α' to be chosen so that $\alpha\eta_{(j)}^{j-1} \in E_{(j)}$. Write $\alpha = (\alpha_1, \cdots, \alpha_n)$, $\alpha'' = (\alpha_1'', \cdots, \alpha_n'')$. Since $\alpha \in E_{(j)}$, we have $\alpha_1 = \alpha_2 = \cdots = \alpha_{n/e_i}$, $\alpha_{n/e_i} + 1 = \cdots = \alpha_{2n/e_i}$, etc.; we also require that $\alpha_1'' = \cdots = \alpha_{n/e_i}''$, $\alpha_{n/e_i+1}'' = \cdots = \alpha_{2n/e_i}$, etc. (We might make $\alpha_j'' = 0$ if $j > n/e_i$, for instance.) This restricts α', of course, but it turns out that if $\alpha = 0$, then $\chi_0\chi_{\alpha'}$ and χ_0 are conjugate; therefore we need only one α' for each α (and we want $\alpha' = 0$ for $\alpha = 0$). That gives us C_{j-1} extensions of χ to H_0^{j-1}. Notice that $\chi_{\alpha'}$ is then invariant under $m_n^1(e_i)$ and the permutation matrices in $W_{(j)}$; hence so is χ.

We now need to

(i) produce $\eta_{(j-1)}$ and $E_{(j-1)}$;
(ii) verify (1)–(12);
(iii) prove that these extensions are nonconjugate.

Steps (i) and (ii) involve some work; (iii) is not so bad.

(i): This is done in two steps. First we find an η_0 generating A_e such that $\chi^{\eta_0} = \chi$; then we modify η_0 so that we get something nicely embedded. (The proof in [6] is similar, but is arranged slightly differently.)

The first step is straightforward: $\chi(\eta_{(j)}y\eta_{(j)}^{-1}y^{-1}) = 1$ if $y \in H_0^j$ or if $y \in H_0^{j-1} \cap G_{(j)}$. (We have $\chi = \chi_0\chi_{\alpha'}$. Since $\chi_0^{\eta(j)} = \chi_0$, we need only show that $\chi_{\alpha'}(\eta_{(j)}y\eta_{(j)}^{-1}y^{-1}) = 1$. Since $\chi_{\alpha'}(1+\gamma\eta_{(j)}^{j-1}) = \psi\circ Tr^{(e_i)}(\alpha^{\sigma^{j-1}}\gamma)$ if $\gamma \in m_n^1(e_i)$, this follows from Lemma 8.5.) We set

$$\eta_0 - \eta_{(j)}(1 + \delta_{i-1}\eta_{i-1}^{t_i-j+1}) \cdots (1 + \delta_h\eta_h^{t_{h+1}-j+1}),$$

where $\delta_\ell \in m_n^1(e_\ell)$, and use (12) inductively (as in §4) to make $\chi^{\eta_0} = \chi$. Then η_0^{n/e_i} commutes with $m_n^1(e_i)$. It follows that if P is the permutation matrix (described in §6) such that conjugation by P rearranges the rows and columns in the order

$$(1, e_i + 1, \cdots, n - e_i + 1, 2, \cdots, n - e_i + 2, \cdots, e_i - 1, \cdots, n - 1),$$

then $P\eta_0^{n/e_i}P^{-1}$ is a diagonal block matrix $(\xi_1, \cdots, \xi_{n/e_i})$, where each ξ_h is an $e_i \times e_i$ matrix. Let ξ_0 be the matrix such that $P\xi_0P^{-1}$ is the diagonal block matrix $(\xi_1, I, I, \cdots, I)$, $\eta_{(j-1)}$ the matrix such that $P\eta_{(j-1)}P^{-1}$ is the block matrix

$$\begin{bmatrix} 0 & I & 0 & & 0 \\ \vdots & 0 & I & \ddots & \\ \vdots & & \ddots & \ddots & \ddots \\ 0 & & & \ddots & I \\ \xi_1 & 0 & \cdots & \cdots & 0 \end{bmatrix}.$$

Then ξ_0 commutes with χ (because η_0^{n/e_i} does), and $\eta_{(j-1)}$ is ξ_0 multiplied by a permutation matrix in $W_{(j)}$. Hence $\eta_{(j-1)}$ commutes with χ.

Furthermore, $F[\eta_{(j-1)}^{n/e_i}] = E_{(j-1)}$ is nicely embedded, by the construction. Lemma 8.2 now shows that $K_e \cap G_{(j-1)}$ commutes with χ. Since ξ_0 and the permutation matrices in $W_{(j)}$ generate $W_{(j-1)}$, it follows that $G_{(j-1)}$ commutes with χ.

(ii): Thanks to the work done in (i), properties (1)–(9) are easy to check, and it is also not hard to check that J_0^{j-1} commutes with χ. (The calculation is like that in §4.) We need the converses in (10) and (11), since (12) is automatic in this case.

For (10), we reason as in §4. Let $w \in K_e^j$ commute with χ on H_0^{j-1}. Then $w \in J_0^j$, and, dividing by an element known to commute with χ, $w = 1 + \delta_{t_i-j+1}\eta_{i-1}^{t_i-j+1} + \cdots = w_1 w_0$, $w_0 = 1 + \delta_{t_i-j+1}\eta_{i-1}^{t_i-j+1}$ and $w_1 \in K_e^{t_i-j+1} \cap J_0$. Restrict attention to elements in $H_0^j(H_0^{j-1} \cap G_{i-1})$. On these elements $\chi^{w_1} = \chi$, and we must therefore have $\chi^{w_0} = \chi$. For elements $1 + \gamma_{j-1}\eta_{(j)}^{t_i-j+1}$, (12) and Lemma 8.5 say that $w_0 \equiv$ an element in $M_{(t_i)}^{\sim} \bmod K_e^{t_i-j+2}$. But (12) and (6) now imply that $w_0 \equiv$ an element of $G_{(j-1)} \bmod K_e^{t_i-j+2}$. Dividing by this element, we get

$$w = 1 + \delta_{t_{i-1}-j+2}\eta_{i-1}^{t_i-j+2} + \cdots .$$

Dividing by another element known to commute with χ, we get

$$w = 1 + \delta_{t_i-j+1}\eta_{i-2}^{t_{i-1}-j+1} + \cdots ;$$

we repeat the same argument (now on elements in $H_0^j(H_0^{j-1} \cap G_{(i-2)})$), much as in §4. An induction gives (10).

The reasoning in (11) is similar, but messier. The proof is again inductive. To begin with, if $\chi^x = \chi$ on their common domain, then $x \in J_1^j W_{(j)} J_1^j$; from Lemma 8.10, this is also $J_1^j W_{(j-1)} J_1^j$. The second step of the induction is typical of the general step; here is a description. Assume that we have $x = k_1 b k_2$, where $k_1, k_2 \in (J_0^j \cap K_e^{t_{i-1}-j+1})(G_{i-1} \cap K_e^{t_i-j+1}) \cdots (Z_e K_e \cap G_{(j)})$ and $b \in W_{(j-1)}$. Dividing by elements known to fix χ, we may assume that $k_1, k_2 \in J_0^j \cap K_e^{t_{i-1}-j+1}$. Now restrict attention to $H_0^j(H_0^{j-1} \cap G_{i-2})$. On this group, $J_1^j \cap K_e^{t_{i-1}-j+2}$ commutes with χ; we may therefore assume that $k_h = 1 + \delta_h \eta_{i-2}^{t_{i-1}-j+1}$. We know that $\chi^{k_1} = \chi\chi_1$, $\chi^{k_2^{-1}} = \chi\chi_2$, where χ_1, χ_2 are trivial on H_0^j. Thus we have $(\chi\chi_1)^b = \chi\chi_2$ on their common domain.

The idea now is as follows: Lemma 8.10 lets us write $b = b'k_0$, where k_0 normalizes $J_0^j \cap K_e^{t_{i-1}-j+1}$ and $b' \in W_{i-2}$. We will construct a character $\chi^{\sim}$ on $(H_0^{j-1} \cap G_{i-2})$ such that $(\chi^{\sim})^{k_1 b' k_2} = \chi^{\sim}$ on their common domain and Lemma 8.12 applies to $\chi^{\sim}$. This means that $k_1 b' k_2 = k_1' b' k_2'$, with k_1', k_2' elements of $(K_e^{t_{i-1}-j+2} \cap G_{i-2})(K_e^{t_{i-1}-j+1} \cap G_{(t_i-1)}^{\sim})$. Hence $k_1 b k_2 = k_1 b' k_2 (k_2^{-1} k_0 k_2) = k_1' b' k_2' (k_2^{-1} k_0 k_2)$; since $k_2^{-1} k_0 k_2 = k_0 k_3$ and $k_2' k_0 = k_0 k_2''$, with $k_2'', k_3 \in (K_e^{t_{i-1}-j+2} \cap G_{i-2})(K_e^{t_{i-1}-j+1} \cap G_{(t_i-1)})$, a bit of algebra says that we may assume that $k_1', k_2' \in J_1^{j-1}(J_1^j \cap K_e^{t_{i-2}-j+1})$. This will extend the induction and prove (11).

We know that $(\chi\chi_1)^b = \chi\chi_2$ on elements $x \in H_0^{j-1} \cap G_{i-2}$ with bxb^{-1} or $b^{-1}xb \in H_0^{j-1}$. Consider $\chi|_{H_0^{t_i-1} \cap G_{i-2}}$. From (12), $\chi = \chi_{0,i-1}\chi_{1,i-1}$ there, where $\chi_{0,i-1}$ extends to a character on G_{i-2} and $\chi_{1,i-1}$ is a character commuting with $G^{\tilde{}}_{(t_i-1)}$ (the group corresponding to $E^{\tilde{}}_{(t_i-1)}$ in (12)). Therefore each can be extended to $H_0^{j-1} \cap G_{i-2}$ so that $G^{\tilde{}}_{(t_i-1)}$ commutes with both and G_{i-2} commutes with the first. Call the product of these extensions $\chi^{\tilde{}}$. It is not hard to check that

$$(\chi^{\tilde{}})^{k_1} = \chi^{\tilde{}}\chi_1, \ (\chi^{\tilde{}})^{k_2^{-1}} = \chi^{\tilde{}}\chi_2,$$

since, e.g., $(\chi^{\tilde{}})^{k_1}$ depends only on $\chi^{\tilde{}}|_{H_0^{t_i-1}} = \chi|_{H_0^{t_i-1}}$. Because $\chi_1^{k_0} = \chi_1$ and $\chi_2^{k_0} = \chi_2$, we have $(\chi_1)^b = (\chi_1)^{b'}, (\chi_2)^b = (\chi_2)^{b'}$. Therefore $(\chi^{\tilde{}}\chi_1)^{b'} = \chi^{\tilde{}}\chi_2$ on their common domain; because $\chi_{0,i-1}$ is a character on G_{i-1}, Lemma 8.12 applies. Since this means that $(\chi^{\tilde{}})^{k_1 b' k_2} = \chi^{\tilde{}}$, (11) is proved.

(iii): The proof of nonconjugacy is similar. An $x \in Z_e K_e$ that conjugates χ to another χ' must fix χ on H_0^j; by induction, $x \in J_0^j/J_0^{j-1}$. A calculation using Lemma 8.5 then shows that $\chi^x = \chi$ on $G_{(j-1)} \cap H_0^{j-1}$. Since the χ were chosen to be distinct on this group, distinct χ are nonconjugate in $Z_e K_e$. A similar argument works for $x \in G$. We must have $x \in J_0^j W_{(j)} J_0^j$. Say that $x = k_1 b k_2$, as before. Since k_1 and k_2 fix χ on $G_{(j-1)} \cap H_0^{j-1}$, we must have $\chi^b = \chi'$ on $(G_{(j-1)} \cap H_0^{j-1}) \cap b^{-1}(G_{(j-1)} \cap H_0^{j-1})b$. Write $\chi = \chi_0\chi_{\alpha'}, \chi' = \chi_0\chi_{\beta'}$. Since $\chi_0^b = \chi$, we have $\chi_{\alpha'}^b = \chi_{\beta'}$ on their common domain. From Lemma 8.11, $\alpha\eta_{(j)}^{1-j}$ and $\beta_{(j)}^{1-j}$ are conjugate in $G_{(j)}$. Since both are central, $\alpha = \beta$.

Case 3c: $j - 1 = t_{i+1}$. We can again write χ on H_0^{j-1} as $\chi_0\chi_{\alpha'}$, with $\alpha' = Tr_{e_h}(\alpha'')$ so that $\alpha\eta_{(j)}^{1-j}$ generates a field extension $E^{\tilde{}}_{(j)}$ of $E_{(j-1)}$ with $e(E^{\tilde{}}_{(j)}/F) = e_{i+1}$ and $f(E^{\tilde{}}_{(j)}/F) = f_{i+1}$ (= 1, for us). Write $\alpha = (\alpha_1, \cdots, \alpha_n)$, $\alpha'' = (\alpha_1'', \cdots, \alpha_n'')$; we require that $\alpha_1'' = \cdots = \alpha''_{n/e_{i+1}}$, $\alpha''_{n/e_{i+1}+1} = \cdots = \alpha''_{2n/e_{i+1}}$, and so on. We choose one α in each conjugacy class of elements $\alpha\eta_{(j)}^{1-j}$, and one α'' for each such α; that gives C_{j-1} extensions of χ. We set $E_{(j)} = E_i$, $M_{(j)} = M_i$, $G_{(j)} = G_i$, and $\eta_{(j)} = \eta_i$. We also extend the definition of H_0, because $H_0^{t_{i+1}'}$ is now defined.

We first need $\eta_{(j-1)}$. A calculation like that for case 3 of §4 (using Lemma 8.7) shows that if $\delta_0\eta_i^a(1+\delta_1\eta_i+\cdots)$ commutes with χ on $G_i \cap K_e^{j-1}$, then $\delta_0 \in m_n^1(e_{i+1})$. We can choose δ_0 so that $(\delta_0\eta_i)^{1-j} = \alpha\eta_i^{1-j}$ and $\delta_0 \in m_n^1(e_{i+1})$. Then we can use (12) and the same reasoning as in 3b to produce first η_0 and then $\eta_{(j)}$.

Most of properties (1)–(12) are now verified as in 3b, with slight changes. In (12), we set $\alpha_{i+1} = \alpha$ and define $E^{\tilde{}}_{(j)}$ as above; $\chi_0 = \chi_{0,i+1}$ and $\chi_{\alpha''} =$

$\chi_{1,i+1}$. One implication in each of (10) and (11) is the same as in 3b. In (10), to prove that if $w \in Z_e K_e$ commutes with χ, then $w \in J_0^{j-1}$, we write $w = w_1 w$, where $w_1 \in K_e^1$ and $w_0 = \delta\eta_i^a$. For $y = 1 + \gamma\eta_i^{j-1}$ with $\gamma \in m_n^1(e_i)$, we have $\chi^{w_1}(y) = \chi(y)$; thus we need $\chi^{w_0}(y) = y$. A calculation done to compute $\eta_{(j-1)}$ shows that $\delta \in m_n^1(e_{i+1})$. Therefore $w = \delta\eta_{(j)}^a w_1'$, where $w_1' \in K_e^1$. Since $\delta\eta_{(j)}^a$ is known to commute with χ, we may delete it. Thus we may assume that $w \in K_e^1$ and w commutes with χ. The rest of the argument is as in 3b. In (11), too, the first step is slightly different from that in 3(b). Suppose that x commutes with χ. By the inductive hypothesis, $x = k_1 b k_2$ with $k_1, k_2 \in J_1^j$ and $b \in W_i$. Consider elements $y \in G_i \cap K_e^{j-1}$. By hypothesis, $\chi^{k_2^{-1}}(y) = \chi^{k_1 b}(y)$ (if both sides are defined); but $\chi^{k_2^{-1}}(y) = \chi(y)$ and $\chi^{k_1}(y) = \chi(y)$ for these elements, by an application of Lemma 8.2. Thus $\chi^b(y) = \chi(y)$ if $y, byb^{-1} \in \operatorname{Dom}\chi$. Lemma 8.13 implies that b commutes with $E_{(j-1)}^{-}$. From Lemma 8.10, $b = b_1 k$, where $k \in J_1^j$ and $b_1 \in W_{(j-1)}$. Now we continue as in 3b.

The proof that the different extensions are nonconjugate is also like that of 3b. Let $\chi = \chi_0\chi_{\alpha'}$, $\chi^{\#} = \chi_0\chi_{\beta'}$ be two extensions, and let $\chi_{\alpha'}, \chi_{\beta'}$ be determined on $K_e^{j-1} \cap G_i$ by $\alpha, \beta \in m_e^{e_i}(f_i)$ respectively. If $\chi^x = \chi^{\#}$ on their common domain, then $x \in J_1^j W_i J_1^j$ by (11) of the inductive hypothesis. Write $x = k_1 b k_2$. Then J_1^j fixes χ on $K_e^{j-1} \cap G_i$, so that $\chi^b = \chi^{\#}$ on $K_e^{j-1} \cap G_i \cap b^{-1}(K_e^{j-1} \cap G_i)b$. Since $\chi_0^b = \chi$, we have $\chi_{\alpha'}^b = \chi_{\beta'}$ there. Lemma 8.11 implies $\alpha\eta_i^{1-j}$ and $\beta\eta_i^{1-j}$ are conjugate. We chose only one extension from each such conjugacy class; thus $\alpha = \beta$, $\alpha' = \beta'$, and $\chi = \chi^{\#}$.

And that gives the construction down to H_0.

10. Completing the Construction.

Once we have H_0, the main work is done. If $r_0 = r$, then $G_{r_0} = E_{r_0}^{\times}$. We extend χ to a character on $H_0 E_{r_0}^{\times} = H$, making sure that $\chi(\varpi_F I) = 1$; that can be done in $e\prod_{j=0}^{t_r-1} C_j$ ways. If $r_0 < r$, then Lemmas 8.4 and 8.2 ensure that χ extends to a character (also called χ) on $H_0(G_{r_0} \cap Z_e K_e)$. We then invoke the inductive hypothesis that the construction works for GL_m if $m < n$ to pick pairs (H_1, ρ^*) corresponding to triples

$$(s_{r_0+1}/f_{r_0}, e_{r_0+1}/e_{r_0}, f_{r_0+1}/f_{r_0}), \ldots, (s_r/s_{r_0}, e_r/e_{r_0}, f_r/f_{r_0})$$

for G_{r_0}. Extend ρ^* to $H_0 H_1$ by making it trivial on H_1, and let $H = H_0 H_1$, $\rho = \chi \otimes \rho^*$. If $s_r > 0$, ρ^* is a character; if $s_r = 0$, ρ^* is a multiple of a character on $H \cap K_e^1$. We also have $H = K_e^{t_1'}(K_e^{t_2'} \cap G_1) \cdots (K_e^{t_r'} \cap G_{r-1})G_r$; notice that $G_r = E_r^{\times}$, so that $H \subseteq Z_e K_e$.

As with division algebras, we extend ρ to a projective representation on $J = K_e^{t_1''}(K_e^{t_2''} \cap G_1) \cdots (K_e^{t_r''} \cap G_{r-1})G_r$ and tensor with a Weil representation to get an (ordinary) irreducible representation ρ_1 on J. (The details are given in the appendix to [6].) It is not hard to establish inductively that for $x \in Z_e K_e$, ρ^x and ρ intertwine $\Leftrightarrow x \in J$. That ρ_1 induces to an irreducible of G (necessarily supercuspidal) and that distinct ρ give distinct ρ_1 and distinct irreducibles is trickier than in the division algebra case, but the argument was worked out earlier, in [12]. Because $\pi = \mathrm{Ind}_J^G \rho_1$ is irreducible and J is compact mod the center Z of G, π is supercuspidal by standard results. That completes the outline of the proof of Theorem 7.1.

III. Completeness

11. Applying the Matching Theorem.

The final part of the program is to prove that the previous construction gives all supercuspidals (up to tensoring by characters). The result we need is the following theorem of [9] (see also [18]):

11.1. Matching Theorem. *There is a 1–1 correspondence $\pi \leftrightarrow \pi'$ between discrete series representations of $GL_n(F)$ and irreducible unitary representations of $D_n^\times(F)$ that satisfies:*

1. *If $\pi \leftrightarrow \pi'$, then π, π' agree on $F^\times$ (i.e., there is a character ψ of $F^\times$ with $\pi(x) = \psi(x)I$, $\pi'(x) = \psi(x)I$ for all $x \in F^\times$);*
2. *If $\pi \leftrightarrow \pi'$ and χ is a character of $F^\times$, then $\pi \otimes (\chi \circ \mathrm{Det}) \leftrightarrow \pi' \otimes (\psi \circ N)$ (N = reduced norm map);*
3. *If $\pi \leftrightarrow \pi'$, then the ε-factors agree up to sign (in particular, π and π' have the same conductoral exponent);*
4. *For appropriate choice of Haar measure, if $\pi \leftrightarrow \pi'$, then π, π' have the same formal degree.*

For a reductive group G, let $\hat{G}_{\mathrm{disc}}$ = set of (equivalence classes of) irreducible discrete series representations, $\hat{G}_{\mathrm{sup}}$ = set of (equivalence classes of) irreducible supercuspidals, $\hat{G}_{\mathrm{sp}}$ = complement of $\hat{G}_{\mathrm{sup}}$ in $\hat{G}_{\mathrm{disc}}$. From [22], we know:

11.2. Theorem. *There is a natural 1–1 correspondence*

$$(GL_n(F))\hat{}_{\mathrm{sp}} \leftrightarrow \bigcup_{m|n, m<n} (GL_m(F))\hat{}_{\mathrm{sup}}.$$

Let $\pi' \in (D_n^\times)\hat{}$ be associated with data $(s_1, e_1, f_1), \cdots, s_r, e_r, f_r))$, and assume that char $F = 0$. The ε-factors are computed in [10]; they depend on various choices of characters, but the conductoral exponent depends in a simple way on s_r. The same is true for elements of $GL_m(F)\hat{}_{\mathrm{sup}}$ associated with the same data (where $e_r f_r = m$), as is shown in [2], and [10] (together with the construction in [22]) shows how to associate the ε-factors of corresponding representations under the correspondence of Theorem 11.2.

The general idea of the proof of completeness is now easy to explain. From (1) and (2) of the Matching Theorem, we may restrict attention to representations trivial on ϖ_F. For given M, there are only finitely many such representations with conductoral exponent less than or equal to M, and we need only match these up in some 1–1 manner. Now consider the representations of $D_n^\times$ with data $(s_1, e_1, f_1), \cdots, (s_r, e_r, f_r)$ and $e_r f_r = m < n$. These correspond 1–1 to representations of $D_m^\times$ with data $(s_1 m/n, e_1, f_1), \cdots, (s_r m/n, e_r, f_r)$. Assume inductively that these correspond 1–1 to the supercuspidal representations of $GL_m(F)$ with the same data; then by Theorem 11.2 we get a 1–1 correspondence between representations of $D_n^\times$ with data $(s_1, e_1, f_1), \cdots, (s_r, e_r, f_r)$ such that $e_r f_r = m < n$, and representations in $(GL_n(F))^\wedge_{\mathrm{sp}}$. The conductoral exponents agree, and we get all representations in $GL_n(F)^\wedge_{\mathrm{sp}}$ corresponding to those in $(D_n^x)^\wedge$ with $e_r f_r < n$ if we tensor with appropriate characters.

The representations still not accounted for are those of $GL_n(F)^\wedge_{\mathrm{sup}}$ and those of $(D_n^\times)^\wedge$ with data such that $e_r f_r = n$. But we have constructed an equal number on both sides, and their conductoral exponents agree. We now tensor with characters (using (2) of Theorem 11.1), and the induction step is checked.

The case of char $F \neq 0$ could probably be done by similar calculations, but it is just as easy to use [14] and establish a correspondence between supercuspidal representations of $GL_n(F)$ with given conductor and those of $GL_n(F[\sqrt[m]{p}])$ (where F is an unramified extension of $\mathbb{Q}_p$ and m is large) with the same conductor. For more information, see §11 of [6].

One obvious question is whether the above correspondence is in any sense the correspondence of the Matching Theorem. Calculations in [5] and [6] show that the above correspondence matches formal degrees for the supercuspidal representations, and it is shown in [8] that if $p \nmid n$, the elements π, π' corresponding under the matching theorem have $e_r = e'_{r'}$, $f_r = f'_{r'}$ (where data corresponding to π' has primes). It seems likely that this last result holds for all n, but the proof may require some new ideas.

Appendix

In this appendix, we compute the cocycle for the projective representations $\sigma'_1, \sigma'_2, \ldots$ mentioned in §5. The computation is like that in the Appendix to [6], but is somewhat simpler. Assume inductively that we have an irreducible representation σ_{i-1} on $J_0(J \cap K^{s'_i})$ with the following properties:

a. $\sigma_{i-1}\big|_{J\cap K^{s'_{r_0}}}$ is a multiple of χ.
b. Write $\chi\big|_{s'_i} = \chi_0\chi_1$, where χ_0 extends $\chi_{0,i}$ as in Lemma 4.1, and assume s_i/e_{i-1} even. Write $t = s_i/(2e_{i-1})$, $s = s_i/2 = s''_i$. For $\alpha_1, \alpha_2 \in k_{n/e_{i-1}}$, $B(\alpha_1, \alpha_2) = \chi_1((1+\alpha_1\eta^t_{i-1}), (1+\alpha_2\eta^t_{i-1})) = \chi_1(1 + (\alpha_1^{\sigma^s}\alpha_2 - \alpha_2^{\sigma^s}\alpha_1)\eta^{2t}_{i-1})$ is an alternating bilinear form with radical k_{n/e_i}.
c. χ_0 extends to a character (also to be called χ_0) of J^s_i.

We want to create a representation σ_i on $J_0(J \cap K^{s'_{i+1}})$ with similar properties. If s_i/e_{i-1} is odd, then $J_0(J \cap K^{s'_i+1}) = J_0(J \cap K^{s'_i})$, and there is nothing to prove. We therefore assume that s_i/e_{i-1} is even. Choose a set of coset representatives V for $k_{n/e_{i-1}}/k_{n/e_i}$; we may assume that these elements form a k_{n/e_i}-space. We actually need to work with the coset representatives of V in D, but there should be no confusion if we identify the two. Write $\eta = \eta_{r_0}$, $H_i = H(J \cap K_e^{s'_i})$, $J_i = H(J \cap K_e^{s})$, $H_i^j = H_i \cap K^j$, $J_i^j = J_i \cap K^j$. Coset representatives for J_i^s/H_i^s are given by elements $1 + \alpha\eta_i^t$, where $\alpha \in V$, and coset representatives for $J_i/J_i^1 \cong H_i/H_i^1$ are given by elements $\delta\eta^h$, $h \in \mathbb{Z}$ and $\delta \in k^{\times}_{n/e_{r_0}} = k_0^{\times}$ (for brevity).

Notice also that:

d. $\chi_0^{\delta} = \chi_0$ for $\delta \in k_0^{\times}$, while $\chi_0^{\eta^{-h}}\chi_0^{-1} = \psi^{(h)}$ is a character on $H_i^s/H_i^{s_i}$. The $\psi^{(h)}$ are fixed by $k_0^{\times}$ because η^{-h} normalizes $k_0^{\times}$.

e. If $x = 1 + \alpha\eta_i^t$ is one of the coset representatives for J_i^s/H_i^s, then $\sigma_i^x(y) = \sigma_i(y)$ for all $y \in H_i^1$. (The point is we have operator equality, not just equivalence; the reason is that $\sigma(xyx^{-1}y^{-1}) = I$.)

We now extend σ_i to a projective representation ρ_i on J_i by

$$\rho_i(x(1+\alpha\eta_i^t)) = \chi_0(1+\alpha\eta_i^t)\sigma_i(x), \quad x \in H_i,\ \alpha \in V.$$

We need to check that this is a projective representation, but the following computation will take care of that. For $y_1, y_2 \in H_i^1$, we compute:

$$\begin{aligned}(A.1)\qquad & \rho_i(y_1\delta_1\eta^g(1+\alpha_1\eta_i^t))\rho_i(y_2\delta_2\eta^h(1+\alpha_2\eta_i^t)) \\ & \cdot[\rho_i(y_1\delta_1\eta^g(1+\alpha_1\eta_i^t)y_2\delta_2\eta^h(1+\alpha_2\eta_i^t))]^{-1} \\ & = \chi_0(1+\alpha_1\eta_i^t)\chi_0(1+\alpha_2\eta_i^t) \\ & \cdot\sigma_i(y_1\delta_1\eta^g y_2\delta_2\eta^h)\sigma_i(y_1\delta_1\eta^g y_2\delta_2\eta^h)^{-1} \\ & \cdot[\rho_i((y_2\delta_2\eta^h)^{-1}\cdot(1+\alpha_1\eta_i^t)y_2\delta_2\eta^h(1+\alpha_2\eta_i^t)]^{-1}\end{aligned}$$

Write $y_2^{-1}(1+\alpha_1\eta_i^t)y_2 = z_0(1+\alpha_1\eta_i^t)$, with $z_0 \in H_i^{s+1}$. As noted earlier, $\chi(z_0) = 1$, so that $\sigma_i(z_0) = I$. Since $\delta_2\eta^h$ commutes with χ on H_i^{t+1}, we can omit z_0 from our calculations. Therefore

$$\begin{aligned}&\rho_i((y_2\delta_2\eta^h)^{-1}(1+\alpha_1\eta_i^t)y_2\delta_2\eta^h(1+\alpha_2\eta_i^t)) \\ &= \rho_i([\delta_2\eta^h]^{-1}(1+\alpha_1\eta_i^t)\delta_2\eta^h(1+\alpha_2\eta_i^t)).\end{aligned}$$

Write $[\delta_2\eta^h]^{-1}(1+\alpha_1\eta_i^t)\delta_2\eta^h = u(\delta_2, h; \alpha_1)(1+\alpha_3\eta_i^t)$, with $u(\delta_2, h, \alpha_1) \in H_i^{s'_i}$ and $\alpha_3 = \delta_2^{-1}(\alpha_1\delta_2)^{\sigma^{hf_0}} \in V$ (here, $f_0 = f_{r_0}$). Then the last expression is

$$\sigma_i(u(\delta_2, h; \alpha_1))\rho_i((1+\alpha_3\eta_i^t)(1+\alpha_2\eta_i^t)),$$

which is scalar. Therefore (A.1) is scalar and ρ_i is projective. Write

$$(1+\alpha_3\eta_i^t)(1+\alpha_2\eta_i^t) = (1+(\alpha_2+\alpha_3)\eta_i^t)z(\alpha_2,\alpha_3), \quad z(\alpha_2,\alpha_3) \in H_i^t.$$

Then

$$\begin{aligned}\rho_i((1+\alpha_3\eta_i^t)(1+\alpha_2\eta_i^t) &= \chi_0[(1+(\alpha_2+\alpha_3)\eta_i^t)z(\alpha_2,\alpha_3)]\chi_1[z(\alpha_2,\alpha_3)]I \\ &= \rho_i(1+\alpha_3\eta_i^t)\rho_i(1+\alpha_2\eta_i^t)\chi_1[z(\alpha_2,\alpha_3)] \\ &= \chi_0(1+\alpha_1\eta_i^t)\chi_0(1+\alpha_2\eta_i^t)\chi_1[z(\alpha_2,\alpha_3)]\psi^{(h)}(1+\alpha_1\eta^t).\end{aligned}$$

After combining terms, we get the scalar in (A.1) to be

$$[\chi_1(z(\alpha_2,\alpha_3))\psi^{(h)}(1+\alpha_1\eta_i^t)\chi(u(\delta_2,h;\alpha_1))]^{-1}.$$

Write $\xi(\delta_1,g,\alpha_1;\delta_2,h,\alpha_2)=\psi^{(h)}(1+\alpha_1\eta_i^t)\chi(u(\delta_2,h;\alpha_1))$; here, (δ,g,α) is shorthand for $\delta\eta^g(1+\alpha\eta_i^t)$. Since $z(\alpha_2,\alpha_3)$ is a cocycle, ξ is also a cocycle.

We first show that ξ is independent of δ_1 and δ_2. The independence of δ_1 is obvious. For the other part, write $\delta_2\eta^h=\eta^h\varepsilon$, $\varepsilon\in k_0$. Then

$$\begin{aligned}(\delta_2\eta^h)^{-1}(1+\alpha_1\eta_i^t)\delta_2\eta^h &= \varepsilon^{-1}[\eta^{-h}(1+\alpha_1\eta_i^t)\eta^h]\varepsilon \\ &= \varepsilon^{-1}[u(1,h;\alpha_1)(1+\alpha_1^{\sigma^{th}}\eta_i^t)]\varepsilon.\end{aligned}$$

Because $(\delta_2\eta^h)^{-1}(1+\alpha_1\eta_i^t)\delta_2\eta^h=u(\delta_2,h;\alpha_1)(1+\alpha_3\eta_i^t)$, it follows that $\varepsilon^{-1}u(1,h;\alpha_1)\varepsilon=u(\delta_2,h;\alpha_1)$. Since $\chi_1^\varepsilon=\chi_1$, then $\chi_1(u(\delta_2,h;\alpha_1))=\chi_1(u(1,h;\alpha_1))$; this proves the claim.

Write $C(g,\alpha_1;h,\alpha_2)=\xi(1,g,\alpha_1;1,h,\alpha_2)$. This is independent of g and α_2, and we sometimes indicate this by writing it as $\chi^*(h,\alpha_1)$. The cocycle condition gives

$$\begin{aligned}\chi^*(h,\alpha_1+\alpha_2) &= C(g,\alpha_1+\alpha_2;h,\alpha_3) \\ &= C(g,\alpha_1;0,\alpha_2)C((g,\alpha_1\cdot(0,\alpha_2);h,\alpha_3) \\ &= C(g,\alpha_1;(0,\alpha_2)\cdot(h,\alpha_3))C(0,\alpha_2;h,\alpha_3) \\ &= \chi^*(h,\alpha_1)\chi^*(h,\alpha_2).\end{aligned}$$

Similarly,

$$\chi^*(g,\alpha)\chi^*(h,\alpha^g)=\chi^*(g+h,\alpha),$$

where α^g satisfies $\eta^{-g}(1+\alpha\eta_i^t)\eta^g\equiv 1+\alpha^g\eta_i^t \bmod H_i^{s+1}$. So if we write

$$\chi^*(g,\alpha)=\varphi(g)(\alpha),\quad \varphi:\mathbb{Z}\mapsto V^{\hat{}},$$

then φ is a 1-cocycle. C is a coboundary if φ is, since if $\varphi(g)=\mu^g/\mu$ ($\mu\in V^{\hat{}}$), then $C(g,\alpha_1;h,\alpha_2)=\nu((g,\alpha_1)\cdot(h,\alpha_2))/\nu(g,\alpha_1)\nu(h,\alpha_2)$ when $\nu(g,\alpha)=\mu(\alpha)$.

Since $\eta^{e_{r_0}}$ is a central element in $D^\times$ (mod H_i^1), it is easy to check that the order of φ divides e_{r_0}. $V^{\hat{}}$ is a p-group, and standard theory implies that the order of φ divides p. So by the above remarks, C is a coboundary if $p \nmid e_{r_0}$. If $p \mid e_{r_0}$, we need to consider the inverse of the rest of the cocycle,

$$C_0(\delta_1\eta^g(1+\alpha_1\eta_i^t), \delta_2\eta^h(1+\alpha_2\eta_i^t)) = \chi_1(z(\alpha_2, \alpha_3)),$$

α_3 as defined earlier. This is the cocycle for the Weil representation, and $C_0(1+\alpha_1\eta_i^t,\ 1+\alpha_2\eta_i^t) = B(\alpha_1, \alpha_2)$ is a nondegenerate bilinear form on $V \times V$. Hence there exists β with $C_0(1+\beta\eta_i^t, 1+\alpha\eta_i^t) = \varphi(1)(\alpha)^{-1}$, for all $\alpha \in V$. Replace η by $\eta_* = \eta(1+\alpha_0\eta^t)$; then

$$C_0C(\delta_1\eta^g(1+\alpha_1\eta_i^t), \delta_2\eta^h(1+\alpha_2\eta_i^t)) = C_0(\delta_1\eta_*^g(1+\alpha_1\eta_i^t), \delta_2\eta_*^h(1+\alpha_2\eta_i^t)).$$

That is, the cocycle $(C_0C)^{-1}$ is one for the Weil representation, but we need to use η_* instead of η to generate the cyclic group. Another way to say this is that we are applying an outer automorphism to J_i/J_i^1. This outer automorphism depends only on χ, and not on all of σ_i.

We now tensor ρ_i with a restriction of the Weil (or oscillator) representation W to get an ordinary representation on J_i. Since W is trivial on H_i^1, the induction hypotheses apply to $\rho_i \otimes W = \sigma_{i+1}$. Since W is irreducible on J_i^s/H_i^s, any operator intertwining σ_{i+1} with itself must be of the form $A \otimes I$; as σ_i is irreducible on H_i, A must be a multiple of I. Therefore σ_{i+1} is irreducible. This extends the induction.

References

1. F. Bruhat and J. Tits, *Groupes réductifs sur un corps local I*, Publ. Math. IHES **41** (1972), 5–252.

2. C. Bushnell, *Hereditary orders, Gauss sums, and supercuspidal representations of* GL_N, J. reine angew. Math. **375/376** (1977), 184–210.

3. H. Carayol, *Représentations cuspidale du groupe linéaire*, Ann. Sci. École Norm. Sup. (4) **17** (1984), 191–225.

4. L. Corwin, *Representations of division algebras over local fields*, Advances in Math. **13** (1974), 249–257.

5. ———, *The unitary dual for the multiplicative group of arbitrary division algebras over local fields*, J. Am. Math. Soc. **2** (1989), 565–598.

6. ———, *A construction of the supercuspidal representations of* $GL_n(F)$, *F p-adic*, Trans. Amer. Math. Soc. (to appear).

7. L. Corwin and R. Howe, *Computing characters of tamely ramified p-adic division algebras*, Pac. J. Math. **73** (1977), 461–477.

8. L. Corwin, A. Moy, and P. J. Sally, Jr., *Degrees and formal degrees for division algebras and* GL_n *over a p-adic field*, Pac. J. Math. **141** (1990), 21–45.

9. P. Deligne, D. Kazhdan, and M.-F. Vigneras, *Représentations des algébres centrales simples p-adiques*, Représentations des groupes réductifs sur un corps local, Hermann, Paris, 1984, pp. 33–117.

10. R. Godement and H. Jacquet, *Zeta functions of Simple Algebras*, Lecture Notes in Math., Vol. 462, Springer-Verlag, Berlin-New York, 1975.

11. R. Howe, *Representation theory for division algebras over local fields (tamely ramified case)*, Bull. Amer. Math. Soc. **77** (1971), 1063–1066.

12. ———, *Tamely ramified supercuspidal representations of GL_n*, Pac. J. Math. **73** (1977), 365–381.

13. N. Iwahori and H. Matsumoto, *On some Bruhat decompositions and the structure of the Hecke ring of $\mathfrak{p}$-adic Chevalley groups*, Publ. Math. IHES **25** (1965), 5–48.

14. D. Kazhdan, *Representations of groups over closed local fields*, J. D'Analyse Math. **47**, 175–179.

15. H. Koch, *Eisensteinsche Polynomfolgen und Arithmetik in Divisionsalgebren über lokalen Körpern*, Math. Nachr. **104** (1981), 229–251.

16. P. Kutzko and D. Manderscheid, *On intertwining operators for $GL_N(F)$, F a non-archimedean p-adic field*, Duke Math. J. **57** (1988), 275–293.

17. A. Moy, *Local constants and the tame Langlands correspondence*, Amer. J. Math. **108** (1986), 863–930.

18. J. Rogawski, *Representations of $GL(n)$ and division algebras over a p-adic field*, Duke Math. J. **50** (1983), 161–196.

19. A. Silberger, *Introduction to Harmonic Analysis on Reductive p-adic Groups*, Mathematical Notes, Vol. 23, Princeton Univ. Press, Princeton, 1979.

20. T. Springer, *Characters of special groups*, Semisimple Algebraic Groups and Related Finite Groups, Lecture Notes in Math., Vol. 131, Springer-Verlag, Berlin-New York, 1970, pp. 121–166.

21. A. Weil, *Basic Number Theory*, Springer-Verlag, Berlin-New York, 1967.

22. A. V. Zelevinskii, *Induced representations of reductive p-adic groups, II. On irreducible representations of $GL(n)$*, Ann. Sci. École Norm. Sup. (4) **13** (1980), 165–211.

DEPARTMENT OF MATHEMATICS, RUTGERS UNIVERSITY, NEW BRUNSWICK, NJ 08903

A REMARK ON THE DUNKL DIFFERENTIAL–DIFFERENCE OPERATORS

GERRIT J. HECKMAN

University of Nijmegen

§1. INTRODUCTION

Let E be a Euclidean vector space of dimension n with inner product $(\cdot,\cdot)$. For each $\alpha \in E$ with $(\alpha,\alpha) = 2$ we write

$$r_\alpha(\lambda) = \lambda - (\alpha,\lambda)\alpha, \ \lambda \in E \tag{1.1}$$

for the orthogonal reflection in the hyperplane perpendicular to α.

Definition 1.1. A normalized root system R in E is a finite set of non-zero vectors in E, normalized by $(\alpha,\alpha) = 2$ for all $\alpha \in R$, such that $r_\alpha(\beta) \in R$ for all $\alpha, \beta \in R$.

Let $R \subset E$ be a normalized root system. We write $W = W(R)$ for the group generated by the reflections r_α, $\alpha \in R$. Denote by $\mathbb{C}[E]$ the algebra of $\mathbb{C}$-valued polynomial functions on E. For $w \in W$, $\xi \in E$, and $\alpha \in R$ introduce the operators

$$w,\ \partial_\xi,\ \Delta_\alpha : \mathbb{C}[E] \longrightarrow \mathbb{C}[E] \quad \text{by} \tag{1.2}$$

$$(wp)(\lambda) = p(w^{-1}\lambda), \tag{1.3}$$

$$(\partial_\xi p)(\lambda) = \frac{d}{dt}\{p(\lambda + t\xi)\}_{t=0}, \tag{1.4}$$

$$(\Delta_\alpha p)(\lambda) = \frac{p(\lambda) - p(r_\alpha\lambda)}{(\alpha,\lambda)}. \tag{1.5}$$

Remark 1.2. The operators Δ_α, $\alpha \in R$, were studied by Bernstein, Gel'fand and Gel'fand and are related to the Schubert cells and the cohomology of G/P [BGG]. They are the infinitesimal analogues of the Demazure operators [De1,2].

For a fixed generic $\lambda \in E$ let $R_+ = \{\alpha \in R : \ (\alpha,\lambda) > 0\}$ be a positive subsystem of R.

Definition 1.3. Suppose $(k_\alpha)_{\alpha\in R}$ is a collection of complex numbers such that $k_{w\alpha} = k_\alpha$ for all $w \in W$ and for all $\alpha \in R$. For each $\xi \in E$ the operator D_ξ defined by

$$D_\xi = \partial_\xi + \sum_{\alpha\in R_+} k_\alpha(\alpha,\xi)\Delta_\alpha : \mathbb{C}[E] \longrightarrow \mathbb{C}[E] \tag{1.6}$$

is called a Dunkl differential-difference operator.

Remark 1.4. It is easy to see that D_ξ is independent of the choice of the positive subsystem $R_+ \subset R$. If we write $q_\alpha = e^{2\pi i k_\alpha}$ then one can think of the operator D_ξ as a q-analogue (corresponding to the case $k_\alpha \to 0$) of the directional derivative ∂_ξ. We also write $D_\xi = D_\xi(k)$ to indicate the dependence on

$$k \in K = \{k = (k_\alpha)_{\alpha\in R} \in \mathbb{C}^R : \ k_{w\alpha} = k_\alpha \ \forall w \in W, \ \forall \alpha \in R\}.$$

Theorem 1.5. *(Dunkl [Du]) The relation $D_\xi D_\eta = D_\eta D_\xi$ is valid for all $\xi, \eta \in E$.*

Let $\mathbb{C}[E^*]$ be the symmetric algebra on E. For $\pi \in \mathbb{C}[E^*]$ we write ∂_π when we think of π as a constant coefficient differential operator on E (rather than as a polynomial function on E^*). In view of Theorem 1.5 the constant coefficient differential operator ∂_π has a well defined q-analogue

$$D_\pi : \mathbb{C}[E] \longrightarrow \mathbb{C}[E] \tag{1.8}$$

defined for a monomial $\pi = \xi_1^{d_1} \ldots \xi_n^{d_n}$ by

$$D_\pi = D_\pi(k) = D_{\xi_1}^{d_1} \ldots D_{\xi_n}^{d_n} \tag{1.9}$$

and extended by linearity.

Theorem 1.6. *(Dunkl [Du]) Suppose $\xi_1, \ldots, \xi_n$ is an orthonormal basis for E. The q-analogue of the Laplacian is given by*

$$\sum_{j=1}^n D_{\xi_j}^2 = \sum_{j=1}^n \partial_{\xi_j}^2 + 2 \sum_{\alpha\in R_+} k_\alpha \frac{1}{(\alpha,\cdot)} \{\partial_\alpha - \Delta_\alpha\}. \tag{1.7}$$

In Section 2 we review the proofs of both theorems as given by Dunkl.

We write $\mathbb{C}[E]^W$ and $\mathbb{C}[E^*]^W$ for the spaces of W-invariants in $\mathbb{C}[E]$ and $\mathbb{C}[E^*]$ respectively. We denote by $\mathbb{A}$ the associative algebra of endomorphisms of $\mathbb{C}[E]$ generated by (multiplication by) $(\xi,\cdot)$ and D_η for $\xi, \eta \in E$. Let $\mathbb{A}^W = \{D \in \mathbb{A} : \ wD = Dw \ \forall w \in W\}$ be the subalgebra of W-invariant operators in $\mathbb{A}$, and denote by

$$\mathrm{Res}(D) : \mathbb{C}[E]^W \longrightarrow \mathbb{C}[E]^W, \ D \in \mathbb{A}^W \tag{1.10}$$

the restriction of D to $\mathbb{C}[E]^W$. Clearly $\mathrm{Res} : \mathbb{A}^W \to \mathrm{End}(\mathbb{C}[E]^W)$ is a homomorphism of algebras. Since $wD_\xi w^{-1} = D_{w\xi} \ \forall w \in W, \forall \xi \in E$, we have $D_\pi \in \mathbb{A}^W \ \forall \pi \in \mathbb{C}[E^*]^W$.

Theorem 1.7. *Suppose by the Chevalley theorem that*

$$\mathbb{C}[E]^W = \mathbb{C}[p_1, \dots, p_n],$$

with $p_1, \dots, p_n$ homogeneous of degrees $d_1 \leq \dots \leq d_n$. Then the set

$$\{\mathrm{Res}(D_\pi) : \pi \in \mathbb{C}[E^*]^W\} \tag{1.11}$$

is a commuting family of differential operators in the Weyl algebra

$$\mathbb{C}[k, p_1, \dots, p_n, \frac{\partial}{\partial p_1}, \dots, \frac{\partial}{\partial p_n}]$$

containing the operator

$$\mathrm{Res}(\sum_{j=1}^n D_{\xi_j}^2) = \sum_{j=1}^n \partial_{\xi_j}^2 + 2 \sum_{\alpha \in R_+} k_\alpha \frac{1}{(\alpha, \cdot)} \partial_\alpha. \tag{1.12}$$

Remark 1.8. The proof of this theorem is trivial. However it can be reformulated as the complete integrability for the generalized non-periodic Calogero-Moser system (both on the quantum mechanical level of differential operators and on the classical mechanical level of symbols). For root systems R of type A the complete integrability of the Calogero-Moser system was first established by Moser by realizing the system as a Lax pair [Mo]. The method of Moser was extended by Olshanetsky and Perelomov to cover the root systems R of classical type [OP]. In the crystallographic case, where $(\alpha, \beta)^2 \in \mathbb{Z}\ \forall \alpha, \beta \in R$, the above theorem had been obtained earlier by Opdam using transcendental methods [HO, He1, Op1-2, He2].

Suppose $S \subset R$ is a set of roots in R invariant under W. Let $S_+ = S \cap R_+$ and put

$$p_S(\cdot) = \prod_{\alpha \in S_+} (\alpha, \cdot) \quad \in \mathbb{C}[E], \tag{1.13}$$

$$\pi_S = \prod_{\alpha \in S_+} \alpha \quad \in \mathbb{C}[E^*]. \tag{1.14}$$

Clearly we have

$$w p_S = \chi(w) p_S, \quad w \pi_S = \chi(w) \pi_S \quad \text{for all } w \in W \tag{1.15}$$

for some one dimensional character $\chi = \chi_S$ of W. Conversely, every $p \in \mathbb{C}[E]$ which satisfies the relation

$$wp = \chi(w) p \quad \text{for all } w \in W$$

is divisible in $\mathbb{C}[E]$ by p_S. Although $p_S^{-1} D_{\pi_S}(k)$ need not be an endomorphism of $\mathbb{C}[E]$, it follows that $p_S^{-1} D_{\pi_S}(k)(p) \in \mathbb{C}[E]^W \ \ \forall p \in \mathbb{C}[E]^W$, and hence

$$G(1_S, k) := \mathrm{Res}(p_S^{-1} D_{\pi_S}(k)) \in \mathrm{End}(\mathbb{C}[E]^W) \tag{1.16}$$

is a well defined endomorphism of $\mathbb{C}[E]^W$. We also write

$$G(-1_S, k) := \mathrm{Res}(D_{\pi_S}(k - 1_S) \cdot p_S) \in \mathrm{End}(\mathbb{C}[E]^W) \tag{1.17}$$

where $k - 1_S \in K$ is the multiplicity function defined by $(k - 1_S)_\alpha = k_\alpha - 1$ for $\alpha \in S$ and $(k - 1_S)_\alpha = k_\alpha$ for $\alpha \in R \backslash S$.

Theorem 1.9. *The operators (1.16) and (1.17) are differential operators in the Weyl algebra* $\mathbb{C}[k, p_1, \ldots, p_n, \frac{\partial}{\partial p_1}, \ldots, \frac{\partial}{\partial p_n}]$ *and satisfy the shift relations*

$$G(1_S, k)\mathrm{Res}(D_\pi(k)) = \mathrm{Res}(D_\pi(k+1_S))G(1_S, k), \tag{1.18}$$

$$G(-1_S, k)\mathrm{Res}(D_\pi(k)) = \mathrm{Res}(D_\pi(k-1_S))G(-1_S, k) \tag{1.19}$$

for all $\pi \in \mathbb{C}[E^*]^W$, *where* $(k \pm 1_S)_\alpha = k_\alpha \pm 1$ *for* $\alpha \in S$ *and* $(k \pm 1_S)_\alpha = k_\alpha$ *for* $\alpha \in R \backslash S$.

The proofs of both Theorem 1.7 and 1.9 will be given in Section 3.

Remark 1.10. In the terminology of Opdam the operator (1.16) is a raising operator and the operator (1.17) a lowering operator for the commuting family (1.11). In the crystallographic case the above theorem was obtained by Opdam [Op2].

Recall Macdonald's (infinitesimal) constant term conjecture:

$$\int_E \prod_{\alpha \in R_+} |(\alpha, \lambda)|^{2s} d\gamma(\lambda) = \prod_{j=1}^{n} \frac{(sd_j)!}{s!} \quad \text{for all } \mathcal{R}(s) > 0, \tag{1.20}$$

where $d\gamma(\lambda) = (2\pi)^{-\frac{n}{2}} e^{-\frac{1}{2}(\lambda,\lambda)} d\lambda$ is the Gaussian measure on E [Ma]. The same arguments as given in [Op3, Section 6] show that the evaluation of this integral is equivalent to

$$G(-1, k)(1) = |W| \cdot \prod_{i=1}^{n} \prod_{j=1}^{m_i} (d_i k - j), \tag{1.21}$$

where $-1 = -1_R$ and $k = k_\alpha$ for all $\alpha \in R$. This latter formula is in turn related to the normalization of the "multivariable Bessel function associated with R" at $\xi = 0$. This normalization problem has been analyzed by Opdam, and the desired formula (1.21) can be obtained [Op4]. One can further proceed as in [Op3, Section 7] to compute the Bernstein-Sato polynomial of the discriminant without the crystallographic restriction, in accordance with a conjecture of Yano and Sekiguchi [YS].

§2. The Dunkl differential-difference operators

Using the bracket $[\cdot, \cdot]$ for the commutator of endomorphisms of $\mathbb{C}[E]$, for $\xi, \eta \in E$ we write

$$[D_\xi, D_\eta] = I + II + III, \tag{2.1}$$

where

$$I = [\partial_\xi, \partial_\eta] = 0, \tag{2.2}$$

$$II = \sum_{\alpha \in R_+} k_\alpha \{(\alpha, \xi)[\Delta_\alpha, \partial_\eta] + (\alpha, \eta)[\partial_\xi, \Delta_\alpha]\}, \tag{2.3}$$

$$III = \sum_{\alpha, \beta \in R_+} k_\alpha k_\beta (\alpha, \xi)(\beta, \eta)[\Delta_\alpha, \Delta_\beta]. \tag{2.4}$$

Lemma 2.1. *For $\xi \in E$, $\alpha \in R$ we have*

$$[\partial_\xi, \Delta_\alpha] = \frac{(\alpha, \xi)}{(\alpha, \cdot)}\{r_\alpha \partial_\alpha - \Delta_\alpha\}. \tag{2.5}$$

Proof. Using the definition $\Delta_\alpha = \frac{1}{(\alpha,\cdot)}(1 - r_\alpha)$ we get

$$\begin{aligned}[\partial_\xi, \Delta_\alpha] &= [\partial_\xi, \frac{1}{(\alpha, \cdot)}](1 - r_\alpha) + \frac{1}{(\alpha, \cdot)}[\partial_\xi, 1 - r_\alpha] \\ &= -\frac{(\alpha, \xi)}{(\alpha, \cdot)^2}(1 - r_\alpha) + \frac{1}{(\alpha, \cdot)} r_\alpha(\partial_\xi - \partial_{r_\alpha \xi}) \\ &= -\frac{(\alpha, \xi)}{(\alpha, \cdot)}\Delta_\alpha + \frac{(\alpha, \xi)}{(\alpha, \cdot)} r_\alpha \partial_\alpha.\end{aligned}$$

□

Using (2.5) the second term (2.3) can be rewritten as

$$II = \sum_{\alpha \in R_+} k_\alpha \frac{(\alpha, \xi)(\alpha, \eta)}{(\alpha, \cdot)}\{r_\alpha \partial_\alpha - \Delta_\alpha\}(-1 + 1) = 0. \tag{2.6}$$

The third term (2.4) can be written as

$$III = \sum_{\alpha, \beta \in R_+} k_\alpha k_\beta \{(\alpha, \xi)(\beta, \eta) - (\alpha, \eta)(\beta, \xi)\}\Delta_\alpha \Delta_\beta. \tag{2.7}$$

To complete the proof of Theorem 1.5 we need only verify the vanishing of this third term.

Proposition 2.2. *Suppose $B(\cdot, \cdot)$ is a bilinear form on E such that*

(2.8) $\quad B(r_\alpha \lambda, r_\alpha \mu) = B(\mu, \lambda)$ *for all* $\lambda, \mu \in E$ *and* $\alpha \in R \cap \operatorname{span}\langle \lambda, \mu \rangle$.

If $w \in W$ is a pure rotation (i.e. dim *$Im(w - \mathrm{Id}) = 2$), then*

$$\sum_{\alpha, \beta \in R_+,\ r_\alpha r_\beta = w} k_\alpha k_\beta B(\alpha, \beta) \frac{1}{(\alpha, \cdot)(\beta, \cdot)} = 0 \tag{2.9}$$

and

$$\sum_{\alpha, \beta \in R_+,\ r_\alpha r_\beta = w} k_\alpha k_\beta B(\alpha, \beta)\Delta_\alpha \Delta_\beta = 0. \tag{2.10}$$

Proof. Using the definition $\Delta_\alpha = \frac{1}{(\alpha,\cdot)}(1 - r_\alpha)$ the left hand side of (2.10) can be written as a sum of the following three terms:

$$A = \sum k_\alpha k_\beta B(\alpha, \beta) \frac{1}{(\alpha, \cdot)(\beta, \cdot)}, \tag{2.11}$$

$$B = -\sum k_\alpha k_\beta B(\alpha,\beta)\left\{\frac{1}{(\alpha,\cdot)(r_\alpha\beta,\cdot)}r_\alpha + \frac{1}{(\alpha,\cdot)(\beta,\cdot)}r_\beta\right\}, \tag{2.12}$$

$$C = \sum k_\alpha k_\beta B(\alpha,\beta)\frac{1}{(\alpha,\cdot)(r_\alpha\beta,\cdot)}r_\alpha r_\beta, \tag{2.13}$$

with the summations over the same index set as in (2.9) and (2.10).

Let $S = R \cap \operatorname{Im}(w - \mathrm{Id})$ be the normalized root system of $W(S)$, the largest dihedral group containing w. If $w = r_\alpha r_\beta$ then, for $\gamma \in S$, we have $r_\gamma w r_\gamma = w^{-1}$ and hence $r_{r_\gamma\alpha} r_{r_\gamma\beta} = r_\beta r_\alpha$. We claim that $r_\gamma A = A$ for all $\gamma \in S$. Indeed we have

$$\begin{aligned} r_\gamma A &= \sum_{\alpha,\beta\in R_+,\ r_\alpha r_\beta = w} k_\alpha k_\beta B(\alpha,\beta)\frac{1}{(r_\gamma\alpha,\cdot)(r_\gamma\beta,\cdot)} \\ &= \sum_{\alpha,\beta\in r_\gamma R_+,\ r_\beta r_\alpha = w} k_\alpha k_\beta B(r_\gamma\alpha,r_\gamma\beta)\frac{1}{(\alpha,\cdot)(\beta,\cdot)} \\ &= \sum_{\alpha,\beta\in r_\gamma R_+,\ r_\beta r_\alpha = w} k_\alpha k_\beta B(\beta,\alpha)\frac{1}{(\alpha,\cdot)(\beta,\cdot)} \\ &= A \end{aligned}$$

since the summation in (2.9) is independent of the choice of R_+. Let $S_+ = R_+ \cap S$ and put $p_S = \prod_{\alpha\in S_+} (\alpha,\cdot)$. Then p_S transforms under the group $W(S)$ according to the sign character; conversely, every polynomial in $\mathbb{C}[E]$ which transforms under $W(S)$ according to the sign character is divisible in $\mathbb{C}[E]$ by p_S. Now observe that $p_S A \in \mathbb{C}[E]$ transforms under $W(S)$ according to the sign character. Hence $A \in \mathbb{C}[E]$. Since A is homogeneous of degree minus two we have $A = 0$. This proves (2.9).

Since $w = r_\alpha r_\beta = r_{r_\alpha\beta} r_\alpha$ and $B(\alpha,\beta) = B(r_\alpha\beta, r_\alpha\alpha) = -B(r_\alpha\beta,\alpha)$, the vanishing of term (2.12) is clear. For (2.13) we write $C = -Aw = 0$. □

Lemma 2.3. *For fixed ξ and η in E the bilinear form on E defined by*

$$B(\lambda,\mu) = (\lambda,\xi)(\mu,\eta) - (\lambda,\eta)(\mu,\xi) \tag{2.14}$$

satisfies condition (2.8).

Proof. Clearly $B(\mu,\lambda) = -B(\lambda,\mu)$ is an alternating form. For $\lambda \in E$, $\lambda \neq 0$, we write $\lambda' = \sqrt{2}|\lambda|^{-1}\lambda$ and obtain

$$B(r_{\lambda'}\lambda, r_{\lambda'}\mu) = B(-\lambda, \mu - (\lambda',\mu)\lambda') = B(-\lambda,\mu) = B(\mu,\lambda).$$

Hence, for generic λ, $\mu \in E$ and for all $\nu \in \operatorname{span}\langle\lambda,\mu\rangle$ such that $(\nu,\nu) = 2$, continuity shows that

$$B(r_\nu\lambda, r_\nu\mu) = B(\mu,\lambda). \quad \square$$

The proof of Theorem 1.5 now follows by regrouping the terms in (2.7) as a sum over $\{\alpha, \beta \in R_+ : r_\alpha r_\beta = w\}$, where $w \in W$ runs over the pure rotations in W, and by applying (2.10).

The proof of Theorem 1.6 is just an easy calculation, as we now show.

$$\begin{aligned}\sum_{j=1}^{n} D_{\xi_j}^2 &= \sum_{j=1}^{n} (\partial \xi_j + \sum_{\alpha \in R_+} k_\alpha (\alpha, \xi_j) \Delta_\alpha)^2 \\ &= \sum_{j=1}^{n} \Big\{ \partial_{\xi_j}^2 + \sum_{\alpha \in R_+} k_\alpha (\alpha, \xi_j)(\partial_{\xi_j} \Delta_\alpha + \Delta_\alpha \partial_{\xi_j}) + \sum_{\alpha, \beta \in R_+} k_\alpha k_\beta (\alpha, \xi_j)(\beta, \xi_j) \Delta_\alpha \Delta_\beta \Big\} \\ &= \sum_{j=1}^{n} \partial_{\xi_j}^2 + \sum_{\alpha \in R_+} k_\alpha (\partial_\alpha \Delta_\alpha + \Delta_\alpha \partial_\alpha) + \sum_{\alpha, \beta \in R_+} k_\alpha k_\beta (\alpha, \beta) \Delta_\alpha \Delta_\beta .\end{aligned}$$

The third term vanishes by Proposition 2.2 and because $\Delta_\alpha^2 = 0$. Using Lemma 2.1 we get

$$\begin{aligned}\partial_\alpha \Delta_\alpha + \Delta_\alpha \partial_\alpha &= [\partial_\alpha, \Delta_\alpha] + 2 \Delta_\alpha \partial_\alpha \\ &= \frac{(\alpha, \alpha)}{(\alpha, \cdot)} \Big\{ r_\alpha \partial_\alpha - \Delta_\alpha \Big\} + \frac{2}{(\alpha, \cdot)} (1 - r_\alpha) \partial_\alpha \\ &= \frac{2}{(\alpha, \cdot)} \Big\{ \partial_\alpha - \Delta_\alpha \Big\}.\end{aligned}$$

§3. The Opdam shift operators

Recall that $D \in \mathrm{End}(\mathbb{C}[p_1, \dots, p_m])$ is a differential operator of degree $\leq d$ if and only if

$$\mathrm{ad}(p)^{d+1}(D) = 0 \quad \text{for all } p \in \mathbb{C}[p_1, \dots, p_n]. \tag{3.1}$$

Hence (1.11), (1.16) and (1.17) are seen to be differential operators from

$$\mathrm{ad}(p)(D_\xi) = \mathrm{ad}(p)(\partial_\xi) = -\partial_\xi(p) \tag{3.2}$$

$$\mathrm{ad}(p)^2(D_\xi) = 0 \tag{3.3}$$

for all $p \in \mathbb{C}[E]^W$, and $\xi \in E$. Hence Theorem 1.7 is a consequence of Theorems 1.5 and 1.6.

Theorem 3.1. *The q-analogue of the Laplacian is*

$$\mathrm{Res}(p_S^{-1} \circ \Big\{ \sum_{j=1}^{n} D_{\xi_j}^2(k) \Big\} \circ p_S) = \mathrm{Res}\Big(\sum_{j=1}^{n} D_{\xi_j}^2 (k + 1_S) \Big). \tag{3.4}$$

Proof. We first observe that the left hand side of (3.4) is a well-defined endomorphism of $\mathbb{C}[E]^W$. Then apply Theorem 1.6 and evaluate each of the three resulting terms. The first term yields

$$\begin{aligned} p_S^{-1} \circ \sum_{j=1}^n \partial_{\xi_j}^2 \circ p_S &= \sum_{j=1}^n \partial_{\xi_j}^2 + 2 \sum_{\alpha \in S_+} \frac{1}{(\alpha, \cdot)} \partial_\alpha + p_S^{-1} (\sum_{j=1}^n \partial_{\xi_j}^2)(p_S) \\ &= \sum_{j=1}^n \partial_{\xi_j}^2 + 2 \sum_{\alpha \in S_+} \frac{1}{(\alpha, \cdot)} \partial_\alpha. \end{aligned}$$

The second term becomes

$$\begin{aligned} p_S^{-1} &\circ \Big\{ 2 \sum_{\alpha \in R_+} k_\alpha \frac{1}{(\alpha, \cdot)} \partial_\alpha \Big\} \circ p_S \\ &= 2 \sum_{\alpha \in R_+} k_\alpha \frac{1}{(\alpha, \cdot)} \partial_\alpha + p_S^{-1} \cdot \Big(2 \sum_{\alpha \in R_+} k_\alpha \frac{1}{(\alpha, \cdot)} \partial_\alpha \Big)(p_S) \\ &= 2 \sum_{\alpha \in R_+} k_\alpha \frac{1}{(\alpha, \cdot)} \partial_\alpha + 2 \sum_{\alpha \in R_+, \beta \in S_+} k_\alpha \frac{(\alpha, \beta)}{(\alpha, \cdot)(\beta, \cdot)} \\ &= 2 \sum_{\alpha \in R_+} k_\alpha \frac{1}{(\alpha, \cdot)} \partial_\alpha + 2 \sum_{\beta \in S_+} k_\beta \frac{(\beta, \beta)}{(\beta, \cdot)^2} + 2 \sum_{\substack{\alpha \in R_+, \beta \in S_+ \\ \alpha \neq \beta}} k_\alpha \frac{(\alpha, \beta)}{(\alpha, \cdot)(\beta, \cdot)} \\ &= 2 \sum_{\alpha \in R_+} k_\alpha \frac{1}{(\alpha, \cdot)} \partial_\alpha + 2 \sum_{\beta \in S_+} k_\beta \frac{2}{(\beta, \cdot)^2} \end{aligned}$$

by the same argument as in the proof of Proposition 2.2. Finally, the third term becomes

$$\begin{aligned} p_S^{-1} \circ \Big\{ 2 \sum_{\alpha \in R_+} k_\alpha \frac{1}{(\alpha, \cdot)} \Delta_\alpha \Big\} \circ p_S &= 2 \sum_{\alpha \in R_+} k_\alpha \frac{1}{(\alpha, \cdot)^2} \{1 - p_S^{-1} \circ r_\alpha \circ p_S\} \\ &= 2 \sum_{\alpha \in R_+} k_\alpha \frac{1}{(\alpha, \cdot)^2} \{1 - \chi_S(r_\alpha) r_\alpha\} \\ &= 2 \sum_{\alpha \in S_+} k_\alpha \frac{1}{(\alpha, \cdot)^2} \{1 + r_\alpha\} + 2 \sum_{\alpha \in R_+ \setminus S_+} k_\alpha \frac{1}{(\alpha, \cdot)} \Delta_\alpha \\ &= 2 \sum_{\alpha \in S_+} k_\alpha \frac{2}{(\alpha, \cdot)^2} - 2 \sum_{\alpha \in S_+} k_\alpha \frac{1}{(\alpha, \cdot)} \Delta_\alpha + 2 \sum_{\alpha \in R_+ \setminus S_+} k_\alpha \frac{1}{(\alpha, \cdot)} \Delta_\alpha. \end{aligned}$$

Taking all three terms together yields

$$\begin{aligned} p_S^{-1} \circ \Big\{ \sum_{j=1}^n D_{\xi_j}^2(k) \Big\} \circ p_S = \sum_{j=1}^n \partial_{\xi_j}^2 &+ 2 \sum_{\alpha \in R_+} k_\alpha \frac{1}{(\alpha, \cdot)} \partial_\alpha + 2 \sum_{\alpha \in S_+} \frac{1}{(\alpha, \cdot)} \partial_\alpha \\ &+ 2 \sum_{\alpha \in S_+} k_\alpha \frac{1}{(\alpha, \cdot)} \Delta_\alpha - 2 \sum_{\alpha \in R_+ \setminus S_+} k_\alpha \frac{1}{(\alpha, \cdot)} \Delta_\alpha. \quad \square \end{aligned}$$

Corollary 3.2. *The following shift relations are valid:*

$$G(1_S,k)\mathrm{Res}\left(\sum_{j=1}^{n} D^2_{\xi_j}(k)\right) = \mathrm{Res}\left(\sum_{j=1}^{n} D^2_{\xi_j}(k+1_S)\right) G(1_S,k) \tag{3.5}$$

$$G(-1_S,k)\mathrm{Res}\left(\sum_{j=1}^{n} D^2_{\xi_j}(k)\right) = \mathrm{Res}\left(\sum_{j=1}^{n} D^2_{\xi_j}(k-1_S)\right) G(-1_S,k). \tag{3.6}$$

Proof. Indeed we have

$$\begin{aligned}
\mathrm{Res}\Big(p_S^{-1}D_{\pi_S}(k)\Big)\mathrm{Res}\Big(\sum_{j=1}^{n} D^2_{\xi_j}(k)\Big) &= \mathrm{Res}\Big(\sum_{j=1}^{n} p_S^{-1}D_{\pi_S}(k)D^2_{\xi_j}(k)\Big)\\
&= \mathrm{Res}\Big(\sum_{j=1}^{n} p_S^{-1}D^2_{\xi_j}(k)D_{\pi_S}(k)\Big)\\
&= \mathrm{Res}\Big(\sum_{j=1}^{n} p_S^{-1}D^2_{\xi_j}(k)p_S\Big)\mathrm{Res}\Big(p_S^{-1}D_{\pi_S}(k)\Big)\\
&= \mathrm{Res}\Big(\sum_{j=1}^{n} D^2_{\xi_j}(k+1_S)\Big)\mathrm{Res}\Big(p_S^{-1}D_{\pi_S}(k)\Big)
\end{aligned}$$

which proves relation (3.5). Relation (3.6) is proved similarly. □

Theorem 3.3. *Let E, H, and F be the endomorphisms of $\mathbb{C}[E]$ defined by*

$$E = \frac{1}{2}\sum_{j=1}^{n}(\xi_j,\cdot)^2, \tag{3.7}$$

$$H = \sum_{j=1}^{n}(\xi_j,\cdot)\partial_{\xi_j} + \Big(\frac{n}{2} + \sum_{\alpha\in R_+} k_\alpha\Big), \tag{3.8}$$

$$F = -\frac{1}{2}\sum_{j=1}^{n} D^2_{\xi_j}. \tag{3.9}$$

These operators satisfy the commutation relations of $sl(2)$:

$$[H,E] = 2E,\ [H,F] = -2F,\ [E,F] = H. \tag{3.10}$$

Proof. The Euler operator $\sum_{j=1}^{n}(\xi_j,\cdot)\partial_{\xi_j}$ acts as multiplication by d on the space of homogeneous polynomials in $\mathbb{C}[E]$ of degree d. Hence the commutation relations $[H,E] = 2E$ and $[H,F] = -2F$ merely state the fact that E and F are homogeneous of degree plus two and minus two respectively.

Since $[p, \Delta_\alpha] = 0$ for all $p \in \mathbb{C}[E]^W$ and $\alpha \in R$, we obtain

$$[E, D_\xi] = [E, \partial_\xi] = -(\xi, \cdot) \text{ for all } \xi \in E, \tag{3.11}$$

and therefore

$$\begin{aligned}
[E,F] &= -\frac{1}{2}\sum_{j=1}^{n}[E, D_{\xi_j}^2] = \frac{1}{2}\sum_{j=1}^{n}\{(\xi_j,.)D_{\xi_j} + D_{\xi_j}(\xi_j,.)\} \\
&= \sum_{j=1}^{n}(\xi_j,.)D_{\xi_j} + \frac{1}{2}\sum_{j=1}^{n}[D_{\xi_j},(\xi_j,.)] \\
&= \sum_{j=1}^{n}(\xi_j,.)D_{\xi_j} + \frac{n}{2} + \frac{1}{2}\sum_{j=1}^{n}\sum_{\alpha\in R_+} k_\alpha(\alpha,\xi_j)[\Delta_\alpha,(\xi_j,.)] \\
&= \sum_{j=1}^{n}(\xi_j,.)\partial_{\xi_j} + \sum_{\alpha\in R_+} k_\alpha(\alpha,\cdot)\Delta_\alpha + \frac{n}{2} + \sum_{\alpha\in R_+} k_\alpha r_\alpha \\
&= \sum_{j=1}^{n}(\xi_j,.)\partial_{\xi_j} + (\frac{n}{2} + \sum_{\alpha\in R_+} k_\alpha).
\end{aligned}$$

Here we have used that, for $\xi \in E$,

$$\begin{aligned}
[\Delta_\alpha,(\xi,\cdot)] &= -\frac{1}{(\alpha,\cdot)}[r_\alpha,(\xi,\cdot)] \\
&= -\frac{1}{(\alpha,\cdot)}\{(r_\alpha\xi,\cdot) - (\xi,\cdot)\}r_\alpha = (\alpha,\xi)r_\alpha.
\end{aligned}$$

□

Proposition 3.4. *The inner product $(\cdot,\cdot)$ on E yields an isomorphism between $\mathbb{C}[E]$ and $\mathbb{C}[E^*]$. If $p \in \mathbb{C}[E]$ is homogeneous of degree d, and if $\pi \in \mathbb{C}[E^*]$ is the element corresponding to p under the isomorphism, then*

$$D_\pi = (-1)^d \frac{1}{d!} \operatorname{ad}(F)^d(p). \tag{3.12}$$

Proof. Clearly $\operatorname{ad}(H)D_\pi = -dD_\pi$ and, by Theorem 1.5, $\operatorname{ad}(F)D_\pi = 0$. Using (3.11) and induction on d (assuming π to be a monomial as in (1.9) with $d = d_1 + \cdots + d_n$) it is easy to see that $(-1)^d(1/d!)\operatorname{ad}(E)^d(D_\pi) = p$ and hence $\operatorname{ad}(E)^{d+1}(D_\pi) = 0$. By the standard representation theory of $sl(2)$ we obtain (3.12). □

Corollary 3.5. *For $\pi \in \mathbb{C}[E^*]^W$ we have*

$$\operatorname{Res}\left(p_S^{-1} \circ D_\pi(k) \circ p_S\right) = \operatorname{Res}\left(D_\pi(k + 1_s)\right). \tag{3.13}$$

Proof. This is easily derived from Theorem 3.1 and Proposition 3.4. □

The proof of Theorem 1.9 now goes along the same lines as the proof of Corollary 3.2.

Remark 3.6. This type of argument—using a copy of $sl(2)$ to reduce the computation of higher order operators to those of a second order operator—goes back to Harish-Chandra [Ha].

Acknowledgements. The author would like to thank the organizers of the conference for their invitation to deliver this paper, and William Barker in particular for the hospitality extended while at Bowdoin College.

References

[BGG] I. N. Bernstein, I. M. Gel'fand, and S. I. Gel'fand, *Schubert cells and the cohomology of G/P*, Russ. Math. Surveys **28** (1973), 1–26.

[De1] M. Demazure, *Désingularisation des variétés de Schubert généralisés*, Ann. Sc. Éc. Norm. Sup. **7** (1974), 53–88.

[De2] ———, *Une nouvelle formule des caractères*, Bulletin Sc. Math. **98** (1974), 163–172.

[Du] C. F. Dunkl, *Differential-difference operators associated to reflection groups*, Trans. AMS **311** (1989), 167–183.

[Ha] Harish-Chandra, *Differential operators on a semisimple Lie algebra*, Amer. J. Math. **79** (1957), 87–120; also in the Collected Works, Vol. 2, 243–276.

[HO] G. J. Heckman and E. M. Opdam, *Root systems and hypergeometric functions I*, Comp. Math. **64** (1987), 329–352.

[He1] G. J. Heckman, *Root systems and hypergeometric functions II*, Comp. Math. **64** (1987), 353–373.

[He2] ———, *Hecke algebras and hypergeometric functions*, Inv. Math. **100** (1990), 403–417.

[Ma] I. G. Macdonald, *Some conjectures for root systems*, Siam J. Math. Analysis **13** (1982), 988–1007.

[Mo] J. Moser, *Three integrable systems connected with isospectral deformation*, Adv. Math. **16** (1975), 197–220.

[OP] M. A. Olshanetsky and A.M. Perelomov, *Completely integrable systems connected with semisimple Lie algebras*, Inv. Math. **37** (1976), 93–108.

[Op1-2] E. M. Opdam, *Root systems and hypergeometric functions III, IV*, Comp. Math. **67** (1988), 21–49 and 191–209..

[Op3] ———, *Some applications of hypergeometric shift operators*, Inv. Math. **98** (1989), 1–18.

[Op4] ———, *Dunkl operators, Bessel functions and the discriminant for a finite Coxeter group*, in preparation.

[YS] T. Yano and J. Sekiguchi, *The microlocal structure of weighted homogeneous polynomials associated with Coxeter systems I, II*, Tokyo J. Math. **2** (1979), 193–219; *II* **4** (1981), 1–34.

University of Nijmegen, Mathematisch Instituut, Toernooiveld, 6525 ED Nijmegen, The Netherlands

E-mail: Heckman @ sci.kun.nl

INVARIANT DIFFERENTIAL OPERATORS AND WEYL GROUP INVARIANTS

SIGURDUR HELGASON

Massachusetts Institute of Technology

1. INTRODUCTION

In this research announcement we describe the relationship between the algebra $\mathbf{Z}(G)$ of bi-invariant differential operators on a simple noncompact Lie group G and the algebra $\mathbf{D}(G/K)$ of invariant differential operators on the symmetric space G/K associated with G (cf. §4). The natural map $\mu : \mathbf{Z}(G) \longrightarrow \mathbf{D}(G/K)$ turns out to be surjective except in exactly four cases. These cases involve the exceptional groups and for them the relationship between $\mathbf{Z}(G)$ and $\mathbf{D}(G/K)$ is quite complicated. However for all cases of G/K, each $D \in \mathbf{D}(G/K)$ is the "ratio" of two $\mu(Z_1)$ and $\mu(Z_2)$ (with $Z_1, Z_2 \in \mathbf{Z}(G)$). See Theorem 4.1.

The question is motivated partly by the theory of generalized spherical functions and partly by an intertwining property of the Radon transform on G/K relative to the horocycles. This is explained in §2. In §3 we state without proof some applications of finite-dimensional representations to the geometry of the space of horocycles. I am indebted to H. Schlichtkrull for helpful suggestions.

As usual, $\mathbb{R}$ and $\mathbb{C}$ denote the fields of real and complex numbers respectively, $\mathbb{Z}$ the ring of integers, and $\mathbb{Z}^+ = \{n \in \mathbb{Z} : n \geq 0\}$. If X is a manifold we write $\mathcal{D}(X)$ for $C_c^\infty(X)$, and $\mathcal{E}(X)$ for $C^\infty(X)$. If G is a Lie group, Ad denotes its adjoint representation.

2. TRANSMUTATION OPERATORS

The Radon transform on $\mathbb{R}^n$ associates to a function f on $\mathbb{R}^n$ its integral $\hat{f}$ over each hyperplane in the space. Formally we write

$$\hat{f}(\omega, p) = \int_{(x,\omega)=p} f(x)dm(x), \tag{1}$$

where $\omega \in \mathbb{R}^n$ is a unit vector, $p \in \mathbb{R}$ is arbitrary, and $(\ ,\ )$ denotes the inner product. Here dm is the Euclidean measure on the hyperplane $(x, \omega) = p$. The transform (1) has the elementary property

$$(Lf)^\wedge = \square \hat{f}, \tag{2}$$

where L is the Laplacian on $\mathbb{R}^n$ and $\square = d^2/dp^2$. If f is a radial function then $\hat{f}(\omega, p)$ is independent of ω so we write

$$f(x) = F(|x|), \quad \hat{f}(\omega, p) = \hat{F}(p), \tag{3}$$

where the functions F and $\hat{F}$ are even. Then (1) takes the form

$$\hat{F}(p) = \Omega_{n-1} \int_0^\infty F\left((p^2 + t^2)^{\frac{1}{2}}\right) t^{n-2} dt, \tag{4}$$

where Ω_{n-1} is the area of the unit sphere in $\mathbb{R}^{n-1}$, and (2) becomes

$$\left(\left(\frac{d^2}{dr^2} + \frac{n-1}{r}\frac{d}{dr}\right) F\right)^\wedge = \frac{d^2}{dp^2}\hat{F}. \tag{5}$$

Using the well-known range characterization of $\mathcal{D}(\mathbb{R}^n)^\wedge$ and denoting by $\mathcal{D}_{\text{even}}(\mathbb{R})$ the space of even functions in $\mathcal{D}(\mathbb{R})$ we can thus state the following result.

Theorem 2.1. *The operator $A : F \longrightarrow \hat{F}$ is a bijection of $\mathcal{D}_{\text{even}}(\mathbb{R})$ onto itself which satisfies the identity*

$$A \circ \left(\frac{d^2}{dr^2} + \frac{n-1}{r}\frac{d}{dr}\right) = \frac{d^2}{dp^2} \circ A.$$

This means that A is a "transmutation operator," converting the singular operator in (5) into the constant coefficient operator d^2/dp^2 (cf. [DL] for more general results).

The formulas in (2) and (5) have analogs for symmetric spaces $X = G/K$, where G is a connected noncompact semisimple Lie group with finite center and K a maximal compact subgroup. Fix an Iwasawa decomposition $G = KAN$ (A abelian, N nilpotent). The analogs for X of the hyperplanes in $\mathbb{R}^n$ are the *horocycles,* that is, the orbits in X of the groups gNg^{-1} conjugate to N. The Radon transform $f \longrightarrow \hat{f}$ is defined by

$$\hat{f}(\xi) = \int_\xi f(x) dm(x), \tag{6}$$

where ξ is a horocycle and dm the volume element on it. The group G permutes the horocycles ξ transitively and the space Ξ of all horocycles ξ is naturally identified with G/MN, where M is the centralizer of A in K. Let $\mathbf{D}(G/K)$ and $\mathbf{D}(G/MN)$ denote the algebras of G-invariant differential operators on X and Ξ, respectively. Both of these algebras are commutative. Let $\mathbf{D}(A)$ denote the algebra of invariant differential operators on A. Given $U \in \mathbf{D}(A)$ we consider the operator D_U on $\mathcal{E}(G/MN)$ defined by

$$(D_U\phi)(kaMN) = U_a(\phi(kaMN))), \quad k \in K, a \in A \tag{7}$$

We then have the following result ([H1]) analogous to (2).

Theorem 2.2.

(i) *The mapping* $U \to D_U$ *is an isomorphism of* $\mathbf{D}(A)$ *onto* $\mathbf{D}(G/MN)$.

(ii) *Given* $D \in \mathbf{D}(G/K)$ *there exists a* $\hat{D} \in \mathbf{D}(G/MN)$ *such that*

$$(Df)^\wedge = \hat{D}\hat{f}, \qquad f \in \mathcal{D}(X). \tag{8}$$

If $f \in \mathcal{E}(G/K)$ is K-invariant let $\bar{f}$ denote the restriction $f|A\cdot o$, o being the origin in X. Given a differential operator D on X let $\Delta(D)$ denote its radial part, that is, the (singular) differential operator on A (which we identify with $A \cdot o$) such that

$$\Delta(D)(\bar{f}) = (Df)^- \qquad \text{for } f \ K\text{-invariant}. \tag{9}$$

For example, if X is the hyperbolic plane $\mathbf{H}^2 = \mathbf{SU}(1,1)/\mathbf{SO}(2)$ we have the well-known expression

$$\Delta(L_{H^2}) = \frac{d^2}{dr^2} + \coth r \frac{d}{dr}. \tag{10}$$

Let $\mathcal{D}_W(A)$ denote the space of Weyl group invariants in $\mathcal{D}(A)$. Then $f \longrightarrow \bar{f}$ is a bijection of the space $\mathcal{D}_K(X)$ of K-invariants in $\mathcal{D}(X)$ onto $\mathcal{D}_W(A)$. In analogy with (3) we can write (with $F \in \mathcal{D}_W(A)$, $\hat{F} \in \mathcal{D}(A)$),

$$f(ka\cdot o) = F(a)\ , \quad \hat{f}(kaMN) = \hat{F}(a), \ k \in K, a \in A. \tag{11}$$

Let log denote the inverse of the exponential mapping for A (as a Lie group), let ρ denote half the sum of the restricted roots with multiplicity, and put $e^\rho(a) = e^{\rho(\log a)}$, $a \in A$. Then (8) and the Paley-Wiener theorem for (6) imply the following symmetric space analog for Theorem 2.1.

Theorem 2.3. *The mapping* $\mathcal{A} : F \longrightarrow e^\rho \hat{F}$ *is a bijection of* $\mathcal{D}_W(A)$ *onto itself which satisfies the identity*

$$\mathcal{A} \circ \Delta(D) = (e^\rho \hat{D} \circ e^{-\rho}) \circ \mathcal{A}, \qquad D \in \mathbf{D}(G/K). \tag{12}$$

The operators $\hat{D}$ and $e^\rho \hat{D} \circ e^{-\rho}$ are constant coefficient differential operators on the Euclidean space A. Writing $\hat{D}$ as a polynomial $\hat{D} = P(H_1, \dots, H_\ell)$ in a basis (H_i) of the Lie algebra of A we have

$$e^\rho \hat{D} \circ e^{-\rho} = P(H_1 - \rho(H_1), \dots, H_\ell - \rho(H_\ell)).$$

Thus (12) shows that $\mathcal{A}$ is a transmutation operator for all the singular operators $\Delta(D)$ simultaneously, that is, $\mathcal{A}$ converts them all into differential operators with constant coefficients. This was used in [H2] to prove that each $D \in \mathbf{D}(G/K)$ has a fundamental solution.

3. Finite-dimensional Representations

Let $\mathfrak{g} = \mathfrak{k} + \mathfrak{a} + \mathfrak{n}$ be the Lie algebra decomposition corresponding to $G = KAN$ and let $\langle , \rangle$ denote the Killing form of $\mathfrak{g}$. Let $\Sigma = \Sigma(\mathfrak{g}, \mathfrak{a})$ denote the set of restricted roots and Σ^+ the subset of positive ones corresponding to $\mathfrak{n}$. If $g \in G$ let $H(g) \in \mathfrak{a}$ be determined by $g \in K \exp H(g) N$.

Let $\ell = \operatorname{rank} G/K$ and consider linear forms $\mu_1, \ldots, \mu_\ell$ on $\mathfrak{a}$, each satisfying the integrality condition

$$\frac{\langle \mu_i, \alpha \rangle}{\langle \alpha, \alpha \rangle} \in \mathbb{Z}^+ \qquad \text{for } \alpha \in \Sigma^+. \tag{1}$$

We extend $\mathfrak{a}$ to a Cartan subalgebra $\mathfrak{h}$ of $\mathfrak{g}$ and extend each μ_i to a linear form on the complexification $\mathfrak{h}^{\mathbb{C}}$ of $\mathfrak{h}$ by making it 0 on $\mathfrak{h} \cap \mathfrak{k}$. Then there is an irreducible representation π_i of G with highest weight μ_i on the complex finite-dimensional vector space V_i. By [H3], this π_i has an MN-fixed vector $\mathrm{e}_i \neq 0$ and a K-fixed vector $v_i \neq 0$. Let π be the direct sum representation $\pi = \pi_1 \oplus \cdots \oplus \pi_\ell$ of G on the direct sum $V = \bigoplus_j V_j$ and put

$$\mathrm{e} = \mathrm{e}_1 + \cdots + \mathrm{e}_\ell, \qquad v = v_1 + \cdots + v_\ell.$$

We also consider the contragredient representations $\pi_1', \ldots, \pi_\ell'$ and the corresponding MN-fixed vectors $e_1', \ldots e_\ell'$. Let W denote the annihilator

$$W = \{w \in V : \ e_i'(w) = 0 \qquad \text{for } 1 \leq i \leq \ell\}.$$

The following description of the concepts of §2 inside the vector space V is established in [H6]. A special form of (i) is also observed in [E].

Theorem 3.1. *Assume the linear forms μ_i, $1 \leq i \leq \ell$, are linearly independent. Then the following statements hold:*

(i) *The maps*

$$gK \longrightarrow \pi(g)v \quad , \quad gMN \longrightarrow \pi(g)\mathrm{e}$$

of X and Ξ into V are both injective and imbed X and Ξ as G-orbits in V.

(ii) *Under the imbedding in (i) the horocycle space Ξ is the cone*

$$\{\pi(k)(s_1\mathrm{e}_1 + \cdots + s_\ell \mathrm{e}_\ell) : \quad \textit{all} \quad s_i > 0,\ k \in K\}.$$

(iii) *Viewing, by (i), X as a subset of V, the horocycle $\pi(N)v$ is the intersection $X \cap (v + W)$.*

Since G permutes the horocycles transitively, each of them is, by (iii), a plane section with X.

4. Bi-invariant Operators

Let $\mathbf{Z}(G)$ denote the (commutative) algebra of differential operators on G invariant under all left and right translations. For $D \in \mathbf{Z}(G)$ define $D_1 \in \mathbf{D}(G/K)$ and $D_2 \in \mathbf{D}(G/MN)$ by

$$(D_1 f) \circ p = D(f \circ p), \quad (D_2\phi) \circ \pi = D(\phi \circ \pi)$$

for $f \in \mathcal{D}(X)$ and $\phi \in \mathcal{D}(\Xi)$, where $p : G \longrightarrow G/K$ and $\pi : G \longrightarrow G/MN$ are the natural maps. Then an easy general argument [H4] (valid even for K and MN arbitrary closed subgroups) shows that

$$(D_1 f)^{\wedge} = D_2 \hat{f}. \tag{1}$$

One might then wonder if (8) §2 is a special case; in other words, if we put

$$\mathbf{Z}(G/K) = \{D_1 \ : \ D \in \mathbf{Z}(G)\},$$

is the subalgebra $\mathbf{Z}(G/K) \subset \mathbf{D}(G/K)$ actually equal to $\mathbf{D}(G/K)$? The question was once mentioned to me by Harish-Chandra in a different context: for certain generalizations of spherical functions it is most natural to work with $\mathbf{Z}(G/K)$, but from a different viewpoint it is most natural to work with $\mathbf{D}(G/K)$. The above question as to whether $\mathbf{Z}(G/K)$ equals $\mathbf{D}(G/K)$ is therefore natural and important.

An affirmative answer to the question has appeared in the literature (cf. [Be], pp. 243, 248). However, as we shall see below, the situation is more complicated.

Theorem 4.1. *Let G/K be a symmetric space of the noncompact type. Then for each $D \in \mathbf{D}(G/K)$ there exist $Z_1, Z_2 \in \mathbf{Z}(G/K)$ such that*

$$Z_1 D = Z_2 \ \text{ and } Z_1 \neq 0.$$

However, one cannot in general take Z_1 to be a constant. More precisely, the following result holds.

Theorem 4.2. *Suppose G/K is an irreducible symmetric space of the noncompact type. Then $\mathbf{Z}(G/K) = \mathbf{D}(G/K)$ except for the four cases*

$$G/K = E_6/\mathbf{SO}(10)\mathbb{R},\ E_6/F_4,\ E_7/E_6\mathbb{R},\ E_8/E_7\mathbf{SU}(2). \tag{2}$$

We shall just indicate the proof, the full details of which are given in [H5]. We have the inclusions $\mathfrak{a} \subset \mathfrak{h} \subset \mathfrak{g}$ and $\mathfrak{a}^{\mathbb{C}} \subset \mathfrak{h}^{\mathbb{C}} \subset \mathfrak{g}^{\mathbb{C}}$, superscript denoting complexification. Let $\Delta = \Delta(\mathfrak{g}^{\mathbb{C}}, \mathfrak{h}^{\mathbb{C}})$ denote the set of roots of $\mathfrak{g}^{\mathbb{C}}$ with respect to $\mathfrak{h}^{\mathbb{C}}$. With a bar denoting restriction to $\mathfrak{a}$ we have $\Sigma = \{\bar{\alpha} : \alpha \in \Delta, \bar{\alpha} \neq 0\}$. We give Δ an order compatible with that of Σ

and split the set $\Delta^+ = \{\alpha \in \Delta : \alpha > 0\}$ into subsets $\Delta^+ = P_0 \cup P$, where $P_0 = \{\alpha \in \Delta^+ : \bar{\alpha} = 0\}$ and $P = \Delta^+ - P_0$. We put

$$\rho_o = \frac{1}{2}\sum_{\alpha\in P_0} \alpha, \quad \rho = \frac{1}{2}\sum_{\alpha\in P} \bar{\alpha}$$

so that ρ is the same as defined in §2. Let $W = W(\Sigma)$ and $\tilde{W} = W(\Delta)$ be the Weyl groups of the root systems Σ and Δ, respectively. Then W acts on $\mathfrak{a}, \mathfrak{a}^{\mathbf{C}}, (\mathfrak{a}^{\mathbf{C}})^*$ (asterisk denoting dual) and $\tilde{W}$ acts on $\mathfrak{h}, \mathfrak{h}^{\mathbf{C}}$ and $(\mathfrak{h}^{\mathbf{C}})^*$. Let $I(\mathfrak{a}^{\mathbf{C}})$ and $I(\mathfrak{h}^{\mathbf{C}})$ denote the corresponding W and $\tilde{W}$ invariants, respectively, in the symmetric algebras $S(\mathfrak{a}^{\mathbf{C}})$ and $S(\mathfrak{h}^{\mathbf{C}})$. We view $(\mathfrak{a}^{\mathbf{C}})^*$ as a subset of $(\mathfrak{h}^{\mathbf{C}})^*$ by taking each $\lambda \in (\mathfrak{a}^{\mathbf{C}})^*$ to be identically zero on $\mathfrak{h}^{\mathbf{C}} \cap \mathfrak{k}^{\mathbf{C}}$. Let $\tilde{W}_0$ be the translation of $\tilde{W}$ by ρ_0, that is, the group of transformations

$$\mu \longrightarrow s\mu + s\rho_o - \rho_o, \qquad \mu \in (\mathfrak{h}^{\mathbf{C}})^*,$$

as s runs through $\tilde{W}$. The corresponding algebra $I_o(\mathfrak{h}^{\mathbf{C}})$ of invariants consists of the functions $\mu \longrightarrow P(\rho_o + \mu)$ on $(\mathfrak{h}^{\mathbf{C}})^*$ as P runs through $I(\mathfrak{h}^{\mathbf{C}})$.

Using Fourier analysis on G/K it is not difficult to prove the following result where ϕ_λ denotes the usual spherical function

$$\phi_\lambda(gK) = \int_K e^{(i\lambda-\rho)(H(gk))}dk, \quad \lambda \in \mathfrak{a}^*_{\mathbf{C}}.$$

Lemma 4.3. *Let $D_1, D_2 \in \mathbf{D}(G/K)$ and let $c_1(\lambda), c_2(\lambda)$ be defined by*

$$D_1\phi_\lambda = c_1(\lambda)\phi_\lambda, \qquad D_2\phi_\lambda = c_2(\lambda)\phi_\lambda.$$

If $c_1(\lambda) = c_2(\lambda)$ for all $\lambda \in \mathfrak{a}^*_{\mathbf{C}}$, then $D_1 = D_2$.

On the other hand

$$D\phi_\lambda = \Gamma(D)(i\lambda)\phi_\lambda, \qquad Z\phi_\lambda = \gamma(Z)(\rho_o + i\lambda)\phi_\lambda,$$

where $\Gamma : \mathbf{D}(G/K) \longrightarrow I(\mathfrak{a}^{\mathbf{C}})$ and $\gamma : \mathbf{Z}(G) \longrightarrow I(\mathfrak{h}^{\mathbf{C}})$ are the Harish-Chandra isomorphisms [HC1], [HC2]. We then obtain the following result.

Proposition 4.4. *The following properties of the symmetric space G/K are equivalent:*

(i) $\mathbf{Z}(G/K) = \mathbf{D}(G/K)$.

(ii) *The restriction from $(\mathfrak{h}^{\mathbf{C}})^*$ to $(\mathfrak{a}^{\mathbf{C}})^*$ maps $I_o(\mathfrak{h}^{\mathbf{C}})$ onto $I(\mathfrak{a}^{\mathbf{C}})$.*

In a similar fashion Theorem 4.1 is reduced to the statement that if J_o denotes the image of $I_o(\mathfrak{h}^{\mathbf{C}})$ under the restriction from $(\mathfrak{h}^{\mathbf{C}})^*$ to $(\mathfrak{a}^{\mathbf{C}})^*$ then the quotient field $\mathbb{C}(J_o)$ equals the quotient field $\mathbb{C}(I(\mathfrak{a}^{\mathbf{C}}))$. This last statement is proved by studying the homomorphisms of J_0 into $\mathbb{C}$.

Since the translation by ρ_o does not preserve homogeneity of polynomials on $(\mathfrak{h}^{\mathbf{C}})^*$, criterion (ii) is hard to use directly for proving Theorem 4.2. Fortunately it can be stated in a simpler form.

Proposition 4.5. *Assume G is simple. Denoting the restriction from $(\mathfrak{h}^{\mathbf{C}})^*$ to $(\mathfrak{a}^{\mathbf{C}})^*$ by a bar, the following conditions are equivalent:*

(i) $I_o(\mathfrak{h}^{\mathbf{C}})^- = I(\mathfrak{a}^{\mathbf{C}})$,
(ii) $I(\mathfrak{h}^{\mathbf{C}})^- = I(\mathfrak{a}^{\mathbf{C}})$.

The implication (ii) $\Longrightarrow$ (i) is proved by simple induction on the degree (cf. [F-J], Theorem 6.4). For the converse we resort to classification. The algebras $I(\mathfrak{h}^{\mathbf{C}})$ and $I(\mathfrak{a}^{\mathbf{C}})$ have homogeneous algebraically independent generators $j_1, \ldots, j_r$ and $i_1, \ldots, i_\ell$ with degrees denoted by $n_1, \ldots, n_r$ and $m_1, \ldots, m_\ell$. In all but the four exceptional cases (2) the m_i appear among the n_j. For the four cases (2) (in Cartan's notation EIII, EIV, EVII, EIX) we have

$$EIII: (n_1, \ldots, n_r) = (2,5,6,8,9,12), \qquad (m_1, \ldots, m_\ell) = (2,4)$$
$$EIV: (n_1, \ldots, n_r) = (2,5,6,8,9,12), \qquad (m_1, \ldots, m_\ell) = (2,3)$$
$$EVII: (n_1, \ldots, n_r) = (2,6,8,10,12,14,18), \qquad (m_1, \ldots, m_\ell) = (2,4,6)$$
$$EIX: (n_1, \ldots, n_r) = (2,8,12,14,18,20,24,30), (m_1, \ldots, m_\ell) = (2,6,8,12)$$

Because of the algebraic independence we see that in each of these cases i_2 (of respective degree $4,3,4,6$) does not belong to $I(\mathfrak{h}^{\mathbf{C}})^-$. Thus $I(\mathfrak{h}^{\mathbf{C}})^-$ is not equal to $I(\mathfrak{a}^{\mathbf{C}})$. That $I_o(\mathfrak{h}^{\mathbf{C}})^-$ is also different from $I(\mathfrak{a}^{\mathbf{C}})$ is a consequence of the following lemma kindly communicated to me by David Vogan. An exhaustive study of such anomalies has been done by Bien [Bi].

Lemma 4.6. *(Vogan) For each of the four cases in (2) there exist dominant integral functions $\lambda \neq \mu$ on $\mathfrak{h}^{\mathbf{C}}$ such that $\lambda, \mu \in \mathfrak{a}^*$ and such that $\lambda + \rho_o$ and $\mu + \rho_o$ are conjugate by $\tilde{W}$.*

For all other cases of G/K one has $I(\mathfrak{h}^{\mathbf{C}})^- = I(\mathfrak{a}^{\mathbf{C}})$ by explicit verification for each case (cf. [H5]). For G/K classical this was done earlier in [H2]. For the exceptional spaces G/K we make extensive use of Lee's paper [L]. Proposition 4.5 follows from these results, and thus so does Theorem 4.2.

Remark. For the exceptional cases (2) the relationship between $\mathbf{Z}(G/K)$ and $\mathbf{D}(G/K)$ is quite complicated. Consider, for example, the case EVII where both j_2 and i_3 have degree 6. Here $i_3 \neq \bar{j}_2$ whereas i_3 is a rational function of $\bar{j}_1$, $\bar{j}_2$, $\bar{j}_3$, and $\bar{j}_4$. Thus, in Theorem 4.1, if D is the sixth degree generator in $\mathbf{D}(G/K)$, Z_2 has degree at least 10.

References

[Be] F. A. Berezin, *Laplace operators on semisimple Lie groups*, Amer. Math. Soc. Transl. **21** (1962), 239–339.

[Bi] F.V, Bien, *$\mathcal{D}$-modules and spherical representations*, Math. Notes, Princeton Univ. Press, Princeton, N. J., 1990.

[DL] J. Delsarte and J.L. Lions, *Moyennes Généralisées*, Comm. Math. Helv. **33** (1959), 59–69.

[E] L. Ehrenpreis, *The use of partial differential equations for the study of group representations*, Proc. Symp. Pure Math. Vol. XXVI, Amer. Math. Soc. 1973, 317–320.

[F-J] M. Flensted-Jensen, *Spherical functions on a real semisimple Lie group. A method of reduction to the complex case*, J. Funct. Anal. **30** (1978), 106–146.

[HC1] Harish-Chandra, *The characters of semisimple Lie groups*, Trans. Amer. Math. Soc. **83** (1956) 98–163.

[HC2] ________ , *Spherical functions on a semisimple Lie group I*, Amer. J. Math. **801** (1958), 241–310.

[H1] S. Helgason, *Duality and Radon transform for symmetric spaces*, Amer. J. Math. **85** (1963), 667–692.

[H2] ________ , *Fundamental solutions of invariant differential operators on symmetric spaces*, Amer. J. Math. **86** (1964), 565–601.

[H3] ________ , *Radon-Fourier transforms on symmetric spaces and related group representations*, Bull. Amer. Math. Soc. **71** (1965), 757–763.

[H4] ________ , *Some results on Radon transforms, Huygens' principle and X-ray transforms*, Contemp. Math. **63** (1987), 151–177.

[H5] ________ , *Some results on invariant differential operators on symmetric spaces*, Amer. J. Math. (to appear).

[H6] ________ , *Geometric Analysis on Symmetric Spaces*, monograph in preparation.

[L] C.Y. Lee, *Invariant polynomials of Weyl groups and applications to the centers of universal enveloping algebras*, Can. J. Math. XXVI (1974), 583–592.

DEPARTMENT OF MATHEMATICS, MASSACHUSETTS INSTITUTE OF TECHNOLOGY, ROOM 2-182, CAMBRIDGE, MA 02139

THE SCHWARTZ SPACE OF A GENERAL SEMISIMPLE LIE GROUP

REBECCA A. HERB

University of Maryland

1. INTRODUCTION

Let G be a connected semisimple Lie group. The tempered spectrum of G consists of families of representations unitarily induced from cuspidal parabolic subgroups. Each family is parameterized by unitary characters of a θ-stable Cartan subgroup. The Schwartz space $\mathcal{C}(G)$ is a space of smooth functions decreasing rapidly at infinity and satisfying the inclusions $C_c^\infty(G) \subset \mathcal{C}(G) \subset L^2(G)$. The Plancherel theorem expands Schwartz class functions in terms of the distribution characters of the tempered representations. Very roughly, we can write $f(x) = \sum_H f_H(x)$, $f \in \mathcal{C}(G), x \in G$, where

$$f_H(x) = \int_{\hat{H}} \Theta(H:\chi)(R(x)f)m(H:\chi)d\chi.$$

Here the summation is over a full set of representatives H of conjugacy classes of Cartan subgroups of G, $\Theta(H:\chi)$ is the distribution character of the representation corresponding to $\chi \in \hat{H}$, $R(x)f$ is the right translate of f by $x \in G$, and $m(H:\chi)d\chi$ is Plancherel measure.

When G has finite center, the Plancherel formula was proven by Harish-Chandra in [**HC**1,2,3]. Along the way he constructed wave packets of Eisenstein integrals which are Schwartz class, and showed for every K-finite $f \in \mathcal{C}(G)$ that each function f_H occuring in the Plancherel formula decomposition of f is a finite sum of such wave packets coming from the H-series of tempered representations. In particular, this shows that each f_H is an element of $\mathcal{C}(G)$. Thus, for the finite center case, the work of Harish-Chandra gives a complete description of the K-finite functions in the Schwartz space in terms of their Fourier transforms.

The Plancherel formula for general connected semisimple Lie groups, that is, those with possibly infinite centers, was proven by the author together with J. Wolf in [**HW** 1,2]. The proof is independent of Harish-Chandra's result and bypasses the technical machinery of Eisenstein integrals, c-functions, and wave packets. Instead it uses explicit character

Partially supported by NSF grant DMS-88-02586 .

formulas and Fourier inversion of orbital integrals. This technique of proof was first used by P. Sally and G. Warner for rank one linear groups [**SW**] and was extended to linear groups of arbitary rank by the author [**H** 1,2,3]. Although this proof provides a quick route to the Plancherel formula, it leaves open the Paley-Wiener problem of characterizing Schwartz functions in terms of their Fourier transforms. In this paper I would like to describe what is known about this problem for general semisimple groups and illustrate the results in the example of the universal covering group of $SL(2,\mathbb{R})$. Details can be found in [**H** 4,5, **HW** 3,4,5].

2. General Results

Let G be a connected reductive group and K a maximal relatively compact subgroup, i.e., the center Z of G is contained in K and K/Z is a maximal compact subgroup of G/Z. Then $K = K_1 \times V$ where K_1 is compact and V is a vector group central in K. Of course, if G is semisimple with finite center, then $V = \{0\}$, while if G is simple and has infinite center, then V is one-dimensional. Let $\mathfrak{v}$ denote the real Lie algebra of V. Then for all $h \in i\mathfrak{v}^*$, e^h is a one-dimensional unitary character of K, trivial on K_1. Thus every K-type $\tau \in \hat{K}$ lies in a continuous family $\{\tau \otimes e^h : h \in i\mathfrak{v}^*\}$.

Let $P = MAN$ be the Langlands decomposition of a cuspidal parabolic subgroup of G. Let $H = TA$ be a Cartan subgroup so that $T \subset M \cap K$ is a relatively compact Cartan subgroup of M. Then $T \subset K$, implying that e^h is a character of T for all $h \in i\mathfrak{v}^*$. Thus each $\chi \in \hat{T}$ lies in a continuous family $\{\chi \otimes e^h : h \in i\mathfrak{v}^*\}$. Fix $\chi \in \hat{T}$ and let $\pi(T : h)$ denote the relative discrete series representation of M parameterized by $\chi \otimes e^h$. Define

$$\begin{aligned} \mathcal{D} \;=\; \{h \in i\mathfrak{v}^* \;&:\; \text{the Harish-Chandra parameter of } \pi(T : h) \\ &\text{lies in the same Weyl chamber as that of } \pi(T : 0)\}. \end{aligned}$$

Then $\mathcal{D}$ is a convex open set and $\{\pi(T : h) : h \in \mathcal{D}\}$ is a continuous family of relative discrete series representations of M. For example, if $M = G$ is simple and has infinite center so that V is one-dimensional, then $\mathcal{D}$ is an interval and is unbounded just in case the corresponding representations are holomorphic or anti-holomorphic relative discrete series. Now $\{\pi(H : h : \nu) = Ind^G_{MAN}(\pi(T : h) \otimes e^{i\nu} \otimes 1) : h \in \mathcal{D}, \nu \in \mathfrak{a}^*\}$ is a continuous family of tempered representations of G. (Note that in general these representations are not all distinct since, in addition to the usual Weyl group identifications, in the h variable they depend only on the restriction of e^h to T.) The characters vary smoothly in $h \in \mathcal{D}$ and $\nu \in \mathfrak{a}^*$ and extend holomorphically to $h \in \mathfrak{v}^*_{\mathbb{C}}$ and $\nu \in \mathfrak{a}^*_{\mathbb{C}}$. Further, for fixed $\tau \in \hat{K}$ the multiplicity of $\tau \otimes e^h$ in $\pi(H : h : \nu)$ is independent of $h \in \mathcal{D}$ and $\nu \in \mathfrak{a}^*$.

We obtain spherical functions of matrix coefficients for this family by using the Harish-Chandra Eisenstein integral construction. Thus we first fix a double representation (τ_1, τ_2) of K on a finite-dimensional vector space W. For any $h \in \mathfrak{v}^*_{\mathbb{C}}$, $(\tau_{1,h}, \tau_{2,h})$ is the double representation of K on W

given by $\tau_{i,h} = \tau_i \otimes e^h$ for $i = 1, 2$. As in [**HW** 3,4,5], we can now define smooth functions $F : \mathfrak{v}^*_{\mathbb{C}} \times M \to W$ so that:

(i) F is a holomorphic function of $h \in \mathfrak{v}^*_{\mathbb{C}}$;
(ii) for all $k_1, k_2 \in K \cap M, m \in M$, and $h \in \mathfrak{v}^*_{\mathbb{C}}$,

$$F(h : k_1 m k_2) = \tau_{1,h}(k_1) F(h : m) \tau_{2,h}(k_2);$$

(iii) for every $h \in \mathcal{D}$ and $w^* \in W^*$, $m \to < F(h : m), w^* >$ is a matrix coefficient of the representation $\pi(T : h)$;
(iv) when $\mathcal{D}$ is unbounded, F has polynomial growth for $h \in \mathcal{D}$.

If $x \in G$, use $G = KMAN$ to express $x = k(x)m(x)\exp H(x)n(x)$ and extend F to G by $F(h : x) = \tau_{1,h}(k(x))F(h : m(x))$. The Eisenstein integral is then defined by

$$E(P : F : h : \nu : x) = \int_{K/Z} F(h : xk)\tau_{2,h}(k^{-1})e^{(i\nu-\rho)H(xk)}d(kZ),$$

where $\rho(H)$ is 1/2 the trace of $\text{ad}(H)$ on $\mathfrak{n}$.

By a wave packet of Eisenstein integrals we mean a function

$$\Phi_\alpha(x) = \int_{\mathcal{D}\times\mathfrak{a}^*} E(P : F : h : \nu : x)\alpha(h : \nu)m(h : \nu)dhd\nu.$$

Here α is a jointly smooth function on $\text{cl}(\mathcal{D}) \times \mathfrak{a}^*$, with all derivatives rapidly decreasing at infinity, and $m(h : \nu)dhd\nu$ is the Plancherel measure. Note that Φ_α takes values in the finite-dimensional vector space W. To obtain scalar-valued wave packets take $w^* \in W^*$ and set $\phi_\alpha(x) = < \Phi_\alpha(x), w^* >$.

Theorem 1 **(Harish-Chandra).** *Suppose the semisimple part of G has finite center. Then each ϕ_α constructed as above is a Schwartz class function on G. Further, if $f \in \mathcal{C}(G)$ is K-finite, then f is a finite sum of wave packets of this type.*

When the semisimple part of G has infinite center, there are no K-finite functions in the Schwartz space. However we can generalize the notion of K-finiteness in a natural way by saying $f \in \mathcal{C}(G)$ is K-compact if its K-types lie in a compact subset of $\hat{K}$. The K-compact functions are dense in $\mathcal{C}(G)$. Now the second part of Harish-Chandra's theorem has a direct analog in the general case.

Theorem 2. *Suppose $f \in \mathcal{C}(G)$ is K-compact. Then f is a finite sum of wave packets ϕ_α as above.*

The first part of Harish-Chandra's theorem requires more serious modification in the infinite center case. In general, the individual wave packets occuring in the decomposition of a Schwartz function will not themselves be Schwartz class functions. This is because of interference between different series of tempered representations which occurs any time the limits of discrete series occurring in the decompositon of a reducible principal series representation are actual limits along a continuous family of relative discrete series representations. In order that an individual wave packet be Schwartz class it is necessary to make stronger assumptions on the function α to guarantee that the wave packet avoids interference at these limits of discrete series and reducible principal series.

The limits of discrete series from the chamber associated with $\mathcal{D}$ correspond to elements h in bdry($\mathcal{D}$), the boundary of cl($\mathcal{D}$). In order to avoid interference at these points we require that α have zeroes of infinite order along bdry($\mathcal{D}$) in the sense that α and all its derivatives in h and ν vanish on bdry$(\mathcal{D}) \times \mathfrak{a}^*$.

The reducible principal series representations which overlap in a non-trivial way with other series are exactly those for which the Plancherel function $m(h : \nu)$, which is separately smooth in both variables, fails to be jointly smooth. In order to avoid interference at these points we need to assume that $(h,\nu) \rightarrow E(P : F : h : \nu : x)\alpha(h : \nu)m(h : \nu)$ is jointly smooth on $\mathcal{D} \times \mathfrak{a}^*$. Since $E(P : F)$ is jointly smooth and α is always assumed to be jointly smooth, this is a restriction on α only at those points where $m(h : \nu)$ fails to be jointly smooth.

Theorem 3. *Let ϕ_α be a wave-packet as above. Then $\phi_\alpha \in \mathcal{C}(G)$ if and only if α has zeroes of infinite order along the walls of $\mathcal{D}$ and*

$$(h,\nu) \rightarrow E(P : F : h : \nu : x)\alpha(h : \nu)m(h : \nu)$$

is jointly smooth on $\mathcal{D} \times \mathfrak{a}^*$.

In order to characterize which sums of non-Schwartz class wave packets patch together to form a Schwartz class function it is necessary to look at matching conditions. The conditions above which guarantee that a single wave packet is Schwartz will be a special case of these matching conditions. That is, they will be the conditions necessary for a wave packet to match with the zero function from all other series.

It has been known for some time that matching conditions would be necessary for Schwartz functions in the infinite center case. This was first pointed out to me in 1984 by D. Miličić. He and H. Kraljević discovered this phenomenon when they were studying the Fourier transform for the C^*-algebra of the universal covering group of $SL(2,\mathbb{R})$. [**KM**]

The matching conditions for Schwartz functions are consequences of the following character identity which is stated in its simplest form here. (For the more general statement and details see [**H** 4].) Assume G is a connected reductive group with rank $G =$ rank K, and B a maximal relatively compact Cartan subgroup of G. Let β be a singular imaginary root of $(\mathfrak{g}, \mathfrak{b})$. In order to have non-trivial matching conditions we assume β is not orthogonal to $\mathfrak{v}$. (Note $\mathfrak{v} \subset \mathfrak{b}$ since V is central in K.) Let $C^{\pm}$ be two Weyl chambers of $i\mathfrak{b}^*$ which are separated only by the hyperplane corresponding to $\beta = 0$. Let $H = TA$ be the Cartan subgroup of G obtained by the Cayley transform c with respect to β and let $P = MAN$ be a corresponding cuspidal parabolic subgroup. Let C' be the Weyl chamber of it^* such that $\tau|_t \in C'$ for all $\tau \in C^{\pm}$.

Now fix $\chi \in \hat{B}$ and let $\Theta(B : C^{\pm} : h)$ denote the continuous family of coherently continued relative discrete series characters parameterized by $\chi \otimes e^h$ and $C^{\pm}$. Thus if the Harish-Chandra parameter $\tau(h)$ of the relative discrete series representation $\pi(B : h)$ lies in $C^{\pm}$, then $\Theta(B : C^{\pm} : h)$ is its character. $T \subset B$ in this case so we can define $\chi' \in \hat{T}$ by restricting χ to T. Let $\Theta(T : C' : h)$ denote the coherently continued relative discrete series character of M corresponding to C' and $\chi' \otimes e^h$, and for $\nu \in \mathfrak{a}^*$ define

$$\Theta(H : C' : h : \nu) = \mathrm{Ind}_{MAN}^{G}(\Theta(T : C' : h) \otimes e^{i\nu} \otimes 1).$$

We pick the linear functionals $h_1 \in i\mathfrak{b}^*$ and $\nu_1 \in \mathfrak{a}^*$ such that

$$2 < h_1, \beta > / < \beta, \beta >= 1 \text{ and } 2 < \nu_1, {}^c\beta > / <{}^c\beta, {}^c\beta >= 1.$$

Theorem 4. *Let G' denote the regular set of G, and suppose that $\tau(h)$ is orthogonal to β when $h = h_0$. Then, for any integer $k \geq 0$ and $x \in G'$, the following character identity is valid:*

$$(d/ds)^k_{s=0}\Theta(B : C^+ : h_0 + sh_1 : x) + (d/ds)^k_{s=0}\Theta(B : C^- : h_0 + sh_1 : x)$$
$$= (\partial/\partial s \pm i\partial/\partial t)^k_{s=t=0}\Theta(H : C' : h_0 + sh_1 : t\nu_1 : x).$$

In the case $k = 0$ (i.e., no derivatives), this theorem is Schmid's identity [S]. The general case of the theorem is an easy consequence of a more general statement of Schmid's identity.

The above character identity leads to matching conditions for the Fourier transforms of Schwartz functions as follows. Suppose $\tau(h_0)$ is semi-regular, orthogonal to a unique singular imaginary root β, and an element of the intersection of the closures of $C^{\pm}$. Then we can assume that, for s sufficiently small, $\tau(h_0 + sh_1)$ lies in C^+ when $s > 0$ and lies in C^- when $s < 0$. Thus $\Theta(B : C^+ : h_0 + sh_1)$ is equal to the tempered character $\Theta(B : h_0 + sh_1)$ for small $s > 0$, and $\Theta(B : C^- : h_0 + sh_1) = \Theta(B : h_0 + sh_1)$ for small $s < 0$. Further, $\Theta(H : C' : h_0+sh_1 : t\nu_1)$ is equal to the tempered character $\Theta(H : h_0 + sh_1 : t\nu_1)$ for all small s and all (real) t. By integrating against $f \in \mathcal{C}(G)$ we obtain the following theorem.

Theorem 5. *Let $f \in \mathcal{C}(G)$. Then for any integer $k \geq 0$,*

$$(\lim_{s\downarrow 0} + \lim_{s\uparrow 0})(d/ds)^k\Theta(B : h_0 + sh_1 : R(x)f) =$$
$$(\partial/\partial s \pm i\partial/\partial t)^k_{s=t=0}\Theta(H : h_0 + sh_1 : t\nu_1 : R(x)f).$$

Theorem 5 can be generalized to the case where $\tau(h_0)$ lies on more than one root hyperplane. However, in this case the identities for all roots with respect to which $\tau(h_0)$ is singular have to be combined in order to obtain an identity which involves only tempered characters.

Finally, character identities for limits of discrete series can easily be induced to prove more general identities involving any tempered representations induced from limits of discrete series. These induced versions of the character identities should be sufficient to give all matching conditions on Schwartz functions. However, at this time it has not yet been proven that these matching conditions are sufficient to guarantee that a finite sum of wave packets patch together to form a Schwartz function.

3. Universal Covering Group of $SL(2,\mathbb{R})$

Let $\bar{G} = SL(2,\mathbb{R})$. Let $\bar{B}$ be the compact Cartan subgroup of $\bar{G}$. Then $\bar{B}$ is a circle, and so the discrete series representations of $\bar{G}$ are parameterized by the integers. For $n \in \mathbb{Z} - \{0\}$, let $\Theta(\bar{B} : n)$ denote the character of the corresponding discrete series representation and let $\Theta(\bar{B} : \pm : n)$ denote the coherent continuations of discrete series characters. Then the two limits of discrete series have characters $\Theta(\bar{B} : \pm : 0)$. Let $\bar{H}$ be the non-compact Cartan subgroup of $\bar{G}$. Then $\bar{H} = \{\pm I\} \times A$, where A is a one-dimensional vector group, and the principal series representations correspond to characters of $\bar{H}$. It will be convenient to write their characters as $\Theta(\bar{H} : \epsilon : \nu)$ where $\nu \in \mathbb{C}$, $\epsilon = 0$ gives the non-spherical principal series, and $\epsilon = 1$ gives the spherical principal series. These characters are tempered just when $\nu \in \mathbb{R}$. Further, $\Theta(\bar{H} : 0 : 0)$ is the character of the only reducible principal series representation, and $\Theta(\bar{H} : 0 : 0) = \Theta(\bar{B} : + : 0) + \Theta(\bar{B} : - : 0)$.

The Plancherel formula for $\bar{G}$ can be written as $f(x) = f_{\bar{B}}(x) + f_{\bar{H}}(x)$ where:

$$f_{\bar{B}}(x) = c_{\bar{B}} \sum_{n \in \mathbb{Z}} \Theta(\bar{B} : n)(R(x)f)|n|;$$

$$f_{\bar{H}}(x) = c_{\bar{H}} \sum_{\epsilon=0,1} \int_{-\infty}^{+\infty} \Theta(\bar{H} : \epsilon : \nu)(R(x)f)\nu \sinh \pi\nu(\cosh \pi\nu - (-1)^{\epsilon})^{-1} d\nu.$$

Note that the limits of discrete series don't occur in the formula for $f_{\bar{B}}$— hence there are no matching conditions between $f_{\bar{B}}$ and $f_{\bar{H}}$.

Let G be the universal covering group of $SL(2,\mathbb{R})$ with covering map $\pi : G \to \bar{G}$. Then $B = \pi^{-1}(\bar{B}) \cong \mathbb{R}$ is the relatively compact Cartan subgroup of G. With these definitions, the characters of the relative discrete series of G can be parameterized as $\Theta(B : h)$, $h \in \mathbb{R} - \{0\}$, and the coherent continuations of relative discrete series characters as $\Theta(B : \pm : h)$, $h \in \mathbb{R}$. Thus, for every $x \in G'$, the function $h \to \Theta(B : \pm : h)(x)$ is smooth, and the characters of the limits of discrete series are actual limits along these smooth families of characters.

Let $H = \pi^{-1}(\bar{H})$. Then $H = TA$ where $T = \mathbb{Z}$ is the (infinite) center of G. The characters of the principal series of G can be parameterized as $\Theta(H : h : \nu)$ where $h \in \mathbb{R}$ and $\nu \in \mathbb{C}$. These characters are tempered for $\nu \in \mathbb{R}$ and depend only on the coset of h in $\mathbb{R}/2\mathbb{Z}$.

The general form of Schmid's identity for this case says that

$$\Theta(B:+:h)+\Theta(B:-:h)=\Theta(H:h:-ih)$$

for all $h \in \mathbb{R}$. Now since both sides are smooth functions of h, for all $k \geq 0$ we have

$$(d/dh)^k_{h=0}\Theta(B{:}+{:}h)+(d/dh)^k_{h=0}\Theta(B{:}-{:}h)=(d/dh)^k_{h=0}\Theta(H{:}h{:}-ih).$$

This identity can be written completely in terms of tempered characters as

$$(\lim_{h\downarrow 0}+\lim_{h\uparrow 0})(d/dh)^k\Theta(B:h)=(\partial/\partial h-i\partial/\partial\nu)^k_{h=\nu=0}\Theta(H:h:\nu).$$

Now, since $\Theta(H:h:\nu)$ is an even function of ν,

$$(\partial/\partial h-i\partial/\partial\nu)^k_{h=\nu=0}\Theta(H:h:\nu)=(\partial/\partial h+i\partial/\partial\nu)^k_{h=\nu=0}\Theta(H:h:\nu).$$

Integrating against $f \in \mathcal{C}(G)$ we obtain a special case of Theorem 5.

Theorem 6.

$$(\lim_{h\downarrow 0}+\lim_{h\uparrow 0})(d/dh)^k\Theta(B:h:R(x)f)=$$
$$(\partial/\partial h\pm i\partial/\partial\nu)^k_{h=\nu=0}\Theta(H:h:\nu:R(x)f).$$

The Plancherel formula for G can be written as $f(x)=f_B(x)+f_H(x)$ where:

$$f_B(x)=c_B\int_{-\infty}^{+\infty}\Theta(B:h)(R(x)f)|h|dh;$$

$$f_H(x)=\int_{-1}^{+1}\int_{-\infty}^{+\infty}\Theta(H:h:\nu)(R(x)f)\nu\sinh\pi\nu(\cosh\pi\nu-\cos\pi h)^{-1}d\nu dh.$$

Here Theorem 6 gives an infinite collection of matching conditions between the terms occuring in f_B and f_H.

Suppose that f is a single Schwartz class wave packet. In the case that f is a wave packet of relative discrete series matrix coefficients corresponding to the Weyl chamber with, for example, $h>0$, we will have

$$\Theta(B:h)(R(x)f)=0 \text{ for all } h<0, \text{ and}$$
$$\Theta(H:h:\nu)(R(x)f)=0 \text{ for all } h,\ \nu.$$

Thus, by Theorem 6,

$$\lim_{h\downarrow 0}(d/dh)^k\Theta(B:h:R(x)f)=0$$

for all $k\geq 0$. That is, $\Theta(B:h:R(x)f)$ has a zero of infinite order along the wall of the Weyl chamber. Thus the wave packet must satisfy the extra requirement of Theorem 3 at the limit of discrete series.

In the case that f is a wave packet of Eisenstein integrals coming from the principal series, $\Theta(B:h)(R(x)f)=0$ for all h so that

$$(\partial/\partial h \pm i\partial/\partial\nu)^k_{h=\nu=0}\Theta(H:h:\nu:R(x)f)=0$$

for all $k\geq 0$. This implies that $h^2+\nu^2$ divides $\Theta(H:h:\nu)(R(x)f)$. But this, in turn, implies that $\Theta(H:h:\nu)(R(x)f)m(h:\nu)$ is jointly smooth at $(h,\nu)=(0,0)$ since $m(h:\nu)=\nu\sinh\pi\nu(\cosh\pi\nu-\cos\pi h)^{-1}$ acts like $\nu^2(h^2+\nu^2)^{-1}$ near $(0,0)$. Thus the wave packet satisfies the extra requirement of Theorem 3 at the reducible principal series.

References

[HC1] Harish-Chandra, *Harmonic analysis on real reductive groups I*, J. Funct. Anal. **19** (1975), 104–204.

[HC2] ———, *Harmonic analysis on real reductive groups II*, Inv. Math. **36** (1976), 1–55.

[HC3] ———, *Harmonic analysis on real reductive groups III*, Annals of Math. **104** (1976), 117–201.

[H1] R. Herb, *Fourier inversion and the Plancherel theorem for semisimple Lie groups*, Amer. J. Math. **104** (1982), 9–58.

[H2] ———, *Fourier inversion and the Plancherel theorem* (Proc. Marseille Conf., 1980), Lecture Notes in Math., vol. 880, Springer-Verlag, Berlin and New York, 1981, pp. 197–210.

[H3] ———, *The Plancherel theorem for semisimple groups without compact cartan subgroups* (Proc. Marseille Conf., 1982), Lecture Notes in Math., vol. 1020, Springer- Verlag, Berlin and New York, 1983, pp. 73–79.

[H4] ———, *The Schwartz space of a general semisimple Lie group II: Wave packets associated to Schwartz functions*, Trans. AMS. (to appear).

[H5] ———, *The Schwartz space of a general semisimple Lie group III: c-functions*, Advances in Math. (to appear).

[HW1] R. Herb, J. Wolf, *The Plancherel theorem for general semisimple groups*, Compositio Math. **57** (1986), 271–355.

[HW2] ———, *Rapidly decreasing functions on general semisimple groups*, Compositio Math. **58** (1986), 73–110.

[HW3] ———, *Wave packets for the relative discrete series I: The holomorphic case*, J. Funct. Anal. **73** (1987), 1–37.

[HW4] ———, *Wave packets for the relative discrete series II: The non-holomorphic case*, J. Funct. Anal. **73** (1987), 38–106.

[HW5] ———, *The Schwartz space of a general semisimple group I: Wave packets of Eisenstein integrals,*, Advances in Math. **80** (1990), 164–224.

[KM] H. Kraljević, D. Miličić, *The C^*-algebra of the universal covering group of $SL(2,\mathbb{R})$*, Glasnik Mat. Ser. III 7 **27** (1972), 35–48.

[SW] P. Sally, G. Warner, *The Fourier transform on semisimple Lie groups of real rank one*, Acta Math. **131** (1973), 1–26.

[S] W. Schmid, *Two character identities for semisimple Lie groups* (Proc. Marseille Conf., 1976) Lecture Notes in Math., vol. 587, Springer-Verlag, Berlin-New York, 1977.

Dept. of Mathematics, University of Maryland, College Park, MD 20742

E-mail: rah@emmy.umd.edu

INTERTWINING FUNCTORS AND IRREDUCIBILITY OF STANDARD HARISH-CHANDRA SHEAVES

DRAGAN MILIČIĆ
University of Utah

INTRODUCTION

Let $\mathfrak{g}$ be a complex semisimple Lie algebra and σ an involution of $\mathfrak{g}$. Denote by $\mathfrak{k}$ the fixed point set of this involution. Let K be a connected algebraic group and φ a morphism of K into the group $G = \mathrm{Int}(\mathfrak{g})$ of inner automorphisms of $\mathfrak{g}$ such that its differential is injective and identifies the Lie algebra of K with $\mathfrak{k}$. Let X be the flag variety of $\mathfrak{g}$, i.e. the variety of all Borel subalgebras in $\mathfrak{g}$. Then K acts algebraically on X, and it has finitely many orbits which are locally closed smooth subvarieties. The typical situation is the following: $\mathfrak{g}$ is the complexification of the Lie algebra of a connected real semisimple Lie group G_0 with finite center, K is the complexification of a maximal compact subgroup of G_0, and σ the corresponding Cartan involution.

Let $\mathfrak{h}$ be the (abstract) Cartan algebra of $\mathfrak{g}$, Σ the root system in $\mathfrak{h}^*$, and Σ^+ the set of positive roots determined by the condition that the homogeneous line bundles $\mathcal{O}(-\mu)$ on X corresponding to dominant weights μ are positive. For each $\lambda \in \mathfrak{h}^*$, A. Beilinson and J. Bernstein defined a G-homogeneous twisted sheaf of differential operators $\mathcal{D}_\lambda$ on X (compare [1], [5]). For a detailed discussion of their construction see §2 in Schmid's lecture in this volume [11].

Let $\mathcal{M}_{coh}(\mathcal{D}_\lambda, K)$ be the category of coherent $\mathcal{D}_\lambda$-modules on X with algebraic K-action ([5], Appendix). The objects of this category are called *Harish-Chandra sheaves.* Every Harish-Chandra sheaf has finite length, and there is a simple geometric description of irreducible Harish-Chandra sheaves which we shall describe now. Let Q be a K-orbit in X and $i : Q \longrightarrow X$ the natural inclusion. Then $\mathcal{D}_\lambda$ induces a K-homogeneous twisted sheaf of differential operators $\mathcal{D}_\lambda^i$ on Q. Fix $x \in Q$, let $\mathfrak{b}_x$ be the Borel subalgebra corresponding to this point, and define $\mathfrak{n}_x = [\mathfrak{b}_x, \mathfrak{b}_x]$. Let $\mathfrak{c}$ be a σ-stable Cartan subalgebra in $\mathfrak{b}_x$. Then the composition of the canonical maps $\mathfrak{c} \longrightarrow \mathfrak{b}_x/\mathfrak{n}_x \longrightarrow \mathfrak{h}$ is an isomorphism. It induces an isomorphism, called a *specialization*, of the Cartan triple $(\mathfrak{h}^*, \Sigma, \Sigma^+)$ onto the

---

1980 *Mathematics Subject Classification* (1985 *Revision*). 22E47.
Supported in part by NSF Grant DMS 88-02827 .

triple $(\mathfrak{c}^*, R, R^+)$; here R is the root system of the pair $(\mathfrak{g}, \mathfrak{c})$ and R^+ the set of positive roots determined by $\mathfrak{b}_x$. Let ρ be the half-sum of positive roots in Σ^+. If the restriction of the specialization of $\lambda + \rho$ to $\mathfrak{t} = \mathfrak{k} \cap \mathfrak{c}$ is the differential of a character of the identity component of the stabilizer S_x of x in K, there exist K-homogeneous $\mathcal{D}_\lambda^i$-connections on Q—we say that they are compatible with $\lambda + \rho$. Let τ be an irreducible K-homogeneous connection on Q compatible with $\lambda + \rho$. Then its direct image $R^0 i_+(\tau)$ is the *standard Harish-Chandra sheaf* $\mathcal{I}(Q, \tau)$. It is holonomic and therefore of finite length. Moreover, it has a unique irreducible $(\mathcal{D}_\lambda, K)$-submodule $\mathcal{L}(Q, \tau)$. The irreducible objects $\mathcal{L}(Q, \tau)$ exhaust the isomorphism classes of all irreducible objects in the category $\mathcal{M}_{coh}(\mathcal{D}_\lambda, K)$. Therefore, the composition series of standard Harish-Chandra sheaves $\mathcal{I}(Q, \tau)$ consist of modules isomorphic to some $\mathcal{L}(Q', \tau')$ for orbits Q' in the closure of Q and irreducible K-homogeneous connections τ' on Q' compatible with $\lambda + \rho$.

For integral $\lambda \in \mathfrak{h}^*$, the structure of the composition series of these modules is determined by Vogan's version of the Kazhdan-Lusztig conjectures [13]. As in the Verma module case, one should expect that the necessary and sufficient condition for the irreducibility of standard Harish-Chandra sheaves must be a far less deep result than the Kazhdan-Lusztig conjectures. If K is the fixed point set of an involution acting on a covering group of G, such a result is equivalent to the irreducibility theorem of [12]. Although the final result in this case (as in the case of Verma modules) suggests that the irreducibility criterion is completely controlled by SL_2-phenomena, this is not so evident from the existing proofs. The purpose of this paper is to describe the irreducibility result for the general case, and to sketch a proof which is conceptually as simple as in the case of Verma modules. This result is a part of a joint work with Henryk Hecht, Wilfried Schmid and Joseph A. Wolf. The complete details will appear in [6].

1. The basic example

In this section we discuss the simplest case of $\mathfrak{g} = \mathfrak{sl}(2, \mathbb{C})$. In this case the group $\mathrm{Int}(\mathfrak{g})$ of inner automorphisms of $\mathfrak{g}$ can be identified with $\mathrm{PSL}(2, \mathbb{C})$, and we can identify the flag variety X of $\mathfrak{g}$ with the one-dimensional projective space $\mathbb{P}^1$. If we denote by $[x_0, x_1]$ the projective coordinates of $x \in \mathbb{P}^1$, the corresponding Borel subalgebra $\mathfrak{b}_x$ is the Lie subalgebra of $\mathfrak{sl}(2, \mathbb{C})$ which leaves the line x invariant. Let σ be conjugation by $\begin{pmatrix} -1 & 0 \\ 0 & 1 \end{pmatrix}$ in $\mathfrak{g}$. Then $\mathfrak{k}$ is the subalgebra of diagonal matrices in $\mathfrak{g}$.

Let T the one-dimensional torus which stabilizes both $0 = [1, 0]$ and $\infty = [0, 1]$. Its Lie algebra is $\mathfrak{k}$. Hence, K can be an arbitrary n-fold covering of T with covering map φ. The K-orbits in $\mathbb{P}^1$ are $\{0\}$, $\{\infty\}$, and $\mathbb{C}^*$.

First we want to construct a suitable trivializations of $\mathcal{D}_\lambda$ on the open cover of $\mathbb{P}^1$ consisting of $\mathbb{P}^1 - \{0\}$ and $\mathbb{P}^1 - \{\infty\}$. We denote by $\alpha \in \mathfrak{h}^*$ the positive root of $\mathfrak{g}$ and put $\rho = \frac{1}{2}\alpha$ and $t = \alpha\check{}(\lambda)$, where $\alpha\check{}$ is the dual root of α.

Let $\{E, F, H\}$ denote the standard basis of $\mathfrak{sl}(2, \mathbb{C})$:

$$E = \begin{pmatrix} 0 & 1 \\ 0 & 0 \end{pmatrix} \quad F = \begin{pmatrix} 0 & 0 \\ 1 & 0 \end{pmatrix} \quad H = \begin{pmatrix} 1 & 0 \\ 0 & -1 \end{pmatrix}.$$

They satisfy the commutation relations

$$[H, E] = 2E \quad [H, F] = -2F \quad [E, F] = H.$$

Also, H spans the Lie algebra $\mathfrak{k}$. Moreover, we remark that if we specialize at 0, H corresponds to the dual root $\alpha\check{}$, but if we specialize at ∞, H corresponds to the negative of $\alpha\check{}$.

First we discuss $\mathbb{P}^1 - \{\infty\}$. On this set we define the usual coordinate z by $z([1, x_1]) = x_1$. In this way one identifies $\mathbb{P}^1 - \{\infty\}$ with the complex plane $\mathbb{C}$. After a short calculation we get

$$E = -z^2\partial - (t+1)z, \quad F = \partial, \quad H = 2z\partial + (t+1)$$

in this coordinate system. Analogously, on $\mathbb{P}^1 - \{0\}$ with the natural coordinate $\zeta([x_0, 1]) = x_0$, we have

$$E = \partial, \quad F = -\zeta^2\partial - (t+1)\zeta, \quad H = -2\zeta\partial - (t+1).$$

On $\mathbb{C}^*$ these two coordinate systems are clearly related by $\zeta = \frac{1}{z}$. This implies that $\partial_\zeta = -z^2\partial_z$, i. e., on $\mathbb{C}^*$ the second trivialization gives

$$E = -z^2\partial, \quad F = \partial - \frac{1+t}{z} \quad H = 2z\partial - (t+1).$$

Therefore, the first and the second trivialization on $\mathbb{C}^*$ are related by the automorphism of $\mathcal{D}_{\mathbb{C}^*}$ induced by

$$\partial \longrightarrow \partial - \frac{1+t}{z} = z^{1+t}\,\partial\, z^{-(1+t)}.$$

Now we want to analyze the standard Harish-Chandra sheaves attached to the open K-orbit $\mathbb{C}^*$. If we identify K with another copy of $\mathbb{C}^*$, the stabilizer in K of any point in the orbit $\mathbb{C}^*$ is the group M of nth roots of 1. Let η_0 be the trivial representation of M, η_1 the identity representation of M, and $\eta_k = (\eta_1)^k$, $2 \le k \le n-1$, the remaining irreducible representations of the cyclic group M. Denote by τ_k the irreducible K-equivariant connection on $\mathbb{C}^*$ corresponding to the representation η_k of M, and by $\mathcal{I}(\mathbb{C}^*, \tau_k)$ the corresponding standard Harish-Chandra sheaf in $\mathcal{M}_{coh}(\mathcal{D}_\lambda, K)$. To analyze these $\mathcal{D}_\lambda$-modules it is convenient to introduce a trivialization of $\mathcal{D}_\lambda$ on $\mathbb{C}^* = \mathbb{P}^1 - \{0, \infty\}$ such that H corresponds to the differential operator

$2z\partial$ on the orbit $\mathbb{C}^*$ and $T \cong \mathbb{C}^*$ acts on it by multiplication. We obtain this trivialization by restricting the original z-trivialization to $\mathbb{C}^*$ and twisting it by the automorphism

$$\partial \longrightarrow \partial - \frac{1+t}{2z} = z^{\frac{1+t}{2}}\, \partial\, z^{-\frac{1+t}{2}}.$$

This gives a trivialization of $\mathcal{D}_\lambda|\mathbb{C}^*$ which satisfies

$$E = -z^2\partial - \frac{1+t}{2}z, \quad F = \partial - \frac{1+t}{2z}, \quad H = 2z\partial.$$

The global sections of τ_k on $\mathbb{C}^*$ form the linear space spanned by functions $z^{p+\frac{k}{n}}$, $p \in \mathbb{Z}$. To analyze irreducibility of the standard $\mathcal{D}_\lambda$-module $\mathcal{I}(\mathbb{C}^*, \tau_k)$ we have to study its behavior at 0 and ∞. By the preceding discussion, if we use the z-trivialization of $\mathcal{D}_\lambda$ on $\mathbb{C}^*$, $\mathcal{I}(\mathbb{C}^*, \tau_k)$ looks like the $\mathcal{D}_{\mathbb{C}}$-module which is the direct image of the $\mathcal{D}_{\mathbb{C}^*}$-module generated by $z^{\frac{k}{n}-\frac{1+t}{2}}$. This module is clearly reducible if and only if it contains functions regular at the origin, i. e. if and only if $\frac{k}{n} - \frac{1+t}{2}$ is an integer. Analogously, the module $\mathcal{I}(\mathbb{C}^*, \tau_k)|\mathbb{P}^1 - \{0\}$ is reducible if and only if $\frac{k}{n} + \frac{1+t}{2}$ is an integer. Therefore, $\mathcal{I}(\mathbb{C}^*, \tau_k)$ is irreducible if and only if neither $\frac{k}{n} - \frac{1+t}{2}$ nor $\frac{k}{n} + \frac{1+t}{2}$ is an integer.

We can summarize this as follows.

1.1. Lemma. *Let K be the n-fold covering of T, $k \in \{0, 1, \dots, n-1\}$, and $\lambda \in \mathfrak{h}^*$. Then the following conditions are equivalent:*

(i) $\alpha\check{}(\lambda) \notin \{\frac{2k}{n}, -\frac{2k}{n}\} + 2\mathbb{Z} + 1$;
(ii) *the standard module $\mathcal{I}(\mathbb{C}^*, \tau_k)$ is irreducible.*

2. The irreducibility theorem

First we shall formulate the irreducibility result precisely. To do this we must analyze in detail the parametrization of K-homogeneous connections compatible with $\lambda + \rho \in \mathfrak{h}^*$ on a K-orbit Q.

Fix $x \in Q$ and denote by B_x the Borel subgroup of G with Lie algebra $\mathfrak{b}_x$. Then $S_x = \varphi^{-1}(\varphi(K) \cap B_x)$ is the stabilizer of x in K. The Borel subalgebra $\mathfrak{b}_x$ contains a σ-stable Cartan subalgebra $\mathfrak{c}$ and all such Cartan subalgebras in $\mathfrak{b}_x$ are S_x-conjugate [7]. Therefore, Q determines a unique K-conjugacy class of σ-stable Cartan subalgebras of $\mathfrak{g}$.

The involution σ defines an involution on the root system R in $\mathfrak{c}^*$, and its pull-back by the specialization map is an involution σ_Q on the root system Σ which depends only on Q. Therefore we can divide the roots in Σ in the following groups:

$\Sigma_{Q,I} = \{\alpha \in \Sigma \,|\, \sigma_Q\alpha = \alpha\}$ – Q-imaginary roots
$\Sigma_{Q,\mathbb{R}} = \{\alpha \in \Sigma \,|\, \sigma_Q\alpha = -\alpha\}$ – Q-real roots,
$\Sigma_{Q,\mathbb{C}} = \Sigma - (\Sigma_{Q,I} \cup \Sigma_{Q,\mathbb{R}})$ – Q-complex roots.

The Lie algebra $\mathfrak{s}_x = \mathfrak{k} \cap \mathfrak{b}_x$ of S_x is the semidirect product of $\mathfrak{t} = \mathfrak{k} \cap \mathfrak{c}$ with the nilpotent radical $\mathfrak{u}_x = \mathfrak{k} \cap \mathfrak{n}_x$ of $\mathfrak{s}_x$. Let U_x be the unipotent subgroup of K corresponding to $\mathfrak{u}_x$; it is the unipotent radical of S_x. Let C be the torus in G corresponding to $\mathfrak{c}$. Put $T = \varphi^{-1}(\varphi(K) \cap C)$. Then S_x is the semidirect product of T with U_x. A K-homogeneous connection τ on Q compatible with $\lambda + \rho$ determines a finite-dimensional algebraic representation ω of S_x on the geometric fibre $T_x(\tau)$ of τ at x. This representation is trivial on U_x, hence it can be viewed as a representation of the group T. The differential of the representation ω, considered as a representation of $\mathfrak{t}$, is a direct sum of a finite number of copies of the one dimensional representation defined by the restriction of the specialization of $\lambda + \rho$ to $\mathfrak{t}$. What remains to be described is the action of the other components of S_x. The information relevant for determination of irreducibility of standard Harish-Chandra sheaves is determined by the action of the elements which will be described now.

Let α be a Q-real root. Denote by $\mathfrak{s}_\alpha$ the three-dimensional simple algebra generated by the root subspaces corresponding to α and $-\alpha$. Let S_α be the connected subgroup of G with Lie algebra $\mathfrak{s}_\alpha$; it is isomorphic either to $\mathrm{SL}(2, \mathbb{C})$ or to $\mathrm{PSL}(2, \mathbb{C})$. Denote by H_α the element of $\mathfrak{s}_\alpha \cap \mathfrak{c}$ such that $\alpha(H_\alpha) = 2$. Then $m_\alpha = \exp(\pi i H_\alpha) \in G$ satisfies $m_\alpha^2 = 1$. Moreover, $\sigma(m_\alpha) = \exp(-\pi i H_\alpha) = m_\alpha^{-1} = m_\alpha$. Clearly, $m_\alpha = 1$ if $S_\alpha \cong \mathrm{PSL}(2, \mathbb{C})$, and $m_\alpha \neq 1$ if $S_\alpha \cong \mathrm{SL}(2, \mathbb{C})$—in the latter case m_α corresponds to the negative of the identity matrix in $\mathrm{SL}(2, \mathbb{C})$. Let $\mathfrak{k}_\alpha = \mathfrak{s}_\alpha \cap \mathfrak{k}$; it is the Lie algebra of a one dimensional torus K_α in K. Its image $\varphi(K_\alpha)$ in G is a torus in S_α. Therefore, $m_\alpha \in \varphi(K_\alpha)$. The composition of $\varphi : K_\alpha \longrightarrow S_\alpha$ and the covering projection $S_\alpha \longrightarrow \mathrm{Int}(\mathfrak{s}_\alpha)$ is an n-fold covering map between two one dimensional tori. If we identify K_α with $\mathbb{C}^*$, the kernel of this map is isomorphic to $\{e^{\frac{2\pi i p}{n}} \mid 0 \leq p \leq n-1\}$. Let n_α correspond to $e^{\frac{2\pi i}{n}}$ under this isomorphism (there are two possible choices for n_α and they are inverses of each other). Then φ maps n_α to m_α, hence n_α lies in T.

Let

$$D_-(Q) = \{\beta \in \Sigma^+ \cap \Sigma_{Q,\mathbb{C}} \mid -\sigma_Q \beta \in \Sigma^+\}.$$

Then $D_-(Q)$ is the union of $-\sigma_Q$-orbits consisting pairs $\{\beta, -\sigma_Q\beta\}$. Let A be a set of representatives of $-\sigma_Q$-orbits in $D_-(Q)$. Then, for arbitrary Q-real root α, the number

$$\delta_Q(m_\alpha) = \prod_{\beta \in A} e^\beta(m_\alpha)$$

is independent of the choice of A and equal to ± 1.

Following B. Speh and D. Vogan [12][1], we say that τ satisfies the SL_2-*parity condition with respect to the Q-real root* α if the spectrum of the linear transformation $\omega(n_\alpha)$ does not contain $-e^{\pm i\pi\alpha\check{}(\lambda)}\delta_Q(\varphi(n_\alpha))$. Since

[1] In fact, they consider the reducibility condition, while ours is the irreducibility condition.

n_α is determined up to inversion, this condition does not depend on the choice of n_α. Clearly, this condition specializes to the condition of 1.1.(i) in our basic example.

Let Σ_α be the smallest σ_Q-invariant closed root subsystem of Σ containing α. Then $\Sigma_\alpha \cap \Sigma^+$ is a set of positive roots in Σ_α which contains α and $-\sigma_Q\alpha$. Put

$$C_-(Q) = \{\alpha \in D_-(Q) \mid \alpha \text{ is minimal in } \{\alpha, -\sigma_Q\alpha\} \text{ with respect to the ordering of } \Sigma_\alpha\}.$$

Then $C_-(Q)$ contains at least one representative of each $-\sigma_Q$-orbit in $D_-(Q)$. Finally, let

$$\Sigma_\lambda = \{\alpha \in \Sigma \mid \alpha\check{}(\lambda) \in \mathbb{Z}\}$$

be the root subsystem of Σ consisting of all roots integral with respect to λ. We can now state the main result of this paper.

2.1. Theorem. *Let Q be a K-orbit in X, λ an element of $\mathfrak{h}^*$, and τ an irreducible K-homogeneous connection on Q compatible with $\lambda + \rho$. Then the following conditions are equivalent:*

(i) *$C_-(Q) \cap \Sigma_\lambda = \emptyset$, and τ satisfies the SL_2-parity condition with respect to every Q-real root in Σ; and*

(ii) *the standard $\mathcal{D}_\lambda$-module $\mathcal{I}(Q,\tau)$ is irreducible.*

Let $\tilde{G}$ be a covering of G and σ the involution of $\tilde{G}$ determined by the involution σ of $\mathfrak{g}$. Assume that K is the fixed point set of σ in $\tilde{G}$. Then we say that the pair $(\mathfrak{g}, K)$ is *linear*. In this case we have a slightly simpler criterion, which is equivalent to [12].

2.2. Corollary. *Assume that $(\mathfrak{g}, K)$ is a linear pair. Let Q be a K-orbit in X, λ an element of $\mathfrak{h}^*$, and τ an irreducible K-homogeneous connection on Q compatible with $\lambda + \rho$. Then the following conditions are equivalent:*

(i) *$D_-(Q) \cap \Sigma_\lambda = \emptyset$, and τ satisfies the SL_2-parity condition with respect to every Q-real root in Σ; and*

(ii) *the standard $\mathcal{D}_\lambda$-module $\mathcal{I}(Q,\tau)$ is irreducible.*

In the next example we show that, for a pair $(\mathfrak{g}, K)$ which is not linear, the first condition of 2.2 can fail for an irreducible standard module.

2.3. Example. Let G_0 be the universal cover of $\mathrm{SL}(3,\mathbb{R})$, $\mathfrak{g}$ its complexified Lie algebra, K the complexification of a maximal compact subgroup of G_0, and σ the corresponding Cartan involution. Then σ acts on the Lie algebra $\mathfrak{g} = \mathfrak{sl}(3,\mathbb{C})$ of all 3×3 complex matrices of trace zero by $\sigma(A) = -A^t$, where A^t is the transpose of the matrix A.

There are two K-conjugacy classes of σ-stable Cartan subalgebras in $\mathfrak{g}$: (1) the "split" class consisting of Cartan subalgebras on which σ acts as -1, which is represented by the subalgebra of all diagonal matrices in $\mathfrak{g}$, and (2) the "fundamental" class consisting of Cartan subalgebras on which σ acts as a reflection, represented by the Cartan subalgebra

$$\left\{ \begin{pmatrix} a & b & 0 \\ -b & a & 0 \\ 0 & 0 & -2a \end{pmatrix} \middle| \, a, b \in \mathbb{C} \right\}.$$

The flag variety X of $\mathfrak{g}$ is three dimensional, and the only K-orbit attached to the "split" class of σ-stable Cartan subalgebras is the open orbit Q_o.

For any fundamental σ-stable Cartan subalgebra $\mathfrak{c}$, the root system of $(\mathfrak{g}, \mathfrak{c})$ consists of two imaginary roots and four complex roots. Let W_K be the subgroup of the Weyl group of $(\mathfrak{g}, \mathfrak{c})$ consisting of elements induced by the elements of K. Then the order of W_K is equal to 2, and the only nontrivial element of W_K acts as $-\sigma$ on $\mathfrak{c}$. It follows that there are three K-orbits attached to the "fundamental" class of σ-stable Cartan subalgebras. The closed orbit C, which is one dimensional, has the property that two simple roots in Σ^+ are permuted by σ_C. Therefore, both of them are C-complex and their sum is C-imaginary. The remaining two orbits are two dimensional. They correspond to the cases where one simple root is Q-imaginary.

Let x be a point in one of the K-orbits attached to the "fundamental" conjugacy class of σ-stable Cartan subalgebras. Then the stabilizer S_x of that point in K is connected. Therefore, every irreducible algebraic representation of S_x is one dimensional and completely determined by its differential.

Let Q be one of the two dimensional orbits. Let α be the Q-complex simple root and β the Q-imaginary simple root. Then $\gamma = \alpha + \beta$ is the other Q-complex positive root, and $\sigma_Q(\alpha) = -\gamma$. Therefore, $D_-(Q) = \{\alpha, \gamma\}$, but $C_-(Q) = \alpha$.

By the preceding discussion, there is at most one irreducible K-homogeneous connection on Q compatible with $\lambda + \rho \in \mathfrak{h}^*$. It exists if and only if $\beta\check{}(\lambda) \in \frac{1}{2}\mathbb{Z}$. We denote the corresponding standard module by $\mathcal{I}(Q, \lambda)$. Evidently, the quantity $\gamma\check{}(\lambda) = \alpha\check{}(\lambda) + \beta\check{}(\lambda)$, can be an integer without $\alpha\check{}(\lambda)$ being an integer, i.e., $\mathcal{I}(Q, \lambda)$ can be irreducible for λ integral with respect to γ.

3. Intertwining functors

Let θ be a Weyl group orbit in $\mathfrak{h}^*$. We consider the derived category $D^b(\mathcal{U}_\theta)$ of bounded complexes of $\mathcal{U}_\theta$-modules. For each $\lambda \in \mathfrak{h}^*$ we also consider the derived category $D^b(\mathcal{D}_\lambda)$ of bounded complexes of (quasi-coherent) $\mathcal{D}_\lambda$-modules. The derived functor $R\Gamma$ of the functor of global sections Γ maps $D^b(\mathcal{D}_\lambda)$ into $D^b(\mathcal{U}_\theta)$ since its right cohomological dimension is $\leq \dim X$. If λ is regular, this functor is an equivalence of categories

[2]. On the other hand, the localization functor $L\Delta_\lambda$, defined by

$$L\Delta_\lambda(V^\cdot) = \mathcal{D}_\lambda \overset{L}{\otimes}_{\mathcal{U}_\theta} V^\cdot, \quad V^\cdot \in D^b(\mathcal{U}_\theta),$$

maps $D^b(\mathcal{U}_\theta)$ into the derived category $D^-(\mathcal{D}_\lambda)$ of complexes bounded from below for arbitrary λ. If λ is regular, the left cohomological dimension of Δ_λ is finite, and $L\Delta_\lambda$ defines a quasi-inverse of $R\Gamma$. This implies, in particular, that for any two $\lambda, \mu \in \theta$, the categories $D^b(\mathcal{D}_\lambda)$ and $D^b(\mathcal{D}_\mu)$ are equivalent. This equivalence is given by the functor $L\Delta_\mu \circ R\Gamma$ from $D^b(\mathcal{D}_\lambda)$ into $D^b(\mathcal{D}_\mu)$. We now describe another functor, defined in geometric terms, which is (under certain conditions) isomorphic to $L\Delta_\mu \circ R\Gamma$. This is the intertwining functor of Beilinson and Bernstein ([2], [3]; for compete details see [9]).

Define the action of $G = \mathrm{Int}(\mathfrak{g})$ on $X \times X$ by

$$g \cdot (x, x') = (g \cdot x, g \cdot x')$$

for $g \in G$ and $(x, x') \in X \times X$. The G-orbits in $X \times X$ can be parametrized in the following way. First we introduce a relation between Borel subalgebras in $\mathfrak{g}$. Let $\mathfrak{b}$ and $\mathfrak{b}'$ be two Borel subalgebras in $\mathfrak{g}$. Let $\mathfrak{c}$ be a Cartan subalgebra of $\mathfrak{g}$ contained in $\mathfrak{b} \cap \mathfrak{b}'$. Denote by R the root system of $(\mathfrak{g}, \mathfrak{c})$ in $\mathfrak{c}^*$ and by R^+ the set of positive roots determined by $\mathfrak{b}$. This determines a specialization of the Cartan triple $(\mathfrak{h}^*, \Sigma, \Sigma^+)$ into $(\mathfrak{c}^*, R, R^+)$. On the other hand, $\mathfrak{b}'$ determines another set of positive roots in R, which corresponds via this specialization to $w(\Sigma^+)$ for some uniquely determined $w \in W$. The element $w \in W$ doesn't depend on the choice of $\mathfrak{c}$, and we say that $\mathfrak{b}'$ is in *relative position* w with respect to $\mathfrak{b}$.

Define

$$Z_w = \{(x, x') \in X \times X \mid \mathfrak{b}_{x'} \text{ is in the relative position } w \text{ with respect to } \mathfrak{b}_x\}$$

for $w \in W$. Then the map $w \longrightarrow Z_w$ is a bijection of W onto the set of G-orbits in $X \times X$, hence the sets Z_w, $w \in W$, are smooth subvarieties of $X \times X$.

Denote by p_1 and p_2 the projections of Z_w onto the first and second factor in $X \times X$, respectively. Let $\Omega_{Z_w|X}$ be the invertible $\mathcal{O}_{Z_w}$-module of top degree relative differential forms for the projection $p_1 : Z_w \longrightarrow X$. Let $\mathcal{T}_w$ be its inverse. The twisted sheaves of differential operators $\mathcal{D}_{w\lambda}$ and $\mathcal{D}_\lambda$, "pulled back" to Z_w by the projections p_1 and p_2 respectively, determine twisted sheaves $(\mathcal{D}_{w\lambda})^{p_1}$ and $\mathcal{D}_\lambda^{p_2}$ on Z_w ([5], A.1). It is easy to check that they differ by the twist by $\mathcal{T}_w$, i.e.,

$$(\mathcal{D}_{w\lambda})^{p_1} = (\mathcal{D}_\lambda^{p_2})^{\mathcal{T}_w}.$$

Since the morphism $p_2 : Z_w \longrightarrow X$ is a surjective submersion, the inverse image p_2^+ is an exact functor from the category $\mathcal{M}(\mathcal{D}_\lambda)$ of $\mathcal{D}_\lambda$-modules into $\mathcal{M}((\mathcal{D}_\lambda)^{p_2})$. Twisting by $\mathcal{T}_w$ defines an exact functor

$$\mathcal{V} \longrightarrow \mathcal{T}_w \otimes_{\mathcal{O}_{Z_w}} p_2^+(\mathcal{V})$$

from $\mathcal{M}(\mathcal{D}_\lambda)$ into $\mathcal{M}((\mathcal{D}_{w\lambda})^{p_1})$. Therefore, we have a functor

$$\mathcal{V}^{\cdot} \longrightarrow \mathcal{T}_w \otimes_{\mathcal{O}_{Z_w}} p_2^+(\mathcal{V}^{\cdot})$$

from $D^b(\mathcal{D}_\lambda)$ into $D^b((\mathcal{D}_{w\lambda})^{p_1})$. By composing it with the direct image functor

$$Rp_{1+} : D^b((\mathcal{D}_{w\lambda})^{p_1}) \longrightarrow D^b(\mathcal{D}_{w\lambda}),$$

we get the functor from $D^b(\mathcal{D}_\lambda)$ into $D^b(\mathcal{D}_{w\lambda})$ given by the formula

$$LI_w(\mathcal{V}^{\cdot}) = Rp_{1+}(\mathcal{T}_w \otimes_{\mathcal{O}_{Z_w}} p_2^+(\mathcal{V}^{\cdot}))$$

for any $\mathcal{V}^{\cdot} \in D^b(\mathcal{D}_\lambda)$. This is the left derived functor of the functor

$$I_w(\mathcal{V}) = R^0p_{1+}(\mathcal{T}_w \otimes_{\mathcal{O}_{Z_w}} p_2^+(\mathcal{V}))$$

from $\mathcal{M}(\mathcal{D}_\lambda)$ into $\mathcal{M}(\mathcal{D}_{w\lambda})$. It is called *the intertwining functor* (attached to $w \in W$).

Moreover, we have the following basic fact.

3.1. Proposition. *Let $w \in W$ and $\lambda \in \mathfrak{h}^*$. Then LI_w is an equivalence of the category $D^b(\mathcal{D}_\lambda)$ with $D^b(\mathcal{D}_{w\lambda})$.*

Intertwining functors satisfy a natural "product formula." It allows the reduction of the analysis of intertwining functors to the ones attached to simple reflections.

3.2. Proposition. *Let $w, w' \in W$ be such that $\ell(w'w) = \ell(w') + \ell(w)$. Then, for any $\lambda \in \mathfrak{h}^*$, the functors $LI_{w'} \circ LI_w$ and $LI_{w'w}$ from $D^b(\mathcal{D}_\lambda)$ into $D^b(\mathcal{D}_{w'w\lambda})$ are isomorphic; in particular the functors $I_w \circ I_{w'}$ and $I_{w'w}$ from $\mathcal{M}(\mathcal{D}_\lambda)$ into $\mathcal{M}(\mathcal{D}_{w'w\lambda})$ are isomorphic.*

Let $\alpha \in \Sigma$. We say that $\lambda \in \mathfrak{h}^*$ is *α-antidominant* if $\alpha^\check(\lambda)$ is not a strictly positive integer. For any $S \subset \Sigma^+$, we say that $\lambda \in \mathfrak{h}^*$ is *S-antidominant* if it is α-antidominant for all $\alpha \in S$. Put

$$\Sigma_w^+ = \{\alpha \in \Sigma^+ \mid w\alpha \in -\Sigma^+\} = \Sigma^+ \cap (-w^{-1}(\Sigma^+))$$

for any $w \in W$. The name of the functor LI_w comes from the following basic result of Beilinson and Bernstein to which we alluded before.

3.3. Theorem. *Suppose $w \in W$ and $\lambda \in \mathfrak{h}^*$ are Σ_w^+-antidominant and regular. Then LI_w is an equivalence of the category $D^b(\mathcal{D}_\lambda)$ with $D^b(\mathcal{D}_{w\lambda})$, isomorphic to $L\Delta_{w\lambda} \circ R\Gamma$.*

We also have the following estimate for the left cohomological dimension of the intertwining functors.

3.4. Proposition. *Let $w \in W$ and $\lambda \in \mathfrak{h}^*$. Then the left cohomological dimension of I_w is less than or equal to* $\mathrm{Card}(\Sigma_w^+ \cap \Sigma_\lambda)$.

In particular, we have the following consequence which is critical for our argument.

3.5. Corollary. *Let $w \in W$ and $\lambda \in \mathfrak{h}^*$ be such that $\Sigma_w^+ \cap \Sigma_\lambda = \emptyset$. Then*

$$I_w : \mathcal{M}(\mathcal{D}_\lambda) \longrightarrow \mathcal{M}(\mathcal{D}_{w\lambda})$$

is an equivalence of categories and $I_{w^{-1}}$ is its quasi-inverse.

4. A SKETCH OF THE PROOF OF THE IRREDUCIBILITY THEOREM

The idea of the proof of the irreducibility theorem is simple. We shall show that, if a standard Harish-Chandra sheaf is reducible, then there exists an intertwining functor which is an equivalence of categories and which maps the original sheaf into a standard Harish-Chandra sheaf for which the reducibility is obvious.

The standard Harish-Chandra sheaves attached to irreducible K-homogeneous connections on closed K-orbits are obviously irreducible. The analogous remark is all we need in the Verma module case—in this case the orbits in question are just Bruhat cells $C(w)$, $w \in W$, in X. Since the stabilizers of the unipotent radical N of a Borel subgroup of G are always connected, then for each Bruhat cell $C(w)$ and $\lambda \in \mathfrak{h}^*$ there exists a unique standard $\mathcal{D}_\lambda$-module $\mathcal{I}(w,\lambda)$ supported on the closure of $C(w)$. By a direct calculation [9],

$$I_{w^{-1}}(\mathcal{I}(1, w^{-1}\lambda)) = \mathcal{I}(w,\lambda).$$

If I_w satisfies the conditions of 3.5, it is an equivalence of categories and $\mathcal{I}(w,\lambda)$ is irreducible. A slightly more careful argument also implies the necessity of this condition.

In the case of Harish-Chandra modules we cannot reduce the argument to the case of a closed K-orbit. However, we can do the next best thing: we can reduce the argument to the orbits of minimal dimension attached to a particular K-conjugacy class of σ-stable Cartan subalgebras. The orbit Q has the minimal dimension among all K-orbits attached to a particular conjugacy class of σ-stable Cartan subalgebras if and only if the set $D_-(Q)$ is empty.

Assume that $D_-(Q)$ is not empty. Then it contains a simple root α, and this root must be in $C_-(Q)$. Let X_α be the flag variety of all parabolic subalgebras of type α in $\mathfrak{g}$. Denote by p_α the natural projection of X onto X_α. Then $p_\alpha^{-1}(p_\alpha(Q))$ is a K-invariant subset of X which is the union of two K-orbits: the orbit Q, and another orbit Q' which satisfies $\dim Q' = \dim Q - 1$. The orbit Q' is attached to the same conjugacy class of σ-stable Cartan subalgebras as Q, but

$$D_-(Q') = s_\alpha(D_-(Q) - \{\alpha, -\sigma_Q\alpha\}),$$

i.e., $\operatorname{Card} D_-(Q') = \operatorname{Card} D_-(Q) - 2$.

Let τ be an irreducible K-homogeneous connection on Q compatible with $\lambda+\rho$. Then there exists an irreducible K-homogeneous connection τ' on Q', compatible with $s_\alpha\lambda + \rho$, such that the following result holds.[2]

[2]Compare with §6. of Schmid's lecture in this volume [11].

4.1. Lemma.

$$I_{s_\alpha}(\mathcal{I}(Q',\tau')) = \mathcal{I}(Q,\tau).$$

In addition, τ satisfies the SL_2*-parity condition with respect to a Q-real root β if and only if τ' satisfies the* SL_2*-parity condition with respect to the Q'-real root $s_\alpha\beta$.*

As was the case for the previous formula in the Verma module situation, this result is a straightforward consequence of the geometry of K-orbits and the base change [4]. If $\alpha\check{}(\lambda)$ is not an integer, I_{s_α} is an equivalence of categories by 3.5, and $\mathcal{I}(Q,\tau)$ is irreducible if and only if $\mathcal{I}(Q',\tau')$ is irreducible. On the other hand, if $\alpha\check{}(\lambda)$ is an integer, $\mathcal{I}(Q,\tau)$ contains an obvious submodule of local sections "which extend over Q'"—hence $\mathcal{I}(Q,\tau)$ is reducible, and we are done. This inductive argument allows us to eliminate $D_-(Q)$ completely and reduce the discussion to the case of standard Harish-Chandra sheaves attached to orbits of minimal dimension for a given conjugacy class of σ-stable Cartan subalgebras. This procedure is not unique; in some situations we can choose several different simple complex roots to do the reduction of dimension. Also, in each step we "lose" a pair of complex roots from $D_-(Q)$, and the integrality of λ with respect to one of them doesn't necessarily imply the integrality with respect to the other (compare 2.3). Fortunately, the smaller set $C_-(Q)$ has the property that it contains all "relevant" roots, and if it contains a pair $\{\alpha, -\sigma_Q\alpha\}$, the integrality with respect to one of them implies integrality with respect to the other if the SL_2-parity condition holds for all Q-real roots. On the contrary, in the linear case, the integrality of λ with respect to any Q-complex root α implies the integrality with respect to $\sigma_Q\alpha$. This is the reason why we could use the full set $D_-(Q)$ in the corollary to the main theorem.

This reduces the proof to the case of orbits of minimal dimension attached to a particular conjugacy class of σ-stable Cartan subalgebras. Let Q be such K-orbit. Since $D_-(Q)$ is empty in this situation, the set of all positive Q-complex roots is σ_Q-invariant. Therefore, the union of positive roots and Q-real roots is a σ_Q-stable parabolic set of roots, and it determines a generalized flag variety X_Θ for some subset Θ of the set of simple roots in Σ. Let p_Θ be the corresponding natural projection from X onto X_Θ. The projection $p_\Theta(Q)$ of the orbit Q to X_Θ is a closed K-orbit in X_Θ. Therefore, the inverse image $p_\Theta^{-1}(p_\Theta(Q))$ is a smooth closed subvariety in X invariant under the action of K. Let $F = p_\Theta^{-1}(p_\Theta(x))$ be a fibre of the projection p_Θ passing through the point $x \in Q$. Let $\mathfrak{p}$ be the parabolic subalgebra determined by $p_\Theta(x)$ and $\mathfrak{g}_\Theta$ the Levi factor of $\mathfrak{p}$ which contains the σ-stable Cartan subalgebra $\mathfrak{c}$. Clearly $\mathfrak{g}_\Theta$ is σ-stable. Let K_Θ be the centralizer of the center of $\mathfrak{g}_\Theta$ in K. Then K_Θ acts on $\mathfrak{g}_\Theta$ by automorphisms and its Lie algebra is identified with $\mathfrak{k}_\Theta = \mathfrak{k} \cap \mathfrak{g}_\Theta$. The map $\mathfrak{b} \longmapsto \mathfrak{b} \cap \mathfrak{g}_\Theta$ defines an isomorphism of F with the flag variety X_Θ of $\mathfrak{g}_\Theta$. The map $Q' \longmapsto Q' \cap F$ defines a bijection between the K-orbits in

$p_\Theta^{-1}(p_\Theta(Q))$ and K_Θ-orbits in X_Θ. A more careful analysis shows that an appropriate derived functor of the "restriction" to F defines an equivalence of the full subcategory of $\mathcal{M}_{coh}(\mathcal{D}_\lambda, K)$ consisting of modules supported in $p_\Theta^{-1}(p_\Theta(Q))$ with the corresponding category of $\mathcal{D}$-modules on X_Θ and this equivalence maps standard Harish-Chandra sheaves into standard Harish-Chandra sheaves.

Since $Q \cap F$ is dense in F, this reduces the proof to the case of standard modules attached to the open K-orbit Q in the flag variety X. Also, Q is attached to a conjugacy class of Cartan subalgebras on which σ acts as -1, i.e., all roots in Σ are Q-real. Let τ be a K-homogeneous connection on Q. The stabilizer S_x of a point $x \in Q$ is finite. The intersection $\mathfrak{b}_x \cap \sigma(\mathfrak{b}_x)$ is equal to a σ-stable Cartan subalgebra $\mathfrak{c}$ on which σ acts as -1. Let α be a simple root. Then there exists the unique Borel subalgebra $\mathfrak{b}_y$, $y \in X$, containing $\mathfrak{c}$ and in relative position s_α with respect to $\mathfrak{b}_x$. The point y is contained in Q and its stabilizer S_y is equal to S_x. Therefore, the representation ω of S_x in $T_x(\tau)$ determines another K-homogeneous connection τ_α on Q such that the representation of $S_y = S_x$ in $T_y(\tau_\alpha)$ is equal to ω. The next lemma follows again from the analysis of the SL_2-situation and the base change.

4.2. Lemma. *Assume τ satisfies the SL_2-parity condition with respect to a simple root α. Then*

$$I_{s_\alpha}(\mathcal{I}(Q,\tau)) = \mathcal{I}(Q,\tau_\alpha).$$

If we assume in addition that $p = -\alpha\check{}(\lambda)$ is an integer,

$$I_{s_\alpha}(\mathcal{I}(Q,\tau)) = \mathcal{I}(Q,\tau)(p\alpha) = \mathcal{I}(Q,\tau) \otimes_{\mathcal{O}_X} \mathcal{O}(p\alpha).$$

Moreover, if τ satisfies the SL_2-parity condition with respect to a root β, τ_α satisfies the SL_2-parity condition with respect to $s_\alpha\beta$.

Therefore, if $\alpha\check{}(\lambda)$ is not an integer, I_{s_α} is an equivalence of categories by 3.5, and $\mathcal{I}(Q,\tau)$ is irreducible if and only if $\mathcal{I}(Q,\tau_\alpha)$ is irreducible. On the other hand, if $\alpha\check{}(\lambda)$ is an integer, $\mathcal{I}(Q,\tau_\alpha) = \mathcal{I}(Q,\tau)(p\alpha)$ and the same assertion is obvious. Therefore, to check irreducibility, we can freely "move around" λ by the action of the Weyl group.

Assume that the SL_2-parity condition fails for some root β. Then, by applying the intertwining functors, we can assume that it fails for a simple root. In this case the reducibility is obvious: the standard Harish-Chandra sheaf has a submodule of local sections which extend over a K-orbit of codimension one in X. This proves the necessity of the condition. The sufficiency is equally simple. If the parity condition holds for all roots and $\mathcal{I}(Q,\tau)$ is reducible, then $\mathcal{I}(Q,\tau)$ has a nontrivial quotient supported on a closed subvariety of X of codimension ≥ 1. By applying the intertwining functors we can decrease the codimension of the support of this quotient until it reaches 1 [2]. Let Q' be a K-orbit of codimension one in X contained in the support of this quotient. In this case there exists a simple root α such that $p_\alpha^{-1}(p_\alpha(Q'))$ contains the open orbit Q. For an arbitrary point $y \in Q'$, its fibre $F = p_\alpha^{-1}(p_\alpha(y))$ is isomorphic to $\mathbb{P}^1$—the flag variety of $\mathfrak{sl}(2,\mathbb{C})$—and $Q \cap F$ corresponds to $\mathbb{C}^*$. On the other hand, $Q' \cap F$ corresponds to either $\{0\}$ or $\{0,\infty\}$. Since the restriction to F of a standard Harish-Chandra sheaf attached to Q is a standard Harish-Chandra sheaf on $\mathbb{P}^1$ of the type we discussed in 1, it is irreducible by 1.1. But this contradicts the fact that it should have a quotient supported in $\{0,\infty\}$. This completes the proof of the main theorem.

Finally, we would like to make a remark about the relationship of our result with the main result of Speh and Vogan [12]. Their result gives a necessary and sufficient condition for irreducibility of the principal series representations for regular infinitesimal characters and leaves the singular case open. The reason for this is that the Beilinson-Bernstein equivalence of categories fails for singular infinitesimal characters—there exist $\mathcal{D}_\lambda$-modules with no cohomology at all! This allows global sections of reducible standard Harish-Chandra sheaves to be irreducible in some cases. In the case of irreducible unitary principal series representations, I. Mirković proved that they are always global sections of some irreducible standard Harish-Chandra sheaf [10]; this explains relative simplicity of the tempered spectrum of semisimple Lie groups.

A completely analogous argument for irreducibility of standard modules works for the category of generalized Verma modules [9] and the category of Whittaker modules [8].

REFERENCES

1. A. Beilinson, J. Bernstein, *Localisation de $\mathfrak{g}$-modules*, C. R. Acad. Sci. Paris, Ser. I **292** (1981), 15–18.
2. A. Beilinson, J. Bernstein, *A generalization of Casselman's submodule theorem*, Representation Theory of Reductive Groups, Birkhäuser, Boston, 1983, pp. 35–52.
3. A. Beilinson, *Localization of representations of reductive Lie Algebras*, Proceedings of International Congress of Mathematicians, August 16-24, 1983, Warszawa, pp. 699–710.
4. A. Borel et al., *Algebraic $\mathcal{D}$-modules*, Academic Press, Boston, 1987.
5. H. Hecht, D. Miličić, W. Schmid, J. A. Wolf, *Localization and standard modules for real semisimple Lie groups I: The duality theorem*, Inventiones Math. **90** (1987), 297–332.
6. H. Hecht, D. Miličić, W. Schmid, J. A. Wolf, *Localization and standard modules for real semisimple Lie groups II: Irreducibility, vanishing theorems and classification*, in preparation.
7. T. Matsuki, *The orbits of affine symmetric spaces under the action of minimal parabolic subgroups*, J. Math. Soc. Japan **31** (1979), 332–357.
8. D. Miličić, W. Soergel, *Twisted Harish-Chandra sheaves and Whittaker modules*, in preparation.

9. D. Miličić, *Localization and representation theory of reductive Lie groups*, (manuscript), to appear.

10. I. Mirković, *Classification of irreducible tempered representations of semisimple Lie groups*, Ph. D. Thesis, University of Utah, 1986.

11. W. Schmid, *Construction and classification of irreducible Harish-Chandra modules*, in this volume.

12. B. Speh, D. Vogan, *Reducibility of generalized principal series representations*, Acta Math. **145** (1980), 227–299.

13. D. Vogan, *Irreducible characters of semisimple Lie groups III: proof of the Kazhdan-Lusztig conjectures in the integral case*, Inventiones Math. **71** (1983), 381–417.

DEPARTMENT OF MATHEMATICS, UNIVERSITY OF UTAH, SALT LAKE CITY, UT 84112
E-mail: Milicic @ math.utah.edu

FUNDAMENTAL G–STRATA

LAWRENCE MORRIS

Clark University

§1. INTRODUCTION

1.1. Let k be a non-archimedean local field, with valuation ring $\mathcal{O}$ and prime ideal $\mathcal{P}$. Then $\mathcal{O}/\mathcal{P}$ is a finite field with q elements which we denote by $\mathbf{F}_q$.

1.2. Let G be the group of k-valued points of a reductive algebraic group defined over k. Then G has the natural structure of a totally disconnected locally compact Hausdorff topological group.

A representation of G is a pair (π, W) where W is a complex vector space and

$$\pi : G \longrightarrow \mathrm{Aut}_{\mathbf{C}}(W)$$

is a homomorphism from G to the invertible linear operators of W. The representation (π, W) is *smooth* if the stabilizer in G of every vector in V is open. (π, W) is *admissible* if it is smooth, and if for each open compact subgroup $K \subset G$, the space of vectors fixed by K is finite dimensional. Finally, a representation is said to be *irreducible* if it has no proper subrepresentation other than the trivial one.

By a theorem of Jacquet [J], any irreducible smooth representation of G is admissible.

A philosophy that has met with considerable success in recent years is the analysis of irreducible admissible representations of G via restriction to "suitable" open compact subgroups. This approach can be traced back at least to R. Howe's notion of "essential K-types" [H]. One could also compare it with the theory of K-types developed for real Lie groups. In the non-archimedean case, the state of affairs is much more primitive.

In section 2 below we briefly review a "minimal K-type" theorem for the general linear groups. In section 3 we describe some remarkable congruence subgroups for the classical groups. Using these we develop the framework for an analogous theorem for these groups in sections 4 and 5. Finally, in section 6 we illustrate some of the ideas (and difficulties) which lie behind the proof. For details we refer the reader to [M3].

Supported by NSF Grant DMS-8802842 .

§2. $GL(V)$

2.1. Let V be a finite dimensional k-vector space of dimension N, and let $G = \mathbf{G}L(V)$. In this case the relevant open compact subgroups turn out to be certain congruence subgroups.

Let $\mathcal{L}$ be a *lattice chain* in V, i.e., a family of free $\mathcal{O}$-modules (in V), each of rank N, totally ordered by inclusion, and closed under multiplication by elements of k^*. Such a family is countable and is completely described by a "slice":

$$\cdots \supset L_{-1} = \mathcal{P}^{-1}L_{e-1} \supsetneq L_0 \supsetneq \cdots \supsetneq L_{e-1} \supsetneq \mathcal{P}L_0 = L_e \supset \cdots$$

where $e =$ *period* of $\mathcal{L}, 1 \leq e \leq N$.

Let $\mathcal{A} = \{x \in \mathrm{End}_k(V) \mid xL_i \subseteq L_i\}$. This is a hereditary order with Jacobson radical

$$\mathcal{B} = \{x \in \mathrm{End}_k(V) \mid xL_i \subseteq L_{i+1}, \quad i \in \mathbb{Z}\}.$$

Let $P_0 = \mathcal{A}^*$ (units in $\mathcal{A}$) and, for each $n \in \mathbb{N}$, define $P_n = 1 + \mathcal{B}^n$. Then we have the following result.

Proposition.

(a) $P^+ = N_G(P_0)$ *is a compact mod centre subgroup of* G.
(b) P_n *is an open normal subgroup of* P_0.
(c) *The commutator* $(P_m, P_n) \subseteq P_{m+n}$ $(m, n \geq 1)$.
(d) *The group* P_n/P_{n+1} *(which is abelian by (c)) is isomorphic to the (additive) group* $\mathcal{B}^n/\mathcal{B}^{n+1}$, $n \geq 1$.
(e) $P_0/P_1 \simeq \Pi_i^r \mathbf{G}L(V_i)$, *where each* V_i *is an* $\mathbf{F}_q$*-vector space.*

It follows from (d) that, for $n \geq 1$, the Pontrjagin dual of P_n/P_{n+1} can be identified with $\mathcal{B}^{c-(n+1)}/\mathcal{B}^{c-n}$, where c is a suitable integer.

2.2. We remark that in the language of Bruhat-Tits, P_0 corresponds to a parahoric subgroup of G, and the family $\{P_n\}$ corresponds to the filtration given by the standard affine height function associated to P_0. In this form, a version of the above proposition holds for any (connected) reductive group defined over k. (See [P-R].)

2.3 Definition. A G-stratum is a triple $(P,\ n,\ \psi)$ where $P = P_0$ is a group defined as above, n is a non-negative integer, and ψ is an irreducible representation of P_n/P_{n+1}.

2.4. For $n \geq 1$ the remark above on the Pontrjagin dual means that ψ corresponds to an element $b + \mathcal{B}^{c-n}, b \in \mathcal{B}^{c-(n+1)}$. Explicitly, let Ω be an additive character of k with conductor $\mathcal{O}$. Then $\psi(x) = \Omega(\mathrm{trace}((x-1)b))$, where trace is matrix trace; in this case $c = 1 - e$, where e is the period of the lattice chain $\mathcal{L}$. We write $\psi_b = \psi$ to mean that ψ corresponds to $\bar{b} = b + \mathcal{B}^{c-n}$.

2.5. We say that (P, n, ψ) is *fundamental* if $n = 0$ and ψ is a cuspidal representation of P_0/P_1, or if $n \geq 1$ and $\psi = \psi_b$ where $b + \mathcal{B}^{c-n}$ contains no nilpotent elements. (Note: $b + \mathcal{B}^{c-n} \subseteq \mathrm{End}_k(V)$.)

2.6. Suppose now that (π, W) is an irreducible admissible representation of G. We say π *contains* (P, n, ψ) if ψ occurs in the restriction of π to P_n.

The following beautiful result was conjectured by Allen Moy. A proof, as well as a brief history, can be found in [B]; see also [H-M].

Theorem. *(π, W) always contains fundamental G-strata.*

In fact, given $P = P_0$, define $f(P, \pi)$ to be the least integer f such that $\pi \mid P_{f+1}$ contains the trivial character. There is always a G-stratum $(P, f(P, \pi), \psi)$ contained in π. Now suppose $f(P, \pi) \geq 1$ for all such P. Choose P so that $f(P, \pi)/e(P) \leq f(P', \pi)/e(P')$ for any other P'. (Here $e(P)$ is the period of the filtration associated to $P : 1 \leq e(P) \leq N$.) The theorem above can then be refined in this situation to say that *any G-stratum $(P, f(P, \pi), \psi)$ occurring in π is fundamental.*

Similarly, if there is a P for which $f(P, \pi) = 0$, choose one such that $e(P) \geq e(P')$ for all P' with $f(P', \pi) = 0$. Then any G-stratum $(P, 0, \psi)$ occurring in π is fundamental.

2.7. Theorem 2.6 provides the basis for a much deeper analysis of the admissible dual of G. For more details, see the lectures of Bushnell and Kutzko which appear elsewhere in these proceedings.

§3. Classical Groups

3.1. One would like to have results similar to Theorem 2.6 for other reductive groups G. In particular, one would like to be able to

(1) find suitable families $\{P_n\}_{n \geq 0}$,
(2) describe P_n/P_{n+1} in some nice way, and
(3) find analogues of (fundamental) G-stratum, "normalized valuation" $f/e, \ldots$

The first problem that occurs with this strategy is that the obvious candidates $\{P_n\}_{n \geq 0}$ *do not suffice*. Namely, one cannot hope to understand the representation theory of G using only the canonical filtrations of 2.2.

There are several reasons for this:

- Allen Moy has explicitly listed "minimal K-types" for small rank groups (see, e.g., [Mo]). In doing these calculations he was compelled to produce ad hoc "non standard" filtrations at some stage.
- In [M1], [M2] the author associates irreducible supercuspidal representations of G (G a classical group) to "tame" cuspidal data. Part of this data typically includes a tamely ramified compact maximal torus T in G. Part of the construction involves (a family of)

parahoric subgroups and filtrations which reflect the ramified arithmetic of T. These filtrations are rarely "standard". One would expect such filtrations (among others) to be included in a description of G-strata.

3.2. Suppose $\sigma_0 : k \longrightarrow k$ is an involution (possibly trivial) with a fixed field k_0. From now on *we assume that the characteristic of* $\mathbf{F}_q$ *is not* 2.

Let V be a finite dimensional k-vector space and

$$f : V \times V \longrightarrow k$$

a non-degenerate (skew) hermitian form:

$$\epsilon\sigma_0 f(w, \lambda v) = f(\lambda v, w) = (\sigma_0\lambda) f(v, w) .$$

Here $\epsilon = \epsilon(f) \in \{+1, -1\}$.

We write G for the group of isometries of f. It is the group of k_0-valued points of a reductive group defined over k_0.

3.3. If $L \subseteq V$ is an $\mathcal{O}$-lattice, then one defines the *complementary lattice*

$$L^\# = \{v \in V \mid f(v, L) \subseteq \mathcal{O}\} .$$

If $\mathcal{L}$ is a lattice chain, we set

$$\mathcal{L}^\# = \{L^\# \mid L \in \mathcal{L}\} .$$

This is again a lattice chain, which we call the *dual chain*.

We now make the following assumptions.

(a) The collection $\mathcal{L} \cup \mathcal{L}^\#$ is again a lattice chain (in particular it is "self-dual").

(b) Define an *overlapping sequence* in $\mathcal{L} \cup \mathcal{L}^\#$ to be a sequence of successive lattices of the form

$$N_k \supsetneqq N_{k+1} \supsetneqq \cdots \supsetneqq N_{k+l} \supsetneqq N_{k+l+1}$$

where $N_k, N_{k+l+1} \notin \mathcal{L} \cap \mathcal{L}^\#$, and

$$N_{k+i} \in \mathcal{L} \cap \mathcal{L}^\#, \qquad 1 \leq i \leq l .$$

We assume

C(i) $L \in \mathcal{L} \backslash \mathcal{L}^\#$ implies its successor is in $\mathcal{L}^\#$, and similarly for $M \in \mathcal{L}^\# \backslash \mathcal{L}$.

C(ii) The overlapping sequences all have the form $N_k \in \mathcal{L}$, $N_{k+l+1} \in \mathcal{L}^\#$, or they all have the form $N_k \in \mathcal{L}^\#$, $N_{k+l+1} \in \mathcal{L}$.

We call a pair of lattice chains $(\mathcal{L}, \mathcal{M})$ a *C-chain* if $\mathcal{M} = \mathcal{L}^{\#}$ and $\mathcal{L} \cup \mathcal{M}$ satisfies the conditions (a) and (b) above. We assume that matters have been arranged so that all overlapping sequences as above have the form $N_k \in \mathcal{L}$, $N_{k+l+1} \in \mathcal{L}^{\#} = \mathcal{M}$. As we shall see shortly, to each C-chain one can associate a congruence filtration with some remarkable properties.

3.4. C-chains exists in abundance. We give two examples:

(i) Suppose $\mathcal{L}$ is a self-dual lattice chain. Then $(\mathcal{L}, \mathcal{L})$ is a C-chain.
(ii) If T is a tamely ramified compact maximal torus in G, then T gives rise to a C-chain. We refer the reader to [M1] for the construction. (The set of such C-chains is a proper subset of the set of all C-chains.)

3.5. Suppose $(\mathcal{L}, \mathcal{M})$ is a C-chain. There are associated hereditary orders $\mathcal{A}_{\mathcal{L}}, \mathcal{A}_{\mathcal{M}}$, with Jacobson radicals $\mathcal{B}_{\mathcal{L}}, \mathcal{B}_{\mathcal{M}}$. If σ is the involution on $\mathrm{End}_k(V)$ associated to f, then we have $\sigma\mathcal{A}_{\mathcal{L}} = \mathcal{A}_{\mathcal{M}}$, etc. Set $\mathcal{A} = \mathcal{A}_{\mathcal{L}} \cap \mathcal{A}_{\mathcal{M}} = \mathcal{A}_{\mathcal{L}\cup\mathcal{M}}$.

We construct a special family of $(\mathcal{A}, \mathcal{A})$-bimodules as follows. Define $\mathcal{B}_0 = \mathcal{A}_{\mathcal{L}} \cap \mathcal{A}_{\mathcal{M}}$, and for each $i \in \mathcal{Z}$ let

$$\mathcal{B}_{2i} = \mathcal{B}_{\mathcal{L}}^{i} \cap \mathcal{B}_{\mathcal{M}}^{i} \quad \text{and} \quad \mathcal{B}_{2i+1} = \mathcal{B}_{\mathcal{L}}^{i+1} + \mathcal{B}_{\mathcal{M}}^{i+1} .$$

Let $e = \mathcal{O}$-period of $\mathcal{L}$ ($=\mathcal{O}$-period of $\mathcal{M}$) so that $\mathcal{B}_{\mathcal{L}}^{e} = \mathcal{P}\mathcal{A}_{\mathcal{L}}$.

Proposition.

(a) $\mathcal{B}_{\mathcal{L}\cup\mathcal{M}} = \mathcal{B}_{\mathcal{L}} + \mathcal{B}_{\mathcal{M}}$.
(b) $\mathcal{B}_{\mathcal{L}}^{i} + \mathcal{B}_{\mathcal{M}}^{i} \supset \mathcal{B}_{\mathcal{L}}^{i} \cap \mathcal{B}_{\mathcal{M}}^{i} \supset \mathcal{B}_{\mathcal{L}}^{i+1} + \mathcal{B}_{\mathcal{M}}^{i+1}$ *for each* $i \in \mathcal{Z}$.
(c) *Set* $(\mathcal{B}_{2i})^* = \{x \in \mathrm{End}_k(V) \mid \mathrm{trace}(x\mathcal{B}_{2i}) \subseteq \mathcal{O}\}$. *Then*

$$\mathcal{P}^2(\mathcal{B}_{2(e-i)})^* = \mathcal{B}_{\mathcal{L}}^{i+1} + \mathcal{B}_{\mathcal{M}}^{i+1}.$$

Corollary. *The infinite sequence*

$$\cdots \supset \mathcal{A} \supset \mathcal{B} \supset \mathcal{B}_{\mathcal{L}} \cap \mathcal{B}_{\mathcal{M}} \supset \mathcal{B}_{\mathcal{L}}^{2} + \mathcal{B}_{\mathcal{M}}^{2} \supset \cdots$$

is a periodic lattice chain of $(\mathcal{A}, \mathcal{A})$*-bimodules in* $\mathrm{End}_k(V)$*, self dual with respect to trace, and stable under* σ.

A similar statement holds for skew elements in the context of Lie algebras.

3.6. Set $P = \mathcal{A}^* \cap G = \mathcal{A} \cap G$, and for each positive integer n, define

$$P_n = \{x \in P \mid x - 1 \in \mathcal{B}_n\} .$$

Note that $\overline{\mathcal{A}} = \mathcal{A}/\mathcal{B}_{\mathcal{L}\cup\mathcal{M}}$ inherits an involution $\overline{\sigma}$ from that on $\mathcal{A}$. Set

$$N(\overline{\mathcal{A}}, \overline{\sigma}) = \{x \in \overline{\mathcal{A}} \mid x\overline{\sigma}x = 1\}.$$

The following analogue of Proposition 2.1 is then obtained.

Theorem.

(a) *For each $n > 0$, P_n is a normal subgroup of P, and the commutator subgroup (P_m, P_n) lies in P_{m+n}.*
(b) *The natural map $P \longrightarrow N(\overline{\mathcal{A}}, \overline{\sigma})$ is surjective with kernel P_1.*
(c) *For each n there exists a bijection $\mathcal{B}_n^- \longrightarrow P_n$ given by*

$$x \longmapsto (x-1)(1+x)^{-1}$$

($\mathcal{B}_n$ is stable under σ; we let $\mathcal{B}_n^-$ denote the subspace of skew elements)
(d) *If $2i \geq j \geq i \geq 1$, there is an isomorphism of abelian groups induced by $x \longmapsto x-1$:*

$$P_i/P_j \longrightarrow \mathcal{B}_i^-/\mathcal{B}_j^-$$

3.7. The preceding results allow one to give a convenient description of the Pontrjagin dual $(P_i/P_j)^\wedge$ (i, j as in the theorem). Indeed, let $\Omega : k_0^+ \longrightarrow S^1$ be a character with conductor $\mathcal{O}_0$, let e_0 be the ramification degree of k over k_0, and set $c = 1 - 2ee_0$. Also let tr denote the composite trace map

$$\mathrm{End}_k(V) \longrightarrow k \longrightarrow k_0$$

Proposition 3.5 and Theorem 3.6 then imply the following result.

Corollary. *There are natural isomorphisms*

$$(\mathcal{B}_{c-j}/\mathcal{B}_{c-i})^- \simeq \mathcal{B}_{c-j}^-/\mathcal{B}_{c-i}^- \longrightarrow (P_i/P_j)^\wedge .$$

The isomorphism on the right is given by

$$\overline{b} = b + \mathcal{B}_{c-i}^- \longmapsto (p \longmapsto \Omega(tr(b(p-1)))) .$$

We write ψ_b for the character given by the coset $b + \mathcal{B}_{c-i}^-$; similarly we write δ_ψ for the coset corresponding to ψ.

3.8 Example. Let $\mathcal{L}$ be a self dual lattice chain, and consider the C-chain $(\mathcal{L}, \mathcal{L})$ (cf 3.4). Then $\mathcal{B}_{\mathcal{L}}^i = \mathcal{B}_{\mathcal{M}}^i$, so $\mathcal{B}_{2i} = \mathcal{B}_{\mathcal{L}}^i = \mathcal{B}_{\mathcal{M}}^i$ and

$$\mathcal{B}_{2i+1} = \mathcal{B}_{\mathcal{L}}^{i+1} + \mathcal{B}_{\mathcal{M}}^{i+1} = \mathcal{B}_{\mathcal{L}}^{i+1} = \mathcal{B}_{\mathcal{M}}^{i+1} = \mathcal{B}_{2(i+1)} .$$

Thus in this case one recovers the filtration that arises from powers of the Jacobson radical.

§4. G-Strata

4.1. As one might expect, the notion of a G-stratum is a bit more subtle in this framework; it involves a certain non-degeneracy condition which needs a few preliminaries. To avoid some minor technical complications in what follows, we shall exclude the case where k is a quadratic ramified extension over k_0.

4.2. There is always a (unique) self dual "slice" for $\mathcal{L} \cup \mathcal{M}$ of the form

$$\supset N_{r-1}^{\#} \supset \cdots \supset N_0^{\#} \supset N_0 \supset \cdots \supset N_{r-1} \supset \mathcal{P}N_{r-1}^{\#}$$

where $N_i^{\#}$ denotes the complementary $\mathcal{O}$-lattice (with respect to f) of N_i, and possibly $N_0^{\#} = N_0, \quad N_{r-1} = \mathcal{P}N_{r-1}^{\#}$.

Let $\overline{N}$ denote the direct sum of successive quotients in this slice. If s is the period of $\mathcal{L} \cup \mathcal{M}$, then $\overline{N}$ is a $\mathbb{Z}/s\mathbb{Z}$ graded $\mathbf{F}_q$-space, naturally equipped with a non-degenerate sesquilinear form

$$\overline{f} : \overline{N} \times \overline{N} \longrightarrow \mathbf{F}_q \ .$$

Define $\overline{U}$ to be the direct sum of those A/B in $\overline{N}$ where $A \in \mathcal{L}$, and let $\overline{V}$ be the obvious complement to $\overline{U}$ in $\overline{N}$.

Lemma.

(a) *$\overline{U}$ is a $\mathbb{Z}/e\mathbb{Z}$–graded $\mathbf{F}_q$–space (e the period of $\mathcal{L}$).*
(b) $\overline{f} = \overline{f}_{\overline{U}} \oplus \overline{f}_{\overline{V}}$.
(c) *There is a natural map $\phi_j : \mathcal{B}_{2j} \longrightarrow \mathrm{End}(\overline{U})_{j(\mathrm{mod}\ e)}$ which preserves skew elements (cf. (b)). If $\mathcal{K}_{2j} = ker\phi_j$, then $\mathcal{K}_{2j}$ is an $(\mathcal{A},\mathcal{A})$-bimodule and $\mathcal{K}_{2j} \supset \mathcal{B}_{2j+1}$.*

Define

$$(\mathcal{K}_{2j})^{\#} = \{x \in \mathrm{End}_k(V) \mid tr_0(x\mathcal{K}_{2j}) \subseteq \mathcal{O}_0\},$$

where tr_0 is the composition

$$\mathrm{End}_k(V) \underset{\text{Trace}}{\longrightarrow} K \underset{tr_{k/k_0}}{\longrightarrow} k_0 \ .$$

Then $(\mathcal{K}_{2j})^{\#}$ is also a (σ–stable) $(\mathcal{A},\mathcal{A})$-bimodule.

4.3. By duality (c.f. 3.7) one obtains

$$(\mathcal{B}_{2j}/\mathcal{K}_{2j})^{\wedge} \simeq (\mathcal{K}_{2j})^{\#}/\mathcal{B}_{2j}^{\#} = (\mathcal{K}_{2j})^{\#}/\mathcal{B}_{c-2j}$$

(note that $c - 2j = 1 - 2ee_0 - 2j$ is odd).

For $n > 0$ define $\mathcal{J}_{2n} = (\mathcal{K}_{c-2n-1})^{\#}$. Then $\mathcal{B}_{2n} \supset \mathcal{J}_{2n} \supset \mathcal{B}_{2n+1}$.

Lemma.

(a) $\mathcal{J}_{2n}$ *is a* σ*-stable two sided ideal in* $\mathcal{A}$.

(b) $\mathrm{J}_{2n} = \{x \in P \mid x - 1 \in \mathcal{J}_{2n}\}$ *is a normal subgroup of* P.

4.4. Let ψ_b be a character of $\mathcal{B}^-_{2n-1}/\mathcal{B}^-_{2n+1}$. We say that ψ_b is *non-degenerate* if $\psi_b \neq 1$ implies $\psi_b \mid \mathcal{J}^-_{2n} \neq 1$. (This can easily be interpreted in terms of cosets.) By a *G-stratum* we mean either:

(a) a quadruple $((\mathcal{L}, \mathcal{M}), n, \psi_b, \psi)$ where $(\mathcal{L}, \mathcal{M})$ is a C-chain, n is a positive integer, ψ_b is a non-degenerate character of $\mathcal{B}^-_{2n-1}/\mathcal{B}^-_{2n+1}$, and ψ is a character of $\mathcal{J}^-_{2n}/\mathcal{B}^-_{2n+1}$ such that $\psi_b \mid \mathcal{J}^-_{2n} = \psi$; or

(b) a pair $((\mathcal{L}, \mathcal{M}), \psi)$ where $(\mathcal{L}, \mathcal{M})$ is a C-chain and ψ is an irreducible representation of $(P/P_1)^0$. Here $(P/P_1)^0$ denotes the identity component of the finite group of Lie type P/P_1.

Remarks.

(i) Note that $\mathcal{J}^-_{2n}/\mathcal{B}^-_{2n+1} \subseteq \mathcal{B}^-_{2n}/\mathcal{B}^-_{2n+1}$ and $\mathcal{B}^-_{2n}/\mathcal{B}^-_{2n+1} \simeq P_{2n}/P_{2n+1}$. Hence $\mathcal{B}^-_{2n}/\mathcal{B}^-_{2n+1}$ is an abelian group.

(ii) If $n \geq 2$, then $\mathcal{B}^-_{2n-1}/\mathcal{B}^-_{2n+1} \simeq P_{2n-1}/P_{2n+1}$, and a G-stratum can be described by a non-degeneracy condition on a character ψ_b of P_{2n-1}/P_{2n+1}.

4.5. Consider $\tilde{\mathcal{A}} = \oplus_{i \in \mathbb{Z}} \mathcal{B}_{2i}/\mathcal{B}_{2i+2}$. This is a $\mathbb{Z}$-graded $\mathbf{F}_q$-algebra equipped with an involution. By periodicity it can be "collapsed" to a $\mathbb{Z}/e\mathbb{Z}$-graded algebra $\overline{\mathcal{A}}$ which is isomorphic as a vector space to $\bigoplus_{i=0}^{e-1} \mathcal{B}_{2i}/\mathcal{B}_{2i+2}$. (If k is quadratic ramified over k_0, more care must be exercised here). In any case, one can speak of nilpotent elements in $\tilde{\mathcal{A}}$ or $\overline{\mathcal{A}}$.

4.6. Let ψ_b be a character of $\mathcal{B}^-_{2n-1}/\mathcal{B}^-_{2n+1}$ represented by a coset

$$b + \mathcal{B}^-_{-2ee_0-2n+2} \in \mathcal{B}^-_{-2ee_0-2n}/\mathcal{B}^-_{-2ee_0-2n+2}\,.$$

We say that ψ_b is *nilpotent* if the coset above is nilpotent.

4.7. A G-stratum $((\mathcal{L}, \mathcal{M}), n, \psi_b, \psi)$ (respectively, $((\mathcal{L}, \mathcal{M}), \psi)$) is said to be *fundamental* if ψ_b is not nilpotent (respectively, if ψ is a cuspidal representation of $(P_0/P_1)^0$).

§5. The Main Theorem

5.1. Let π be an irreducible smooth representation of G. We say that π *contains* the G-stratum $((\mathcal{L}, \mathcal{M}), n, \psi_b, \psi)$ (respectively, $((\mathcal{L}, \mathcal{M}), \psi)$) if $\pi \mid \mathrm{J}_{2n}$ contains ψ (respectively, $\pi \mid P$ contains ψ).

5.2 Theorem. *If π is an irreducible smooth representation of G, then π always contains a fundamental G-stratum.*

5.3. The basic idea in the proof of Theorem 5.2 is the following. If $b+\mathcal{B}_{2j+2}^-$ is a nilpotent coset of $\mathcal{B}_{2j}^-/\mathcal{B}_{2j+2}^- \simeq (\mathcal{B}_{2j}/\mathcal{B}_{2j+2})^-$, then it gives rise to a skew nilpotent element $\overline{b}$ in $\mathrm{End}(\overline{U})_{j(\mathrm{mod}\ e)}$ via the map ϕ_j of 4.2(c). Such an element $\overline{b}$ gives rise to a homogeneous graded flag of "weight" spaces in $\overline{U}$: if $\overline{b}^d = 0, \overline{b}^{d-1} \neq 0$, then there is a flag

$$\overline{U} = \overline{U}(1-d) \supset \overline{U}(2-d) \supset \cdots \supset \overline{U}(d-1) \supsetneq 0$$

such that, for each λ,

$$\overline{U}(\lambda) = \bigoplus_{i \in \mathbb{Z}/e\mathbb{Z}} \overline{U}_i(\lambda)$$

Moreover $\overline{U}(\lambda)^{\perp} = \overline{U}(1-\lambda)$ if "$\perp$" denotes orthogonal complement with respect to the form $\overline{f}_{\overline{U}}$. (This is a graded version of Jordan normal form.)

Thus, given a nilpotent ψ_b, one produces a homogeneous flag associated to it. The strategy is then to show that starting from this flag, one can find a particular subsequence of the family $\{\overline{U}_i(\lambda)\}$ which "lifts" to give a C-chain $(\mathcal{L}', \mathcal{M}')$, where the $\mathcal{O}$-period of $\mathcal{L}'$ is e', with the following properties:

(i) $b + \mathcal{B}_{2j+2}^- \subseteq \mathcal{B}'^-_{2j'}$, and
(ii) $j'/e' > j/e$.

One then shows that since π is admissible one can eventually produce G-strata for which it is impossible to "raise the normalized level" as in (ii) above. This is similar to the strategy used in the proof of Theorem 2.6.

5.4. There are several difficulties that one encounters in attempting to implement the ideas of 5.3. First, the subsequence of $\{\overline{U}_i(\lambda)\}$, and its "dual" family, only give rise to a C-chain with the properties (i) and (ii) above when the coset $b+\mathcal{B}_{2j+2}$ is of a certain type; this ultimately accounts for the non-degeneracy condition in the definition of G-strata. Second, the coset $b + \mathcal{B}'^-_{2j'+2}$ may not be of this same type: one must modify it slightly. Third, one has to show that the pair $(\mathcal{L}', \mathcal{M}')$ is a C-chain.

6. Two Examples

6.1. In this section we shall provide two examples which illustrate some of the ideas and difficulties mentioned in sections 5.3 and 5.4.

6.2. Consider the case where V is a 4 dimensional k-vector space equipped with a non-degenerate alternating form f. Assume that e_1, e_2, e_3, e_4 is a basis of V with respect to which f has the matrix

$$\begin{pmatrix} 0 & 0 & 0 & 1 \\ 0 & 0 & 1 & 0 \\ 0 & -1 & 0 & 0 \\ -1 & 0 & 0 & 0 \end{pmatrix}.$$

Let

$$b = \begin{pmatrix} 0 & 0 & 0 & 1 \\ 0 & 0 & 0 & 0 \\ 0 & 0 & 0 & 0 \\ 0 & 0 & 0 & 0 \end{pmatrix},$$

and consider the standard lattice chain $\mathcal{L} : \cdots \supset L \supset \mathcal{P}L \supset \cdots$, where $L = \mathcal{O}e_1 \oplus \mathcal{O}e_2 \oplus \mathcal{O}e_3 \oplus \mathcal{O}e_4$. We view $\mathcal{L}$ as a C-chain $(\mathcal{L}, \mathcal{L})$. The corresponding hereditary order is $M_4(\mathcal{O})$, with Jacobson radical $M_4(\mathcal{P})$. Moreover $b \in M_4(\mathcal{O})$: the normalized level of b is $0/1$.

Let $\overline{b}$ denote the image of b in $M_4(\mathbf{F}_q)$ acting on $\overline{V} = L/\mathcal{P}L$. By the Jordan normal form for symplectic nilpotent elements, one has the flag

$$\overline{V} = \overline{V}(-1) \supsetneq \overline{V}(0) \supsetneq \overline{V}(1) \supsetneq (0),$$

where $\overline{V} =< \overline{e_1} >$, $\overline{V}_i(0) =< \overline{e_1}, \overline{e_2}, \overline{e_3} >$. In this case $\mathcal{L}$ is given by the slice

$$\cdots \supset L \supset \mathcal{O}e_1 \oplus \mathcal{P}e_2 \oplus \mathcal{P}e_3 \oplus \mathcal{P}e_4 \supset \mathcal{P}L \supset \cdots$$

and $\mathcal{M}'$ is given by the slice

$$\cdots \supset L \supset \mathcal{O}e_1 \oplus \mathcal{O}e_2 \oplus \mathcal{O}e_3 \oplus \mathcal{P}e_4 \supset \mathcal{P}L \supset \cdots .$$

The new level of b is equal to $1/2 > 0/1$. This example occurs in [Mo]. (If one pursues this once more, one obtains the C-chain $(\mathcal{L}'', \mathcal{L}'')$:

$$\begin{aligned} \mathcal{L}'' \cdots &\supset \mathcal{P}^{-1}e_1 \oplus \mathcal{O}e_2 \oplus \mathcal{O}e_3 \oplus \mathcal{O}e_4 \\ &\supset \mathcal{O}e_1 \oplus \mathcal{O}e_2 \oplus \mathcal{O}e_3 \oplus \mathcal{P}e_4 \\ &\supset \mathcal{O}e_1 \oplus \mathcal{P}e_2 \oplus \mathcal{P}e_3 \oplus \mathcal{P}e_4 \\ &\supset \cdots \end{aligned}$$

and b has level $1/1 > 1/2$.)

6.3. Let V be the 10 dimensional k-space with basis $e_1, \cdots, e_{10}$ and alternating form defined by $f(e_i, e_{11-i}) = 1 = -f(e_{11-i}, e_i)$ for $1 \leq i \leq 5$, and $f(e_i, e_j) = 0$ otherwise. Let x be the element of $\mathrm{End}_k(V)$ which acts on this basis in the following way:

$$\begin{gathered} e_{10} \longmapsto e_9 \longmapsto e_2 \longmapsto e_1 \longmapsto 0, \\ e_8 \longmapsto e_6 \longmapsto e_4 \longmapsto 0, \\ e_7 \longmapsto e_5 \longmapsto e_3 \longmapsto 0. \end{gathered}$$

Consider the C-chain $(\mathcal{L}, \mathcal{M})$ which can be described as follows

$$\begin{aligned} M_0 &= \mathcal{O}e_1 \oplus \cdots \oplus \mathcal{O}e_9 \oplus \mathcal{P}e_{10}, \\ M_1 &= \mathcal{O}e_1 \oplus \cdots \oplus \mathcal{O}e_6 \oplus \mathcal{P}e_7 \oplus \cdots \oplus \mathcal{P}e_{10}, \\ M_2 &= \mathcal{O}e_1 \oplus \mathcal{P}e_2 \oplus \cdots \oplus \mathcal{P}e_{10}, M_3 = \mathcal{P}M_0 . \end{aligned}$$

Then $\mathcal{L} \cup \mathcal{M}$ is described by the slice

$$\cdots \supset M_0 \supsetneq L_0 \supsetneq M_1 \supsetneq L_1 \supsetneq M_2 \supsetneq L_2 \supsetneq \mathcal{P}M_0 \supset \cdots$$

where $L_2/\mathcal{P}M_0 =< \overline{e_1}, \overline{e_{10}} >$, $L_1/M_2 =< \overline{e_2} >$, and $L_0/M_1 =< \overline{e_9} >$. Then one finds that $x \in \mathcal{B}_\mathcal{L} \cap \mathcal{B}_\mathcal{M} = \mathcal{B}_2$, so x has normalized valuation 1/3.

At the next stage one finds the chain $(\mathcal{L}', \mathcal{M}')$ where $M_0' = M_1$, $L_0' = L_1$, and $M_1' = L_1'$ is described by $M_1''/\mathcal{P}M_0' =< \overline{e_1}, \overline{e_7}, \overline{e_8}, \overline{e_9} >$. We have

$$\cdots \supset M_0' \supset L_0' \supset M_1' = L_1' \supset \mathcal{P}M_0' \supset \cdots$$

with successive quotients

$$M_0'/L_0' =< \overline{e_5}, \overline{e_6} >,$$
$$L_0'/M_1' =< \overline{e_2}, \overline{e_3}, \overline{e_4}, \overline{e_{10}} > \text{, and}$$
$$M_1'/\mathcal{P}M_0' =< \overline{e_1}, \overline{e_7}, \overline{e_8}, \overline{e_9} > .$$

If x is replaced by the element x' defined by

$$x' : e_{10} \longmapsto e_9 \longmapsto e_2 \longmapsto e_1 \longmapsto 0 ,$$

and $x'e_j = 0$ for other j, then one finds that $x' \in \mathcal{B}'_{\mathcal{L}'} \cap \mathcal{B}'_{\mathcal{M}'}$, with normalized valuation $1/2 > 1/3$.

This is quite typical of what happens in general.

References

[B] C. Bushnell, *Hereditary orders, Gauss sums, and supercuspidal representations of GL_N*, J. Reine Angew. Math. **375/6** (1987), 184–210.

[H] R. Howe, *Some qualitative results on the representation theory of GL_N over a p-adic local field*, Pac. J. Math. **73** (1977), 479–538.

[HM] R. Howe and A. Moy, *Minimal K-types for GL_N over a p-adic field*, Astérisque, Nos. 171–172 (1989), 257–273.

[J] H. Jacquet, *Sur les représentations des groupes réductifs p-adiques*, C. R. Acad. Sci., Paris, Ser. A **280** (1975), 1271–1272.

[M1] L. Morris, *Tamely ramified supercuspidal representations of the classical groups I: filtrations*, preprint.

[M2] ———, *Tamely ramified supercuspidal representations of the classical groups II: representation theory*, preprint.

[M3] ———, *Fundamental G-strata for classical groups*, preprint.

[Mo] A. Moy, *Representations of GSp_4 over a p-adic field I, II*, Comp. Math. **66** (1988), 237–284; *II*, ibid **66** (1988), 285–328.

[PR] G. Prasad and M. S. Raghunathan, *Topological central extensions of semisimple groups over local fields I*, Ann. Math. **119** (1984), 143–201.

Department of Mathematics/Computer Science, Clark University, Worcester, MA 01610-1477

CONSTRUCTION AND CLASSIFICATION OF IRREDUCIBLE HARISH–CHANDRA MODULES

WILFRIED SCHMID

Harvard University

§1 INTRODUCTION.

Typically, irreducible Harish-Chandra modules are constructed not directly, but as unique irreducible submodules, or unique irreducible quotients, of so-called standard modules. As in some other contexts, standard modules are obtained by cohomological constructions which tend to be "easy" on the level of Euler characteristic. For certain values of the parameters in these constructions, there is a vanishing theorem; standard modules arise when the vanishing theorem applies.

Standard modules have been used to classify irreducible Harish-Chandra modules by Langlands, Vogan-Zuckerman, and Beilinson-Bernstein. Langlands' standard modules are parabolically induced from discrete series or "limits of discrete series" representations [11,8], which in turn may be viewed as cohomologically induced [16]. Vogan-Zuckerman, in effect, reverse the process; their standard modules are cohomologically induced from principal series representations [20]. Beilinson-Bernstein, finally, associate standard modules to $K_{\mathbb{C}}$-orbits in the flag variety by a one-step process [1]. For the sake of completeness, I ought to mention also Vogan's classification by lowest K-type [19]; it does not rely on standard modules, but is conceptually close to the classification of Vogan-Zuckerman, and has also been deduced from the Beilinson-Bernstein classification [4].

The separate induction steps of both Langlands and Vogan-Zuckerman can be combined into a single procedure: Zuckerman's derived functors attach Harish-Chandra modules—among them the standard modules of [11] and [20]—to G-orbits in the flag variety. Though purely algebraic, this construction was motivated by a then hypothetical geometric construction, recently carried out by Wolf and myself [17,18], which is the subject of Wolf's lecture. In a somewhat different sprit, Hecht-Taylor also relate Harish-Chandra modules and G-orbits in the flag variety [7].

Supported in part by a John Simon Guggenheim Memorial Fellowship and NSF Grant DMS-87-01578.

In these lectures I shall show how one can pass back and forth between the various standard modules, in terms of their geometric realizations[1]. This is joint work with Hecht, Miličić and Wolf—full details will appear as part of [6]. I am indebted to Dragan Miličić for pointing out several inaccuracies in the original version of this manuscript.

Here is a brief summary of the contents of the various sections: I begin with a discussion of infinitesimally equivariant line bundles and twisted sheaves of differential operators in §2. Section three recalls Beilinson-Bernstein's $\mathcal{D}$-module construction of standard Harish-Chandra modules and gives an alternative description in terms of local cohomology. Standard modules arising from G-orbits are the subject of §4. The duality theorem of [5] provides the bridge between modules corresponding to G-orbits on the one hand, and to $K_{\mathbb{C}}$-orbits on the other. Section five contains the statement of the duality theorem, along with a sketch of a proof which is identical to that in [5] in a technical sense, but avoids the machinery of $\mathcal{D}$-module theory. The different types of standard modules are identified with each other in §6; this amounts to "independence of the polarization", in the language of geometric quantization, under appropriate hypotheses. I conclude with the example of $\mathrm{Sl}(2,\mathbb{C})$, in §7.

§2 Equivariant line bundles and TDO's.

Throughout these lectures, G denotes a semisimple Lie group. I assume that G is connected and linear—solely for ease of exposition: arguments that are fairly standard by now make it a simple matter to extend all methods and results to much larger classes of groups; cf. Appendix B in [5]. In particular, G has a complexification $G_{\mathbb{C}}$. I fix a maximal compact subgroup $K \subset G$. The choice of K determines a Cartan involution θ and Cartan decomposition $\mathfrak{g} = \mathfrak{k} \oplus \mathfrak{p}$ of the complexified Lie algebra $\mathfrak{g}$ of G.

Let X be the flag variety of $\mathfrak{g}$, i.e., the variety of Borel subalgebras. The group $G_{\mathbb{C}}$ acts transitively on X, via Ad. This action identifies the quotients $\mathfrak{b}/[\mathfrak{b},\mathfrak{b}]$, for the various Borel subalgebras $\mathfrak{b} \subset \mathfrak{g}$. In other words, $\mathfrak{b}/[\mathfrak{b},\mathfrak{b}]$ is the fibre at $\mathfrak{b} \in X$ of a canonically trivial vector bundle. By definition, its generic fibre $\mathfrak{h}$ is the "abstract Cartan algebra" for $\mathfrak{g}$. The "abstract root system" Φ of $\mathfrak{h}$ comes equipped with a distinguished positive root system Φ^+: a root α is negative if $[\mathfrak{b},\mathfrak{b}]$ contains the α-root space[2]. Thus any concrete Cartan subalgebra $\mathfrak{h}_1$ of $\mathfrak{g}$ becomes canonically isomorphic to $\mathfrak{h}$ as soon as one specifies a positive root system $\Phi^+(\mathfrak{h}_1,\mathfrak{g})$ in the root system $\Phi(\mathfrak{h}_1,\mathfrak{g})$.

Temporarily I make the identification $X \cong G_{\mathbb{C}}/B$, where B is the stabilizer of some base point $\mathfrak{b} \in X$. The differentials of the holomorphic characters of the Borel subgroup B form a lattice Λ in the linear dual $\mathfrak{h}^*$

[1] Identifications between certain standard modules, by algebraic arguments, are part of Vogan's proof of the Kazhdan-Lusztig conjectures in the integral case [21].

[2] Caution: [5,6,14] use the opposite convention.

of $\mathfrak{h} \cong \mathfrak{b}/[\mathfrak{b},\mathfrak{b}]$ – $\Lambda \subset \mathfrak{h}^*$ is the weight lattice of $G_{\mathbb{C}}$. For each $\lambda \in \Lambda$, the character e^λ of B associates a holomorphic line bundle $\boldsymbol{L}_\lambda \to X$ to the principal bundle $B \to G_{\mathbb{C}} \to G_{\mathbb{C}}/B \cong X$. Since $G_{\mathbb{C}}$ operates on the principal bundle, it operates on the associated line bundle: $\boldsymbol{L}_\lambda$ is a $G_{\mathbb{C}}$-equivariant holomorphic line bundle—a holomorphic line bundle, equipped with an action of $G_{\mathbb{C}}$ compatible with the $G_{\mathbb{C}}$-action on X. Every $G_{\mathbb{C}}$-equivariant line bundle over X is of this type, for a unique $\lambda \in \Lambda$. Under the resulting parametrization of the $G_{\mathbb{C}}$-equivariant line bundles by the weight lattice Λ, the canonical line bundle corresponds to -2ρ, where ρ denotes the half sum of the positive roots:

$$\boldsymbol{L}_{-2\rho} \cong \text{ canonical bundle of } X. \tag{2.1}$$

Since X is projective, the line bundles $\boldsymbol{L}_\lambda$ have uniquely determined algebraic structures. According to the Borel-Weil theorem, $\boldsymbol{L}_\lambda$ is ample in the sense of algebraic geometry precisely when the weight λ is dominant and regular.

I shall need the notion of infinitesimally equivariant line bundle over a real or complex homogeneous space. Let J be a Lie group with Lie algebra $\mathfrak{j}$, and $J_0 \subset J$ a closed subgroup. I write $\mathcal{O}$ for the sheaf of real analytic functions on J/J_0 or, if J/J_0 comes equipped with a J-invariant complex structure[3], for the sheaf of holomorphic functions. Note that the derivative of the J-action on J/J_0 turns $\mathfrak{j}$ into a Lie algebra of real analytic, respectively holomorphic, vector fields. Let $U \subset J/J_0$ be an open subset, $\boldsymbol{L} \to U$ a real analytic, respectively holomorphic, line bundle, and $\mathcal{O}(\boldsymbol{L})$ the sheaf of real analytic, respectively holomorphic, sections. The structure of $\mathfrak{j}$-equivariant line bundle for $\boldsymbol{L}$ consists of an action of the Lie algebra $\mathfrak{j}$ on the stalks of $\mathcal{O}(\boldsymbol{L})$, such that: (i) $\mathfrak{j}$ acts continuously, relative to the sheaf topology for $\mathcal{O}(\boldsymbol{L})$ and the discrete topology for $\mathfrak{j}$; (ii) the map $\mathcal{O} \otimes \mathcal{O}(\boldsymbol{L}) \to \mathcal{O}(\boldsymbol{L})$, given by multiplication of sections by functions, obeys Leibnitz' rule. Equivalently, the Lie algebra $\mathfrak{j}$ acts on the space of sections of $\mathcal{O}(\boldsymbol{L})$ over any open subset of U, compatibly with restriction to smaller subsets, and again subject to Leibnitz' rule for multiplication by functions. Globally defined J-equivariant line bundles $\boldsymbol{L} \to J/J_0$ are $\mathfrak{j}$-equivariant in this sense: the derivative of the J-action satisfies both (i) and (ii). One can show that conversely every globally defined $\mathfrak{j}$-equivariant line bundle over a simply connected quotient J/J_0 is globally equivariant, if not with respect to J, then at least with respect to the universal covering group $\tilde{J}$.

Locally, any $\mathfrak{j}$-equivariant line bundle $\boldsymbol{L} \to U \subset J/J_0$ can be raised to fractional powers. To see this, fix a point $x \in U$ and complex constant c. Choose a nowhere vanishing section s of $\boldsymbol{L}$, defined on some simply connected neighborhood U_x of x. For $Z \in \mathfrak{j}$, Zs is another section of $\boldsymbol{L}$ over U_x, hence $Zs = fs$, for some function f on U_x. Set $Z(s^c) = c(Zs)s^{c-1} = cfs^c$; then

[3] for example, if J is a complex Lie group and J_0 a closed complex Lie subgroup

$L^c \to U_x$, the line bundle with generating section s^c, becomes $\mathfrak{j}$-equivariant. Different choices of the local generating section s lead to the same L^c, up to isomorphism within the class of $\mathfrak{j}$-equivariant line bundles. It is an exercise, which will be skipped here, to define L^c over the germ of a neighborhood of x in a way which is formally independent of arbitrary choices. Germs of $\mathfrak{j}$-equivariant line bundles at x can also be tensored. If we regard tensoring as "addition", and raising to the c-th power as "multiplication by c", the isomorphism classes of germs of $\mathfrak{j}$-equivariant line bundles at x form a complex vector space.

To simplify the next statement, I suppose that J and J_0 are complex Lie groups. Thus $\mathfrak{j}$ and $\mathfrak{j}_0$, the Lie algebra of J_0, are complex Lie algebras.

(2.2) Lemma. *The vector space of isomorphism classes of germs of* $\mathfrak{j}$-*equivariant holomorphic line bundles at the identity coset in* J/J_0 *is isomorphic to the dual space of* $\mathfrak{j}_0/[\mathfrak{j}_0, \mathfrak{j}_0]$.

Proof. The problem is local and does not change if J_0 is replaced by its identity component. I therefore suppose that J_0 is connected. Let L be the germ of a $\mathfrak{j}$-equivariant holomorphic line bundle at the identity coset, and s the germ of a generating section. The map $Z \mapsto s^{-1}Zs$ defines a 1-cocycle on $\mathfrak{j}$, with values in $\mathcal{O}_0$ = stalk of the structure sheaf; conversely, any such cocycle defines the germ of a $\mathfrak{j}$-equivariant line bundle. If s is modified by an invertible function f, the coboundary of $\log f$ gets added to this cocycle. Thus $H^1(\mathfrak{j}, \mathcal{O}_0)$ parametrizes the germs of $\mathfrak{j}$-equivariant holomorphic line bundles. If one resolves $\mathcal{O}_0$ by the relative de Rham complex of the fibration $J \to J/J_0$, the resulting spectral sequence collapses and gives $H^1(\mathfrak{j}, \mathcal{O}_0) \cong H^1(\mathfrak{j}_0, \mathbb{C}) \cong (\mathfrak{j}_0/[\mathfrak{j}_0, \mathfrak{j}_0])^*$. □

I apply the preceding discussion to the flag variety of $\mathfrak{g}$, viewed as homogeneous space for $G_{\mathbb{C}}$. Each $\mathfrak{b} \in X$ is the Lie algebra of its own stabilizer, and $\mathfrak{b}/[\mathfrak{b}, \mathfrak{b}]$ is canonically isomorphic to $\mathfrak{h}$. Hence, in this situation, (2.2) asserts:

$$\mathfrak{h}^* = \{\text{isomorphism classes of germs of } \mathfrak{g}\text{-equivariant holomorphic line bundles at } \mathfrak{b}\} \tag{2.3}$$

I let $L_{\lambda,\mathfrak{b}}$ denote the germ at $\mathfrak{b}$ corresponding to $\lambda \in \mathfrak{h}^*$ via (2.3). The notation is consistent with the parametrization of the globally defined $G_{\mathbb{C}}$-equivariant line bundles by the weight lattice $\Lambda \subset \mathfrak{h}^*$: for each $\lambda \in \Lambda$, $L_{\lambda,\mathfrak{b}}$ is the germ at $\mathfrak{b}$ of the global line bundle L_λ, as can be seen by retracing the proof of (2.2).

There is an obstruction to fitting together the germs $L_{\lambda,\mathfrak{b}}$ for a fixed $\lambda \in \mathfrak{h}^*$, with $\mathfrak{b}$ varying over some open subset $U \subset X$, into a $\mathfrak{g}$-equivariant line bundle L over U; this obstruction lies in the cohomology group $H^2(U, \mathbb{C}^*)$. If the obstruction vanishes, L is determined only up to a flat line bundle over U, and such flat line bundles are parametrized by $H^1(U, \mathbb{C}^*)$. In the particular situation of interest here, both the obstruction and the ambiguity can be made more explicit, as I shall describe next.

Let J be an algebraically defined subgroup of either the complex algebraic group $G_{\mathbb{C}}$, or of its real form G, and $\mathfrak{j}$ the Lie algebra of J. The two examples that will be important are $J = G$ and $J = K_{\mathbb{C}}$ (= complexification of the maximal compact subgroup $K \subset G$). Let $S \subset X$ be a J-orbit in the flag variety. I shall be concerned with $\mathfrak{g}$-equivariant holomorphic line bundles $\boldsymbol{L} \to U$, defined over the germ U of a neighborhood of S in X, with the additional datum of a J-equivariant structure for the restriction of $\boldsymbol{L}$ to S. The two equivariant structures are required to be compatible, in the sense that the derivative of the J-action on the restriction of $\boldsymbol{L}$ to S agrees with the $\mathfrak{j}$-action coming from the inclusions $\mathfrak{j} \subset \mathfrak{g}$, $S \subset X$. I shall refer to this set of hypotheses by calling $\boldsymbol{L}$ "a $(\mathfrak{g}, J)$-equivariant holomorphic line bundle". As homogeneous space for J, S can be identified with J/J_0, where J_0 = isotropy subgroup of J at some $\mathfrak{b}_0 \in S$. Since $\mathfrak{h} \cong \mathfrak{b}_0/[\mathfrak{b}_0, \mathfrak{b}_0]$ and $\mathfrak{h}^* \cong$ annihilator of $[\mathfrak{b}_0, \mathfrak{b}_0]$ in $\mathfrak{b}_0^*$, I may view each $\lambda \in \mathfrak{h}^*$ as a linear function on $\mathfrak{b}_0$.

(2.4) Lemma. *The possible ways of patching together the germs $\boldsymbol{L}_{\lambda,\mathfrak{b}}$, for $\mathfrak{b} \in S$, into a $(\mathfrak{g}, J)$-equivariant line bundle $\boldsymbol{L} \to U$, over the germ U of a neighborhood of S in X, correspond bijectively to those characters of J_0 whose differential agrees with the restriction of λ to $\mathfrak{j}_0 = \mathfrak{j} \cap \mathfrak{b}_0$.*

Proof. J-equivariant line bundles on S are associated to the principal bundle $J_0 \to J \to J/J_0 \cong S$ by a character of J_0. If the differential of this character coincides with the restriction of λ to $\mathfrak{j}_0$, then at each $\mathfrak{b} \in S$ the line bundle in question extends locally to a $\mathfrak{g}$-equivariant holomorphic line bundle, whose germ at $\mathfrak{b}$ equals $\boldsymbol{L}_{\lambda,\mathfrak{b}}$, on a neighborhood $U_{\mathfrak{b}}$ in X, compatibly with the $\mathfrak{j}$-equivariant structure on S. The obstruction to fitting together these various locally defined $\mathfrak{g}$-equivariant line bundles is topological; it vanishes on the germ of a neighborhood of S since the restricted germs on S fit together, by construction, into a J-equivariant line bundle on S. The J-equivariant structure of that line bundle on S eliminates the ambiguity in the process of patching together the germs $\boldsymbol{L}_{\lambda,\mathfrak{b}}$ along S. □

If J is an algebraically defined subgroup of $G_{\mathbb{C}}$, then every J-orbit S in X is locally closed in the Zariski topology, and thus becomes an algebraic variety in its own right. In this situation, when a J-equivariant holomorphic line bundle on S has an algebraic structure, one may ask if its $(\mathfrak{g}, J)$-equivariant extensions can be defined as algebraic line bundles on Zariski neighborhoods of S in X. Simple examples show that this is asking for too much. Note, however, that the weight lattice Λ generates $\mathfrak{h}^*$ as complex vector space, and that consequently every germ $\boldsymbol{L}_{\lambda,\mathfrak{b}}$ is a tensor product of fractional powers of germs, at $\mathfrak{b}$, of globally defined algebraic $G_{\mathbb{C}}$-equivariant line bundles. Every $\mathfrak{g}$-equivariant line bundle $\boldsymbol{L}$ over an open (in the Hausdorff topology) subset U of X can therefore be defined by system of transition functions $f_{\alpha,\beta}$, relative to some open cover $\{U_\alpha\}$ of U, such that the logarithmic derivatives of the $f_{\alpha,\beta}$ are algebraic functions. In particular, $\boldsymbol{L}$ restricts to an algebraic line bundle on any (Zariski) locally closed algebraic subspace $S \subset X$, provided S is entirely contained in U. By definition, the formal neighborhood of S in X is the inverse limit of the jet bundles of the embedding—equivalently, of the quotients $\mathcal{O}_X/(\mathcal{I}_S)^n$, where $\mathcal{I}_S$ denotes the ideal sheaf of S. With $\boldsymbol{L}$, U, and S as before, the tensor product over $\mathcal{O}_X$ of $\mathcal{O}(\boldsymbol{L})$ with these quotients are sheaves of holomorphic sections of algebraic vector bundles on S; the reason is the same as that for the algebraic structure of $\boldsymbol{L}$ on S. I summarize this state of affairs as follows:

(2.5) Lemma. *Let $S \subset X$ be a locally closed algebraic subspace, and $\boldsymbol{L} \to U$ a $\mathfrak{g}$-equivariant holomorphic line bundle, defined on a (Hausdorff) open subset $U \subset X$ which contains S. Then the restriction of $\boldsymbol{L}$ to the formal neighborhood of S in X has a canonical algebraic structure.*

I write $\mathcal{D}_X$ for the sheaf of linear differential operators on X, with holomorphic coefficients. This is a sheaf of algebras with unit, over the sheaf of rings $\mathcal{O}_X$. If $\boldsymbol{L} \to U$ is a holomorphic line bundle over an open subset $U \subset X$, one can "twist $\mathcal{D}_X$ by $\boldsymbol{L}$": $\mathcal{D}(\boldsymbol{L}) = \mathcal{O}(\boldsymbol{L}) \otimes_{\mathcal{O}_X} \mathcal{D}_X \otimes_{\mathcal{O}_X} \mathcal{O}(\boldsymbol{L}^{-1})$ is the sheaf of differential operators which act on holomorphic sections of $\boldsymbol{L}$. By definition, a twisted sheaf of differential operators, or TDO for short, is a sheaf of algebras with unit, over $\mathcal{O}_X$, which is locally isomorphic (as sheaf of $\mathcal{O}_X$-algebras with unit) to $\mathcal{D}_X$. Since line bundles are locally trivial, $\mathcal{D}(\boldsymbol{L})$ is a particular example of TDO. When local trivializations of a line bundle $\boldsymbol{L}$ are related by transition functions $f_{\alpha,\beta}$, the corresponding local trivializations $\mathcal{D}(\boldsymbol{L}) \cong \mathcal{D}_X$ are related by "conjugation by $f_{\alpha,\beta}$", which involves the logarithmic derivatives of $f_{\alpha,\beta}$, but not $f_{\alpha,\beta}$ itself. Thus $\mathcal{D}(\boldsymbol{L}^c)$ is well defined for any holomorphic line bundle $\boldsymbol{L}$ and any complex constant c; all such twists of $\mathcal{D}_X$ by fractional powers of line bundles are again TDO's. Slightly more generally, one can twist $\mathcal{D}_X$ by any tensor product of fractional powers of line bundles.

The locally defined $\mathfrak{g}$-equivariant line bundles $\boldsymbol{L}_\lambda$, with $\lambda \in \mathfrak{h}^*$, were obtained as tensor products of fractional powers of globally defined line bundles. Thus $\mathcal{D}(\boldsymbol{L}_\lambda)$ is a TDO, defined on all of X, for any $\lambda \in \mathfrak{h}^*$, even though $\boldsymbol{L}_\lambda$ itself has global meaning only if the parameter λ lies in the weight lattice $\tilde{\Lambda}$ of the universal covering group $\tilde{G}_{\mathbb{C}}$. These globally defined line bundles $\boldsymbol{L}_\lambda$, $\lambda \in \tilde{\Lambda}$, are algebraic. Since logarithmic derivatives of algebraic functions are again algebraic, it follows that all the $\mathcal{D}(\boldsymbol{L}_\lambda)$, for $\lambda \in \mathfrak{h}^*$, have algebraic structures. The definition of $\mathcal{D}_X$ and the notion of TDO make sense equally in the algebraic category. By what was just said, there exists a TDO in the algebraic category, whose image in the holomorphic category is $\mathcal{D}(\boldsymbol{L}_\lambda)$; this algebraic TDO will be denoted by $\mathcal{D}_{\lambda+\rho}$. The shift by ρ has the usual reason: it makes the labeling compatible with the Harish-Chandra homomorphism.

Conjugation of linear differential operators by invertible functions preserves the degree, hence TDO's have well-defined degree filtrations. The filtrants are locally free $\mathcal{O}_X$-modules of finite rank. It is therefore clear what one should mean by a $G_{\mathbb{C}}$- or $\mathfrak{g}$-action, either in the holomorphic or in the algebraic context. A $G_{\mathbb{C}}$-equivariant structure on a TDO is an action of $G_{\mathbb{C}}$ which preserves multiplication and lies over the translation action on X; a $\mathfrak{g}$-equivariant structure is the infinitesimal analogue of this notion. Every $G_{\mathbb{C}}$-equivariant line bundle $\boldsymbol{L} \to X$ induces a $G_{\mathbb{C}}$-equivariant structure not only on $\mathcal{D}(\boldsymbol{L})$, but even on the twists $\mathcal{D}(\boldsymbol{L}^c)$ by fractional powers of $\boldsymbol{L}$, again because conjugation of a differential operator by a non-zero function f depends only on the logarithmic derivative of f. In particular, all the algebraic TDO's $\mathcal{D}_\lambda$, for $\lambda \in \mathfrak{h}^*$, are $G_{\mathbb{C}}$-equivariant in a natural manner.

Since TDO's are locally trivial, they can be described by a system of transition functions. The logarithmic derivatives of any such system of transition functions define a cohomology class in $H^1(X, \mathcal{Z}^1_X)$, which determines the TDO in question up to isomorphism, with $\mathcal{Z}^1_X$ = sheaf of closed 1-forms on X. Both $\mathcal{Z}^1_X$ and the notion of cohomology are to be taken in the holomorphic or algebraic sense, depending on the setting. This parametrization of TDO's is entirely analogous to the parametrization of line bundles by $H^1(X, \mathcal{O}^*_X)$. If $S \subset X$ is a smooth subvariety, $H^1(X, \mathcal{Z}^1_X)$ maps, by restriction, to $H^1(S, \mathcal{Z}^1_S)$. If the cohomology class of a TDO $\mathcal{D}$ on X restricts to the cohomology class of some TDO $\mathcal{D}_S$ on S, one says that $\mathcal{D}_S$ is obtained from $\mathcal{D}$ by restriction, as sheaf of differential operators. This differs, of course, from the sheaf theoretic restriction of $\mathcal{D}$ to S.

§3 Standard modules attached to $K_{\mathbb{C}}$-orbits

In this section X will always be viewed as complex algebraic variety, sheaves will be defined with respect to the Zariski topology, and $\mathcal{O} = \mathcal{O}_X$ will stand for the sheaf of algebraic functions. I fix a particular linear function $\lambda \in \mathfrak{h}^*$. Let $Z(\mathfrak{g})$ denote the center of the universal enveloping algebra $U(\mathfrak{g})$, and $I_\lambda \subset Z(\mathfrak{g})$ the maximal ideal which corresponds to λ via the Harish-Chandra isomorphism, i.e., the annihilator in $Z(\mathfrak{g})$ of the Verma module with highest weight $\lambda - \rho$. Both I_λ and the associative algebra

$$U_\lambda = U(\mathfrak{g})/U(\mathfrak{g})I_\lambda \tag{3.1}$$

depend on the Weyl group orbit of λ in $\mathfrak{h}^*$, rather than on λ itself.

The derivative of the $G_{\mathbb{C}}$-action on the TDO $\mathcal{D}_\lambda$ defines a Lie algebra homomorphism of $\mathfrak{g}$ into the space of global sections $\Gamma(\mathcal{D}_\lambda)$, hence an algebra homomorphism $\tilde{h} : U(\mathfrak{g}) \to \Gamma(\mathcal{D}_\lambda)$. One can show that $\tilde{h}$ annihilates I_λ, so $\tilde{h}$ induces a morphism

$$h : U_\lambda \to \Gamma(\mathcal{D}_\lambda)\,. \tag{3.2}$$

The following result of Beilinson-Bernstein is the starting point of their construction:

3.3 Theorem [1,13]. *h is an isomorphism.*

A sheaf of D_λ-modules is said to be quasi-coherent if, on some neighborhood of every point, it has a presentation by free sheaves, not necessarily of finite rank. Let $M(\mathcal{D}_\lambda)$ denote the category of sheaves of quasi-coherent $\mathcal{D}_\lambda$-modules, and $M(U_\lambda)$ the category of U_λ-modules—equivalently, the category of $U(\mathfrak{g})$-modules with infinitesimal character χ_λ, using Harish-Chandra's notation. Because of (3.3), it makes sense to tensor U_λ-modules over U_λ with $\mathcal{D}_\lambda$; the resulting sheaves are quasi-coherent, since modules over a ring have presentations by free modules, and tensoring is right exact. Thus tensoring defines a right exact functor between the two categories,

$$\Delta : M(U_\lambda) \to M(\mathcal{D}_\lambda)\,, \quad \Delta V = \mathcal{D}_\lambda \otimes_{U_\lambda} V, \tag{3.4}$$

which Beilinson-Bernstein call the localization functor. The global sections of any $\mathcal{S} \in M(\mathcal{D}_\lambda)$ constitute a module over $\Gamma\mathcal{D}_\lambda \cong U_\lambda$, so

$$\Gamma : M(\mathcal{D}_\lambda) \to M(U_\lambda) \tag{3.5}$$

is a left exact functor between the two categories, in the opposite direction.

I shall call λ "integrally dominant" if $2\frac{(\lambda,\alpha)}{(\alpha,\alpha)} \notin \{-1,-2,-3,\dots\}$ for every positive root α; evidently this is a much weaker notion than dominance of the real part[4] of λ. In particular, every $\lambda \in \mathfrak{h}^*$ is Weyl conjugate to an integrally dominant λ. As usual, λ will be called regular if $(\lambda,\alpha) \neq 0$ for all roots α. The next statement, formally analogous to Cartan's theorems A and B, provides the key to Beilinson-Bernstein's $\mathcal{D}$-module construction of representations:

(3.6) Theorem [1,13]. (A) *If $\lambda \in \mathfrak{h}^*$ is regular and integrally dominant, the global sections of any $\mathcal{S} \in M(\mathcal{D}_\lambda)$ generate all of its stalks.* (B) *If $\lambda \in \mathfrak{h}^*$ is integrally dominant and $p > 0$, $H^p(X,\mathcal{S}) = 0$ for every $\mathcal{S} \in M(\mathcal{D}_\lambda)$.*

As a formal consequence of the theorem, Beilinson-Bernstein deduce:

(3.7) Corollary [1,13]. *For λ regular and integrally dominant, Γ defines an equivalence of categories $M(\mathcal{D}_\lambda) \cong M(U_\lambda)$, with inverse Δ.*

Recall that U_λ depends only on the Weyl group orbit of λ. Thus, when λ is regular, the category $M(U_\lambda)$ can be identified with at least one of the categories $M(\mathcal{D}_\mu)$, for μ in the orbit of λ - possibly more than one because the Weyl group orbit of λ may contain several integrally dominant members. On the level of derived categories, all the $M(\mathcal{D}_\mu)$ parametrized by Weyl conjugates μ of λ become equivalent, again under the assumption of regularity [2]. The situation is more subtle for singular λ, as ought to be expected: $U(\mathfrak{g})$-modules with singular infinitesimal character tend to be more complicated. When λ is singular, there exist sheaves $\mathcal{S} \in$

[4] relative to the real structure on $\mathfrak{h}$ which makes all weights real.

$M(\mathcal{S}_\lambda)$ without cohomology in any degree. It is possible to define a quotient category of $M(D_\lambda)$ in which sheaves without cohomology are set equal to zero. That quotient category is equivalent to $M(U_\lambda)$, provided λ is integrally dominant.

The equivalence of categories persists when additional structures are added on both sides in a compatible fashion. Let $M(U_\lambda, K)$ denote the subcategory of all $(\mathfrak{g}, K)$-modules in $M(U_\lambda)$, and $M(\mathcal{D}_\lambda, K_\mathbb{C})$ the category of $K_\mathbb{C}$-equivariant sheaves in $M(\mathcal{D}_\lambda)$. The functors Γ, Δ relate compatible K-actions on U_λ-modules to $K_\mathbb{C}$-actions, compatible with translation on X, on sheaves of $\mathcal{D}_\lambda$-modules. Thus:

(3.8) $M(U_\lambda, K) \cong M(\mathcal{D}_\lambda, K_\mathbb{C})$ if λ is regular and integrally dominant.

The irreducible objects in $M(U_\lambda, K)$ are precisely the irreducible Harish-Chandra modules with infinitesimal character χ_λ (Harish-Chandra's notation). In view of (3.8), the global sections of any irreducible, $K_\mathbb{C}$-equivariant sheaf of $\mathcal{D}_\lambda$-modules form an irreducible Harish-Chandra module if λ is regular and integrally dominant; every irreducible Harish-Chandra module with infinitesimal character χ_λ arises in this manner. When λ is integrally dominant but singular, irreducible Harish-Chandra modules with infinitesimal character χ_λ can still be realized as spaces of global sections of irreducible sheaves in $M(\mathcal{D}_\lambda, K_\mathbb{C})$; the converse fails, because certain irreducible sheaves have no global sections.

By $K_\mathbb{C}$-equivariance, any $\mathcal{S} \in M(\mathcal{D}_\lambda, K_\mathbb{C})$ is supported on a union of $K_\mathbb{C}$-orbits—necessarily on the closure of a single orbit Q if $\mathcal{S}$ is irreducible. A theorem of Kashiwara [3] asserts that the $\mathcal{D}$-module construction of direct image induces an equivalence between the category of sheaves of $\mathcal{D}_S$-modules on a smooth subvariety $\mathcal{S}$ and the category of sheaves of $\mathcal{D}$-modules on the ambient space, with support in S. Applied to $S = Q$, which is smoothly embedded in $X - \partial X$, this fact implies that every irreducible $\mathcal{S} \in M(D_\lambda, K_\mathbb{C})$ can be obtained as the unique irreducible subsheaf (of $\mathcal{D}_\lambda$-modules) in the direct image of an irreducible, $K_\mathbb{C}$-equivariant sheaf of $\mathcal{D}_\lambda|_Q$-modules, on a uniquely determined $K_\mathbb{C}$-orbit Q; here $\mathcal{D}_\lambda|_Q$ is the restricted sheaf of differential operators, as described at the end of §2. I shall not go into details of this argument, but simply recall the final result.

I keep fixed a particular $\lambda \in \mathfrak{h}^*$ (without any assumption of dominance or regularity), a $K_\mathbb{C}$-orbit $Q \subset X$, and a $K_\mathbb{C}$-equivariant, algebraic line bundle $\boldsymbol{L} \to Q$. The data λ, Q, $\boldsymbol{L}$ must satisfy a compatibility condition: let B denote the isotropy subgroup of $G_\mathbb{C}$ at some $\mathfrak{b} \in Q$, χ the character by which $B \cap K_\mathbb{C}$ acts on the fibre of $\boldsymbol{L}$ at $\mathfrak{b}$; then

(3.9) the differential of χ agrees with $\lambda - \rho$ on $\mathfrak{b} \cap \mathfrak{k}$.

Here $\lambda - \rho$ is viewed as linear function of $\mathfrak{b}$ via $\mathfrak{h}^* \cong (\mathfrak{b}/[\mathfrak{b},\mathfrak{b}])^* \cong$ annihilator of $[\mathfrak{b},\mathfrak{b}]$ in $\mathfrak{b}^*$. To emphasize this hypothesis—which depends only on λ, Q,

$\boldsymbol{L}$, not on the choice of $\mathfrak{b} \in Q$—I write $\boldsymbol{L}_\chi$ from now on. From the definition of the TDO $\mathcal{D}_\lambda$, one sees that (3.9) amounts to a compatibility condition on $\mathcal{D}_\lambda$ and $\mathcal{D}_Q(\boldsymbol{L}_\chi)$ = sheaf of differential operators on Q, twisted by $\boldsymbol{L}_\chi$:

$$\text{(3.10)} \qquad \mathcal{D}_Q(\boldsymbol{L}_\chi) = \text{ restriction of } \mathcal{D}_\lambda \text{ to } Q$$

(restriction as sheaf of differential operators). This is precisely the situation in which the $\mathcal{D}$-module direct image of $\mathcal{O}_Q(\boldsymbol{L}_\chi)$ under the inclusion $j:Q \hookrightarrow x$ becomes a sheaf of $\mathcal{D}_\lambda$-modules on X. The group $K_{\mathbb{C}}$ acts on $\boldsymbol{L}_\chi$, hence on the direct image $j_+\mathcal{O}_Q(\boldsymbol{L}_\chi)$; I shall call

$$\text{(3.11)} \qquad j_+\mathcal{O}_Q(\boldsymbol{L}_\chi) \in M(\mathcal{D}_\lambda, K_{\mathbb{C}})$$

the "standard Beilinson-Bernstein sheaf attached to the data λ, Q, $\boldsymbol{L}_\chi$". Kashiwara's theorem and basic results on $\mathcal{D}$-modules imply:

(3.12) Proposition [1,13]. (a) *$j_+\mathcal{O}_Q(\boldsymbol{L}_\chi)$ is a holonomic sheaf of $\mathcal{D}_\lambda$-modules, and thus has a finite composition series;* (b) *$j_+\mathcal{O}_Q(\boldsymbol{L}_\chi)$ contains a unique irreducible subsheaf (of $\mathcal{D}_\lambda$-modules);* (c) *every irreducible sheaf in $M(\mathcal{D}_\lambda, K_{\mathbb{C}})$ arises in this fashion, from uniquely determined data Q, $\boldsymbol{L}_\chi$.*

(3.13) Corollary.

(a) *The cohomology groups $H^p(X, j_+\mathcal{O}_Q(\boldsymbol{L}_\chi))$ are Harish-Chandra modules with infinitesimal character χ_λ;*
(b) *$H^p(X, j_+\mathcal{O}_Q(\boldsymbol{L}_\chi)) = 0$ for $p > 0$ if λ is integrally dominant;*
(c) *$H^0(X, j_+\mathcal{O}_Q(\boldsymbol{L}_\chi))$ contains a unique irreducible Harish-Chandra submodule if λ is regular and integrally dominant.*

Note that (b), (c) follow directly from (3.6-8) and (3.12), as does (a) when λ satisfies the hypotheses of regularity and integral dominance. This also proves (a) for other choices of λ, either by general cohomological arguments, or by the technique of tensoring with finite dimensional representations which, on the geometric side, amounts to tensoring with an algebraically (but not equivariantly) trivial vector bundle.

In effect, corollary (3.13) is Beilinson-Bernstein's classification of irreducible Harish-Chandra modules: they can be realized as unique irreducible submodules of the modules

$$H^0(X, j_+\mathcal{O}_Q(\boldsymbol{L}_\chi)),$$

with λ integrally dominant—not always with uniquely determined data λ, Q, $\boldsymbol{L}_\chi$, since λ may have other integrally dominant Weyl conjugates, but this failure of uniqueness is easy to control. For singular λ, the unique irreducible subsheaf of $j_+\mathcal{O}_Q(L_\chi)$ may not have any sections; when this happens, $H^0(X, j_+\mathcal{O}_Q(\boldsymbol{L}_\chi))$ does not contribute to the classification. It is possible to pin down these extraneous groups $H^0(X, j_+\mathcal{O}_Q(\boldsymbol{L}_\chi))$; more on that in §6.

Before giving an alternate description of the standard Beilinson-Bernstein modules $H^0(X, j_+\mathcal{O}_Q(\boldsymbol{L}_X))$, I digress briefly on the notion of local cohomology. For the moment, X will denote the underlying topological space of either a complex manifold, in its Hausdorff topology, or of a complex algebraic variety, equipped with the Zariski topology. Let Q be a locally closed subspace of X, and $U \subset X$ an open subset containing Q, such that Q is closed in U. The functor Γ_Q : {sheaves of abelian groups} $\to$ {abelian groups}, which assigns to any sheaf $\mathcal{S}$ on X the space of all global sections over U, with support in Q, does not depend on the particular choice of U. It is left exact, and thus has right derived functors $R^p\Gamma_Q$. These, by definition, are the local cohomology groups along Q; notation: $H^p_Q(X, \bullet)$. They can be computed by applying Γ_Q to any flabby resolution of the sheaf in question, since Γ_Q is exact on flabby sheaves. One can also compute local cohomology from sheaf valued chains with respect to a relative covering of the pair $(U, U - Q)$ by Stein, respectively affine, subsets, in analogy to the Čech characterization of sheaf cohomology [9]. There exists a sheaf version of local cohomology: the family of sheaves $\mathcal{H}^p_Q(\cdot)$ associated to the presheaves $V \mapsto H^p_{Q\cap V}(V, \cdot)$. If Q is closed in X, $H^*_Q(X, \cdot)$ is the abutment of spectral sequence with E_2-term $H^p(X, \mathcal{H}^p_Q(\cdot))$. In particular, again for closed Q, if the local cohomology sheaves $\mathcal{H}^d_Q(\mathcal{S})$ of some sheaf $\mathcal{S}$ vanish in all but one degree d, then$H^p_Q(X, \mathcal{S}) = H^{p-d}(X, \mathcal{H}^d_Q(\mathcal{S}))$.

Now I suppose more specifically that X is a smooth complex algebraic variety, and Q a smooth closed subvariety of codimension d. The description of local cohomology in terms of relative affine open covers shows first of all

$$\mathcal{H}^p_Q(\mathcal{O}_X) = 0 \quad \text{if } p \neq d. \tag{3.14a}$$

The local cohomology sheaves vanish outside of Q, as is clear from their definition. Near any given point $x \in Q$, I can choose local algebraic coordinates $z_1, \ldots, z_{n-d}$, $w_1, \ldots, w_d$ on X, centered at x, such that Q is the intersection of the d divisors $D_j = \{w_j = 0\}$. Then

(3.14b) stalk of $\mathcal{H}^d_Q(\mathcal{O}_X)$ at $x \cong$ space of finite Laurent series in the w_j, with algebraic functions of the z_i as coefficients, modulo linear combinations of Laurent series which involve negative powers of at most $d-1$ of the w_j,

again by computation with relative open covers. At fist glance, this quotient of spaces of Laurent series appears to have no invariant meaning. However, its members can be paired naturally with forms of top degree on X, into forms of top degree on Q, by taking residues along the divisors D_j; this pairing makes it possible to see the effect of a change of variables. Note that differential operators with algebraic coefficients act on Laurent series, so

(3.15) $\mathcal{H}^d_Q(\mathcal{O}_X)$ is a sheaf of $\mathcal{D}_X$- modules.

As such, it coincides with the $\mathcal{D}$-module direct image of $\mathcal{O}_Q$ via the inclusion $j : Q \hookrightarrow X$,

$$\mathcal{H}^d_Q(\mathcal{O}_X) \cong j_+(\mathcal{O}_Q), \tag{3.16}$$

as follows from the definition of $\mathcal{D}$-module direct image. The assertions (3.14-16) imply also the analogous statements about the sheaves of sections, on X and Q respectively, of any algebraic vector bundle.

I apply this discussion to the flag variety X, considered as complex algebraic variety, and the inclusion j of a $K_{\mathbb{C}}$-orbit Q. Since Q need not be closed in X, I take the direct image in two steps: first under the closed embedding of Q in the complement of its boundary ∂Q, then under the inclusion of the open set $X - \partial Q$ in X. For the first step, I appeal directly to (3.16). For inclusion of open subsets, the operation of $\mathcal{D}$-module direct image agrees with direct image in the sense of sheaves; also, the local cohomology sheaves $\mathcal{H}^*_Q(\bullet)$, viewed as sheaves on X, are simply the direct image sheaves of the $\mathcal{H}^*_Q(\bullet)$, now viewed as sheaves on $X - \partial Q$. Implicit in these assertions is the vanishing of the higher direct images—a consequence of the following result of Beilinson-Bernstein:

(3.17) $\qquad K_{\mathbb{C}}$ – orbits in the flag variety are affinely embedded

[5]. Thus (3.16) remains valid even though Q may fail to be closed in X.

If λ and χ are related by (3.9), the $K_{\mathbb{C}}$-equivariant algebraic line bundle $\boldsymbol{L}_\chi \to Q$ extends to a $(\mathfrak{g}, K_{\mathbb{C}})$-equivariant holomorphic line bundle $\boldsymbol{L}_{\chi,\lambda-\rho}$ over the germ of a Hausdorff neighborhood of Q in X, as described by lemma (2.4): at each $\mathfrak{b} \in Q$, the germ of $\boldsymbol{L}_{\chi,\lambda-\rho}$ coincides with $\boldsymbol{L}_{\lambda-\rho,\mathfrak{b}}$, in the notation of §2. According to (2.5), the restriction of $\boldsymbol{L}_{\chi,\lambda-\rho}$ to the formal neighborhood of Q has a natural algebraic structure. Note that the local cohomology sheaves (3.14) are completely determined by the datum of the formal neighborhood of Q in X. Thus it makes sense to twist the isomorphism (3.16) by $\boldsymbol{L}_{\chi,\lambda-\rho}$, although this line bundle may not exist as algebraic object on any Zariski neighborhood of Q:

$$j_+\left(\mathcal{O}_Q\left(\boldsymbol{L}_\chi\right)\right) \cong \mathcal{H}^d_Q\left(\mathcal{O}\left(\boldsymbol{L}_{\chi,\lambda-\rho}\right)\right); \tag{3.18}$$

for the reason just mentioned, the notation $\mathcal{O}(\boldsymbol{L}_{\chi,\lambda-\rho})$ should not be taken literally, of course. It does not matter whether one takes local cohomology along Q on X or on $X - \partial Q$, as follows from the definition of local cohomology. Also, because of (3.17), it makes no difference whether the cohomology groups of $j_+(\mathcal{O}_Q(\boldsymbol{L}_\chi))$ are calculated over X or over $X - \partial Q$. Since Q is closed in $X - \partial Q$, I may use the spectral sequence of local cohomology, to conclude:

$$H^*\left(X, j_+\left(\mathcal{O}_Q\left(\boldsymbol{L}_\chi\right)\right)\right) \cong H^{*+d}_Q\left(X, \mathcal{O}\left(\boldsymbol{L}_{\chi,\lambda-\rho}\right)\right). \tag{3.19}$$

This is the description of the Beilinson-Bernstein modules in terms of local cohomology.

§4 Standard modules attached to G-orbits

Zuckerman's derived functor construction of Harish-Chandra modules, though completely algebraic in character, was originally motivated by geometric considerations. Without this motivation, it may seem artificial to think of the derived functor modules as being associated to G-orbits in the flag variety. I shall therefore digress briefly, before turning to the technical definition.

The data to which standard modules will be attached consist of a G-orbit S in the flag variety X and a G-equivariant C^ω (=real analytic) line bundle $\boldsymbol{L} \to S$. I choose a particular "base point" $\mathfrak{b} \in S$. Like every point of X, $\mathfrak{b}$ is stabilized by a Cartan subgroup $H \subset G$, which is unique up to conjugation by elements of the isotropy subgroup of G at $\mathfrak{b}$. The homogeneous space G/H maps equivariantly onto S, and thus $\boldsymbol{L}$ pulls back to a G-equivariant C^ω line bundle on G/H; for simplicity, the pullback to G/H will be denoted by the same symbol $\boldsymbol{L}$. The complexified Lie algebra of H is a Cartan subalgebra of $\mathfrak{g}$, contained in the Borel subalgebra $\mathfrak{b}$; as such, it is canonically isomorphic to the abstract Cartan algebra $\mathfrak{h}$ (cf. the discussion in §2). I identify the quotient $\mathfrak{g}/\mathfrak{h}$ with the fibre of the complexified tangent bundle of G/H at the identity coset. By translation, the image of $\mathfrak{b}$ in $\mathfrak{g}/\mathfrak{h}$ defines a G-equivariant subbundle $\boldsymbol{I}_\mathfrak{b}$ of the complexified tangent bundle. Its rank is half the (real) dimension of G/H. The fact that $\mathfrak{b}$ is a Lie algebra translates into the statement that $\boldsymbol{I}_\mathfrak{b}$ is involutive, i.e., closed under Poisson bracket. A different choice of base point $\mathfrak{b} \in S$ would have led to the same triple $(G/H, \boldsymbol{L}, \boldsymbol{I}_\mathfrak{b})$, up to isomorphism; conversely, the orbit S and line bundle $\boldsymbol{L} \to S$ can be reconstructed from the triple. In this sense, pairs $(S, \boldsymbol{L})$ and triples $(G/H, \boldsymbol{L}, \boldsymbol{I}_\mathfrak{b})$, modulo the appropriate notion of conjugacy, are equivalent sets of data.

The mechanism of geometric quantization associates a G-invariant complex of $\boldsymbol{L}$-valued differential forms to $(G/H, \boldsymbol{L}, \boldsymbol{I}_\mathfrak{b})$, whose length equals half the dimension of G/H. Pulled back to G, the complex takes the form

$$\{C^\infty(G) \otimes L \otimes \wedge^\bullet(\mathfrak{b}/\mathfrak{h})^*\}^H \, , d_{\mathfrak{b},H}. \tag{4.1}$$

Here L stands for the fibre of $\boldsymbol{L}$ at the identity coset, viewed as $(\mathfrak{b}, H)$-module with trivial $[\mathfrak{b},\mathfrak{b}]$-action; $C^\infty(G)$ is regarded as $(\mathfrak{b}, H)$-module by right translation; the superscript H refers to the space of H-invariants in $\{\dots\}$; and $d_{\mathfrak{b},H}$ is the coboundary operator of relative Lie algebra cohomology for $(\mathfrak{b}, H)$, which commutes with the left translation action of G on $C^\infty(G)$. The complex (4.1) can be tied even more directly to the orbit S: its sheaf version, pushed forward to S by the projection $G/H \to S$, is quasi-isomorphic to the $\bar{\partial}_\mathfrak{b}$ complex on S, with C^∞ coefficients and values in $\boldsymbol{L}$ [17,18].

Kostant proposed study of the complex (4.1) in 1966 [10]. If the cohomology groups are to have representation theoretic significance, one must establish the closed range property for the coboundary operator. Zuckerman had the idea of circumventing this subtle analytic problem by going to an algebraic analogue of the geometric complex, which produces Harish-Chandra modules, rather than representations of G.

According to Matsuki [12], the base point $\mathfrak{b} \in S$ and the Cartan subgroup $H \subset G$ that stabilizes $\mathfrak{b}$ can be chosen so that the Cartan involution θ preserves H. Then

$$H = (H \cap K)A \quad \text{(direct product)}, \quad \text{with } A \cong \mathbb{R}^\ell \tag{4.2}$$

for some integer ℓ. Let $M(\mathfrak{g}, H \cap K)$ and $M(\mathfrak{g}, K)$ denote the categories of, respectively, $(\mathfrak{g}, H \cap K)$- and $(\mathfrak{g}, K)$-modules, and

$$\Gamma^K_{H\cap K} : M(\mathfrak{g}, H \cap K) \to M(\mathfrak{g}, K) \tag{4.3}$$

Zuckerman's functor, which assigns the largest $(\mathfrak{g}, K)$-submodule to any $(\mathfrak{g}, H \cap K)$-module. It is visibly left exact. Since $M(\mathfrak{g}, H \cap K)$ contains sufficiently many injectives, this functor has a family of right derived functors $R^p\Gamma^K_{H\cap K}$. As before, I view the fibre L of the line bundle $\boldsymbol{L}$ at the base point $\mathfrak{b}$ as $(\mathfrak{b}, H)$-module. The "produced module"

$$\text{pro}^{\mathfrak{g},H\cap K}_{\mathfrak{b},H\cap K}(\boldsymbol{L}) = (H \cap K)\text{-finite part of } \text{Hom}_{\mathfrak{b}}(U(\mathfrak{g}), L) \tag{4.4}$$

belongs to the category $M(\mathfrak{g}, H \cap K)$. By definition,

$$M^p(S, \boldsymbol{L}) = R^p\Gamma^k_{H\cap K}(\text{pro}^{\mathfrak{g},H\cap K}_{\mathfrak{b},H\cap K}(L)) \tag{4.5}$$

is the p-th Zuckerman module attached to $(S, \boldsymbol{L})$.

To justify the phrase "attached to $(S, \boldsymbol{L})$", I note that the module (4.4) has a natural resolution in the category $M(\mathfrak{g}, H \cap K)$, namely

$$0 \to \text{pro}^{\mathfrak{g},H\cap K}_{\mathfrak{b},H\cap K}(L) \to \text{pro}^{\mathfrak{g},H\cap K}_{\mathfrak{h},H\cap K}(L) \to \text{pro}^{\mathfrak{g},H\cap K}_{\mathfrak{h},H\cap K}(L \otimes (\mathfrak{b}/\mathfrak{h})^*) \to \dots, \tag{4.6}$$

which has geometric meaning: its p-th term is the space of $\boldsymbol{L}$-valued p-forms in the complex (4.1), not with C^∞ coefficients, of course, but with formal power series coefficients at the identity coset. The resolving modules are injective in the category $M(\mathfrak{k}, H \cap K)$, hence $\Gamma^K_{H\cap K}$-acyclic.[5] Thus $M^p(S, \boldsymbol{L})$ is the p-th cohomology group of

$$\Gamma^K_{H\cap K}\left\{\text{pro}^{\mathfrak{g},H\cap K}_{\mathfrak{h},H\cap K}(L \otimes \wedge^\bullet(\mathfrak{g}/\mathfrak{h})^*)\right\}, \tag{4.7}$$

[5] Injectivity in $M(\mathfrak{k}, H \cap K)$ follows from the fact that $U(\mathfrak{g})$ is free over $U(\mathfrak{a}) \otimes U(\mathfrak{k})$, by Poincaré-Birkhoff-Witt, as left $\mathfrak{a}$- and right $\mathfrak{k}$-module [5]; injectivity in $M(\mathfrak{k}, H \cap K)$ implies acyclicity with respect to the functor (4.3): by its very definition, this functor, followed by the forgetful functor from $M(\mathfrak{g}, K)$ to $M(\mathfrak{k}, K)$, is sensitive only to the $(\mathfrak{k}, H \cap K)$-structure of any module to which it is applied.

the complex of K-finite $\boldsymbol{L}$-valued forms at the identity coset in G/H with formal power series coefficients.

An aside for the sake of completeness: it now appears that the coboundary operator of the C^∞ complex (4.1) does not have closed range in general. Instead one should take the hyperfunction analogue of (4.1), whose cohomology groups are Fréchet globalizations of the Zuckerman modules; alternatively, these cohomology groups can be realized as the local cohomology along S of a $\mathfrak{g}$-equivariant holomorphic extension of $\boldsymbol{L}$ to a neighborhood of S in X [17,18]. This again connects the Zuckerman modules $M^p(S,\boldsymbol{L})$ to the orbit S and the G-equivariant line bundle $\boldsymbol{L}$.

I recall that I have identified the universal Cartan algebra $\mathfrak{h}$ with the complexified Lie algebra of the concrete Cartan subgroup $H \subset G$. Since any complex linear function on $\mathfrak{h}$ is completely determined by its restriction to the (real) Lie algebra of H, lemma (2.4) guarantees that the G-equivariant C^ω line bundle $\boldsymbol{L} \to S$ extends uniquely to a $(\mathfrak{g}, G)$-equivariant holomorphic line bundle over the germ of a neighborhood of S in X. Let χ denote the character by which H acts on the fibre L of $\boldsymbol{L}$ at the identity coset and define $\lambda \in \mathfrak{h}^*$ by

$$\lambda = \text{(differential of } \chi) + \rho. \tag{4.8}$$

From now on I shall write $\boldsymbol{L}_\chi$ instead of $\boldsymbol{L}$, and I shall use the same symbol to refer also to the unique $(\mathfrak{g}, G)$-equivariant holomorphic extension.

The standard modules of the Langlands and Vogan-Zuckerman classifications occur among the modules $M^p(S,\boldsymbol{L}_\chi)$. To see this, I suppose first of all that the Cartan subgroup H splits over $\mathbb{R}$. In this situation, $\mathfrak{b}$ is defined over $\mathbb{R}$ and has a Borel subgroup MAN of G as normalizer; here $M = H \cap K$ is finite, $A =$ identity component of H, and N has complexified Lie algebra $[\mathfrak{b},\mathfrak{b}]$. Thus $S \cong G/MAN$ as homogeneous space for G. The complex (4.1) is now the relative de Rham complex of the fibration $G/H \to S \cong G/MAN$, and (4.7) is its analogue with K-finite formal power series coefficients. The Poincaré lemma remains valid in the setting of formal power series, hence

$$M^p(S,\boldsymbol{L}_\chi) = \begin{cases} K\text{-finite part of } \operatorname{Ind}_{MAN}^G(\chi) & \text{if } p = 0 \\ 0 & \text{if } p \neq 0 \end{cases} \tag{4.9}$$

(induction without normalization).

Next, I consider the opposite extreme, of a compact Cartan subgroup H. All roots are then imaginary, so $\mathfrak{b}$ and its complex conjugate $\bar{\mathfrak{b}}$ intersect exactly in $\mathfrak{h}$. In particular, G/H coincides with the orbit S, which is open in X, as can be seen by counting dimensions. Lemma (2.4) puts a uniquely determined G-invariant holomorphic structure on the G-equivariant line bundle $\boldsymbol{L}_\chi \to S$, and (4.1) can be identified with the Dolbeault complex of $\boldsymbol{L}_\chi$-valued forms on S. Whenever λ is anti-dominant and regular, the Dolbeault cohomology vanishes in all but one degree; in the remaining degree, it is a Frechét realization of the discrete series representation π_λ (Harish-Chandra's notation) [16]. The complex (4.7)—i.e., the complex of K-finite Dolbeault forms at the identity coset in $S \cong G/H$ with formal power series coefficients—computes the infinitesimal version of the Dolbeault cohomology:

$$
\begin{aligned}
(4.10) \qquad & M^p(S, \boldsymbol{L}_\chi) = 0 \quad \text{if } p \neq s =_{\text{def}} \frac{1}{2} \dim_{\mathbb{R}} K/H, \quad \text{and} \\
& M^s(S, \boldsymbol{L}_\chi) = \text{ Harish-Chandra module of } \pi_\lambda, \\
& \qquad \text{provided } \lambda \text{ is anti-dominant and regular}
\end{aligned}
$$

[20]. The vanishing statement in (4.10) persists if λ is singular but still anti-dominant; in that case, $M^s(S, \boldsymbol{L}_\lambda)$ either reduces to zero, or belongs to the "limits of the discrete series".

I continue to assume that the θ-stable Cartan subgroup $H \subset G$ stabilizes $\mathfrak{b} \in S$. The G-orbit S will be called "maximally real", or *mr* for short, if $\mathfrak{b} \cap \bar{\mathfrak{b}}$ has maximal possible dimension compared to all Borel subalgebras that are normalized by H. Similarly S is "maximally imaginary", or *mi*, if $\mathfrak{b} \cap \bar{\mathfrak{b}}$ has minimal possible dimension. This terminology is borrowed from geometric quantization: $\mathfrak{b}$ defines an invariant polarization $\boldsymbol{I}_{\mathfrak{b}}$ for G/H, as was described earlier; $\boldsymbol{I}_{\mathfrak{b}}$ is maximally real or maximally imaginary among all invariant polarizations precisely when the orbit S has that property. Note that S is simultaneously of type *mr* and of type *mi* if H splits over $\mathbb{R}$, or if H is compact.

Now I suppose that S is maximally real. The centralizer of A (recall (4.2)!) factors as MA, where M is θ-stable and contains $H \cap K$ as compact Cartan subgroup. Let N be the unipotent radical of the stabilizer of $\mathfrak{b}$ in G; then MA normalizes N, and MAN is the Langlands decomposition of a cuspidal parabolic subgroup P of G. In this situation $S \cong G/HN$, and hence S fibres G-equivariantly over G/P, which is a closed G-orbit in a generalized flag variety; the fibre $S_M \cong MA/H$ lies as open M-orbit in the flag variety of $\mathfrak{m}$ = complexified Lie algebra of M. Zuckerman's functor (4.3) satisfies "induction by stages", in the sense that

$$
(4.11) \qquad \Gamma^K_{H \cap K} = \Gamma^K_{M \cap K} \Gamma^{M \cap K}_{H \cap K},
$$

which leads to a Grothendieck spectral sequence for the higher derived functors. Thus, combining the arguments which led to (4.9) and (4.10), one finds

$$
\begin{aligned}
(4.12) \qquad & M^p(S, \boldsymbol{L}_\chi) = 0 \quad \text{if } p \neq s =_{\text{def}} \frac{1}{2} \dim_{\mathbb{R}} M \cap K/H \cap K, \quad \text{and} \\
& M^s(S, \boldsymbol{L}_\chi) = \operatorname{Ind}^G_P \left(M^s(S_M, \boldsymbol{L}_\chi|_{S_M}) \right), \text{ provided } \lambda \text{ is} \\
& \qquad \text{anti-dominant with respect to all imaginary roots.}
\end{aligned}
$$

Here induction refers to ordinary parabolic induction (without normalization) of Harish-Chandra modules, and $M^s(S_M, \boldsymbol{L}_\chi|_{S_M})$ denotes the Harish-Chandra module for MA associated to the M-orbit S_M and the MA-equivariant line bundle obtained by restricting $\boldsymbol{L}_\chi$ to $S_M \cong MA/H$. According to (4.10), applied to the M-orbit S_M, this inducing module belongs to the discrete series or the limits of the discrete series for M, or vanishes if λ happens to be singular with respect to a compact imaginary simple root. The modules (4.12), corresponding to the various maximally real G-orbits in X and to parameters λ which satisfy an additional negativity condition with respect to the restricted roots on A, are precisely the standard modules of Langlands' classification [11].

One can argue similarly if the orbit S is maximally imaginary. I write L for the centralizer of $H \cap K$ in G. Its complexified Lie algebra $\mathfrak{l}$ normalizes $\mathfrak{u} = [\mathfrak{b}, \mathfrak{b}] \cap \theta[\mathfrak{b}, \mathfrak{b}]$, and $\mathfrak{q} = \mathfrak{l} \oplus \mathfrak{u}$ is a θ-stable parabolic subalgebra of $\mathfrak{g}$. Now S fibres G-equivariantly over G/L, which embeds as open G-orbit in the generalized flag variety whose points are the Aut($\mathfrak{g}$)-conjugates of $\mathfrak{q}$. The fibre $S_L \cong L/H$ lies as closed L-orbit in the flag variety of $\mathfrak{l}$. Taken modulo the central subgroup $H \cap K$, L contains A as split Cartan subgroup. Thus (4.9) applies to the L-orbit S_L : $M^p(S_L, \boldsymbol{L}_\chi|_{S_L})$ vanishes in every non-zero degree p, and is a principal series module in degree zero. An application of the Grothendieck spectral sequence derived from (4.11), with L in place of M, shows:

$$(4.13) \qquad M^p\left(S, \boldsymbol{L}_\chi\right) = R^p\Gamma^K_{L\cap K}\left(\mathrm{pro}^{\mathfrak{g}, L\cap K}_{\mathfrak{q}, L\cap K}\left(M^0\left(S_L, \boldsymbol{L}_\chi|_{S_L}\right)\right)\right).$$

Under an appropriate negativity condition on λ, which will be described more closely in §6, these modules vanish in all but one degree s; in degree $p = s$, they are the standard modules that enter the Vogan-Zuckerman classification [20].

§5 The duality theorem

Both G and $K_\mathbb{C}$ contain K, so K operates on the intersection of any G-orbit S with any $K_\mathbb{C}$-orbit Q. The two orbits are dual to each other (in the sense of Matsuki) if the intersection $S \cap Q$ consists of exactly one K-orbit. According to Matsuki [12], this notion of duality establishes a bijection between the set of G-orbits and the set of $K_\mathbb{C}$-orbits; it is referred to as "duality" because it reverses the order relations on the two sets which are defined by containment of one orbit in the closure of another.

The duality also carries over to the geometric data that enter the constructions of representations attached to the orbits. Let S and Q be dual orbits for G and $K_\mathbb{C}$, $\boldsymbol{L}_S \to S$ a G-equivariant C^ω line bundle, $\boldsymbol{L}_Q$ a $(\mathfrak{g}, K_\mathbb{C})$-equivariant algebraic line bundle over the formal neighborhood of Q in X. The discussion in §2 shows that $\boldsymbol{L}_S$ extends uniquely to a $(\mathfrak{g}, G)$-equivariant holomorphic line bundle on the germ of a neighborhood of S; similarly $\boldsymbol{L}_Q$

extends uniquely to a $(\mathfrak{g}, K_{\mathbb{C}})$-equivariant holomorphic line bundle on the germ of a (Hausdorff) neighborhood of Q. Both extensions restrict to $(\mathfrak{g}, K)$-equivariant line bundles on the germ of a neighborhood of the K-orbit $S \cap Q$. I shall say that the duality between S and Q relates the bundles $\boldsymbol{L}_S$, L_Q if their extensions agree as $(\mathfrak{g}, K)$-equivariant holomorphic line bundles near $S \cap Q$.

(5.1) Lemma. *The preceding definition establishes a bijection between G-equivariant C^ω line bundles $\boldsymbol{L}_S \to S$ and $(\mathfrak{g}, K_{\mathbb{C}})$-equivariant algebraic line bundles $\boldsymbol{L}_Q$ on the formal neighborhood of Q.*

Proof. Let B denote the isotropy subgroups of $G_{\mathbb{C}}$ at some reference point $\mathfrak{b} \in S \cap Q$. Matsuki's proof of the duality shows that $\mathfrak{b}$ is normalized by a θ-stable Cartan subgroup $H \subset G$. Then $H = (H \cap K)A$, as in (4.2), with $A \cong \mathbb{R}^\ell$. I identify the complexified Lie algebra of H with the universal Cartan algebra $\mathfrak{h} \cong \mathfrak{b}/[\mathfrak{b}, \mathfrak{b}]$. On the $\mathbb{R}$-linear span of Φ in $\mathfrak{h}^*$, the Cartan involution θ is the negative of complex conjugation with respect to the real structure defined by G. This implies:

$$(5.2) \qquad \begin{aligned} &H, H \cap K, (H \cap K)_{\mathbb{C}} \text{ are Levi factors for,} \\ &\text{respectively}, G \cap B, K \cap B, \text{ and } K_{\mathbb{C}} \cap B; \end{aligned}$$

here $(H \cap K)_{\mathbb{C}}$ denotes the complexification of $H \cap K$. In particular χ, the character by which H acts on the fibre of $\boldsymbol{L}_S$ at $\mathfrak{b}$, completely determines the G-equivariant line bundle $\boldsymbol{L}_S$; thus $\boldsymbol{L}_S = \boldsymbol{L}_\chi$, in the notation of §4. I define $\lambda \in \mathfrak{h}^*$ by the identity (4.9). By (2.4), the holomorphic extension of $\boldsymbol{L}_\chi$ has germ $\boldsymbol{L}_{\lambda-\rho, \mathfrak{b}}$ at $\mathfrak{b}$. But then $\boldsymbol{L}_{\lambda-\rho, \mathfrak{b}}$ must also be the germ of $\boldsymbol{L}_Q$ at $\mathfrak{b}$ if $\boldsymbol{L}_Q$ is related to $\boldsymbol{L}_\chi$ by the duality between S and Q. Again by (2.4), the one remaining datum that pins down the $(\mathfrak{g}, K_{\mathbb{C}})$-equivariant structure of $\boldsymbol{L}_Q$ is the character ϕ by which $K_{\mathbb{C}} \cap B$ acts on the fibre at $\mathfrak{b}$. Like any holomorphic character of $K_{\mathbb{C}} \cap B$, ϕ is determined by its restriction to the Levi factor $(H \cap K)_{\mathbb{C}}$, and thus even by its restriction to the real form $H \cap K$ of $(H \cap K)_{\mathbb{C}}$. Since $\boldsymbol{L}_Q$ and $\boldsymbol{L}_\chi$ are assumed to agree as K-equivariant line bundles along $S \cap Q$, ϕ and χ must restrict to the same character on $H \cap K$. The compatibility condition on ϕ and λ which (2.4) requires is then automatically satisfied, by (4.9). Thus $\boldsymbol{L}_S = \boldsymbol{L}_\chi$ determines, and is determined by, $\boldsymbol{L}_Q$. □

For the statement of the duality theorem, I consider a pair of dual orbits S, Q, and line bundles $\boldsymbol{L}_S$, $\boldsymbol{L}_Q$ as above, which are related by the duality. From §3, I recall that the single datum of the $(\mathfrak{g}, K_{\mathbb{C}})$-equivariant line bundle $\boldsymbol{L}_Q$ is equivalent to the datum of its restriction to Q, together with a TDO $\mathcal{D}_\lambda$ whose restriction to Q acts on $\mathcal{O}_Q(\boldsymbol{L}_Q)$. This latter pair of data, in turn, associates the Beilinson-Bernstein modules $H^p(X, j_+(\mathcal{O}_Q(\boldsymbol{L}_Q)))$ to the orbit Q. I define

$$(5.3) \qquad s = \dim_{\mathbb{R}}(S \cap Q) - \dim_{\mathbb{C}} Q,$$

and $\boldsymbol{K}_X$ = canonical bundle of X.

(5.4) Theorem [5]. *There exists a natural* $(\mathfrak{g}, K)$*-invariant pairing*

$$H^p\left(X, j_+\left(\mathcal{O}_Q\left(\boldsymbol{L}_Q\right)\right)\right) \times M^{s-p}\left(S, \boldsymbol{K}_X \otimes \boldsymbol{L}_S^{-1}\right) \to \mathbb{C},$$

which exhibits the two factors as dual to each other in the category of Harish-Chandra modules.

The remainder of this section is devoted to a sketch of the proof of (5.4). I shall freely use the notation of the proof of (5.1). Let $\mathfrak{b}_{\mathrm{opp}}$ denote the Borel subalgebra which contains the complexified Lie algebra $\mathfrak{h}$ of H, but is opposed to $\mathfrak{b}$. The "variety of ordered Cartans",

$$(5.5)\qquad \begin{aligned} Y &= \{(\mathfrak{b}_1, \mathfrak{b}_2) \in X \times X \mid \mathfrak{b}_1 + \mathfrak{b}_2 = \mathfrak{g}\} \\ &= \text{open orbit of } G_{\mathbb{C}} \text{ acting diagonally in } X \times X, \end{aligned}$$

passes through the point $(\mathfrak{b}, \mathfrak{b}_{\mathrm{opp}})$, at which $G_{\mathbb{C}}$ has isotropy subgroup $H_{\mathbb{C}}$. Thus $Y \cong G_{\mathbb{C}}/H_{\mathbb{C}}$ is the quotient of two reductive algebraic groups, hence a smooth affine variety. The projection π of Y onto the first factor of $X \times X$ exhibits Y as fibre bundle over X,

$$(5.6)\qquad \begin{array}{ccc} Y & \cong & G_{\mathbb{C}}/H_{\mathbb{C}} \\ \downarrow{\scriptstyle\pi} & & \downarrow \\ X & \cong & G_{\mathbb{C}}/B \end{array}, \qquad \text{with fibre } \pi^{-1}(\mathfrak{b}) \cong B/H_{\mathbb{C}} \cong [\mathfrak{b}, \mathfrak{b}].$$

The crux of the proof of (5.4) is to reinterpret both ingredients of the pairing as cohomology groups of complexes on Y.

I let $\tilde{S}$ and $\tilde{Q}$ denote, respectively, the G-orbit and the $K_{\mathbb{C}}$-orbit of the identity coset in $Y \cong G_{\mathbb{C}}/H_{\mathbb{C}}$. Then $\tilde{S}$ lies over S, $\tilde{Q}$ lies over Q, $\tilde{S} \cong G/H$, and $\tilde{Q} \cong K_{\mathbb{C}}/(K \cap H)_{\mathbb{C}}$; cf. (5.2). Restricted to $\tilde{Q}$, π has fibre $K_{\mathbb{C}} \cap B / K_{\mathbb{C}} \cap H_{\mathbb{C}} \cong \mathfrak{k} \cap [\mathfrak{b}, \mathfrak{b}]$. A small calculation in the Lie algebra shows

$$(5.7)\qquad s = \dim(\mathfrak{k} \cap [\mathfrak{b}, \mathfrak{b}]).$$

Thus, with $n = \dim[\mathfrak{b}, \mathfrak{b}] = \dim X$,

$$(5.8)\qquad \begin{aligned} &\text{the fibres of} \quad \pi : \tilde{Q} \to Q \quad \text{have codimension } n - s \\ &\quad \text{in the fibres of} \quad \pi : Y \to X. \end{aligned}$$

As in §3, d will stand for the codimension of Q in X; because of (5.8), $\tilde{Q}$ has codimension $n + d - s$ in Y. More notation: $\boldsymbol{T}^*_{Y|X}$ is the relative cotangent bundle of the fibration (5.6)—i.e., the dual of the bundle of (holomorphic) tangent vectors along the fibres. The coboundary operator of the relative de Rham complex turns

$$(5.9)\qquad \mathcal{H}^{n+d-s}_{\tilde{Q}}\left(\mathcal{O}_Y\left(\pi^* \boldsymbol{L}_Q \otimes \wedge^{\bullet} \boldsymbol{T}^*_{Y|X}\right)\right)$$

into a complex of sheaves.

(5.10) Lemma. *The cohomology sheaves of the complex (5.9) vanish in all degrees other than $n-s$. The direct image via π of the cohomology sheaf in degree $n-s$ coincides with $\mathcal{H}^d_Q(\mathcal{O}_X(\boldsymbol{L}_Q))$, and its higher derived images are zero.*

Proof. The problem is local with respect to the base. Thus I may replace $\boldsymbol{L}_Q$ by the trivial line bundle. Let F be the fibre of (5.6) over a generic point of Q, $\Omega^\bullet_F$ the sheaf of algebraic differential forms on F. Since $F \cong \mathbb{C}^n$ and $F \cap \widetilde{Q} \cong \mathbb{C}^s$,

$$H^p\left(H^{n-s}_{F\cap Q}(F,\Omega^\bullet_F)\right) = \begin{cases} \mathbb{C} & \text{if } p = n-s, \\ 0 & \text{otherwise,} \end{cases}$$

as follows from a direct computation, using the explicit description (3.14) of local cohomology. This statement, with local cohomology for X along Q as coefficients, translates into the lemma. □

I combine lemma (5.10) with the Leray spectral sequence for the fibration $\pi : \widetilde{Q} \to Q$:

$$\begin{aligned}
H^p(X, j_+\mathcal{O}_Q(\boldsymbol{L}_Q)) &\cong H^p(X-\partial Q, j_+\mathcal{O}_Q(\boldsymbol{L}_Q)) && \text{because of (3.17)} \\
&\cong H^p\left(X-\partial Q, \mathcal{H}^d_Q(\mathcal{O}_X(\boldsymbol{L}_Q))\right) && \text{by (3.18)} \\
&\cong H^p\left(Q, \mathcal{H}^d_Q(\mathcal{O}_X(\boldsymbol{L}_Q))\right) && \mathcal{H}^d_Q(\cdot) \text{ supported on } Q \cup \partial Q \\
&\cong \mathbb{H}^{n+d-s}\left(\widetilde{Q}, \mathcal{H}^{n+p-s}_{\widetilde{Q}}\left(\mathcal{O}_Y\left(\pi^*\boldsymbol{L}_Q \otimes \wedge^\bullet \boldsymbol{T}^*_{Y|X}\right)\right)\right) \\
& && \text{by (5.10) and Leray;}
\end{aligned}$$

in the last line, $\mathbb{H}^{n+p-s}(\dots)$ refers to the hypercohomology of the complex of sheaves (5.9). But $\widetilde{Q} \cong K_{\mathbb{C}}/(K \cap H)_{\mathbb{C}}$ is a quotient of reductive algebraic groups, hence affine, so cohomology on $\widetilde{Q}$ vanishes in non-zero degrees. I conclude:

$$\text{(5.11)} \qquad \begin{aligned} &H^p(X, j_+(\mathcal{O}_X(\boldsymbol{L}_Q))) = (n-s+p)\text{-th cohomology group of} \\ &\quad \text{the complex } \; H^0\left(\widetilde{Q}, \mathcal{H}^{n+d-s}_{\widetilde{Q}}\left(\mathcal{O}_Y\left(\pi^*\boldsymbol{L}_Q \otimes \wedge^\bullet \boldsymbol{T}^*_{Y|X}\right)\right)\right). \end{aligned}$$

This is the description of the Beilinson-Bernstein modules that will make the duality visible.

I recall the interpretation of the complex (4.7)—which computes the Zuckerman modules $M^p(S,\boldsymbol{L})$—as a complex of $\boldsymbol{L}$-valued K-finite formal power series at the identity coset in the real analytic manifold G/H. According to the discussion below (5.6), this manifold can be identified with $\widetilde{S}$, and has Y as complexification. Hence (4.7) may be viewed as a complex of $K_{\mathbb{C}}$-finite, holomorphic formal power series at the identity coset in $Y \cong G_{\mathbb{C}}/H_{\mathbb{C}}$. $K_{\mathbb{C}}$-finite formal power series extend to $\widetilde{Q}$, the $K_{\mathbb{C}}$-orbit of the identity coset, as objects that are algebraic[6] along $\tilde{Q}$ but formal in directions normal to $\tilde{Q}$; in other words, to algebraic sections over the formal neighborhood of $\tilde{Q}$ in Y. The $H_{\mathbb{C}}$-module $\wedge^{\bullet}(\mathfrak{b},\mathfrak{h})^*$, which appears in the complex (4.7), is the fibre of $\wedge^{\bullet}\boldsymbol{T}^*_{Y|X}$ at the identity coset in $Y \cong G_{\mathbb{C}}/H_{\mathbb{C}}$, and the coboundary operator of (4.7) corresponds to the differential in the relative de Rham complex for $Y \to X$, since both are given by the same universal formula. Hence (4.7)—more precisely, the complex on $\tilde{Q}$ with which it has been identified—is the relative de Rham complex, over the formal neighborhood of $\tilde{Q}$. I apply these considerations with $\boldsymbol{K}_X \otimes \boldsymbol{L}_S^{-1}$ in place of $\boldsymbol{L}$:

(5.12) $M^p\left(S, \boldsymbol{K}_X \otimes \boldsymbol{L}_S^{-1}\right) =$ p-th cohomology of the relative de Rham complex over the formal neighborhood of $\tilde{Q}$, with values in $\pi^*(\boldsymbol{K}_X \otimes \boldsymbol{L}_Q^{-1})$.

On the right hand side of this equality, everything takes place in the algebraic context. To explain the switch from $\boldsymbol{L}_S$ to $\boldsymbol{L}_Q$, I note that the former is defined near S, so $\pi^*\boldsymbol{L}_S$ need not be defined near $\tilde{Q}$. In the passage from K-finite formal power series on $\tilde{S}$ to $K_{\mathbb{C}}$-finite algebraic section on the formal neighborhood of $\tilde{Q}$, the coefficient line bundle must also be extended $K_{\mathbb{C}}$-equivariantly. But $\boldsymbol{L}_S$ and $\boldsymbol{L}_Q$ agree as $(\mathfrak{g}, K)$-equivariant holomorphic line bundles near $S \cap Q$, and K is a real form of $K_{\mathbb{C}}$, hence $\pi^*\boldsymbol{L}_Q$ is the $K_{\mathbb{C}}$-equivariant extension of $\pi^*\boldsymbol{L}_S$ from the germ of a neighborhood of $\tilde{S} \cap \tilde{Q}$ to the formal neighborhood of $\tilde{Q}$.

The description (3.14) of the local cohomology sheaves exhibits them not only as sheaves of $\mathcal{O}$-modules, but even as sheaves of modules over the sheaf of algebraic functions on the formal neighborhood of the submanifold in question. This, coupled with contraction of $\pi^*\boldsymbol{L}_Q$ against $\pi^*\boldsymbol{L}_Q^{-1}$ and the wedge product pairing

$$\begin{aligned} \wedge^p \boldsymbol{T}^*_{Y|X} \times \pi^*\boldsymbol{K}_X \otimes \wedge^{n-p}\boldsymbol{T}^*_{Y|X} &\to \pi^*\boldsymbol{K}_X \otimes \wedge^n \boldsymbol{T}^*_{Y|X} \\ &= \wedge^n\left(\pi^*\boldsymbol{T}^*_X\right) \otimes \wedge^n \boldsymbol{T}^*_{Y|X} \cong \wedge^{2n}\boldsymbol{T}^*_Y = \boldsymbol{K}_Y, \end{aligned} \tag{5.13}$$

defines a natural pairing of the complexes in formulas (5.11) and (5.12) into $H^0(\tilde{Q}, \mathcal{H}^{n+d-s}_{\tilde{Q}}(\Omega^{2n}_Y))$. As was mentioned in §3, there exists a morphism

$$\mathcal{H}^{n+d-s}_{\tilde{Q}}\left(\Omega^{2n}_Y\right) \to \Omega^{n-d+s}_{\tilde{Q}} \qquad (n-d+s = \dim \tilde{Q}), \tag{5.14}$$

which, in terms of the description (3.14) of local cohomology, is given by taking residues along the divisors D_j. Algebraic forms of top degree on

[6] $K_{\mathbb{C}}$ acts algebraically and transitively on $\tilde{Q}$, so $K_{\mathbb{C}}$-finite holomorphic functions, section, etc. on $\tilde{Q}$ are automatically algebraic.

$\tilde{Q} \cong K_{\mathbb{C}}/(K \cap H)_{\mathbb{C}}$ can be integrated over the cycle $\tilde{Q} \cap \tilde{S} \cong K/K \cap H$, so $H^0(\tilde{Q}, \mathcal{H}^{n+d-s}_{\tilde{Q}}(\Omega^{2n}_Y))$ maps to $\mathbb{C}$ via (5.14) and integration. At this point, the complexes appearing in (5.11) and (5.12) pair into $\mathbb{C}$, which induces a pairing

$$H^p\left(X, j_+\left(\mathcal{O}_X\left(\boldsymbol{L}_Q\right)\right)\right) \times M^{s-p}\left(S, \boldsymbol{K}_X \otimes \boldsymbol{L}_S^{-1}\right) \to \mathbb{C}; \tag{5.15}$$

note that the shift of indices in (5.11), (5.13) accounts for the indexing in (5.15).

The pairing of complexes that induces (5.15) is put together from various natural intermediate pairings, is therefore natural itself, hence $(\mathfrak{g}, K)$-invariant. It remains to be shown that this pairing of complexes identifies each as the dual, in the $(\mathfrak{g}, K)$-category, of the other. Both complexes carry $K_{\mathbb{C}}$-invariant geometric filtrations: in one case, the increasing filtration induced by the total number of poles in the description (3.14) of local cohomology; in the other, the decreasing filtration defined by the order of vanishing along $\tilde{Q}$. As is clear from the definition, the pairing between the two complexes respects these filtrations. Thus it suffices to verify the non-degeneracy of the pairing on the graded level. The graded pieces of the complexes consist of spaces of algebraic sections of $K_{\mathbb{C}}$-equivariant vector bundles of finite rank, over $\tilde{Q}$; at this point, the non-degeneracy follows from a simple application of the (algebraic) Peter Weyl-theorem.

§6 Independence of the polarization

We saw in §4 that the standard modules of Langlands' classification of irreducible Harish-Chandra modules are attached to maximally real G-orbits, whereas the modules of the Vogan-Zuckerman classification correspond to maximally imaginary G-orbits. In each case, there are restrictions on the parameter λ which

(i) imply vanishing of the modules in all but one degree;
(ii) insure the existence of a unique irreducible submodule or quotient module;
(iii) make the resulting correspondence between geometric data and irreducible modules bijective.

I shall call a $K_{\mathbb{C}}$-orbit Q maximally real (or *mr* for short), respectively maximally imaginary (or *mi*), if the dual G-orbit S has this property. Then, by the duality theorem, the duals of the two classes of standard modules are attached to $K_{\mathbb{C}}$-orbits of type *mr* or *mi*. All $K_{\mathbb{C}}$-orbits, on the other hand, enter the Beilinson-Bernstein classification, though proportionately fewer parameters λ contribute to the classification for any one orbit.

All three classifications schemes classify the same irreducible Harish-Chandra modules, of course, but there is no obvious reason *a priori* why the standard modules in the three schemes must be the same (or dual to each other). It turns out that standard modules attached to neighboring orbits coincide if the parameters satisfy a simple condition—this is what I mean by "independence of the polarization". To explain the terminology, I recall that G-orbits in the flag variety encode the datum of an invariant polarization $I_{\mathfrak{b}}$ for quotients G/H, as described at the beginning §4; under appropriate hypotheses, the modules arising from a homogeneous line bundle $\boldsymbol{L} \to G/H$ and choice of invariant polarization depend only on the line bundle, not on the polarization. Coincidences between standard modules attached to neighboring orbits can be chained together, to identify the standard modules of the three classifications with each other.

Independence of the polarization can be established in the setting of either G-orbits or $K_{\mathbb{C}}$-orbits: one implies the other via the duality theorem. Beilinson and Bernstein have given a general functorial construction which relates the localization functors (3.4) for Weyl-conjugate parameters λ [2]. The effect of these so-called intertwining functors on standard modules is studied in [6], to prove independence of the polarization, as well as for other purposes[7]. In this section I shall give a more concrete version of essentially the same argument: for each pair of neighboring orbits, one fibres equivariantly over the other; the Leray spectral sequence of the fibration brings about isomorphisms between standard modules attached to the two orbits.

In preparation for a precise statement, I consider a particular $K_{\mathbb{C}}$-orbit Q. Each $\mathfrak{b}$ in $Q \cap S$, the intersection of Q with the dual G-orbit S, is normalized by a unique θ-stable Cartan subgroup $H \subset G$. I use the natural identifications $\mathfrak{h} \cong \mathfrak{b}/[\mathfrak{b},\mathfrak{b}] \cong$ complexified Lie algebra of G to transfer the Cartan involution and complex conjugation (with respect to the real structure of G) to the universal Cartan algebra $\mathfrak{h}$. These transferred operations remain constant as $\mathfrak{b}$ varies over $Q \cap S$, because K acts transitively on the intersection, and are therefore canonically attached to the orbit Q. I write θ_Q and τ_Q for the Cartan involution and complex conjugation which Q induces on $\mathfrak{h}$. Note that the root system of $\mathfrak{g}$ has a universal incarnation $\Phi \subset \mathfrak{h}^*$. According to my earlier definition, the "universal positive root system" $\Phi^+ \subset \Phi$ is chosen so that $\mathfrak{h} \cong \mathfrak{b}/[\mathfrak{b},\mathfrak{b}]$ acts by negative roots on the derived algebra $[\mathfrak{b},\mathfrak{b}]$ of any $\mathfrak{b} \in X$. A root $\alpha \in \Phi$ will be called real, imaginary, or complex with respect to Q, depending on whether $\tau_Q\alpha = \alpha$, $\tau_Q\alpha = -\alpha$, or $\tau_Q\alpha \neq \pm\alpha$. This can also be described in terms of the action of θ_Q : all roots are purely τ_Q-imaginary on the $(+1)$-eigenspace of θ_Q, and τ_Q-real on (-1)-eigenspace, hence

$$\theta_Q\alpha = -\tau_Q\alpha, \quad \text{for every root } \alpha. \tag{6.1}$$

In particular, α is a complex root precisely when $\theta_Q\alpha \neq \pm\alpha$.

Each simple root $\alpha \in \Phi^+$ determines a generalized flag variety X_α and $G_{\mathbb{C}}$-equivariant fibration

$$P_\alpha : X \to X_\alpha, \quad \text{with fibre } \mathbb{P}^1; \tag{6.2}$$

[7] See also Miličić's lecture in this volume [14].

p_α sends any $\mathfrak{b} \in X$ to the parabolic subalgebra obtained by adjoining the α-root space to $\mathfrak{b}$. The image $p_\alpha(Q)$ of a $K_\mathbb{C}$-orbit $Q \subset X$ is obviously a $K_\mathbb{C}$-orbit in X_α, so $p_\alpha^{-1}(Q)$ consists of one or several $K_\mathbb{C}$-orbits, Q among them. It is not difficult to analyze the various possible configurations [6,14], but only the case of a Q-complex simple root α will be relevant for the purposes of these lectures.

(6.3) Lemma. *Let $\alpha \in \Phi^+$ be a Q-complex, simple root. Then:*

(a) *$p_\alpha^{-1}p_\alpha(Q)$ is the union of Q and exactly one other $K_\mathbb{C}$-orbit Q_α;*
(b) *$\theta_{Q_\alpha} = s_\alpha \theta_Q s_\alpha$ ($s_\alpha =$ reflection about the root hyperplane $\alpha^\perp$);*
(c) *α is complex also with respect to the orbit Q_α;*
(d) *one of the roots $\theta_Q\alpha$, $\theta_{Q_\alpha}(\alpha)$ is positive, the other negative. Reverse the labeling of Q, Q_α, if necessary, so that $\theta_Q\alpha$ becomes positive. In that case:*
(e) *on Q_α, p_α restricts to a fibration $p_\alpha : Q_\alpha \to p_\alpha(Q)$ with fibre $\mathbb{C}$; on Q, p_α induces a biregular morphism $Q \tilde{\to} p_\alpha(Q)$.*

Proof. As in the earlier discussion, I fix $\mathfrak{b} \in Q \cap S$, and identify the universal Cartan algebra $\mathfrak{h}$ with the complexified Lie algebra of the θ-stable Cartan subgroup $H \subset G$ that normalizes $\mathfrak{b}$. I pick n_α in the normalizer of H in $G_\mathbb{C}$, a representative of the Weyl reflection s_α about the root α. Then $\mathfrak{b}_\alpha = \operatorname{Ad} n_\alpha(\mathfrak{b})$ is also normalized by H, and $p_\alpha(\mathfrak{b}_\alpha) = p_\alpha(\mathfrak{b})$. A small calculation shows $\dim(\mathfrak{k} \cap \mathfrak{b}_\alpha) = \dim(\mathfrak{k} \cap \mathfrak{b}) \pm 1$; the positive sign applies if and only if $\theta_Q\alpha$ is negative. Thus Q_α, the $K_\mathbb{C}$-orbit through $\mathfrak{b}_\alpha$, also lies over $p_\alpha(Q)$, but is distinct from Q. I suppose for the moment $\theta_Q\alpha > 0$. Let $X \in \mathfrak{k} \cap \mathfrak{b}$ be complementary to $\mathfrak{k} \cap \mathfrak{b}_\alpha$, and let $N_X \subset K_\mathbb{C}$ denote the one-parameter subgroup generated by X. Another small calculation shows that the N_X-orbit of $\mathfrak{b}$ covers the entire fibre of p_α over $p_\alpha(\mathfrak{b})$, except for the point $\mathfrak{b}$ itself. This implies (a) and (e). Now I drop the hypothesis $\theta_Q\alpha > 0$. The complexified Lie algebra of H is naturally isomorphic both to $\mathfrak{b}/[\mathfrak{b},\mathfrak{b}]$ and to $\mathfrak{b}_\alpha/[\mathfrak{b}_\alpha,\mathfrak{b}_\alpha]$, and these in turn are both isomorphic to the universal Cartan algebra $\mathfrak{h}$. The resulting identifications between $\mathfrak{h}$ and the complexified Lie algebra of H differ by s_α, since s_α relates the positive root systems which correspond to $\mathfrak{b}$ and $\mathfrak{b}_\alpha$, respectively. This implies (b). But then $\theta_Q\alpha \neq \pm\alpha$ is equivalent to $\theta_{Q_\alpha}(\alpha) \neq \pm\alpha$, so (c) follows from (b). Similar reasoning, using the simplicity of α, shows that the roots $\theta_Q\alpha$ and $\theta_{Q_\alpha}(\alpha)$ have opposite signs, as asserted by (d). Finally, if $\theta_Q\alpha$ is negative, I reverse the roles of Q and Q_α. This has the effect of making $\theta_Q\alpha$ positive, and thus leads to the situation that was treated already. □

I consider a pair of $K_\mathbb{C}$-orbits Q, Q_α, which are related as in the statement of the lemma. The composition of $p_\alpha : Q_\alpha \to p_\alpha(Q)$ with the biregular morphism $p_\alpha(Q) \cong Q$ defines a $K_\mathbb{C}$-equivariant fibration

$$p : Q_\alpha \to Q, \quad \text{with fibre } \mathbb{C}. \tag{6.4}$$

I choose base points $\mathfrak{b} \in Q$, $\mathfrak{b}_\alpha \in Q_\alpha$, such that $p(\mathfrak{b}_\alpha) = \mathfrak{b}$, and denote the isotropy subgroups of $G_{\mathbb{C}}$ at these two points by B, B_α, respectively. Then, because of the equivariance of p,

$$K_{\mathbb{C}} \cap B \supset K_{\mathbb{C}} \cap B_\alpha. \tag{6.5}$$

Let $\boldsymbol{L}_\chi \to Q$ be a $K_{\mathbb{C}}$-equivariant line bundle, associated to an algebraic character χ of $K_{\mathbb{C}} \cap B$ and $\lambda \in \mathfrak{h}^*$ a linear function which is related to χ by the compatibility condition (3.9), i.e.,

$$d\chi = \text{ restriction of } \lambda - \rho \text{ to } \mathfrak{k} \cap \mathfrak{b}, \tag{6.6}$$

via the identification $\mathfrak{h}^* \cong (\mathfrak{b}/[\mathfrak{b}, \mathfrak{b}])^* \cong$ annihilator of $[\mathfrak{b}, \mathfrak{b}]$ in $\mathfrak{b}^*$. As discussed in §3, these are the data that enter the construction of the Beilinson-Bernstein modules $H^p(X, j_+\mathcal{O}_Q(\boldsymbol{L}_\chi))$.

The pullback $p^*\boldsymbol{L}_\chi \to Q_\alpha$ is associated also to the character χ or more precisely, its restriction to $K_{\mathbb{C}} \cap B_\alpha$ via the containment (6.5). However, the analogue of (6.6), with $\mathfrak{b}_\alpha$ in place of $\mathfrak{b}$, fails because the natural isomorphisms $\mathfrak{h} \cong \mathfrak{b}/[\mathfrak{b}, \mathfrak{b}]$, $\mathfrak{h} \cong \mathfrak{b}_\alpha/[\mathfrak{b}_\alpha, \mathfrak{b}_\alpha]$ do not agree on $\mathfrak{k} \cap \mathfrak{b}_\alpha$: if we identify the abstract Cartan algebra $\mathfrak{h}$ with a concrete Cartan subalgebra of $\mathfrak{g}$ which is contained in both $\mathfrak{b}$ and $\mathfrak{b}_\alpha$, then the two isomorphisms $\mathfrak{h} \cong \mathfrak{b}/[\mathfrak{b}, \mathfrak{b}]$, $\mathfrak{h} \cong \mathfrak{b}_\alpha/[\mathfrak{b}_\alpha, \mathfrak{b}_\alpha]$ are related by the reflection s_α about the simple root α. Thus, on $\mathfrak{k} \cap \mathfrak{b}_\alpha$, $d\chi$ agrees not with $\lambda - \rho$ via $\mathfrak{h}^* \cong$ annihilator of $[\mathfrak{b}_\alpha, \mathfrak{b}_\alpha]$ in $\mathfrak{b}_\alpha^*$, but with $s_\alpha(\lambda - \rho) = s_\alpha\lambda - \rho + \alpha$. The relative cotangent bundle $\boldsymbol{T}^*_{X|X_\alpha}$ of the fibration (6.1) is a $G_{\mathbb{C}}$-equivariant line bundle, defined on all of X and modeled on the $(-\alpha)$-root space; hence

$$\boldsymbol{T}^*_{X|X_\alpha} \cong \boldsymbol{L}_{-\alpha}, \tag{6.7}$$

in the notation of §2. Let ϕ denote the algebraic character by which the isotropy subgroup $K_{\mathbb{C}} \cap B_\alpha \subset K_{\mathbb{C}}$ at $\mathfrak{b}_\alpha$ acts on the fibre of $\boldsymbol{T}^*_{X|X_\alpha} \otimes p^*\boldsymbol{L}_\chi$. To simplify the notation, I set

$$\boldsymbol{L}_\phi = \boldsymbol{T}^*_{X|X_\alpha} \otimes p^*\boldsymbol{L}_\chi. \tag{6.8}$$

According to the earlier discussion, the $K_{\mathbb{C}}$-equivariant line bundle $\boldsymbol{L}_\phi \to Q_\alpha$ and the linear functional $s_\alpha\lambda \in \mathfrak{h}^*$ satisfy the compatibility condition

$$d\phi = \text{ restriction of } s_\alpha\lambda - \rho \text{ to } \mathfrak{k} \cap \mathfrak{b}_\alpha, \tag{6.9}$$

where I now identify $\mathfrak{h}^*$ with the annihilator of $[\mathfrak{b}_\alpha, \mathfrak{b}_\alpha]$ in $\mathfrak{b}_\alpha^*$. The direct image of $\mathcal{O}_{Q_\alpha}(\boldsymbol{L}_\phi)$ via the inclusion $j_\alpha : Q_\alpha \hookrightarrow X$ is therefore a sheaf of $\mathcal{D}_{s_\alpha\lambda}$-modules, and the Beilinson- Bernstein modules $H^p(X, j_\alpha(\mathcal{O}_{Q_\alpha}(\boldsymbol{L}_\phi)))$ are modules over the same quotient $U_{s_\alpha\lambda} \cong U_\lambda$ of $U(\mathfrak{g})$ as those attached to the data $(Q, \boldsymbol{L}_\chi, \lambda)$.

The next statement makes precise what I mean by "independence of the polarization":

(6.10) Lemma. *If $2\frac{(\lambda,\alpha)}{(\alpha,\alpha)} \notin \{-1,-2,\dots\}$, then there exist natural isomorphisms*

$$H^*(X,(j_\alpha)_+(\mathcal{O}_{Q_\alpha}(\boldsymbol{L}_\phi))) \cong H^*(X,j_+\mathcal{O}_Q(\boldsymbol{L}_\chi))).$$

Proof. I use the description (3.19) of the Beilinson-Bernstein modules. Let d denote the codimension of Q, as in §3; then Q_α has codimension $d-1$. Because of (3.17), the local cohomology sheaves along a $K_{\mathbb{C}}$-orbit have the same cohomology on the orbit itself as on the ambient space X. Thus

$$\begin{aligned} &H^*\left(X,j_+\left(\mathcal{O}_Q\left(\boldsymbol{L}_Q\right)\right)\right) \cong H^*\left(Q,\mathcal{H}^d_Q\left(\mathcal{O}\left(\boldsymbol{L}_{\chi,\lambda-\rho}\right)\right)\right), \quad \text{and} \\ &H^*\left(X,(j_\alpha)_+\left(\mathcal{O}_{Q_\alpha}\left(\boldsymbol{L}_\phi\right)\right)\right) \cong H^*\left(Q_\alpha,\mathcal{H}^{d-1}_{Q_\alpha}\left(\mathcal{O}\left(\boldsymbol{L}_{\phi,s_\alpha\lambda-\rho}\right)\right)\right). \end{aligned} \tag{6.11}$$

The lemma now follows from the Leray spectral sequence of the fibration (6.4), provided one knows

$$p_*\mathcal{H}^{d-1}_{Q_\alpha}\left(\mathcal{O}\left(\boldsymbol{L}_{\phi,s_\alpha\lambda-\rho}\right)\right) \cong \mathcal{H}^d_Q\left(\mathcal{O}\left(\boldsymbol{L}_{\chi,\lambda-\rho}\right)\right); \tag{6.12}$$

the higher derived images vanish since p has affine fibres. In proving (6.12), I may dispense with the $K_{\mathbb{C}}$-action because the problem is local with respect to the base. Let $\mu \in \mathfrak{h}^*$ be the projection of $\lambda-\rho$ into the hyperplane $\alpha^\perp$. Then $\boldsymbol{L}_{\mu,\mathfrak{b}}$, as defined in §2, is equivariantly trivial along the fibres of the fibration (6.2), and consequently extends to a $\mathfrak{g}$-equivariant algebraic line bundle over a Zariski neighborhood of $p_\alpha^{-1}p_\alpha(\mathfrak{b})$. By construction, both $\boldsymbol{L}_{\chi,\lambda-\rho}$ and $\boldsymbol{L}_{\phi,s_\alpha\lambda-\rho}$ agree with this line bundle, up to fractional powers of $\boldsymbol{L}_\alpha$. Thus, twisting both sides of (6.12) by the reciprocal line bundle, I can replace $\boldsymbol{L}_{\chi,\lambda-\rho}$ by a fractional power $\boldsymbol{L}_\alpha^c$ and $\boldsymbol{L}_{\phi,s_\alpha\lambda-\rho}$ by $\boldsymbol{L}_\alpha^{c-1}$, without loss of generality. This reduces (6.12) to a problem on the fibre, with local cohomology along $p_\alpha(Q)$ in X_α playing the role of coefficients. On the fibre, the statement comes down to a simple fact about Verma modules for $\mathfrak{sl}_2$: an irreducible Verma module can be realized either as a space of sections of an $\mathfrak{sl}_2$-equivariant line bundle over the open Bruhat cell, or as local cohomology of the inverse line bundle, tensored with the canonical bundle, along the complement of the open cell. This now implies (6.12), and therefore the lemma. □

Define a partial order on the set of $K_{\mathbb{C}}$-orbits as follows: in the situation of lemma (6.3),

$$Q < Q_\alpha \quad \text{if} \quad \theta_Q(\alpha) \quad \text{is positive}; \tag{6.13}$$

by definition, $<$ is the minimal partial ordering compatible with (6.13). Note that $Q_1 < Q_2$ implies $\dim Q_1 < \dim Q_2$.

(6.14) Lemma. *The following are equivalent:* (i) *a $K_{\mathbb{C}}$-orbit Q is minimal, respectively maximal, in the order $<$;* (ii) *Q is of type mi, respectively mr;* (iii) *$\theta_Q\alpha$ is positive, respectively negative, for every Q-complex root $\alpha \in \Phi^+$;* (iv) $\theta_Q\alpha$ is positive, respectively negative, for every Q-complex, simple root α.

Proof. As was remarked earlier, every $\mathfrak{b}$ in the intersection $Q \cap S$ of a $K_{\mathbb{C}}$-orbit Q with the dual G-orbit S is normalized by a θ-stable Cartan subgroup $H \subset G$. If $\mathfrak{b}$ maximizes the dimension of $\mathfrak{b} \cap \bar{\mathfrak{b}}$ among all Borel subalgebras normalized by H—i.e., if S is maximally real—τ_Q sends every positive Q-complex root to another positive root, and conversely; equivalently, by (6.1), θ_Q makes all positive Q-complex roots negative. I claim: if $\theta_Q\alpha < 0$ for every Q-complex simple root α, then $\theta_Q\alpha < 0$ whenever $\alpha \in \Phi^+$ is Q-complex. To see this, I write α as an integral linear combination of simple roots. In the expression for $\theta_Q\alpha$, at least one real or complex simple root must occur with non-zero coefficient. also, θ_Q acts as multiplication by -1 on real roots and leaves imaginary roots alone. The claim follows. Thus, if Q fails to be maximally real, there exists at least one simple complex root α, such that $\theta_Q\alpha$ is positive. But then $Q < Q_\alpha$, so Q is not maximal in the order. On the other hand, if Q is not maximal in the order, it must be expressible as $(Q_1)_\alpha$ as in (6.3), where now α is simple, complex with respect to the orbit Q_1, and $\theta_{Q_1}\alpha > 0$. Then, by (6.3b,c), α is complex also with respect to Q, and $\theta_Q\alpha < 0$; in particular, Q cannot be of type *mr*. The other set of assertions is treated similarly. □

To get from a standard module in the Beilinson-Bernstein classification to a Langlands standard module, I start with data $(Q_0, \chi_0\lambda_0)$, subject to the compatibility condition (6.6). To simplify the discussion, I assume

$$\mathrm{Re}\,(\lambda_0, \alpha) \geq 0 \quad \text{for every} \quad \alpha \in \Phi^+; \tag{6.15}$$

though more restrictive than integral dominance, this hypothesis suffices to realize all irreducible Harish-Chandra modules, since every Weyl group orbit in $\mathfrak{h}^*$ contains some λ_0 which satisfies it. I shall inductively construct sets of data (Q_k, χ_k, λ_k), $0 \leq k \leq m$, again subject to (6.6), such that

(6.16)

(a) $\mathrm{Re}\,(\lambda_k, \alpha) \geq 0$ if $\alpha \in \Phi^+$ is Q_k-real or Q_k-imaginary;

(b) $\mathrm{Re}\,(\lambda_k, \alpha) \geq 0$ if $\alpha \in \Phi^+$ is Q_k-complex and $\theta_{Q_k}\alpha \in \Phi^+$;

(c) $\mathrm{Re}\,(\lambda_k - \theta_{Q_k}\lambda_k, \alpha) \geq 0$ if $\alpha \in \Phi^+$ is Q_k-complex and $\theta_{Q_k}\alpha \in -\Phi^+$;

(d) $Q_{k-1} < Q_k$, and $H^*\left(X, (j_{k-1})_+ \mathcal{O}_{Q_{k-1}}\left(\mathcal{L}_{\chi_{k-1}}\right)\right)$

$$\cong H^*\left(X, j_{k+}\mathcal{O}_{Q_k}\left(\mathcal{L}_{\chi_k}\right)\right).$$

For $k = 0$, (6.15) implies (a-c), and (d) hold vacuously.

(6.17) Lemma. *Either Q_m is of type mr, or the inductive construction can be continued.*

Proof. I suppose for the moment that there exists a simple root α, such that (i) α is Q_m-complex, (ii) $\theta_{Q_m}\alpha \in \Phi^+$, (iii) $\mathrm{Re}(\lambda_m - \theta_{Q_m}\lambda_m, \alpha) \leq 0$. With this choice of α, I let Q_m play the role of Q in lemmas (6.3), (6.10), and set $Q_{m+1} = Q_\alpha$, $\chi_{m+1} = \phi$, $\lambda_{m+1} = s_\alpha\lambda_m$; the hypothesis of (6.10) follows from (6.16b). Then, for $k = m+1$, (6.16d) holds by construction. According to (6.3), $\theta_{Q_{m+1}}$ is the s_α-conjugate of θ_{Q_m}, so (6.16a-c) at the $(m+1)$st step follow from the corresponding statements at the m-th step except possibly (c) for the simple root α. But that comes down to property (iii) of α. It remains to be shown that α with the required properties exists unless Q_m is of type mr. If not, $\mathrm{Re}(\lambda_m - \theta_{Q_m}\lambda_m, \alpha) = \mathrm{Re}(\lambda_m, \alpha - \theta_{Q_m}\alpha) > 0$ for every Q_m-complex simple root such that $\theta_{Q_m}\alpha \in \Phi^+$. On the other hand, this expression is non-negative if $\alpha \in \Phi^+$ is Q_m-complex and $\theta_{Q_m}\alpha \in -\Phi^+$ (by (6.16c)), or if $\alpha \in \Phi^+$ is Q_m-real (by (6.16a), because $\theta_{Q_m}\alpha = -\alpha$), and vanishes for any Q_m-imaginary α (because $\theta_{Q_m}\alpha = \alpha$). Thus the quantity $\mathrm{Re}(\lambda_m - \theta_{Q_m}\lambda_m, \alpha)$ is non-negative for every $\alpha \in \Phi^+$. Since it behaves skew-symmetrically with respect to θ_{Q_m}, this expression must then be zero if $\alpha \in \Phi^+$ is Q_m-complex and $\theta_{Q_m}\alpha \in \Phi^+$. Since it was assumed to be strictly positive if α is Q_m-complex, simple, and $\theta_{Q_m}\alpha \in \Phi^+$, there are no such roots α. Therefore, according to lemma (6.14), Q_m is of type mr. □

In view of the lemma, the inductive ladder terminates at an orbit Q_m of type mr. To simplify matters, I change notation from Q_m, χ_m, λ_m to Q, χ, λ. These quantities satisfy

$$(6.18)\qquad \begin{aligned} &(\lambda, \alpha) \geq 0 \quad \text{for every } Q\text{-imaginary root}^8\ \alpha \in \Phi^+ \\ &\mathrm{Re}(\lambda - \theta_Q\lambda, \alpha) \geq 0 \quad \text{for every } Q\text{-real or } Q\text{-complex } \alpha \in \Phi^+, \end{aligned}$$

as follows from (6.16): $Q = Q_M$ is of type mr, so θ_Q send all Q-complex roots $\alpha \in \Phi^+$ to negative roots, and θ_Q acts as multiplication by -1 on Q-real roots. By induction, the standard module attached to the data (Q_0, χ_0, λ_0) has been identified with a module attached to the mr orbit Q,

$$(6.19)\qquad H^0\left(X, (j_0)_+ \mathcal{O}_{Q_0}\left(\boldsymbol{L}_{\chi_0}\right)\right) \cong H^0\left(X, j_+\mathcal{O}_Q\left(\boldsymbol{L}_\chi\right)\right);$$

the higher cohomology vanishes because of Beilinson-Bernstein's vanishing theorem (3.6B). The duality theorem (5.4) exhibits this module as the dual of a derived functor module

$$(6.20)\qquad H^0\left(X, j_+\mathcal{O}_Q\left(\boldsymbol{L}_\chi\right)\right) \cong M^s\left(S, \boldsymbol{L}_S\right)^*.$$

[8] (λ,α) is real for Q-imaginary roots α since $e^{\lambda-\rho}$ lifts to the character χ on the toroidal part of any Cartan subgroup of G which normalizes any fixed base point $\mathfrak{b} \in Q$.

Here S is the maximally real G-orbit dual to Q, and $\mathbf{L}_S \to S$ the C^ω G-equivariant line bundle whose inverse, tensored with the canonical bundle of X, corresponds to (Q, χ, λ) in the manner described by lemma (5.1).

Let $H = TA$ be the θ-stable Cartan subgroup of G which stabilizes some base point $\mathfrak{b} \in S \cap Q$, with $T = H \cap K$ and $A =$ split part of H. Then

(6.21)
$$\begin{aligned} &H \text{ acts on the fibre of } \mathbf{L}_S \text{ by a character } \psi \text{ such that} \\ &-\lambda = (\text{differential of } \psi) + \rho; \end{aligned}$$

the two shifts by ρ, here and in (6.6), compensate for the appearance of the canonical bundle in the duality theorem. I now apply (4.12), with $-\lambda$ in place of λ : $M^s(S, \mathbf{L}_S)$ is parabolically induced from the module which belongs to the discrete series, to the limits of the discrete series, or vanishes, depending on whether $-\lambda$ is regular with respect to all imaginary roots, regular with respect to all compact imaginary simple roots, or singular with respect to some compact imaginary simple root. I exclude the latter possibility by requiring (λ, α) to be nonzero for α simple and compact imaginary with respect to S. The real part Re ν of the inducing parameter $\nu = \frac{1}{2}(-\lambda + \theta\lambda)$ on the Lie algebra of A is anti-dominant with respect to the positive restricted roots[9], as follows from (6.18).

To make the connection with the Langlands classification, I break up the induction process into two steps: first from MA in the notation of (4.12), to $M_1A_1 =$ centralizer of Re ν, then from M_1A_1 to G. This has the effect of making the real part of the inducing parameter at the second step strictly anti-dominant with respect to the restricted roots on A_1. The result of the first induction step is a tempered (modulo the split part of the center) Harish-Chandra module U of M_1A_1, unitarily induced from MA. Thus, if U is irreducible, $M^s(S, \mathbf{L}_S)$ is one of the standard modules in the Langlands classification. The other case, of a reducible intermediate module U, happens precisely when the unique irreducible subsheaf of $(j_0)_+\mathcal{O}_{Q_0}(\mathbf{L}_{\chi_0})$ fails to have non-zero global sections. In this situation $H^0(X, (j_0)_+\mathcal{O}_{Q_0}(\mathbf{L}_{\chi_0})$ is one of the extraneous standard modules which does not contribute to the Beilinson-Bernstein classification. In effect, the question of whether or not the unique irreducible subsheaf of $(j_0)_+\mathcal{O}_{Q_0}(\mathbf{L}_{\chi_0})$ has non-zero sections is equivalent to the classification of tempered irreducible Harish-Chandra modules. It can be settled by arguments similar to those described in these lecture [4,6,15], thus leading to geometric proofs of the results of Knapp-Zuckerman [8]—proofs that apply equally to non-linear groups G.

The inductive passage from Beilinson-Bernstein standard modules to Langlands standard modules can be reversed. In this way, uniqueness of the realization of an irreducible Harish-Chandra module as quotient of a Langlands standard module becomes equivalent to uniqueness of the Beilinson-Bernstein realization—to the extent that the latter is un-ambiguous, of

[9]Caution: in this paper, I work with the sign convention opposite to the usual one; the unipotent radical of the inducing parabolic subgroup corresponds to negative roots.

course. This is relatively easy to see in the case of regular infinitesimal character, but requires some combinatorial effort in the singular case, for the usual reason.

The dual of any parabolically induced module is again induced, with dual inducing data. Also, modules belonging to the discrete series or limits of the discrete series are dual to modules of the same type. Thus, if $M^s(S, \boldsymbol{L}_S)$ is a standard module in the Langlands classification, I can appeal to (4.12), dualize the induction process, use (4.12) once more, then apply the duality theorem (5.4), to obtain an isomorphism

$$H^0\left(X, \tilde{j}_+ \mathcal{O}_{\tilde{Q}}(\boldsymbol{L}_{\tilde{\chi}})\right) \cong M^s(S, \boldsymbol{L}_S)\,; \tag{6.22}$$

here $(\tilde{Q}, \tilde{\chi}, \tilde{\lambda})$ is a set of Beilinson-Bernstein data subject to the usual compatibility condition (6.6), and $\tilde{j}$ the inclusion map of $\tilde{Q}$ into X. Tracing through the two dualities, one finds

$$\begin{aligned} &(\tilde{\lambda}, \alpha) \geq 0 \quad \text{if } \alpha \in \Phi^+ \text{ is } \tilde{Q}\text{-imaginary}, \\ &\mathrm{Re}(\tilde{\lambda} - \theta_{\tilde{Q}}\tilde{\lambda}, \alpha) \leq 0 \quad \text{for every } \tilde{Q}\text{-real or } \tilde{Q}\text{-complex } \alpha \in \Phi^+. \end{aligned} \tag{6.23a}$$

The Matsuki correspondence relates $\tilde{Q}$ to the complex conjugate (relative to the real structure defined by G) of the G-orbit S. Since S is maximally real, so is its complex conjugate, and hence also the dual $K_{\mathbb{C}}$-orbit:

$$\tilde{Q} \text{ is of type } mr. \tag{6.23b}$$

In view of (4.12), (a) and (b) imply

$$H^p\left(X, \tilde{j}_+ \mathcal{O}_{\tilde{Q}}(\boldsymbol{L}_{\tilde{\chi}})\right) = 0 \quad \text{if } p > 0. \tag{6.23c}$$

Standard modules of the Langlands classification have unique irreducible quotients. Thus, by (6.22), $H^0(X, \tilde{j}_+ \mathcal{O}_{\tilde{Q}}(\boldsymbol{L}_{\tilde{\chi}}))$ contains a unique irreducible quotient, whereas the sheaf of $\mathcal{D}_{\tilde{\lambda}}$-modules $\tilde{j}_+ \mathcal{O}_{\tilde{Q}}(\boldsymbol{L}_{\tilde{\chi}})$ has a unique irreducible subsheaf. No contradiction here: the equivalence of categories (3.8) depends on integral dominance of the parameter λ.

The passage from the standard module $H^0(X, \tilde{j}_+ \mathcal{O}_{\tilde{Q}}(\boldsymbol{L}_{\tilde{\chi}}))$ to a standard Vogan Zuckerman module is similar to the earlier inductive argument: beginning with $(Q_0, \chi_0, \lambda_0) = (\tilde{Q}, \tilde{\chi}, \tilde{\lambda})$, I construct sets of data (Q_k, χ_k, λ_k), $0 \leq k \leq m$, subject to the usual condition (6.6), such that

(6.24)

(a) $(\lambda_k, \alpha) \geq 0$ if $\alpha \in \Phi^+$ is Q_k-imaginary,

(b) $\mathrm{Re}(\lambda_k, \alpha) \leq 0$ if $\alpha \in \Phi^+$ is Q_k-real,

(c) $\mathrm{Re}\left(\lambda_k - \theta_{Q_k}\lambda_k, \alpha\right) \leq 0$ if $\alpha \in \Phi^+$ is Q_k-complex and $\theta_{Q_k}\alpha \in -\Phi^+$,

(d) $\mathrm{Re}\left(\lambda_k + \theta_{Q_k}\lambda_k, \alpha\right) \geq 0$ if $\alpha \in \Phi^+$ is Q_k-complex and $\theta_{Q_k}\alpha \in \Phi^+$,

(e) $Q_{k-1} > Q_k$, and $H^*\left(X, (j_{k-1})_+ \mathcal{O}_{Q_{k-1}}\left(\boldsymbol{L}_{\chi_{k-1}}\right)\right)$

$$\cong H^*\left(X, j_{k+} \mathcal{O}_{Q_k}(\boldsymbol{L}_{\chi_k})\right).$$

For $k = 0$, (a)–(d) are implied by (6.23) in conjunction with (6.14), and (e) holds vacuously.

(6.25) Lemma. *Either Q_m is of type mi, or the inductive construction can be continued.*

I omit the proof, since it is entirely analogous to that of (6.17). According to the lemma, the induction ends with an orbit Q_m of type *mi*. I change notation from (Q_m, χ_m, λ_m) to (Q, χ, λ); then

$$(6.26)\qquad \begin{aligned} &(\lambda, \alpha) \geq 0 \quad \text{for every } Q\text{-imaginary root } \alpha \in \Phi^+, \\ &\mathrm{Re}\,(\lambda + \theta_Q \lambda, \alpha) \geq 0 \quad \text{for every } Q\text{-complex } \alpha \in \Phi^+, \\ &\mathrm{Re}(\lambda, \alpha) \leq 0 \quad \text{for every } Q\text{-real root } \alpha \in \Phi^+. \end{aligned}$$

The induction identifies the standard Langlands module (6.22) with a module attached to the *mi* orbit Q; it also carries along the vanishing theorem (6.23c):

$$(6.27)\qquad \begin{aligned} &\text{(a) } H^0\left(X, j_+ \mathcal{O}_Q(\boldsymbol{L}_\chi)\right) \cong H^0\left(X, \tilde{j}_+ \mathcal{O}_{\tilde{Q}}(\boldsymbol{L}_{\tilde{\chi}})\right), \\ &\text{(b) } H^p\left(X, j_+ \mathcal{O}_Q(\boldsymbol{L}_\chi)\right) = 0 \quad \text{if } p > 0. \end{aligned}$$

The duality theorem and the discussion at the end of Section 4 show that $H^0(X, j_+\mathcal{O}_Q(\boldsymbol{L}_\chi))$ is dual to a cohomologically induced module, induced from a principal series module of a split group.

Vogan, in his book [20], constructs standard modules by cohomological induction from principal series modules of quasi-split groups; he also remarks that the same modules arise by induction from split groups—the equivalence of the two closely related pictures becomes visible by carrying out the cohomological induction process in stages. Vogan chooses principal series modules of quasi-split (rather than split) groups as basic building blocks because this makes it possible to set up a bijection between lowest K-types, of the inducing module on the one hand, and the induced module on the other. It might be illuminating to explain the reason for this choice in geometric terms, along the lines of Chang's proof of Vogan's lowest K-type theorem [4].

Except for the slight discrepancy between induction from split or quasi-split groups, I have shown the original Langlands standard module (6.22) to be dual to a standard module of the Vogan- Zuckerman classification; (6.27) translates into the vanishing theorem that accompanies the classification. As in the earlier argument (6.16-17), the inductive ladder (6.24-25) can be reversed, to pass back and forth between all three classification schemes.

§7 An example: $\mathrm{SL}(2, \mathbb{C})$

The differences between the various classifications of irreducible Harish-Chandra modules become apparent only if G contains Cartan subgroups that are neither compact nor split. Thus $\mathrm{Sl}(2,\mathbb{R})$ is not a good example for the subject of these lectures: all classification schemes associate the discrete series to the conjugacy class of compact Cartan subgroups, and the principal series to the class of split Cartans. The differences do show up for $\mathrm{Sl}(2,\mathbb{C})$, however, as I shall try to explain in this section.

I view $G = \mathrm{Sl}(2\mathbb{C})$ as real Lie group, with maximal compact subgroup $K = SU(2)$. These groups have complexified Lie algebras

$$(7.1)\qquad \begin{aligned} \mathfrak{g} &= \mathfrak{sl}(2,\mathbb{C}) \oplus \mathfrak{sl}(2,\mathbb{C}), \\ \mathfrak{k} &= \text{diagonal in } \mathfrak{sl}(2,\mathbb{C}) \oplus \mathfrak{sl}(2,\mathbb{C}) \cong \mathfrak{sl}(2,\mathbb{C}). \end{aligned}$$

Thus θ, the Cartan involution with respect to K and $\tau =$ complex conjugation with respect to the real structure of G, are given by

$$(7.2)\qquad \begin{aligned} \theta(Z_1, Z_2) &= (Z_2, Z_1), \\ & \qquad\qquad\qquad (Z_1, Z_2 \in \mathfrak{sl}(2,\mathbb{C})) \\ \tau(Z_1, Z_2) &= (\sigma Z_2, \sigma Z_1); \end{aligned}$$

here $\sigma : \mathfrak{sl}(2,\mathbb{C}) \to \mathfrak{sl}(2,\mathbb{C})$ denotes conjugation with respect to $\mathfrak{su}(2)$, i.e., σZ equals the ordinary complex conjugate of $-Z^t$. The flag variety of $\mathfrak{g} = \mathfrak{sl}(2,\mathbb{C}) \oplus \mathfrak{sl}(2,\mathbb{C})$ is the product of the flag varieties of the two summands,

$$(7.3)\qquad X = \mathbb{P}^1 \times \mathbb{P}^1.$$

I identify $\mathbb{P}^1$ with the one point compactification $\mathbb{C} \cup \{\infty\}$ of the complex plane. Both θ and τ operate on X:

$$(7.4)\qquad \begin{aligned} \theta(z_1, z_2) &= (z_2, z_1), \\ & \qquad\qquad\qquad (z_1, z_2 \in \mathbb{P}^1) \\ \tau(z_1, z_2) &= \left(-\frac{1}{\bar z_2}, -\frac{1}{\bar z_1}\right), \end{aligned}$$

with $\bar z =$ ordinary complex conjugate of $z \in \mathbb{P}^1 = \mathbb{C} \cup \{\infty\}$.

On the level of groups, the complexification $K_\mathbb{C}$ of K is itself isomorphic to $\mathrm{Sl}(2,\mathbb{C})$, and lies in $G_\mathbb{C} = \mathrm{Sl}(2,\mathbb{C}) \times \mathrm{Sl}(2,\mathbb{C})$ as diagonal subgroup. In particular, $K_\mathbb{C}$ has two orbits in $X = \mathbb{P}^1 \times \mathbb{P}^1$:

$$(7.5)\qquad \begin{aligned} Q_0 &= \text{diagonal in } \mathbb{P}^1 \times \mathbb{P}^1, \\ Q_1 &= \text{complement of the diagonal}. \end{aligned}$$

These are dual to the two G-orbits

$$(7.6)\qquad \begin{aligned} S_0 &= \left\{(z_1, z_2) \in \mathbb{P}^1 \times \mathbb{P}^1 \,|\, z_1 \bar z_2 \neq -1\right\}, \\ S_1 &= \left\{(z_1, z_2) \in \mathbb{P}^1 \times \mathbb{P}^1 \,|\, z_1 \bar z_2 = -1\right\}; \end{aligned}$$

cf. (7.4). Indeed, $Q_0 \cap S_0$ and $Q_1 \cap S_1$ are single K-orbits, as required by the definition of Matsuki's duality, both isomorphic to $\mathbb{P}^1$.

The abstract root system $\Phi \subset \mathfrak{h}^*$ is of type $A_1 \times A_1$. There are two positive[10] roots α, β, each of them simple. I write θ_0, θ_1 for the Cartan involutions on the abstract Cartan algebra $\mathfrak{h}$, corresponding to the two orbits Q_0, Q_1, as described in §6. A small calculation shows

$$\theta_0\alpha = \beta, \quad \theta_1\alpha = -\beta. \tag{7.7}$$

Thus, according to (6.14),

(7.8) $\qquad Q_0, S_0$ are of type mi, and Q_1, S_1 of type mr.

Up to conjugacy, G contains a unique Cartan subgroup $H = TA$, which is connected. It follows that both G and $K_{\mathbb{C}}$ intersect the stabilizer $B \subset G_{\mathbb{C}}$ of any particular $\mathfrak{b} \in X$ in a connected subgroup. This simplifies the assertion of lemma (2.4): a $(\mathfrak{g}, G)$-equivariant holomorphic line bundle over the germ of a neighborhood of a G-orbit, or a $(\mathfrak{g}, K_{\mathbb{C}})$- equivariant algebraic line bundle over the formal neighborhood of a $K_{\mathbb{C}}$-orbit, is completely determined by its germ $\boldsymbol{L}_{\lambda,\mathfrak{b}}$ at any point $\mathfrak{b}$ of the orbit in question. The intersection $G \cap B$, or $K_{\mathbb{C}} \cap B$, contains a conjugate of the toroidal subgroup T of H, so (2.4) forces an integrality condition on the parameter $\lambda \in \mathfrak{h}^*$. Specifically,

(7.9) $\qquad (\mathfrak{g}, K_{\mathbb{C}})$-equivariant algebraic line bundles over the formal neighborhood of Q_j, $j = 0, 1$, are parametrized by $\{\lambda \in \mathfrak{h}^* \mid \lambda + \theta_j\lambda$ is a half-integral multiple of $\alpha + \theta_j\alpha\}$;

$(\mathfrak{g}, G)$-equivariant line bundles over a neighborhood of the dual G-orbit S_j correspond bijectively to these $(\mathfrak{g}, K_{\mathbb{C}})$-equivariant line bundles, and are therefore parametrized by the same set. If $\lambda \in \mathfrak{h}^*$ satisfies the integrality condition, I denote the corresponding line bundle, over either Q_j or S_j, by $\boldsymbol{L}_\lambda$.

To make the parametrization of representations compatible with their infinitesimal characters, I shift λ by $-\rho$; since ρ lies in the weight lattice, λ satisfies the integrality condition in (7.9) if and only if $\lambda - \rho$ does. Note that S_0 is open in X. Hence, to each line bundle $\boldsymbol{L}_{\lambda-\rho} \to S_0$, the Zuckerman construction associates the algebraic analogue of sheaf cohomology over S_0, with values in $\mathcal{O}(\boldsymbol{L}_{\lambda-\rho})$. Indeed, according to [18],

$$M^p(S_0, \boldsymbol{L}_{\lambda-\rho}) \cong K\text{-finite part of } H^p(S_0, \mathcal{O}(\boldsymbol{L}_{\lambda-\rho}))\,. \tag{7.10}$$

Since S_1 is maximally real, the Zuckerman modules $M^p(S_1, \mathcal{O}(\boldsymbol{L}_{\lambda-\rho})$, for λ as in (7.9), are parabolically induced—in this case from a minimal parabolic

[10] with respect to the intrinsic order; cf. §2.

subgroup of G, which is quasi-split. In the notation of (4.12), $s = 0$ and $G/MAN \cong S_1$ so

$$\begin{aligned} &M^0(S_1, \boldsymbol{L}_{\lambda-\rho}) \cong \text{ space of } K\text{-finite } C^\infty \text{ sections of } \boldsymbol{L}_{\lambda-\rho} \to S_1, \\ &M^p(S_1, \boldsymbol{L}_{\lambda-\rho}) = 0 \quad \text{if } p \neq 0, \end{aligned} \tag{7.11}$$

for any λ that satisfies the integrality condition (7.9).

To simplify various formulas, I shall make no notational distinction between the inclusions of the two $K_{\mathbb{C}}$-orbits in X: both will be denoted by j. The orbit Q_1 lies in X as Zariski open, affine subset. Hence, if $\boldsymbol{L}_{\lambda-\rho} \to Q_1$ is a $(\mathfrak{g}, K_{\mathbb{C}})$-equivariant line bundle, the $\mathcal{D}$-module direct image $j_+\mathcal{O}_{Q_1}(\boldsymbol{L}_{\lambda-\rho})$ coincides with the sheaf direct image, the higher direct images vanish, and

(7.12)
$$\begin{aligned} H^p(X, j_+\mathcal{O}_{Q_1}(\boldsymbol{L}_{\lambda-\rho})) &= H^p(Q_1, \mathcal{O}(\boldsymbol{L}_{\lambda-\rho})) \\ &= \begin{cases} \text{space of algebraic sections of } \boldsymbol{L}_{\lambda-\rho} \to Q_1 & \text{if } p = 0, \\ 0 & \text{if } p \neq 0. \end{cases} \end{aligned}$$

Here cohomology is taken in the algebraic sense, as always in the context of the Beilinson-Bernstein construction. Now let $\boldsymbol{L}_{\lambda-\rho}$ be a $(\mathfrak{g}, K_{\mathbb{C}})$-equivariant algebraic line bundle over the formal neighborhood of Q_0. I use (3.18,19) to reinterpret the Beilinson-Bernstein modules as local cohomology groups:

$$\begin{aligned} H^p(X, j_+\mathcal{O}_{Q_0}(\boldsymbol{L}_{\lambda-\rho})) &\cong H^p\left(X, \mathcal{H}^1_{Q_0}(\mathcal{O}(\boldsymbol{L}_{\lambda-\rho}))\right) \\ &\cong H^{p+1}_{Q_0}(X, \mathcal{O}(\boldsymbol{L}_{\lambda-\rho})). \end{aligned} \tag{7.13}$$

Irreducibility of sheaves of $\mathcal{D}$-modules is a local phenomenon. By construction, the sheaf $j_+\mathcal{O}_{Q_0}(\boldsymbol{L}_{\lambda-\rho})$ has irreducible stalks at all points of the orbit Q_0. Since $\partial Q_0 = \emptyset$, the support of $j_+\mathcal{O}_{Q_0}(\boldsymbol{L}_{\lambda-\rho})$ coincides with Q_0, so

$$j_+\mathcal{O}_{Q_0}(\boldsymbol{L}_{\lambda-\rho}) \quad \text{is an irreducible sheaf of } \mathcal{D}_\lambda\text{-modules.} \tag{7.14}$$

According to Beilinson-Bernstein's equivalence of categories,

$$\begin{aligned} &H^0(X, j_+\mathcal{O}_{Q_0}(\boldsymbol{L}_{\lambda-\rho})) \quad \text{is irreducible or vanishes,} \\ &\text{provided } \lambda \text{ is integrally dominant;} \end{aligned} \tag{7.15}$$

in the integrally dominant situation, this module cannot possibly vanish unless the parameter λ is singular, of course.

The integer s of the duality theorem (5.4) equals 1 in the case of the pair of orbits S_0, Q_0, and is zero for the pair S_1, Q_1. Thus

$$\begin{aligned} H^p(X, j_+\mathcal{O}_{Q_0}(\boldsymbol{L}_{\lambda-\rho})) &\cong M^{1-p}(S_0, \boldsymbol{L}_{-\lambda-\rho})^*, \\ H^0(X, j_+\mathcal{O}_{Q_1}(\boldsymbol{L}_{\lambda-\rho})) &\cong M^0(S_1, \boldsymbol{L}_{-\lambda-\rho})^*; \end{aligned} \tag{7.16}$$

note that the two shifts by $-\rho$ compensate for the appearance of the canonical bundle $\boldsymbol{K}_X \cong \boldsymbol{L}_{-2\rho}$ in (5.4). Induced modules are dual to modules induced from dual inducing data. In the case of the modules (7.11) this means

$$(7.17) \qquad M^0(S_1, \boldsymbol{L}_{\lambda-\rho}) \cong M^0(S_1, \boldsymbol{L}_{-\lambda-\rho})^*,$$

and, via the duality (7.16),

$$(7.18) \qquad H^0(X, j_+\mathcal{O}_{Q_1}(\boldsymbol{L}_{\lambda-\rho})) \cong H^0(X, j_+\mathcal{O}_{Q_1}(\boldsymbol{L}_{-\lambda-\rho}))^*.$$

Independence of the polarization is a further source of identifications between the modules (7.12,7.13):

$$(7.19) \qquad \begin{aligned} &H^p(X, j_+\mathcal{O}_{Q_0}(\boldsymbol{L}_{\lambda-\rho})) \cong H^p(X, j_+\mathcal{O}_{Q_1}(\boldsymbol{L}_{s_\alpha\lambda-\rho})) \\ &\qquad\qquad \text{if } 2\frac{(\lambda,\alpha)}{(\alpha,\alpha)} \neq -1, -2, \dots \\ &H^p(X, j_+\mathcal{O}_{Q_0}(\boldsymbol{L}_{\lambda-\rho})) \cong H^p\left(X, j_+\mathcal{O}_{Q_1}\left(\boldsymbol{L}_{s_\beta\lambda-\rho}\right)\right) \\ &\qquad\qquad \text{if } 2\frac{(\lambda,\beta)}{(\beta,\beta)} \neq -1, -2, \dots \end{aligned}$$

The duality (7.16) now leads to analogous isomorphisms among the modules (7.10,11).

I write the parameter λ as a linear combination of the simple roots α, β,

$$(7.20) \qquad \lambda = \ell_\alpha \frac{\alpha}{2} + \ell_\beta \frac{\beta}{2},$$

with complex coefficients ℓ_α, ℓ_β. Then

(7.21) λ is integrally dominant if and only if $\ell_\alpha, \ell_\beta \neq -1, -2, \dots$

In view of (7.7), (7.9),

(7.22) $\boldsymbol{L}_{\lambda-\rho}$ exists as $(\mathfrak{g}, K_\mathbb{C})$-equivariant algebraic line bundle on the formal neighborhood of Q_0, respectively Q_1, precisely when $\ell_\alpha + \ell_\beta \in \mathbb{Z}$, respectively $\ell_\alpha - \ell_\beta \in \mathbb{Z}$.

The same conditions are necessary and sufficient for the existence of $\boldsymbol{L}_{\lambda-\rho}$ as $(\mathfrak{g}, G)$-equivariant holomorphic line bundle on the germ of a neighborhood of S_0, respectively S_1. On the other hand,

(7.23) $\boldsymbol{L}_{\lambda-\rho}$ is a globally defined $G_\mathbb{C}$-equivariant holomorphic line bundle on X if and only if both ℓ_α and ℓ_β are integral, because $\frac{\alpha}{2}$ and $\frac{\beta}{2}$ generate the weight lattice of $G_{\mathbb{C}}$. Hence, if $\boldsymbol{L}_{\lambda-\rho}$ exists as equivariant line bundle near any one of the four orbits (7.5,6), then either both ℓ_α and ℓ_β are integral—in which case the line bundle exists globally on X—or both are non-integral.

I first treat the latter situation, which is the simpler of the two: I suppose both ℓ_α and ℓ_β are non-integral; this makes λ integrally dominant and regular, regardless of the particular values of ℓ_α, ℓ_β. If $\ell_\alpha + \ell_\beta \in \mathbb{Z}$, then $\boldsymbol{L}_{\lambda-\rho}$ exists on the formal neighborhood of Q_0, and

$$
\begin{aligned}
(7.24)\qquad & H^0\left(X, j_+\mathcal{O}_{Q_0}(\boldsymbol{L}_{\lambda-\rho})\right) \cong H^0\left(X, j_+\mathcal{O}_{Q_1}(\boldsymbol{L}_{s_\alpha\lambda-\rho})\right) \\
& \cong H^0\left(X, j_+\mathcal{O}_{Q_1}\left(\boldsymbol{L}_{s_\beta\lambda-\rho}\right)\right) \cong H^0\left(X, j_+\mathcal{O}_{Q_0}(\boldsymbol{L}_{-\lambda-\rho})\right)
\end{aligned}
$$

by (7.19); note: $s_\beta s_\alpha = s_\alpha s_\beta = -1$. No higher cohomology occurs since λ is integrally dominant. According to (7.15) these modules are non-zero (λ is regular!) and irreducible. They can be identified with principal series modules, as follows from(7.11), (7.17), and the second isomorphism in (7.16). Next, still under the hypothesis of non-integrality, I assume $\ell_\alpha - \ell_\beta \in \mathbb{Z}$, so that $\boldsymbol{L}_{\lambda-\rho}$ exists on the formal neighborhood of Q_1. This amounts to replacing λ by $s_\beta\lambda$ in the preceding discussion:

$$
\begin{aligned}
(7.25)\qquad & H^0\left(X, j_+\mathcal{O}_{Q_1}(\boldsymbol{L}_{\lambda-\rho})\right) \cong H^0\left(X, j_+\mathcal{O}_{Q_0}(\boldsymbol{L}_{s_\alpha\lambda-\rho})\right) \\
& \cong H^0\left(X, j_+\mathcal{O}_{Q_0}\left(\boldsymbol{L}_{s_\beta\lambda-\rho}\right)\right) \cong H^0\left(X, j_+\mathcal{O}_{Q_1}(\boldsymbol{L}_{-\lambda-\rho})\right)
\end{aligned}
$$

is non-zero, irreducible, and is isomorphic to a module of the principal series. In particular,

$$(7.26)\qquad j_+\mathcal{O}_{Q_1}(\boldsymbol{L}_{\lambda-\rho}) \text{ is an irreducible sheaf of } \mathcal{D}_\lambda\text{- modules}$$

as follows from the equivalence of categories (3.8).

I recall the description of the irreducible Beilinson-Bernstein sheaves as given in (3.12). Under the standing assumption that both ℓ_α and ℓ_β are non-integral, the category $M(\mathcal{D}_\lambda, K_{\mathbb{C}})$ contains two, one, or no irreducible objects, depending on whether both, one, or neither of the quantities $\ell_\alpha + \ell_\beta$, $\ell_\alpha - \ell_\beta$ are integers; cf. (7.14) and (7.26). Equivalently, there exist two, one, or no irreducible Harish-Chandra modules with infinitesimal character χ_λ. Each can be associated to either of the two $K_{\mathbb{C}}$-orbits, as is apparent from (7.24-25), or to either of the two G-orbits, via the duality (7.16). In all cases, the parameters λ and $-\lambda$ lead to the same module. The Langlands classification chooses the realization $M^0(S_1, \boldsymbol{L}_{\lambda-\rho})$ with $\mathrm{Re}(\lambda - \theta_1\lambda)$ anti-dominant—i.e., $\mathrm{Re}(\ell_\alpha + \ell_\beta) \leq 0$. Cohomological induction à la Vogan Zuckerman[11] realizes the irreducible modules in the form $M^1(S_0, \boldsymbol{L}_{\lambda-\rho})$

[11] In this section, the standard modules of Vogan-Zuckerman are understood to arise by cohomological induction from split, rather than quasi-split, Levi factors; cf. the penultimate paragraph of §6.

with $\mathrm{Re}(\lambda+\theta_0\lambda)$ anti-dominant—which again translates into $\mathrm{Re}(\ell_\alpha+\ell_\beta)\leq 0$. The Beilinson-Bernstein classification, on the other hand, leaves the choice of $K_{\mathbb{C}}$-orbit open because all Weyl translates of λ are integrally dominant.

Now I suppose that λ is integral, in the sense that $\ell_\alpha,\ell_\beta\in\mathbb{Z}$. Then $\boldsymbol{L}_{\lambda-\rho}$ is well defined as $G_{\mathbb{C}}$-equivariant line bundle on all of X. The identification (3.18) of $j_+\mathcal{O}_{Q_0}(\boldsymbol{L}_{\lambda-\rho})$ with the local cohomology sheaf of $\boldsymbol{L}_{\lambda-\rho}$ along S_0 leads to the exact sequence

$$0\to\mathcal{O}_X(\boldsymbol{L}_{\lambda-\rho})\to j_+\mathcal{O}_{Q_1}(\boldsymbol{L}_{\lambda-\rho})\to j_+\mathcal{O}_{Q_0}(\boldsymbol{L}_{\lambda-\rho})\to 0. \tag{7.27}$$

It was noted earlier that $j_+\mathcal{O}_{Q_0}(\boldsymbol{L}_{\lambda-\rho})$ is an irreducible sheaf of $\mathcal{D}_\lambda$- modules; $\mathcal{O}_X(\boldsymbol{L}_{\lambda-\rho})$ is irreducible for even simpler reasons. I appeal to (3.12), to conclude:

(7.28) the category $M(\mathcal{D}_\lambda,K_{\mathbb{C}})$ contains exactly two irreducible objects, namely $\mathcal{O}_X(\boldsymbol{L}_{\lambda-\rho})$ and $j_+\mathcal{O}_{Q_0}(\boldsymbol{L}_{\lambda-\rho})$.

According to Borel-Weyl-Bott,

$$H^p(X,\mathcal{O}_X(\boldsymbol{L}_{\lambda-\rho}))=0\quad\text{for all } p\text{, if }\lambda\text{ is singular.} \tag{7.29a}$$

If λ is regular, I set $k=k(\lambda)=0,1,$ or 2, depending on whether both, one, or none of the integers ℓ_α, ℓ_β are positive; then

$$\begin{aligned}&H^p(X,\mathcal{O}_X(\boldsymbol{L}_{\lambda-\rho}))=0\quad\text{if } p\neq k,\\ &H^k(X,\mathcal{O}_X(\boldsymbol{L}_{\lambda-\rho}))\ \text{ is non-zero, finite dimensional,}\\ &\qquad\text{irreducible, and has infinitesimal character } \chi_\lambda.\end{aligned} \tag{7.29b}$$

Since Q_1 is affine, $\boldsymbol{L}_{\lambda-\rho}$ has an infinite dimensional space of algebraic sections over Q_1. In view of the exact sequence (7.27) and the finite dimensionality of $H^0(X,\mathcal{O}_X(\boldsymbol{L}_{\lambda-\rho}))$,

(7.30) the sheaf $j_+\mathcal{O}_{Q_0}$ has non-zero global sections,

independently of the particular choice of the parameter λ. Thus, according to the Beilinson-Bernstein equivalence of categories[12] and (7.29a), there exist exactly two irreducible modules in $M(U_\lambda,K)$ if λ is regular, but only one if λ is singular.

If λ is not only integral and regular, but also dominant, the two irreducible Harish-Chandra modules with infinitesimal character χ_λ are

$$F_\lambda=H^0(X,\mathcal{O}_X(\boldsymbol{L}_{\lambda-\rho})), \tag{7.31a}$$

[12] in the extended version that applies even when λ is singular; cf. §3.

which is finite dimensional, and

(7.31b) $$P_\lambda = H^0\left(X, j_+\mathcal{O}_{Q_0}(\boldsymbol{L}_{\lambda-\rho})\right),$$

which belongs to the principal series, as follows from (7.19), (7.16-17), and (7.11). They fit into the exact cohomology sequence induced by (7.27),

(7.32) $$0 \to F_\lambda \to H^0\left(X, j_+\mathcal{O}_{Q_1}(\boldsymbol{L}_{\lambda-\rho})\right) \to P_\lambda \to 0;$$

all higher cohomology vanishes because λ is assumed to be dominant:

(7.33) $$H^p\left(X, j_+\mathcal{O}_{Q_0}(\boldsymbol{L}_{\lambda-\rho})\right) = H^p\left(X, j_+\mathcal{O}_{Q_1}(\boldsymbol{L}_{\lambda-\rho})\right) = 0 \quad \text{if } p \neq 0.$$

Note that the exact sequence (7.32) does not split because it does not split on the level of sheaves; cf. (3.12).

To see what happens for parameters λ in the other three Weyl chambers, I apply the change of polarization identities (7.19), still with λ integral, dominant, regular:

(7.34)
$$H^0\left(X, j_+\mathcal{O}_{Q_1}(\boldsymbol{L}_{s_\alpha\lambda-\rho})\right) \cong H^0\left(X, j_+\mathcal{O}_{Q_1}\left(\boldsymbol{L}_{s_\beta\lambda-\rho}\right)\right) \cong P_\lambda,$$
$$H^p\left(X, j_+\mathcal{O}_{Q_1}(\boldsymbol{L}_{s_\alpha\lambda-\rho})\right) = H^p\left(X, j_+\mathcal{O}_{Q_1}\left(\boldsymbol{L}_{s_\beta\lambda-\rho}\right)\right) = 0 \quad \text{if } p \neq 0.$$

Next, I use (7.18) to dualize the exact sequence (7.32), while observing that the two irreducible modules F_λ, P_λ are self-dual. Indeed, their duals are still irreducible, have the same infinitesimal character because λ is Weyl conjugate of $-\lambda$, and F_λ is obviously not isomorphic to P_λ, hence $F_\lambda^* \cong F_\lambda$, $P_\lambda^* \cong P_\lambda$. I conclude that

(7.35a) $$0 \to P_\lambda \to H^0\left(X, j_+\mathcal{O}_{Q_1}(\boldsymbol{L}_{-\lambda-\rho})\right) \to F_\lambda \to 0$$

is exact, but does not split. The higher cohomology vanishes,

(7.35b) $$H^p\left(X, j_+\mathcal{O}_{Q_1}(\boldsymbol{L}_{-\lambda-\rho})\right) = 0 \quad \text{if } p \neq 0,$$

because Q_1 is affine. Change of polarization (7.19), with $-s_\beta\lambda$ and $-s_\alpha\lambda$ in place of λ, gives

(7.36)
$$\begin{aligned} H^p\left(X, j_+\mathcal{O}_{Q_0}(\boldsymbol{L}_{s_\alpha\lambda-\rho})\right) &\cong H^p\left(X, j_+\mathcal{O}_{Q_0}\left(\boldsymbol{L}_{s_\beta\lambda-\rho}\right)\right) \\ &\cong H^p\left(X, j_+\mathcal{O}_{Q_1}(\boldsymbol{L}_{-\lambda-\rho})\right), \quad \text{for all } p \end{aligned}$$

(note: $s_\alpha s_\beta = s_\alpha s_\beta = -1$). Finally,

(7.37)
$$\begin{aligned} &H^0\left(X, j_+\mathcal{O}_{Q_0}(\boldsymbol{L}_{-\lambda-\rho})\right) \cong H^0\left(X, j_+\mathcal{O}_{Q_1}(\boldsymbol{L}_{-\lambda-\rho})\right), \\ &H^1\left(X, j_+\mathcal{O}_{Q_0}(\boldsymbol{L}_{-\lambda-\rho})\right) \cong F_\lambda, \\ &H^p\left(X, j_+\mathcal{O}_{Q_0}(\boldsymbol{L}_{-\lambda-\rho})\right) = 0 \quad \text{for } p \neq 0, 1, \end{aligned}$$

as follows from (7.27), (7.29), (7.35).

I continue to assume that λ is integral and regular, but drop the assumption that it is dominant. For the purpose of comparing the three classifications, it will be convenient to denote the two irreducible Harish-Chandra modules with infinitesimal character χ_λ by F_λ and P_λ, for λ in any of the Weyl chambers. I label the four chambers as I, II, III, IV;

Figure 1.

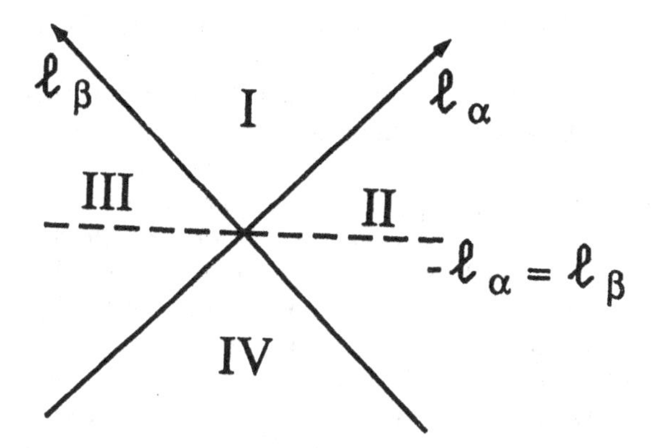

the integer ℓ_α is positive on chambers I and II, whereas ℓ_β is positive on I and III. The Beilinson-Bernstein classification attaches the finite dimensional irreducible module F_λ to the open $K_{\mathbb{C}}$-orbit Q_1, and the other irreducible module P_λ to the closed orbit Q_0, in both cases with λ lying in the dominant Weyl chamber I. Concretely, F_λ is realized as the unique irreducible submodule of the standard module $H^0\left(X, j_+\mathcal{O}_{Q_1}(\boldsymbol{L}_{\lambda-\rho})\right)$ via the exact sequence (7.32), while $P_\lambda = H^0\left(X, j_+\mathcal{O}_{Q_0}(\boldsymbol{L}_{\lambda-\rho})\right)$ coincides with the standard module in which it lies.

The Langlands classification associates both F_λ and P_λ to the closed G-orbit S_1. When λ lies in the anti-dominant chamber IV, the duality (7.16), the self-duality of F_λ and P_λ, and the exact sequence (7.32) exhibit F_λ as the unique irreducible quotient of the standard Langlands module $M^0(S_1, \boldsymbol{L}_{\lambda-\rho})$:

$$0 \to P_\lambda \to M^0(S_1, \boldsymbol{L}_{\lambda-\rho}) \to F_\lambda \to 0 \quad (\ell_\alpha, \ell_\beta < 0) \tag{7.38}$$

is exact and does not split. If λ lies in one of the chambers II or III, $P_\lambda \cong P_\lambda^* \cong M^0(S_1, \boldsymbol{L}_{\lambda-\rho})$ by (7.16) and (7.34). The Langlands classification requires $\lambda - \theta_1\lambda$ to be anti-dominant—in other words, $\ell_\alpha + \ell_\beta$ must be non-positive. Thus P_λ is realized as the full standard module $M^0(S_1, \boldsymbol{L}_{\lambda-\rho})$, with λ in the lower half of either of the chambers II or III. On the line $\ell_\alpha + \ell_\beta = 0$, P_λ is unitarily induced, hence tempered; in this situation the classification leaves open the choice between the two chambers.

Cohomological induction attaches both F_λ and P_λ to the open G-orbit S_0, with λ ranging again over the anti-dominant chamber IV and the bottom halves of II and III. Now the picture is reversed: for λ anti-dominant, $P_\lambda \cong P_\lambda^* \cong M^1(S_0, \boldsymbol{L}_{\lambda-\rho})$ is a full standard module, by (7.16) and (7.31). On the other hand, if λ lies in one of the chambers II or III, then $M^1(S_0, \boldsymbol{L}_{\lambda-\rho})$ contains F_λ as unique irreducible submodule, as follows from the duality (7.16) and (7.35-36):

$$0 \to F_\lambda \to M^1(S_0, \boldsymbol{L}_{\lambda-\rho}) \to P_\lambda \to 0 \quad (\ell_\alpha > 0 > \ell_\beta \text{ or } \ell_\beta > 0 > \ell_\alpha) \tag{7.39}$$

is exact, but does not split. The restriction $\ell_\alpha+\ell_\beta \leq 0$ makes the realization unique, except on the line $\ell_\alpha + \ell_\beta = 0$.

I had remarked in the introduction that standard modules arise in situations when there is a vanishing theorem. The example of $\mathrm{Sl}(2,\mathbb{C})$ illustrates two general phenomena. First, higher cohomology does occur: according to (7.37), $H^1\left(X, j_+\mathcal{O}_{Q_0}\right)(\boldsymbol{L}_{\lambda-\rho}))$ is non-zero for λ in the anti-dominant chamber IV; dually $M^p(S_0,\boldsymbol{L}_{\lambda-\rho}) \neq 0$ for λ in the dominant chamber, in degree $p = 0$—which is different from the generic degree $s = s(S_0) = 1$ for cohomology on S_0. Of course, neither of these two cases contributes to any of the classifications. Secondly, higher cohomology for $K_{\mathbb{C}}$-orbits, and cohomology for G-orbits in degrees other than s, tend to vanish on regions considerably larger than those that enter the various classifications—in the example above, not only on one side of the line $\ell_\alpha + \ell_\beta = 0$, but always on at least three of the four Weyl chambers. In general, vanishing of cohomology on large regions for the parameter λ follows from the vanishing theorem (3.6B) of Beilinson-Bernstein, an obvious vanishing statement for parabolically induced modules (cf. (4.12)), and change of polarization [6].

To finish the example of $\mathrm{Sl}(2,\mathbb{C})$, I must still treat the case of an integral, but singular infinitesimal character χ_λ. This eliminates the irreducible finite dimensional module F_λ. Now the various short exact sequences in the earlier discussion collapse:

$$(7.40)\qquad \begin{gathered} P_\lambda = H^0\left(X, j_+\mathcal{O}_{Q_0}(\boldsymbol{L}_{\lambda-\rho})\right) \cong H^0\left(X, j_+\mathcal{O}_{Q_1}(\boldsymbol{L}_{\lambda-\rho})\right) \cong \\ M^0\left(S_1,\boldsymbol{L}_{\lambda-\rho}\right) \cong M^1\left(S_0,\boldsymbol{L}_{\lambda-\rho}\right) \end{gathered}$$

is nonzero, irreducible, self-dual, and belongs to the principal series, for any integral and singular parameter λ; this is the one and only irreducible Harish-Chandra module with infinitesimal character χ_λ. All other cohomology groups vanish. The Langlands classification attaches P_λ to the closed G-orbit S_0, with λ on the walls of the anti-dominant chamber IV, whereas cohomological induction realizes P_λ on the open G-orbit S_1, again with λ on the walls of chamber IV. The Beilinson-Bernstein classification realizes irreducible Harish-Chandra modules as spaces of sections of irreducible sheaves of $\mathcal{D}_\lambda$-modules. In this particular situation, the sheaf $j_+\mathcal{O}_{Q_1}(\boldsymbol{L}_{\lambda-\rho})$ reduces, so Beilinson Bernstein attach the irreducible module P_λ to the orbit Q_0, with λ dominant—in other words, with λ on the walls of chamber I.

References

1. A. Beilinson and J. Bernstein, *Localization de $\mathfrak{g}$-modules*, C. R. Acad. Sci. Paris, Ser. I **292** (1981), 15–18.
2. ______, *A generalization of Casselman's submodule theorem*, Representation Theory of Reductive Groups (P. C. Trombi, ed.), Progress in Mathematics **40**, Birkhäuser, Boston, 1983, pp. 35–52.
3. A. Borel, et. al., *Algebraic $\mathcal{D}$-modules*, Academic Press, Boston/New York, 1986.

4. J. T. Chang, *Special K-types, tempered characters and the Beilinson-Bernstein realization*, Duke J. Math **56** (1988), 345–383.

5. H. Hecht, D. Miličić, W. Schmid, and J. A. Wolf, *Localization and standard modules for real semisimple Lie groups I: The duality theorem*, Inventiones Math. **90** (1987), 297–332.

6. ———, *Localization and standard modules for real semisimple Lie groups II: Irreducibility, vanishing theorems and classification*, in preparation.

7. H. Hecht and J. L. Taylor, *Analytic localization of representations*, Advances in Math. **79** (1990), 139–212.

8. A. Knapp and G. J. Zuckerman, *Classification of irreducible tempered representations of semisimple Lie groups*, Ann. of Math. **116** (1982), 389–455.

9. H. Komatsu, *Relative cohomology of sheaves of solutions of differential equations*, Hyperfunctions and pseudo-differential equations, Lecture Notes in Mathematics **287**, Springer-Verlag, Berlin/Heidelberg/New York, 1973, pp. 192–261.

10. B. Kostant, *Orbits, symplectic structures, and representation theory*, Proceedings, U.S.-Japan Seminar on Differential Geometry, Kyoto, 1965, p. 71.

11. R. P. Langlands, *On the classification of irreducible representations of real algebraic groups*, mimeographed notes, Institute for Advanced Study, 1973; Reprinted in Representation theory and harmonic analysis on semisimple Lie groups (P. J. Sally, Jr. and D. A. Vogan, eds.), Mathematical Surveys and Monographs, vol. 31, Amer. Math. Soc., Providence, 1989, pp. 101–170.

12. T. Matsuki, *The orbits of affine symmetric spaces under the action of minimal parabolic subgroups*, J. Math. Soc. Japan **31** (1979), 332–357.

13. D. Miličić, *Localization and representation theory of reductive Lie groups*, mimeographed notes (to appear).

14. ———, *Intertwining functors and irreducibility of standard Harish-Chandra sheaves*, in this volume.

15. I. Mirkovic, *Classification of irreducible tempered representations of semisimple Lie groups*, Ph.D. Thesis, Univ. of Utah, Salt Lake City, 1986.

16. W. Schmid, *Homogeneous complex manifolds and representations of semisimple Lie groups*, Ph.D. Thesis, U.C. Berkeley, 1967; Reprinted in Representation theory and harmonic analysis on semisimple Lie groups (P. J. Sally, Jr. and D. A. Vogan, eds.), Mathematical Surveys and Monographs, vol. 31, Amer. Math. Soc., Providence, 1989, pp. 223–286.

17. ———, *Geometric constructions of representations*, Representations of Lie groups, Kyoto, Hiroshima, 1986, Advanced Studies in Pure Mathematics **14**, Kinokuniya, Tokyo, 1988, pp. 349–368.

18. W. Schmid and J. A. Wolf, *Geometric quantization and derived functor modules for semisimple Lie groups*, J. Func. Analysis **90** (1990), 48–112.

19. D. Vogan, *The algebraic structure of representations of semisimple Lie groups I*, Ann. of Math. **109** (1979), 1–60.

20. ———, *Representations of real reductive Lie groups*, Progress in Mathematics, vol. 15, Birkhäuser, Boston, 1981.

21. ———, *Irreducible characters of semisimple Lie groups III: Proof of the Kazhdan-Lusztig conjectures in the integral case*, Invent. Math. **71** (1983), 381–417.

DEPARTMENT OF MATHEMATICS, HARVARD UNIVERSITY, CAMBRIDGE, MASS. 02138

LANGLANDS' CONJECTURE ON PLANCHEREL MEASURES FOR p-ADIC GROUPS

FREYDOON SHAHIDI

Purdue University

1. INTRODUCTION

One of the major achievements of Harish-Chandra was a derivation of the Plancherel formula for real and p-adic groups [9,10]. To have an explicit formula, one will have to compute the measures appearing in the formula; the so called Plancherel measures and formal degrees [12]. (For reasons stemming from L-indistinguishability, we would like to distinguish between the formal degrees for discrete series and the Plancherel measures for non-discrete tempered representations, cf. Proposition 9.3 of [29].) While for real groups the Plancherel measures are completely understood [1, 9, 22], until recently little was known in any generality for p-adic groups [29] (except for their rationality and general form due to Silberger [39]). On the other hand any systematic study of the non-discrete tempered spectrum of a p-adic group would very likely have to follow the path of Knapp and Stein [20, 21] and their theory of R-groups. Since the basic reducibility theorems for p-adic groups are available [40, 41], it is the knowledge of Plancherel measures which would be necessary to determine the R-groups. This is particularly evident from the important and the fundamental work of Keys [16, 17, 18] and the work of the author [29, 30, 31].

On the other hand, based on his results on constant terms of Eisenstein series [23], Langlands conjectured that every Plancherel measure must be a product of certain root numbers with the ratios of the corresponding Langlands L-functions at $s = 1$ and $s = 0$. Otherwise said, he suggested that one must be able to normalize the standard intertwining operators by means of certain local root numbers and L-functions [24]. We refer to the introduction of [29] and to [2, 3, 4, 5] for applications of such normalizations. This was further tested for real groups by Arthur [1]. Since in many instances these local factors (especially L-functions which then determine the poles and zeros of Plancherel measures and thus answer reducibility questions) can be explicitly computed, this leads to explicit formulas for Plancherel measures that are not available from any other method [29, 31,

Partially supported by NSF Grants DMS-8800761 and DMS-9000256.

32]. In fact, except for the the cases coming from minimal parabolics [16, 17, 18], there is not a single example of a Plancherel measure coming from a supercuspidal representation of a non-minimal Levi subgroup of a p-adic group that has been computed in any other way. (We understand that, using Howe-Moy Hecke algebra isomorphisms [11], they can also calculate Plancherel measures in some non-minimal cases.) The main result (Theorem 7.9) of [29] was to prove this conjecture for generic representations of Levi subgroups of quasi-split groups and to set up a program to attack the conjecture in general (Theorem 9.5 of [29] and the conjectural fact that the Plancherel measures are preserved by inner forms [5]). Incidently, we should remark that, although we have not checked carefully, the proof of Proposition 9.6 of [29], when applied to real groups by means of Shelstad's results [38], should basically lead to another proof of Vogan's result on genericity of tempered L-packets for real groups [42].

As a consequence of Theorems 3.5 and 7.9 of [29], a general result (Theorem 8.1 of [29]) was established on reducibility of induced representations from generic supercuspidal representations of maximal Levi subgroups of a quasi-split group. This implied, in particular, that the edge of complementary series (if any) can only have two possible choices, $\tilde{\alpha}$ and $\frac{1}{2}\tilde{\alpha}$, no matter what the group or inducing representation.

The first part of this article (Theorems 3.1, 4.2, and 5.1) is aimed at a survey of these results. In the second part we give three examples. The first two examples, described in Propositions 6.1 and 6.2, determine the reducibility of the representations induced from supercuspidal representations of Levi factors of the Siegel parabolic subgroups of Sp_4, PSp_4, and GSp_4, and the parabolic subgroup of an exceptional split group of type G_2 whose Levi subgroup is generated by the short simple root of G_2. Together with Propositions 8.3 and 8.4 of [29] (also Proposition 5.1 of [43] for GSp_4), this completes the analysis of the unitary duals of all the rank two split p-adic groups supported on their maximal parabolics. Propositions 6.1 and 6.2 both seem to be new.

The paper concludes with explicit formulas for the Plancherel measures for $GL(n)$ and $SL(n)$, presented in Proposition 7.1 and Corollary 7.2, respectively. A formula in terms of certain Rankin-Selberg L-functions for $GL(n)$ [13] for these measures has been in print [32] since 1984. But an explicit expression for the Plancherel measures should make it more convenient to calculate R-groups for $SL(n)$ (Remark 7.3).

2. Notation and Preliminaries

Let F be a non-archimedean field of characteristic zero whose ring of integers and maximal ideal are denoted by O and P, respectively. We shall always fix a uniformizing parameter ϖ, i.e., an element of P such that $P = (\varpi)$. Let q denote the number of elements in the residue field O/P. We use $|\cdot|_F$ to denote the absolute value of F. Then $|\varpi|_F = q^{-1}$.

Let $\mathbf{G}$ be a quasi-split connected reductive algebraic group over F. Fix a Borel subgroup $\mathbf{B}$ of $\mathbf{G}$ over F and write $\mathbf{B} = \mathbf{TU}$, where $\mathbf{T}$ is a maximal torus and $\mathbf{U}$ denotes the unipotent radical of $\mathbf{B}$. For an F-parabolic subgroup $\mathbf{P}$ of $\mathbf{G}$ containing $\mathbf{B}$, let $\mathbf{P} = \mathbf{MN}$ be a Levi decomposition. Then $\mathbf{U} \supset \mathbf{N}$. We use G, B, T, U, P, M, and N to denote the corresponding groups of F-rational points.

If $\mathfrak{a} = \mathrm{Hom}(X(\mathbf{M})_F, \mathbb{R})$ is the real Lie algebra of the maximal split torus $\mathbf{A}$ of the center of $\mathbf{M}$, we use $H_P : \mathbf{M} \to \mathfrak{a}$ to denote the homomorphism of [35,41] defined by

$$q^{\langle \chi, H_P(m) \rangle} = |\chi(m)|_F.$$

Here $X(\mathbf{M})_F$ is the group of F-rational characters of $\mathbf{M}$.

If $\mathbf{A}_0$ is the maximal F-split torus in $\mathbf{T}$, $\mathbf{A}_0 \supset \mathbf{A}$, let ψ be the set of F-roots of $\mathbf{A}_0$. Then $\psi = \psi_+ \cup \psi_-$, where ψ_+ is the set of positive roots of $\mathbf{A}_0$, i.e., those roots generating $\mathbf{U}$. Let $\Delta \subset \psi_+$ be the set of simple roots. Then $\mathbf{M} = \mathbf{M}_\theta$ is generated by a subset θ of Δ.

Given an irreducible admissible representation σ of M and $\nu \in \mathfrak{a}^*_{\mathbb{C}}$, the complex dual of $\mathfrak{a}$, define

$$I(\nu, \sigma) = \operatorname*{Ind}_{MN \uparrow G} \sigma \otimes q^{\langle \nu, H_P(\ ) \rangle} \otimes \mathbf{1}.$$

We use $V(\nu, \sigma)$ to denote the space of $I(\nu, \sigma)$ and we let $I(\sigma) = I(0, \sigma)$ and $V(\sigma) = V(0, \sigma)$.

Let $W(\mathbf{A}_0)$ be the Weyl group of $\mathbf{A}_0$ in $\mathbf{G}$. Fix a $\widetilde{w} \in W(\mathbf{A}_0)$ such that $\widetilde{w}(\theta) \subset \Delta$. Choose a representative $w \in G$ for $\widetilde{w}$. Let $\mathbf{N}_{\widetilde{w}} = \mathbf{U} \cap w\mathbf{N}^- w^{-1}$, where $\mathbf{N}^-$ is the unipotent subgroup of $\mathbf{G}$ opposed to $\mathbf{N}$. Given $f \in V(\nu, \sigma)$, let

$$A(\nu, \sigma, w) f(g) = \int_{N_{\widetilde{w}}} f(w^{-1} n g)\, dn.$$

The integral converges for ν in a certain cone and can be analytically continued to a meromorphic function of ν on all of $\mathfrak{a}^*_{\mathbb{C}}$ (cf. [20, 21, 34, 41]). It intertwines $I(\nu, \sigma)$ with $I(\widetilde{w}(\nu), \widetilde{w}(\sigma))$, where $\widetilde{w}(\sigma)(m') = \sigma(w^{-1} m' w)$, $m' \in M' = wMw^{-1}$. Finally, let $A(\sigma, w) = A(0, \sigma, w)$.

If $\widetilde{w}_0$ is the longest element in the Weyl group of $\mathbf{A}_0$ in $\mathbf{G}$ modulo that of $\mathbf{A}_0$ in $\mathbf{M}$, the Plancherel constant $\mu(\nu, \sigma)$ is defined by

$$A(\nu, \sigma, w_0) A(\widetilde{w}_0(\nu), \widetilde{w}_0(\sigma), w_0^{-1}) = \mu(\nu, \sigma)^{-1} \gamma(G/P)^2.$$

Here

$$\gamma(G/P) = \int_{\overline{\mathbf{N}}_{\widetilde{w}_0}(F)} q^{\langle 2\rho_P, H_P(\overline{n}) \rangle} d\overline{n},$$

where $\overline{\mathbf{N}}_{\widetilde{w}_0} = w_0^{-1} \mathbf{N}_{\widetilde{w}} w_0$. The constant $\mu(\nu, \sigma)$ does not depend on the choice of w_0, nor on that of the defining measures. Also, its dependence on σ is only via the equivalence class of the representation.

Now let ${}^LM$ be the L-group of $\mathbf{M}$. Denote by ${}^L\mathfrak{n}$ the Lie algebra of the L-group ${}^LN$ of $\mathbf{N}$. The group ${}^LM$ acts by adjoint action on ${}^L\mathfrak{n}$. If ${}^L\mathfrak{n}_{\widetilde{w}}$ denotes the Lie algebra of the L-group of $\mathbf{N}_{\widetilde{w}}$, then ${}^L\mathfrak{n}_{\widetilde{w}}$, realized as a subspace of ${}^L\mathfrak{n}$ by $-Ad[w]$, is stable under this adjoint action. Let r and $r_{\widetilde{w}}$ be the adjoint actions of ${}^LM$ on ${}^L\mathfrak{n}$ and ${}^L\mathfrak{n}_{\widetilde{w}}$, respectively.

Let $\rho = \rho_{\mathbf{P}}$ be half the sum of roots whose root spaces generate $\mathbf{N}$. Then, for each α with $X_{\alpha^\vee} \in {}^L\mathfrak{n}$, $\langle 2\rho, \alpha \rangle$ is a positive integer. Let $a_1 < a_2 < \ldots < a_n$ be the distinct values of $\langle 2\rho, \alpha\rangle$. Set

$$V_i = \{X_{\alpha^\vee} \in {}^L\mathfrak{n}_{\widetilde{w}} | \langle 2\rho, \alpha\rangle = a_i\}.$$

Each V_i is invariant under $r_{\widetilde{w}}$. We let $r_{\widetilde{w},i}$ be the restriction of $r_{\widetilde{w}}$ to V_i.

If $\mathbf{P}$ is maximal, then for the non-trivial $\widetilde{w}$ we set $r_{\widetilde{w}} = r$. We then set $r_i = r_{\widetilde{w},i}$; thus $r = \bigoplus_{i=1}^{m} r_i$ with each r_i irreducible (cf. [35]). Let $\alpha \in \Delta$ identify the unique reduced root of $\mathbf{A}$ in $\mathbf{N}$. Set $\widetilde{\alpha} = \langle \rho, \alpha\rangle^{-1}\rho$, an element of $\mathfrak{a}^* = X(\mathbf{M})_F \otimes_{\mathbb{Z}} \mathbb{R}$. Observe that, for each i, $1 \le i \le m$,

$$V_i = \{X_{\beta^\vee} \in {}^L\mathfrak{n} | \langle \widetilde{\alpha}, \beta\rangle = i\},$$

and therefore each V_i is an eigenspace for the action of the connected center of ${}^LM^0$, the connected component of ${}^LM$.

When $F = \mathbb{R}$ we have similar definitions for which we use the same notation.

Fix a non-trivial additive character ψ_F of F. We shall now define a generic character χ of U. Let $\alpha \in \Delta$. If E_α is the smallest extension of F over which the rank one subgroup of $\mathbf{G}$ generated by α splits, we let $\chi | U_\alpha = \psi_F \cdot \mathrm{Tr}_{E_\alpha/F}$. By restriction χ is also a generic character of U^0 which we still denote by χ. Here $\mathbf{U}^0 = \mathbf{U} \cap \mathbf{M}$.

The representation σ is called χ-generic if it can be realized on a space of functions W^0 satisfying $W^0(um) = \chi(u)W^0(m)$. Changing the splitting on $\mathbf{G}$ (or, said in other words, up to L-indistinguishability) every irreducible admissible generic representation is χ-generic with respect to such a χ (cf. [29]). Moreover by Conjecture 9.4 of [29] and Section 6 of [42] every tempered L-packet contains such a representation. Generic representations are thus much more fundamental than once thought.

Next, suppose $\sigma \subset \operatorname{Ind}_{M_\theta N_\theta \uparrow M} \sigma_1 \otimes \mathbf{1}$, where, for each $\theta \subset \Delta$, $\mathbf{M}_\theta\mathbf{N}_\theta$, is a parabolic subgroup of $\mathbf{M}$ and σ_1 is an irreducible admissible representation of M_θ. Let $\theta' = \widetilde{w}(\theta) \subset \Delta$ and fix a reduced decomposition $\widetilde{w} = \widetilde{w}_{n-1} \ldots \widetilde{w}_1$ as in Lemma 2.1.1 of [34]. Then, for each j, there exists a unique root $\alpha_j \in \Delta$ such that $\widetilde{w}_j(\alpha_j) < 0$. For each j, $2 \le j \le n-1$, let $\overline{w}_j = \widetilde{w}_{j-1} \ldots \widetilde{w}_i$. Set $\overline{w}_1 = 1$. Moreover let $\Omega_j = \theta_j \cup \{\alpha_j\}$, where $\theta_1 = \theta$, $\theta_n = \theta'$, and $\theta_{j+1} = \widetilde{w}_j(\theta_j)$, $1 \le j \le n-1$. Then the group $\mathbf{M}_{\Omega_j}$ contains $\mathbf{M}_{\theta_j}\mathbf{N}_{\theta_j}$ as a maximal parabolic subgroup and $\overline{w}_j(\sigma_1)$ is a representation of M_{θ_j}. The L-group ${}^LM_\theta$ acts on V_i. Given an irreducible component of this action, there exists a unique j, $1 \leq j \leq n-1$, which, under $\overline{w}_j$, makes this component equivalent to an irreducible constituent of the action of ${}^L M_j$ on the Lie algebra of ${}^L N_{\theta_j}$ ($\mathbf{M}_{\theta_j}\mathbf{N}_{\theta_j}$ is a maximal parabolic subgroup of $\mathbf{M}_{\Omega_j}$). We denote by $i(j)$ the index of this subspace of the Lie algebra of ${}^L N_{\theta_j}$. Finally let S_i denote the set of all such j's; S_i is, in general, a proper subset of $1 \leq j \leq n-1$. We refer to [36] for several examples.

Langlands' conjecture on Plancherel measures is global by nature. In other words the most important property of the standard normalization is that the normalizing factors can be related globally. For this reason one needs statements about groups over global fields and global L-functions. For this we need further preparation.

Let K be a number field and fix a place v of K. If $\mathbf{G}$ is a quasi-split connected reductive group over K, we use $\mathbf{G} \times_K K_v$ to denote $\mathbf{G}$ as a group over K_v, the completion of K at v. Next suppose $\mathbb{A}_K$ is the ring of adeles of K. Fix a non-trivial character $\psi = \otimes_v \psi_v$ of $\mathbb{A}_K$, trivial on K, and define a character $\chi = \otimes_v \chi_v$ of $\mathbf{U}(\mathbb{A}_K)$ as before, where $\mathbf{U}$ is the unipotent radical of a Borel subgroup of $\mathbf{G}$.

Let $\pi = \otimes_v \pi_v$ be a cusp form on $M = \mathbf{M}(\mathbb{A}_K)$. We shall say π is globally χ-generic if there exists a function φ in the space of π such that

$$\int_{\mathbf{U}^0(K)\backslash \mathbf{U}^0(\mathbb{A}_K)} \varphi(u)\overline{\chi(u)}du \neq 0,$$

where $\mathbf{U}^0 = \mathbf{U} \cap \mathbf{M}$.

Now assume v is such that ψ_v, $\mathbf{G} \times_K K_v$, and π_v are all unramified. If ρ is an analytic representation of ${}^L M$, we define ρ_v by $\rho_v = \rho \cdot \eta_v$, where η_v is the natural map $\eta_v : {}^L M_v \to {}^L M$. Here ${}^L M_v$ is the L-group of $\mathbf{M} \times_K K_v$. Let $L(s, \pi_v, \rho_v)$ be the local Langlands L-function attached to π_v and ρ_v (cf. [7,25]). If S is a finite set of places of K outside of which everything is unramified, we set

$$L_S(s, \pi, \rho) = \prod_{v \notin S} L(s, \pi_v, \rho_v).$$

The first aim of this paper is to define these L-functions at the other places (Section 7 of [29]).

3. The Fundamental Theorem

The standard normalization of intertwining operators as conjectured by Langlands is done by means of certain L-functions and root numbers [24]. Crucial among their properties is that, whenever the representation becomes a local component of a global cusp form, the factors must be those satisfying the corresponding functional equation. In fact this is the only way one can globally relate normalizing factors on different local groups to each other. This is important in many deep applications of the trace formula [2, 3, 4, 19, 30]. To define these local factors, the following theorem was proved in [29] (Theorem 3.5 of [29]).

Theorem 3.1. *Given a local field F of characteristic zero and a quasi-split connected reductive algebraic group* $\mathbf{G}$ *over F containing an F-parabolic subgroup* $\mathbf{P} = \mathbf{MN}$, $\mathbf{N} \subset \mathbf{U}$, *let* $r_{\tilde{w}} = \bigoplus_{i=1}^{m} r_{\tilde{w},i}$ *be the adjoint action of* ${}^L M$ *on* ${}^L \mathfrak{n}_{\tilde{w}}$ *as in Section 2. Then, for every irreducible admissible χ-generic representation σ of M, there exists m complex functions* $\gamma_i(s, \sigma, \psi_F, \tilde{w})$, $s \in \mathbb{C}$, *satisfying the following properties:*

1) If F is archimedean or σ has an Iwahori fixed vector, let $\varphi' : W'_F \to {}^L M$ *be the homomorphism attached to σ, where W'_F is the Deligne-Weil group of F. Denote by* $\varepsilon(s,\ r_{\tilde{w},i} \cdot \varphi',\ \psi_F)$ *and* $L(s,\ r_{\tilde{w},i} \cdot \varphi')$, *the Artin root number and L-function attached to* $r_{\tilde{w},i} \cdot \varphi'$, *respectively. Then*

$$\gamma_i(s, \sigma, \psi_F, \tilde{w}) = \varepsilon(s,\ r_{\tilde{w},i} \cdot \varphi',\ \psi_F) L(1-s,\ \tilde{r}_{\tilde{w},i} \cdot \varphi') / L(s,\ r_{\tilde{w},i} \cdot \varphi').$$

2) For each i, $1 \leq i \leq m$,

$$\gamma_i(s, \sigma, \psi_F, \tilde{w}) \gamma_i(1-s, \tilde{\sigma}, \overline{\psi}_F, \tilde{w}) = 1, \text{ and} \tag{3.1}$$
$$\gamma_i(s, \sigma, \psi_F, \tilde{w}) = \gamma_i(s + s_0, \sigma_0, \psi_F, \tilde{w}),$$

where $\sigma = \sigma_0 \otimes q^{\langle s_0 \tilde{\alpha}, H_P(\ ) \rangle}$ *if F is nonarchimedean.*

3) Inductive property: Suppose $\sigma \subset \operatorname{Ind}_{M_\theta N_\theta \uparrow M} \sigma_1 \otimes \mathbf{1}$, *where* $\mathbf{M}_\theta \mathbf{N}_\theta$ *is a parabolic subgroup of* $\mathbf{M}$ *and σ_1 is an irreducible admissible χ-generic representation of M_θ. Write* $\tilde{w} = \tilde{w}_{n-1} \dots \tilde{w}_1$ *and for each j, $2 \leq j \leq n-1$, let* $\overline{w}_j = \tilde{w}_{j-1} \dots \tilde{w}_1$ *and* $\overline{w}_1 = 1$. *Then for each j, $\overline{w}_j(\sigma_1)$ is a representation of M_{θ_j}. If, for each $j \in S_i$,* $\gamma_{i(j)}(s, \overline{w}_j(\sigma_1), \psi_F, \tilde{w}_j)$, $1 \leq i \leq m$, *denotes the corresponding factor, then*

$$\gamma_i(s, \sigma, \psi_F, \tilde{w}) = \prod_{j \in S_i} \gamma_{i(j)}(s, \overline{w}_j(\sigma_1), \psi_F, \tilde{w}_j).$$

4) Functional equations: Let K be a number field and $\mathbf{G}$ *a quasi-split connected reductive algebraic group over K. Let* $\mathbf{P} = \mathbf{MN}$, $\mathbf{N} \subset \mathbf{U}$, *be a maximal K-parabolic subgroup of* $\mathbf{G}$. *Fix a non-degenerate character* $\chi = \bigotimes_v \chi_v$ *of* $U = \mathbf{U}(\mathbb{A}_K)$, *trivial on K and defined by a non-trivial character* $\psi = \bigotimes_v \psi_v$ *of* $K \backslash \mathbb{A}_K$. *Let* $\pi = \bigotimes_v \pi_v$ *be a globally χ-generic cusp form on* $\mathbf{M}(\mathbb{A}_K)$. *Finally, if r is the adjoint action of ${}^L M$ on ${}^L \mathfrak{n}$, write* $r = \bigoplus_{i=1}^{m} r_i$. *Then* $r_v = \bigoplus_{i=1}^{m} r_{i,v}$ *is the adjoint action of* ${}^L M_v$, *where* $r_v = r \cdot \eta_v$, $r_{i,v} = r_i \cdot \eta_v$, *and* $\eta_v : {}^L M_v \to {}^L M$ *is the natural map. Let S be a finite set of places of K such that, for $v \notin S$,* $\mathbf{G} \times_K K_v$, π_v, *and χ_v are all unramified. Then*

$$L_S(s, \pi, r_i) = \prod_{v \in S} \gamma_i(s, \pi_v, \psi_v) L_S(1-s, \pi, \tilde{r}_i),$$

for every i, $1 \leq i \leq m$, where $\gamma_i(s, \pi_v, \psi_v) = \gamma_i(s, \pi_v, \psi_v, \tilde{w}_0)$.

Moreover, conditions (1), (3), and (4) determine the γ_i uniquely.

Remark 3.2. The factors $\gamma_i(s, \sigma, \psi_F, \tilde{w})$ are all defined locally by means of local coefficients (see the next remark). But to prove their properties and uniqueness one has to employ global methods—more precisely, functional equations. In fact, local proofs are available only for $\mathbf{G} = GL(n)$, and even in that case they are fairly deep and complicated [13, 32]. Thus one must again observe how powerful global methods can be in answering local questions.

Remark 3.3. To make this survey short, we have suppressed the important role played by the theory of local coefficients which was developed in [29, 34, 37].

4. Local Factors and Langlands Conjecture

When $F = \mathbb{R}$ our local factors are those of Artin [27, 37] as defined in [26] (part 1 of Theorem 3.1). In this case the Langlands Conjecture has been verified by Arthur [1]. Therefore for the remainder of this article we shall assume F is non-archimedean.

We first define our local root numbers and L-functions. Start with a maximal parabolic $\mathbf{P}$ and fix an irreducible tempered χ-generic representation σ of M. From now on we use $\gamma(s,\sigma,r_i,\psi_F)$ to denote $\gamma_i(s,\sigma,\psi_F)$, $1 \le i \le m$.

For each i, let $P_{\sigma,i}(t)$ be the unique polynomial satisfying $P_{\sigma,i}(0) = 1$ such that $P_{\sigma,i}(q^{-s})$ has the same zeros as $\gamma(s,\sigma,r_i,\psi_F)$, i.e., $P_{\sigma,i}(t)$ is the unique numerator of $\gamma(s,\sigma,r_i,\psi_F)$ (which is a rational function of q^{-s}) satisfying $P_{\sigma,i}(0) = 1$. Define the L-functions attached to σ, r_i, and $\widetilde{r}_i$ as

$$L(s,\sigma,r_i) = P_{\sigma,i}(q^{-s})^{-1} \quad \text{and} \quad L(s,\sigma,\widetilde{r}_i) = P_{\widetilde{\sigma},i}(q^{-s})^{-1}.$$

They do not depend on ψ_F. Then by (3.1)

$$\gamma(s,\sigma,r_i,\psi_F)L(s,\sigma,r_i)/L(1-s,\sigma,\widetilde{r}_i)$$

is a monomial in q^{-s} which we denote by $\varepsilon(s,\sigma,r_i,\psi_F)$, the root number attached to σ and r_i. Thus

$$\gamma(s,\sigma,r_i,\psi_F) = \varepsilon(s,\sigma,r_i,\psi_F)L(1-s,\sigma,\widetilde{r}_i)/L(s,\sigma,r_i).$$

The definition of L and ε for non-tempered σ is then a consequence of inductive property (3) and the Langlands classification for irreducible admissible representations of p-adic groups (due to Borel-Wallach and Silberger).

The following natural conjecture serves two purposes: On the one hand it provides one of the conditions on normalizing factors demanded by Arthur (Condition R_7 of [1]; it was not among the original conditions conjectured by Langlands in [24]; also see [29]). On the other hand, using inductive property (3), it allows us to prove the multiplicative properties of these factors in general. This is of great interest in the theory of automorphic L-functions. We refer to [36] for an account of this and several examples.

Conjecture 4.1. *If σ is tempered, then each $L(s,\sigma,r_i)$ is holomorphic for $Re(s)>0$.*

It is enough to prove this for σ in the discrete series and $\mathbf{P}$ maximal.

By Proposition 7.2 and 7.3 of [29] the conjecture is a theorem if $m=1$; $m=2$ and

$$L(s,\sigma,r_2)=\prod_j(1-a_j q^{-s})^{-1}\quad(\alpha_j\in\mathbb{C}),$$

(possibly empty) with $|\alpha_j|=1$; m is arbitrary but σ is unitary supercuspidal; and finally $\mathbf{G}$ is a simple classical group, $\mathbf{M}=\mathbf{H}\times GL(n)$ for some classical group H, and σ is at least supercuspidal on one of the factors H or $GL_n(F)$ (Theorem 5.5 of [36]).

Now assume $\mathbf{P}=\mathbf{MN}$, $\mathbf{N}\subset\mathbf{U}$, is any standard parabolic subgroup of $\mathbf{G}$. Fix $\widetilde{w}\in W(\mathbf{A}_0)$ such that $\widetilde{w}(\theta)\subset\Delta$, $\mathbf{P}=\mathbf{P}_\theta$. Let σ be an irreducible unitary χ-generic representation of M. We fix a reduced decomposition $\widetilde{w}=\widetilde{w}_{n-1}\dots\widetilde{w}_1$ and set

$$L(s,\sigma,r_{\widetilde{w}})=\prod_{j=1}^{n-1}\prod_{i=1}^{m_j}L(s,\overline{w}_j(\sigma),r_{\widetilde{w}_j,i})$$

and

$$\varepsilon(s,\sigma,r_{\widetilde{w}},\psi_F)=\prod_{j=1}^{n-1}\prod_{i=1}^{m_j}\varepsilon(s,\overline{w}_j(\sigma),r_{\widetilde{w}_j,i},\psi_F).$$

They are both independent of the decomposition of $\widetilde{w}$.

We shall now normalize the intertwining operator $A(\sigma,w)$ in the way conjectured by Langlands [24]. Let

$$\mathcal{A}(\sigma,w)=\varepsilon(0,\sigma,\widetilde{r}_{\widetilde{w}},\psi_F)L(1,\sigma,\widetilde{r}_{\widetilde{w}})L(0,\sigma,\widetilde{r}_{\widetilde{w}})^{-1}A(\sigma,w),$$

where the right hand side is defined as a limit.

Theorem 4.2. *(Langlands' conjecture). The normalized operator $\mathcal{A}(\sigma,w)$ satisfies:*

a) $\mathcal{A}(\sigma,w_1w_2)=\mathcal{A}(\widetilde{w}_2(\sigma),w_1)\mathcal{A}(\sigma,w_2)$, and

b) $\mathcal{A}(\sigma,w)=\mathcal{A}(\widetilde{w}(\sigma),w^{-1})$, i.e. $\mathcal{A}(\sigma,w)$ is unitary.

Remark 4.3. Theorem 4.2 is clearly equivalent to a formula for Plancherel measures in terms of L-functions and root numbers (Corollary 3.6 of [29]).

5. Reducibility of Induced Representations

One of the consequences of Theorems 3.1 and 4.2 is a general result on reducibility of representations induced from supercuspidal generic representations of Levi factors of maximal parabolic subgroups of any quasi-split group in terms of polynomials $P_{\sigma,i}$, $i=1,2$. More precisely, even the equality (3.1) of Theorem 3.1 allows us to determine the edge of complementary series coming from maximal parabolic subgroups and generic supercuspidal representations of such groups. They turn out to take only two values (if any), $\widetilde{\alpha}$ or $\frac{1}{2}\widetilde{\alpha}$, no matter what the representation and the group are. As mentioned above, such a result is possible only because of the identity (3.1) which provides the only important unknown in the formula for the Plancherel measure obtained by Silberger [39]. This identity is quite deep and is proved by global methods. Its local proof in this generality seems far from reach at present (cf. [13, 32] and the remark at the end of Section 3 of [29] for $GL(n)$). Finally we should remark that in many cases the Levi subgroups are products of A-type groups for which supercuspidal representations are always generic and our assumptions on genericity is automatically satisfied. As a consequence of this, in the next section we shall obtain those parts of the unitary duals of all the rank two split p-adic groups which are supported on their maximal parabolic subgroups.

Let $\mathbf{P}=\mathbf{MN}$ be a maximal parabolic subgroup of a quasi-split connected reductive algebraic group over a p-adic field F. Let σ be an irreducible unitary supercuspidal χ-generic representation of M. Then by Lemma 7.5 of [29], $P_{\sigma,i}\equiv 1$ for $3\leq i\leq m$. Moreover $I(\sigma)$ is irreducible and σ is ramified, i.e. $\widetilde{w}_0(\sigma)\cong\sigma$ if and only if $P_{\sigma,i}(1)=0$ for exactly one of the values $i=1$ or 2 (Corollary 7.6 of [29]). The following is Theorem 8.1 of [29].

Theorem 5.1. *Let $\mathbf{P}=\mathbf{MN}$ be a maximal parabolic subgroup of $\mathbf{G}$, where $\mathbf{G}$ is a quasi-split connected reductive algebraic group over a p-adic field. Let σ be an irreducible unitary supercuspidal χ-generic representation of M. Assume σ is ramified and $I(\sigma)$ is irreducible. Choose a unique i, $i=1$ or 2, such that $P_{\sigma,i}(1)=0$. Then:*

a) *For $0<s<\frac{1}{i}$, $I(s\widetilde{\alpha},\sigma)$ is irreducible and in the complementary series.*
b) *$I(\widetilde{\alpha}/i,\sigma)$ is reducible with a unique χ-generic subrepresentation which is in the discrete series. Its Langlands quotient is never generic. It is a preunitary, non-tempered representation.*
c) *For $s>\frac{1}{i}$, $I(s\widetilde{\alpha},\sigma)$ is irreducible and not in the complementary series.*

If σ is ramified and $I(\sigma)$ is reducible, then no $I(s\widetilde{\alpha},\sigma)$, $s>0$, is preunitary. They are all irreducible. In particular the edge of complementary series (if any) is always either $\widetilde{\alpha}$ or $\frac{1}{2}\widetilde{\alpha}$.

We now state the following corollary, expressing our results in terms of points of reducibility of induced representations.

Corollary 5.2. *Let σ be an irreducible unitary supercuspidal χ-generic representation of M. Then $I(s\widetilde{\alpha},\sigma)$ is irreducible unless some unramified twist of σ is ramified. Assume σ is ramified and $I(\sigma)$ is irreducible. Then the only point of reducibility for $I(s\widetilde{\alpha},\sigma)$ in the region $s>0$ occurs at either $s=1/2$ or $s=1$. If $I(\sigma)$ is reducible, then $s=0$ gives the only point of reducibility.*

Remark 5.3. Both results are valid if σ is generic with respect to any other generic character of U.

6. Examples: Unitary Duals of Rank Two Split Groups

Since unitary duals of all the rank one quasi-split groups are completely determined [17], we can apply Theorem 5.1 to a general rank two split group and obtain that part of the unitary dual which is supported on the maximal parabolic subgroups.

We first let $\mathbf{G}$ be either Sp_4, PSp_4, or GSp_4 over a non-archimedean field F. We let $\mathbf{P} = \mathbf{MN}$ be such that $\mathbf{M}$ is generated by the short simple root α. If $\mathbf{G} = Sp_4$, then $\mathbf{M} = GL_2$. For $\mathbf{G} = PSp_4$, $\mathbf{M} = PGL_2 \times GL_1$. Otherwise it is $GL_2 \times GL_1$. Let σ be an irreducible unitary supercuspidal representation of $M = \mathbf{M}(F)$. Then $\sigma = \sigma_1 \otimes \chi$, where σ_1 is an irreducible unitary supercuspidal representation of $GL_2(F)$ and χ is a character of F^*. We disregard χ if $G = Sp_4(F)$ or $PSp_4(F)$. For $\mathbf{G} = PSp_4$, $\mathbf{M}(F) = GL_2(F)/\{\pm 1\}$ and σ will be a representation of $GL_2(F)$ trivial on $\{\pm 1\}$. Let ω be the central character of σ_1. The element $\widetilde{w}_0$ can be chosen to be $\widetilde{w}_\beta \widetilde{w}_\alpha \widetilde{w}_\beta$, where β is the long simple root. Suppose $\mathbf{G} = Sp_4$ or PSp_4. Then $\widetilde{w}_0(\sigma) \cong \sigma$ if and only if $\widetilde{\sigma}_1 \cong \sigma_1$. In particular, $\omega^2 = 1$. Otherwise, i.e. if $\mathbf{G} = GSp_4$, $\widetilde{w}_0(\sigma) \cong \sigma$ if and only if $\omega = 1$. Thanks are due to David Goldberg for pointing out this difference which was carelessly overlooked in the first version of this paper.

It is instructive to observe that if σ_1 is an irreducible unramified principal series (class one) representation of $GL_2(F)$ defined by the pair (μ_1, μ_2) of unramified characters of F^* and χ is unramified, then the corresponding semi-simple conjugacy class [7,25] in the L-group $GSp_4(\mathbb{C})$ of GSp_4 can be represented by

$$A(I(\sigma)) = \mathrm{diag}(\chi^2(\varpi), \mu_2^{-1}\chi^2(\varpi), \mu_1^{-1}\chi^2(\varpi), \mu_1^{-1}\mu_2^{-1}\chi^2(\varpi)),$$

where ϖ is a uniformizing parameter for F.

However, if $\mathbf{G} = Sp_4$ and $\sigma = \sigma_1$ as above, then ${}^L G = PSp_4(\mathbb{C})$ and

$$\begin{aligned} A(I(\sigma)) = \mathrm{diag}(&(\mu_1\mu_2(\varpi))^{1/2},\ (\mu_1\mu_2^{-1}(\varpi))^{1/2},\\ &(\mu_1^{-1}\mu_2(\varpi))^{1/2},\ (\mu_1^{-1}\mu_2^{-1}(\varpi))^{1/2}) \pmod{\pm 1}, \end{aligned}$$

where the choice of the square root is irrelevant as long as it is consistent for all the entries, i.e. $(\mu_1\mu_2^{-1}(\varpi))^{1/2} = \mu_2^{-1}(\varpi) \cdot (\mu_1\mu_2(\varpi))^{1/2}$ and so on. (That $A(I(\sigma)) = \nu \cdot \mathrm{diag}(\mu_1\mu_2(\varpi), \mu_2(\varpi), \mu_1(\varpi), 1)$ with some $\nu \in \mathbb{C}^*$ follows immediately from definitions, using roots and coroots of Sp_4. To show that $\nu = \chi^2\mu_1^{-1}\mu_2^{-1}(\varpi)$, one uses the decomposition

$$\mathrm{diag}(\eta\alpha, \eta\beta, \beta^{-1}, \alpha^{-1}) = \lambda \cdot \mathrm{diag}(ab, a, b, 1),$$

where $a = \alpha\eta\beta$, $b = \alpha\beta^{-1}$, and $\lambda = \alpha^{-1}$. The character (μ_1, μ_2, χ) of (α, β, η) is then equal to the character (μ_1', μ_2', χ') of (a, b, λ), where $\mu_1' = \chi$, $\mu_2' = \chi\mu_2^{-1}$, and $\chi' = \chi^2\mu_1^{-1}\mu_2^{-1}$. Applying the central cocharacter to

$A(I(\sigma))$ then implies $\nu = \chi^2\mu_1^{-1}\mu_2^{-1}(\varpi)$.) Thus, for example, the adjoint action r of ${}^LM$ on ${}^L\mathfrak{n}$ is simply $r = \rho_2 \oplus \wedge^2\rho_2$, where ρ_2 is the standard representation of $GL_2(\mathbb{C})$. Then

$$P_{\sigma,1}(q^{-s}) = L(s,\sigma_1,\rho_2)^{-1} = 1,$$

while

$$P_{\sigma,2}(q^{-s}) = L(s,\sigma_1,\wedge^2\rho_2)^{-1} = L(s,\omega)^{-1}.$$

We recall that

$$L(s,\omega)^{-1} = 1 - \omega(\varpi)q^{-s}$$

if ω is unramified and is identically one otherwise. Suppose $\mathbf{G} = Sp_4$ or PSp_4. If ω is unramified, $\omega^2 = 1$ with $\omega \neq 1$ implies $\omega(\varpi) = -1$ and therefore $P_{\sigma,2}(1) = 2 \neq 0$. The same is true if ω is ramified. Thus in both cases $I(\sigma)$ is reducible and thus there are no complementary series.

Otherwise, i.e. if $\omega = 1$, then $P_{\sigma,2}(1) = 0$ and therefore $I(\sigma)$ is irreducible. Here we also include GSp_4. Then index i of Theorem 5.1 is $i = 2$. Half the sum of positive roots in $\mathbf{N}$ is $\rho = \frac{3}{2}(\alpha+\beta)$ and therefore $\widetilde{\alpha}$ is in fact equal to $\alpha+\beta$. It is then clear from the definition of H_P that

$$\begin{aligned} q^{\langle s\widetilde{\alpha},H_P(m)\rangle} &= |(\alpha+\beta)(m)|^s \\ &= |\det(m_1)|^s \cdot |\lambda|^s \\ &= |\det(m_1)|^s \cdot |\det(m)|^{s/2}, \end{aligned}$$

where $m = (m_1,\widetilde{\lambda})$ with $m_1 \in GL_2(F)$, $\widetilde{\lambda} = \mathrm{diag}(\lambda,\lambda,1,1)$, $\lambda \in F^*$, and $\det(m) = \lambda^2$ (the determinant as an element in $GL_4(F)$). If $\mathbf{G} = Sp_4$ or PSp_4, we then set $\lambda = 1$. Finally observe that

$$I(s\widetilde{\alpha},\sigma) \cong I((\sigma_1 \otimes \nu^s)\otimes\chi)\otimes|\det(\ )|^s,$$

where $\nu(m_1) = |\det(m_1)|$. But now it follows from Theorem 5.1 that $I((\sigma_1\otimes\nu^s)\otimes\chi)$ is irreducible unless $s = \pm\frac{1}{2}$. Moreover $I((\sigma_1\otimes\nu^s)\otimes\chi)$ has a unique generic special subrepresentation and a unique non-tempered non-generic preunitary quotient. We have thus proved:

Proposition 6.1. a) *Suppose $G = GSp_4(F)$ for F a non-archimedean field. Let α and β be the short and the long simple roots of $\mathbf{G}$, respectively, and denote by $\mathbf{P} = \mathbf{MN}$ the maximal parabolic subgroup in which $\mathbf{M}$ is generated by α, $\mathbf{M} \cong GL_2 \times GL_1$. Fix an irreducible unitary supercuspidal representation $\sigma = \sigma_1 \otimes \chi$ of $M = \mathbf{M}(F)$, where σ_1 is a supercuspidal unitary representation of $GL_2(F)$ with central character ω and χ is a unitary character of F^*. Then $I(\sigma)$ is always irreducible. The representation $I(\sigma_1\nu^s\otimes\chi)$ is reducible if and only if $\omega = 1$ and $s = \pm\frac{1}{2}$, where ν denotes $\nu = |\det(\ )|$ for $GL_2(F)$. The representation $I(\sigma_1\nu^{1/2}\otimes\chi)$ has a unique* *generic special subrepresentation and a unique irreducible preunitary non-tempered non-generic quotient. For $0 < s < 1/2$, all the representations $I(\sigma_1\nu^s \otimes \chi)$ are in the complementary series and $s = 1/2$ is their end point.*

b) For $G = Sp_4(F)$ or $PSp_4(F)$, the representation $I(\sigma)$ is reducible if and only if $\sigma \cong \tilde{\sigma}$ (thus $\omega^2 = 1$) and $\omega \neq 1$. Suppose $\omega = 1$ so that $I(\sigma)$ is irreducible. Then $I(\sigma\nu^s)$ is reducible if and only if $s = \pm 1/2$. The representation $I(\sigma\nu^{1/2})$ has a unique generic special subrepresentation and a unique irreducible preunitary non-tempered non-generic quotient. For $0 < s < 1/2$, all the representations $I(\sigma\nu^s)$ are in the complementary series and $s = 1/2$ is their end point.

c) The Plancherel measure $\mu(s\tilde{\alpha}, \sigma)$ is given by the formula

$$\mu(s\tilde{\alpha}, \sigma) = \gamma(G/P)^2 q^{n(\sigma_1)} \frac{(1 - \omega(\varpi)q^{-2s})(1 - \omega(\varpi)^{-1}q^{2s})}{(1 - \omega(\varpi)q^{-1-2s})(1 - \omega(\varpi)^{-1}q^{-1+2s})}$$

if ω is unramified, and by

$$\mu(s\tilde{\alpha}, \sigma) = \gamma(G/P)^2 q^{n(\sigma_1)+n(\omega)}$$

otherwise. $n(\sigma_1)$ and $n(\omega)$ are the conductors of σ_1 and ω, respectively.

Proof. One only needs prove part c). By Corollary 3.6 of [29] one must calculate

$$\varepsilon(s, \sigma_1, \rho_2, \psi_F)\varepsilon(2s, \omega, \psi_F)\varepsilon(-s, \tilde{\sigma}_1, \rho_2, \overline{\psi}_F)\varepsilon(-2s, \omega^{-1}, \overline{\psi}_F). \tag{6.1.1}$$

Observe that the defining Euclidean measures for intertwining operators are self dual with respect to ψ_F. Therefore, if ψ_F is unramified, then these measures must be the standard ones, i.e. O must have measure one with respect to every direction on the Lie algebra of $\mathbf{N}$. We shall now assume ψ_F is unramified. If one uses [14,45] one immediately sees that

$$\varepsilon(s, \sigma_1, \rho_2, \psi_F) = c(\sigma_1, \psi_F)q^{-n(\sigma_1)s}$$

and

$$\varepsilon(s, \omega, \psi_F) = c(\omega, \psi_F)q^{-n(\omega)s},$$

where $c(\sigma_1, \psi_F)$ and $c(\omega, \psi_F)$ are two non-zero complex numbers. Proposition 7.8 of [29] now implies $c(\tilde{\sigma}_1, \overline{\psi}_F) = \overline{c(\sigma_1, \psi_F)}$. Thus (6.1.1) equals

$$|c(\sigma_1, \psi_F)c(\omega, \psi_F)|^2.$$

By (3.1) this equals $q^{n(\sigma_1)+n(\omega)}$, proving Proposition 6.1. □

Next let $\mathbf{G}$ be a split group of exceptional type G_2. Let $\mathbf{P} = \mathbf{MN}$ be the parabolic subgroup for which $\mathbf{M}$ is generated by the short simple root α. Then $\mathbf{M} = GL(2)$. Let β be the long simple root of $\mathbf{G}$. The isomorphism $\mathbf{M} \cong GL(2)$ is such that $H_\alpha(t) = \mathrm{diag}(t, t^{-1})$, $H_{3\alpha+2\beta}(t) = \mathrm{diag}(t,t)$, $H_\beta(t) = \mathrm{diag}(1,t)$, $H_{3\alpha+\beta} = \mathrm{diag}(t,1)$, $H_{\alpha+\beta}(t) = \mathrm{diag}(t,t^2)$, and $H_{2\alpha+\beta}(t) = \mathrm{diag}(t^2,t)$. Let σ be an irreducible unitary supercuspidal representation of $M = GL_2(F)$. Then $\widetilde{w}_0(\sigma) \cong \sigma$ implies $\sigma \cong \widetilde{\sigma}$.

The adjoint action r of ${}^LM$ on ${}^L\mathfrak{n}$ is $r = \rho_2 \oplus \wedge^2\rho_2 \oplus \rho_2 \otimes \wedge^2\rho_2$, where ρ_2 is the standard representation of $GL_2(\mathbb{C})$. Again

$$P_{\sigma,1}(q^{-s}) = L(s,\sigma,\rho_2)^{-1} = 1$$

and

$$P_{\sigma,2}(q^{-s}) = L(s,\omega)^{-1}.$$

If $\sigma \cong \widetilde{\sigma}$ with $\omega \neq 1$, then $I(\sigma)$ is reducible and there is no more reducibility or complementary series. Now assume $\sigma \cong \widetilde{\sigma}$ and $\omega = 1$. Then $I(\sigma)$ is irreducible. The index i is again $i = 2$. The value of $\widetilde{\alpha}$ is $3\alpha + 2\beta$ and

$$q^{\langle s\widetilde{\alpha}, H_\theta(m)\rangle} = |(3\alpha + 2\beta)(m)|^s.$$

Let $m = \mathrm{diag}(\det m, 1) \cdot m_0$ with $m_0 \in SL_2(F)$. Writing $\mathrm{diag}(\det m, 1) = H_{3\alpha+\beta}(\det m)$ then implies $(3\alpha+2\beta)(m) = \det m$. But now it follows from Theorem 5.1 that $I(\sigma \otimes \nu^s)$ is irreducible unless $s = \pm 1/2$, $\nu = |\det(\ )|$. Moreover $I(\sigma \otimes \nu^{1/2})$ has a unique generic special subrepresentation and a unique irreducible preunitary non-tempered non-generic quotient. Thus:

Proposition 6.2. a) *Let* $\mathbf{G}$ *be an exceptional split group of type* G_2. *Assume the Levi factor* $\mathbf{M}$ *of* $\mathbf{P} = \mathbf{MN}$ *is generated by the short root of* $\mathbf{G}$. *Let* σ *be an irreducible unitary supercuspidal representation of* M. *Then* σ *is ramified if and only if* $\sigma \cong \widetilde{\sigma}$. *Assume* $\sigma \cong \widetilde{\sigma}$ *but* $\omega \neq 1$. *Then* $I(\sigma)$ *is reducible and there are no complementary series. Now suppose* $\sigma \cong \widetilde{\sigma}$ *but* $\omega = 1$. *Then* $I(\sigma)$ *is irreducible. Moreover* $I(\sigma \otimes \nu^s)$ *is irreducible unless* $s = \pm 1/2$. *The representation* $I(\sigma \otimes \nu^{1/2})$ *has a unique generic discrete series subrepresentation and a unique irreducible preunitary non-tempered Langlands quotient. All the representations* $I(\sigma \otimes \nu^s)$, $0 < s < 1/2$, *are in the complementary series and* $s = 1/2$ *is the edge of complementary series.*

b) *The Plancherel measure* $\mu(s\widetilde{\alpha}, \sigma)$ *is given by the formula*

$$\mu(s\widetilde{\alpha},\sigma) = \gamma(G/P)^2 q^{n(\sigma)+n(\sigma\otimes\omega)} \frac{(1-\omega(\varpi)q^{-2s})(1-\omega(\varpi)^{-1}q^{2s})}{(1-\omega(\varpi)q^{-1-2s})(1-\omega(\varpi)^{-1}q^{-1+2s})}$$

if ω *is unramified, and by*

$$\mu(s\widetilde{\alpha},\sigma) = \gamma(G/P)^2 q^{n(\sigma)+n(\omega)+n(\sigma\otimes\omega)}$$

otherwise. Here $n(\sigma)$, $n(\omega)$, *and* $n(\sigma\otimes\omega)$ *are the corresponding conductors.*

Remark 6.3. Together with Propositions 8.3 and 8.4, and Remark 8.5 of [29] (also see [43] for $\mathbf{G} = GSp_4$ with $\mathbf{P}$ equal to the non-Siegel maximal parabolic subgroup of $\mathbf{G}$), this leads to a complete analysis of the unitary duals of all the rank two split groups supported at their maximal parabolic subgroups. The complete unitary dual of $GSp_4(F)$ is the subject matter of a forthcoming paper of M. Tadic, a sequal to his work with P. Sally [28]. Finally we refer to the Corollary to Proposition 6.2 and Proposition 1.1 of [31] for a formula for the Plancherel measure for the exceptional group $\mathbf{G}$ of type G_2 when $\mathbf{M}$ is generated by its long simple root.

Remark 6.4. Although the results of Propositions 6.1 and 6.2 seem to follow the same simple pattern in terms of the inducing representations, this is definitely not the case in general. This is evident from Proposition 8.3 of [29], where $\mathbf{G}$ is of type G_2 and $\mathbf{P}$ is the other maximal parabolic subgroup.

Remark 6.5. It is knowledge of the polynomials $P_{\sigma,1}$ and $P_{\sigma,2}$—or, said in other terms, the L-functions $L(s,\sigma,r_1)$ and $L(s,\sigma,r_2)$—which allows us to obtain such precise results on reducibility of $I(s\widetilde{\alpha},\sigma)$. While there are many cases where these factors are not known, there are instances in which they can be predicted. For example, based on the results of [15], we believe there may be no complementary series coming from the supercuspidal representations of the Levi subgroup $GL_n(F)$ of the group $Sp_{2n}(F)$ whenever $n > 1$ is odd. In other words no reducibility can happen off the unitary axis in this case. This is just an example of a case where both L-functions are identically equal to 1.

Remark 6.6. In general when σ is supercuspidal, the polynomials $P_{\sigma,1}$ and $P_{\sigma,2}$ are such that the operator

$$P_{\sigma,1}(q^{-s})P_{\sigma,2}(q^{-2s})A(s\widetilde{\alpha},\sigma,w_0)$$

is holomorphic and non-zero for all values of s [34]. Therefore local calculation of these polynomials rests upon the knowledge of poles of intertwining operators, a subject in which the method of Olšanskiĭ [46] seems to be useful [47]. One deep and surprising consequence of Theorem 5.1 is that understanding the poles not only determines the reducibility on the unitary axis, but also off it.

7. Explicit Formulas for the Plancherel Measures for $GL(n)$ and $SL(n)$

When $\mathbf{G} = GL(n)$, Langlands' conjecture was proved by the author in [32,33]. Consequently Plancherel measures were given in terms of certain Rankin-Selberg L-functions and root numbers [13] for $GL(n)$. The purpose of this section is to use the results of [13] and [32] to give explicit formulas for the measures.

It is enough to compute Plancherel measures when $\mathbf{P}$ is maximal. Thus let $\mathbf{G}$ equal $GL(n+m)$, where n and m are positive integers, and let $\mathbf{M}$ equal $GL(n) \times GL(m)$. Assume $\sigma = \sigma_1 \otimes \sigma_2$ is a tempered representation of M. Let $L(s,\sigma_1\times\sigma_2)$ and $\varepsilon(s,\sigma_1\times\sigma_2,\psi_F)$ denote the Rankin-Selberg L-function and root number attached to σ_1 and σ_2 by Jacquet, Piatetski-Shapiro, and Shalika in [13], respectively. Then by Theorem 5.1 of [32] and the validity of Conjecture 4.1 in this case,

$$L(s,\sigma_1\times\sigma_2)=L(s,\sigma_1\otimes\sigma_2,\rho_m\otimes\widetilde{\rho}_n)$$

and

$$\varepsilon(s,\sigma_1\times\sigma_2,\psi_F)=\varepsilon(s,\sigma_1\otimes\sigma_2,\rho_m\otimes\widetilde{\rho}_n,\psi_F),$$

where ρ_m and ρ_n are the standard representations of $GL_m(\mathbb{C})$ and $GL_n(\mathbb{C})$, respectively, and the factors on the right are those defined in Section 4. Now, either using our results or Theorem 6.1 of [32], the Plancherel constant $\mu(s\widetilde{\alpha},\sigma_1\otimes\sigma_2)$ satisfies:

$$\mu(s\widetilde{\alpha},\sigma_1\otimes\sigma_2)=\gamma(G/P)^2q^{n(\widetilde{\sigma}_1\times\sigma_2)}\frac{L(1+s,\sigma_1\times\widetilde{\sigma}_2)}{L(s,\sigma_1\times\widetilde{\sigma}_2)}\cdot\frac{L(1-s,\widetilde{\sigma}_1\times\sigma_2)}{L(-s,\widetilde{\sigma}_1\times\sigma_2)},$$

where $n(\widetilde{\sigma}_1\times\sigma_2)$ is an integer defined by

$$\varepsilon(s,\widetilde{\sigma}_1\times\sigma_2,\psi_F)=c(\widetilde{\sigma}_1\times\sigma_2)q^{-n(\widetilde{\sigma}_1\times\sigma_2)s}.$$

The constant $c(\widetilde{\sigma}_1\times\sigma_2)$ is a non-zero complex number. As an example consider the case $n=1$ and $\sigma_2=\mathbf{1}$. Then by Theorem 5.1 of [32], the integer $n(\sigma_1\times\mathbf{1})$ is the conductor of σ_1 (cf.[14]). The purpose of this section is to calculate these L-functions explicitly in order to obtain explicit formulas for $\mu(s\widetilde{\alpha},\sigma_1\otimes\sigma_2)$.

By the product formula for Plancherel measures and Proposition 8.4 of [13], we may assume σ_1 and σ_2 are both in the discrete series. By [6,44], there exist two integers a and t with $at=m$ and an irreducible unitary supercuspidal representation π_0 of $GL_a(F)$ such that if $\pi_i=\pi_0\otimes\nu^{\frac{t+1}{2}-i}$, $1\le i\le t$, then σ_1 is the unique discrete series constituent $\sigma(\pi_1,\ldots,\pi_t)$ of the representation induced from $\pi_1\otimes\pi_2\cdots\otimes\pi_t$. Similarly, choose integers b and u with $bu=n$ and an irreducible unitary supercuspidal representation ρ_0 of $GL_b(F)$ such that $\sigma_2=\sigma(\rho_1,\ldots,\rho_u)$, $\rho_j=\rho_0\otimes\nu^{\frac{u+1}{2}-j}$, $1\le j\le u$.

Assume $n\le m$. Then Theorem 8.2 of [13] implies

$$L(s,\sigma_1\times\sigma_2)=\prod_{j=1}^{u}L(s,\pi_1\times\rho_j).$$

Now assume σ_1 and σ_2 are two irreducible supercuspidal representations of $GL_m(F)$ and $GL_n(F)$, respectively. Using equations (2.4.1) and (2.4.2) and Theorem 2.7 of [13], one sees that $L(s,\sigma_1\times\sigma_2)=1$ unless $\sigma_1\cong\widetilde{\sigma}_2\otimes\nu^{s_0}$ with a complex number s_0 which is not necessarily unique. Then

$$L(s,\sigma_1\times\sigma_2)=L(s+s_0,\widetilde{\sigma}_2\times\sigma_2).$$

Now assume $\sigma_2 \cong \sigma_2 \otimes \eta$, where $\eta \in \hat{F}^*$ is unramified and η denotes $\eta \cdot \det$. Then $\eta^n = 1$. The set of all such η is a cyclic group of order r, $r|n$. This is true because each η is determined by $\eta(\varpi)$ and finite subgroups of $\mathbb{C}^*$ are cyclic. From equation (2.4.1) of [13] it is clear that $L(s, \sigma_2 \times \widetilde{\sigma}_2)^{-1}$ divides $(1 - q^{-ns})$, and therefore poles of $L(s, \sigma_2 \times \widetilde{\sigma}_2)$ are all simple. Moreover, if η is such that $\sigma_2 \cong \sigma_2 \otimes \eta$, then $q^s = \eta(\varpi)$ is in fact a pole of $L(s, \sigma_2 \times \widetilde{\sigma}_2)$. This is clear from the fact that the residue at every such pole is non-zero (see the proof of Proposition 1.2 of [8]). It now follows from the simplicity of the poles of $L(s, \sigma_2 \times \widetilde{\sigma}_2)$ that

$$L(s, \sigma_2 \times \widetilde{\sigma}_2) = (1 - q^{-rs})^{-1},$$

where, as above, r is the order of the cyclic group of unramified characters η satisfying $\sigma_2 \cong \sigma_2 \otimes \eta$.

With notation as in the case of discrete series, our discussion implies

$$L(s, \pi_1 \times \rho_j) = 1$$

unless $\rho_0 \cong \widetilde{\pi}_0 \otimes \nu^{s_0}$ with a pure imaginary number s_0. In this case

$$\rho_j \cong \widetilde{\pi}_1 \otimes \nu^{s_0 + 1/2(t+u) - j}$$

and

$$\widetilde{\rho}_j \cong \widetilde{\pi}_1 \otimes \nu^{-s_0 + 1/2(t-u) - 1 + j}.$$

Therefore

$$L(s, \pi_1 \times \widetilde{\rho}_j) = (1 - q^{rs_0 - r/2(t-u) + r - rj} \cdot q^{-rs})^{-1},$$

while

$$L(s, \widetilde{\pi}_1 \times \rho_j) = (1 - q^{-rs_0 + r/2(t-u) - r + rj} \cdot q^{-rs})^{-1},$$

where r is the order of the cyclic group of unramified characters η satisfying $\pi_0 \cong \pi_0 \otimes \eta$. Observe that s_0 is not unique but q^{-rs_0} is. We thus have:

Proposition 7.1. *Let σ_1 and σ_2 be two discrete series representations of $GL_m(F)$ and $GL_n(F)$, respectively. Choose positive integers a, t, b, and u with $at = m$ and $bu = n$, and irreducible unitary supercuspidal representations π_0 and ρ_0 of $GL_a(F)$ and $GL_b(F)$, respectively, such that $\sigma_1 = \sigma(\pi_1, \ldots, \pi_t)$ and $\sigma_2 = \sigma(\rho_1, \ldots, \rho_u)$, where $\pi_i = \pi_0 \otimes \nu^{\frac{t+1}{2} - i}$, $1 \leq i \leq t$, and $\rho_j = \rho_0 \otimes \nu^{\frac{u+1}{2} - j}$, $1 \leq j \leq u$. Then*

$$\mu(s\widetilde{\alpha}, \sigma_1 \otimes \sigma_2) = \gamma(G/P)^2 q^{n(\widetilde{\sigma}_1 \times \sigma_2)}$$

unless $\rho_0 \cong \pi_0 \otimes \nu^{s_0}$ (and therefore $a = b$) for some pure imaginary number s_0, in which case

$$\mu(s\widetilde{\alpha}, \sigma_1 \otimes \sigma_2) = \gamma(G/P)^2 q^{n(\widetilde{\sigma}_2 \times \sigma_2)}$$
$$\cdot \prod_{j=1}^{u} \frac{(1 - q^{rs_0 - r/2(t-u) + r - jr} \cdot q^{-rs})(1 - q^{-rs_0 + r/2(t-u) - r + jr} \cdot q^{rs})}{(1 - q^{rs_0 - r/2(t-u) - jr} \cdot q^{-rs})(1 - q^{-rs_0 + r/2(t-u) + jr} \cdot q^{rs})},$$

if $n \leq m$. Otherwise, i.e. if $n \geq m$, one must change the role of the triple (b, u, ρ_0) with (a, t, π_0). Here r is the order of the cyclic group of all the unramified characters η satisfying $\pi_0 \cong \pi_0 \otimes \eta$. In particular, if σ_1 and σ_2 are both supercuspidal and $\sigma_2 \cong \sigma_1 \otimes \nu^{s_0}$, then

$$\mu(s\widetilde{\alpha}, \sigma_1 \otimes \sigma_2) = \gamma(G/P)^2 q^{n(\widetilde{\sigma}_1 \times \sigma_2)} \frac{(1 - q^{rs_0} \cdot q^{-rs})(1 - q^{-rs_0} q^{rs})}{(1 - q^{rs_0 - r} \cdot q^{-rs})(1 - q^{-rs_0 + r} \cdot q^{rs})}.$$

Suppose $\mathbf{G} = SL(r)$. Let $\mathbf{P} = \mathbf{MN}$ be a parabolic subgroup of G which we may assume to be standard. Let σ be an irreducible tempered representation of M. Then let $\widetilde{\mathbf{G}} = GL(r)$. There exists a standard parabolic subgroup $\widetilde{\mathbf{P}} = \widetilde{\mathbf{M}}\mathbf{N}$ of $\widetilde{\mathbf{G}}$ such that $\mathbf{M} = \widetilde{\mathbf{M}} \cap \mathbf{G}$. If $\mathbf{P}$ is maximal, then so is $\widetilde{\mathbf{P}}$. By Lemma 1.1 of [30] there exists an irreducible tempered representation $\widetilde{\sigma}$ of $\widetilde{M}$ such that $\sigma \subset \widetilde{\sigma}|M$. Moreover, if $\widetilde{\sigma}_1$ is another such representation of $\widetilde{M}$, then there exists a character $\eta \in \hat{F}^*$ such that $\widetilde{\sigma}_1 \cong \widetilde{\sigma} \otimes \eta$. If σ is in the discrete series, then so is $\widetilde{\sigma}$. Clearly

$$A(\widetilde{\nu}, \widetilde{\sigma}, w)|I(\nu, \sigma) = A(\nu, \sigma, w)$$

and therefore

$$\mu(\nu, \sigma) = \mu(\widetilde{\nu}, \widetilde{\sigma}),$$

where $\widetilde{\nu}$ is any extension of ν from $\mathfrak{a}_{\mathbb{C}}^*$ to $\widetilde{\mathfrak{a}}_{\mathbb{C}}^*$ with obvious notation. We therefore have:

Corollary 7.2. *Let $\mathbf{P} = \mathbf{MN}$ be a standard maximal parabolic subgroup of $SL(r)$. Let σ be a discrete series representation of M. Choose two positive integers m and n, $m + n = r$, such that $\mathbf{M} = GL(r) \cap \widetilde{\mathbf{M}}$, where $\widetilde{\mathbf{M}} = GL(m) \times GL(n)$. Choose a pair of discrete series representations σ_1 and σ_2 of $GL_m(F)$ and $GL_n(F)$ such that $\sigma \subset \widetilde{\sigma}|M$, where $\widetilde{\sigma} = \sigma_1 \otimes \sigma_2$. Then*

$$\mu(s\widetilde{\alpha}, \sigma) = \mu(s\widetilde{\alpha}, \sigma_1 \otimes \sigma_2),$$

where $\mu(s\widetilde{\alpha}, \sigma_1 \otimes \sigma_2)$ are given by the formulas in Proposition 7.1. It is independent of the possible choices of σ_1 and σ_2.

Remark 7.3. Using Proposition 7.1 and Corollary 7.2, it must now be a combinatorial problem to obtain R-groups and therefore determine the number of components of the representation $I(\sigma)$, where σ is in the discrete series. This must take care of the non-discrete part of the tempered spectrum of $SL_r(F)$.

References

1. J. Arthur, *Intertwining operators and residues I: weighted characters*, J. Funct. Anal. **84** (1989), 19–84.

2. ———, *On some problems suggested by the trace formula*, Lie Group Representations II, Lecture Notes in Math, Vol. 1041, Springer-Verlag, Berlin-Heidelberg-New York, 1983, pp. 1–49.

3. ______, *Unipotent automorphic representations: Conjectures*, Orbites Unipotentes et Repŕesentations, II. Groupes p-adic et réels, Société Mathématique de France, Astérisque, Nos. 171–172, 1989, pp. 13–71.

4. ______, *Unipotent automorphic representations: Global motivations*, Proceedings of the Conference on Automorphic Forms, Shimura Varieties, and L-functions, Vol. I (L. Clozel and J.S. Milne, eds.), in *Perspectives in Mathematics*, vol. 10, Academic Press, 1990, pp. 1–75.

5. J. Arthur and L. Clozel, *Simple Algebras, Base Change, and the Advanced Theory of the Trace Formula,* Annals of Math. Studies, Vol. 120, Princeton University Press, Princeton, 1989.

6. I. N. Bernstein and A. V. Zelevinsky, *Induced representations of reductive p-adic groups I,* Ann. Scient. Éc. Norm. Sup. **10** (1977), 441–472.

7. A. Borel, *Automorphic L-functions*, Proc. Sympos. Pure Math., AMS, 33, II (1979), 27–61.

8. S. Gelbart and H. Jacquet, *A relation between automorphic representations of* $GL(2)$ *and* $GL(3)$, Ann. Scient. Éc. Norm Sup. **11** (1978), 471–542.

9. Harish-Chandra, *Harmonic analysis on real reductive groups III. The Maass-Selberg relation and the Plancherel formula*, Annals of Math. **104** (1976), 117–201.

10. ______, *The Plancherel formula for reductive p-adic groups*, reprinted in Collected Papers, Springer Verlag, Berlin-Heidelberg-New York, 1984, pp. 353–367.

11. R. Howe and A. Moy, *Hecke algebra isomorphisms for* GL_n *over a p-adic field*, Jour. of Algebra **131** (1990), 388–424.

12. D. Jabon, D. Keys, and A. Moy, *An explicit Plancherel formula for* $U(2,1)$, preprint.

13. H. Jacquet, I.I. Piatetski-Shapiro, and J.A. Shalika, *Rankin-Selberg convolutions*, Amer. J. Math. **105** (1983), 367–464.

14. ______, *Conducteur des représentations du groupe linéaire*, Math. Ann. **256** (1981), 199–214.

15. H. Jacquet and J.A. Shalika, *Exterior square L-functions*, Proceedings of the Conference on Automorphic Forms, Shimura Varieties, and L-functions, Vol. II (L. Clozel and J.S. Milne, eds.), in *Perspectives in Mathematics*, vol. 11, Academic Press, 1990, pp. 143–226.

16. D. Keys, *On the decomposition of reducible principal series representation of p-adic Chevalley groups*, Pacific J. Math. **101** (1982), 351–388.

17. ______, *Principal series representations of special unitary groups over local field*, Comp. Math. **51** (1984), 115–130.

18. ______, *L-indistinguishability and R-groups for quasi-split groups: Unitary groups of even dimension*, Ann. Scient. Éc. Norm. Sup. **20** (1987), 31–64.

19. D. Keys and F. Shahidi, *Artin L-functions and normalization of intertwining operators*, Ann. Scient. Éc. Norm. Sup. **21** (1988), 67–89.

20. A. W. Knapp and E. M. Stein, *Intertwining operators for semisimple groups*, Annals of Math. **93** (1971), 489–578.

21. ______, *Intertwining operators for semisimple groups, II*, Invent. Math. **60** (1980), 9–84.

22. A. W. Knapp and G. J. Zuckerman, *Classification of irreducible tempered representations of semisimple groups*, Annals of Math. **116** (1982), 389–455.

23. R. P. Langlands, *Euler Products*, Yale University Press, New Haven, 1971.

24. ______, *On the Functional Equations Satisfied by Eisenstein Series*, Lecture Notes in Math., Vol 544, Springer-Verlag, Berlin-Heidelberg-New York, 1976.

25. ———, *Problems in the theory of automorphic forms*, Lec. Notes in Math., Vol. 170, Springer-Verlag, Berlin-Heidelberg-New York, pp. 18–86.

26. ———, *On Artin's L-functions*, Rice University Studies **56** (1970), 23–28.

27. ———, *On the classification of irreducible representations of real algebraic groups*, Representation Theory and Harmonic Analysis on Semisimple Lie Groups (P.J. Sally, Jr. and D.A. Vogan, eds.), Mathematical Surveys and Monographs, AMS, Vol 31, 1989, pp. 101–170.

28. P. J. Sally and M. Tadic, *On representations of non-archimedean symplectic groups*, preprint.

29. F. Shahidi, *A proof of Langlands conjecture on Plancherel measures; complementary series for p-adic groups*, Annals of Math. **132** (1990), 273–330.

30. ———, *Some results on L-indistinguishability for $SL(r)$*, Canadian J. Math. **35** (1983), 1075–1109.

31. ———, *Third symmetric power L-functions for $GL(2)$*, Comp. Math. **70** (1989), 245–273.

32. ———, *Fourier transforms of intertwining operators and Plancherel measures for $GL(n)$*, Amer. J. Math. **106** (1984), 67–111.

33. ———, *Local coefficients and normalization of intertwining operators for $GL(n)$*, Comp. Math. **48** (1983), 271–295.

34. ———, *On certain L-functions*, Amer. J. Math. **103** (1981), 297–356.

35. ———, *On the Ramanujan conjecture and finiteness of poles for certain L-functions*, Annals of Math. **127** (1988), 547–584.

36. ———, *On multiplicativity of local factors*, in Festschrift in honor of I.I. Piatetski-Shapiro, Part II, Israel Mathematical Conference Proceedings **3** (1990), 279–289.

37. ———, *Local coefficients as Artin factors for real groups*, Duke Math. J. **52** (1985), 973–1007.

38. D. Shelstad, *L-indistinguishability for real groups*, Math. Ann. **259** (1982), 385–430.

39. A. Silberger, *Special representations of reductive p-adic groups are not integrable*, Annals of Math. **111** (1980), 571–587.

40. ———, *The Knapp-Stein dimension theorem for p-adic groups*, Proc. Amer. Math. Soc. **68** (1978), 243–246.

41. ———, *Introduction to Harmonic Analysis on Reductive p-adic Groups* Math. Notes, Vol. 23, Princeton University Press, Princeton, 1979.

42. D. Vogan, *Gelfand-Kirillov dimension for Harish-Chandra modules*, Invent. Math. **48** (1978), 75–98.

43. J. L. Waldspurger, *Un exercice sur $GSp(4,F)$ et les représentations de Weil*, Bull. Soc. Math. France **115** (1987), 35–69.

44. A. V. Zelevinsky, *Induced representations of reductive p-adic groups II: on irreducible representations of $GL(n)$*, Ann. Scient. Éc. Norm. Sup. **13** (1980), 165–210.

45. W. Casselman, *On some results of Atkin and Lehner*, Math. Ann. **206** (1973), 311–318.

46. G. I. Olšanskiĭ, *Intertwining operators and complementary series*, Math. USSR Sbornik. **22** (1974), 217–255.

47. B. Tamir, *On L-functions and intertwining operators for unitary groups*, preprint (1990).

DEPARTMENT OF MATHEMATICS, PURDUE UNIVERSITY, WEST LAFAYETTE, IN 47907

TRANSFER AND DESCENT: SOME RECENT RESULTS

DIANA SHELSTAD

Rutgers University

A basic tool for studying the transfer of representations is the dual transfer of orbital integrals. In this paper we report on some recent results for orbital integrals and, in particular, on a descent theorem [LS3].

Throughout this paper F is a local field of characteristic zero with algebraic closure $\overline{F}$ and $\Gamma = \mathrm{Gal}(\overline{F}/F)$; G is a connected reductive group defined over F.

§1 An example

The notions of endoscopy and transfer require some care in formulation, but there is one example where we can give definitions quickly. Suppose H is the (unique up to F-isomorphism) quasi-split inner form of G. Then H is endoscopic for G in the sense of standard endoscopy (where no twisting by an automorphism is specified; see [LS1], [KS]). Up to isomorphism, it is the unique endoscopic group with dimension as great as that of G.

Fix an inner twist $\psi : G \to H$. Although ψ is not defined over F (unless G is quasi-split) we may use it to define a correspondence between points of $G(F)$ and points of $H(F)$. Indeed, ψ induces a bijective map ψ_{conj} from the conjugacy classes in $G(\overline{F})$ to those in $H(\overline{F})$, and because $\psi\sigma(\psi)^{-1}$ is inner, $\sigma \in \Gamma$, ψ_{conj} respects the action of Γ. Thus we have a bijection between the classes defined over F. We shall consider just those elements in G (or H) which are strongly regular in the sense that their centralizers are tori. The conjugacy class in $G(\overline{F})$ of a strongly regular element γ_G of $G(F)$ is defined over F. By a converse theorem of Steinberg for quasi-split groups, the image under ψ_{conj} of this class also contains F-rational elements. We say that any such element γ_H is *an image of* γ_G.

Recall that the stable conjugacy class of a strongly regular element in $G(F)$ (or $H(F)$) consists of the F-rational elements in its $\overline{F}$-conjugacy class. Thus the correspondence (γ_G, γ_H) provides an injective map of the set of stable conjugacy classes of strongly regular elements in $G(F)$ in the stable classes of such elements in $H(F)$ (this map is surjective only if G is quasi-split and thus F-isomorphic to H).

Partially supported by NSF Grant 89-03313.

Because the strongly regular elements in $G(F)$ are dense in $G(F)$ for the topology inherited from F, the correspondence just defined is sufficient to specify *transfer*. Recall that the stable orbital integral of $f \in C_c^\infty(G(F))$ (or $f \in \mathcal{C}(G(F))$) at strongly regular $\gamma \in G(F)$ is

$$\Phi^{st}(\gamma, f) = \sum_{\gamma'} \Phi(\gamma', f), \tag{1.1}$$

where the summation is over representatives γ' for the conjugacy classes in the stable conjugacy class of γ and $\Phi(\gamma', f)$ is the ordinary orbital integral

$$\Phi(\gamma', f) = \int_{G_{\gamma'}(F)\backslash G(F)} f(g^{-1}\gamma' g)\, d\overline{g}.$$

Here $G_{\gamma'} = \mathrm{Cent}(\gamma', G)$ and $d\overline{g}$ is the quotient of some fixed Haar measure on $G(F)$ by a Haar measure on the Cartan subgroup $G_{\gamma'}(F)$. If $\gamma' = g^{-1}\gamma g$, $g \in G(\overline{F})$, then $\mathrm{Int}\, g : G_{\gamma'} \to G_\gamma$ is defined over F: we require that the Haar measures on $G_\gamma(F)$ and $G_{\gamma'}(F)$ be related by transport under $\mathrm{Int}\, g$. The *transfer problem* for (G, H) is to show that for each $f^G \in C_c^\infty(G(F))$ there exists $f^H \in C_c^\infty(H(F))$ such that

$$\Phi^{st}(\gamma_H, f^H) = \begin{cases} \Phi^{st}(\gamma_G, f^G) & \text{if } \gamma_H \text{ is an image of } \gamma_G \\ 0 & \text{if } \gamma_H \text{ is not an image,} \end{cases} \tag{1.2}$$

for all strongly regular elements γ_H in $H(F)$.

Define $\Delta(\gamma_H, \gamma_G)$ for strongly regular $\gamma_H \in H(F)$ and $\gamma_G \in G(F)$ by

$$\Delta(\gamma_H, \gamma_G) = \begin{cases} 1 & \text{if } \gamma_H \text{ is an image of } \gamma_G \\ 0 & \text{otherwise.} \end{cases}$$

Then we may rewrite (1.2) as

$$\Phi^{st}(\gamma_H, f^H) = \sum_{\gamma_G} \Delta(\gamma_H, \gamma_G)\Phi(\gamma_G, f^G) \tag{1.3}$$

for all strongly regular $\gamma_H \in H(F)$. The summation is over representatives γ_G for the conjugacy classes of strongly regular elements in $G(F)$. If (1.2) or (1.3) is true we say that *f^G and f^H have Δ-matching orbital integrals.*

For F archimedean the transfer problem is solved, at least for Schwartz functions [S1] or C_c^∞-functions bifinite under a maximal compact subgroup [CD]. For F nonarchimedean we consider first the problem of *local transfer at the identity*: given $f^G \in C_c^\infty(G(F))$ we are to find $f^H \in C_c^\infty(H(F))$ such that (1.2) holds for strongly regular γ_H near the identity in $H(F)$. This may be reformulated in terms of Shalika germs. For γ strongly regular near the identity, the Shalika germ expansion yields

$$\Phi(\gamma, f) = \sum_{\mathcal{O}} \Gamma_{\mathcal{O}}(\gamma) a_{\mathcal{O}}(f),$$

where the summation is over unipotent conjugacy classes $\mathcal{O}$ in $G(F)$, $\Gamma_{\mathcal{O}}$ is the Shalika germ for $\mathcal{O}$ and $a_{\mathcal{O}}(f)$ is the orbital integral of f along $\mathcal{O}$ (for some normalization of measure). We set

$$\Gamma^{\mathrm{st}}_{\mathcal{O}}(\gamma) = \sum_{\gamma'} \Gamma_{\mathcal{O}}(\gamma'),$$

where the summation is over representatives γ' for the conjugacy classes in the stable conjugacy class of γ; $\Gamma^{\mathrm{st}}_{\mathcal{O}}$ is the *stable germ* for $\mathcal{O}$. For transfer we will use instead the notation γ_G, $\mathcal{O}_G$, and so on. We write $\Gamma^{(H)}_{\mathcal{O}_G}$ for

$$\gamma_H \mapsto \begin{cases} \Gamma^{\mathrm{st}}_{\mathcal{O}_G}(\gamma_G) & \text{if } \gamma_H \text{ is an image of } \gamma_G \\ 0 & \text{if } \gamma_H \text{ is not an image.} \end{cases}$$

Then it follows readily that the pair (G, H) admits local transfer at the identity if and only if

(1.4) *for each unipotent conjugacy class $\mathcal{O}_G$ in $G(F)$, the transferred stable germ $\Gamma^{(H)}_{\mathcal{O}_G}$ is a linear combination of the stable germs $\Gamma^{\mathrm{st}}_{\mathcal{O}_H}$ for $H(F)$.*

For progress on (1.4) see the comments after (2.2).

We shall now assume local transfer at the identity for the centralizers of semisimple elements in $G(F)$ and deduce the full transfer on $G(F)$. The descent argument here is very simple.

First we formulate the assumption precisely. Call a semisimple element ϵ_H in $H(F)$ *an image* of the (semisimple) element ϵ_G in $G(F)$ if there exist a maximal torus T_G over F in G containing ϵ_G and an element x of $H(\overline{F})$ such that ${}_x\psi = \mathrm{Int}\, x \circ \psi : T_G \to H$ is defined over F and carries ϵ_G to ϵ_H. For strongly regular elements this coincides with the earlier notion of image. Applying a lemma of Kottwitz we may further choose x so that $H_{\epsilon_H} = \mathrm{Cent}(\epsilon_H, H)^{\circ}$ is quasi-split. Observe that ${}_x\psi : G_{\epsilon_G} \to H_{\epsilon_H}$ is an inner twist. It is not uniquely determined by ψ but its inner class is determined by the inner class of ψ, i.e., by $\{\mathrm{Int}\, y \circ \psi : y \in H(\overline{F})\}$, and it is only the inner class that matters for the correspondence of F-rational strongly regular points.

Our assumption will be:

(1.5) *for each semisimple element ϵ_H of $H(F)$ the pair $(G_{\epsilon_G}, H_{\epsilon_H})$ admits local transfer at the identity.*

It is then immediate that:

(1.6) *we have local transfer for $(G_{\epsilon_G}, H_{\epsilon_H})$ at any central element of $H_{\epsilon_H}(F)$ and, in particular, at ϵ_H.*

Now we take $f^G \in C_c^\infty(G(F))$ and set

$$\Phi(\gamma_H, f^G) = \sum_{\gamma_G} \Delta(\gamma_H, \gamma_G)\Phi(\gamma_G, f^G).$$

We have to show $\Phi(\cdot\,, f^G)$ is a stable orbital integral on $H(F)$. For this it is sufficient to show that it is *locally* a stable orbital integral on $H(F)$, i.e., for each semisimple $\epsilon_H \in H(F)$ there is $f_{\epsilon_H} \in C_c^\infty(H(F))$ such that

$$\Phi^{st}(\gamma_H, f_{\epsilon_H}) = \Phi(\gamma_H, f^G)$$

for strongly regular γ_H near ϵ_H [LS3, Lemma 2.2.A].

The Harish-Chandra descent for orbital integrals says that near semisimple ϵ_G in $G(F)$ the stable orbital integral $\Phi(\gamma_H, f^G)$ is a sum of stable orbital integrals on the groups $G_{\epsilon'_G}(F)$, with ϵ'_G stably conjugate to ϵ_G. See [LS3, Section 1.5]. From (1.6) we conclude that near each semisimple ϵ_H in $H(F)$, $\Phi(\gamma_H, f^G)$ is a stable orbital integral on $H_{\epsilon_H}(F)$ and hence one on $H(F)$, and transfer for (G, H) follows.

§2 AN OUTLINE OF THE GENERAL SETTING

An *endoscopic group* H for G is given as part of a *set of endoscopic data* for G (see [LS1]). This set also includes a group $\mathcal{H}$ which is "almost" the L-group ${}^LH$ of H and a suitable embedding of $\mathcal{H}$ in ${}^LG$. The group $\mathcal{H}$, rather than ${}^LH$, and the embedding arise naturally in two constructions of endoscopic data: the (T, κ)-construction associated with orbital integrals and the S_ϕ-construction associated with representations. In many cases, $\mathcal{H}$ is isomorphic to ${}^LH$ and transfer involves $H(F)$ itself. In general, however, we take a suitable central extension H_1 of H; in particular, $H_1(F) \to H(F)$ is surjective. Then for transfer we consider representations of $H_1(F)$ which act according to a character λ of $Z_1(F) = \ker(H_1(F) \to H(F))$ specified by our data. On the dual side, we consider the stable orbital integrals of those functions on $H_1(F)$ which transform under translation by $Z_1(F)$ according to the character λ^{-1}.

For the problems of local behavior (around a semisimple point) it makes no difference whether we work on $H_1(F)$ or on $H(F)$. Only in patching together the local results is passage to $H_1(F)$ necessary. This passage affects only a single term Δ_2 in the transfer factor which can be handled quite easily. See §4.4 of [LS1]. To simplify the exposition and save notation we will assume from now on that $\mathcal{H}$ is isomorphic to ${}^LH$ and so take $H_1 = H$.

We follow the steps in Section 1. First, there is a canonical map of semisimple conjugacy classes in $H(\overline{F})$ to such classes in $G(\overline{F})$ and this map respects Galois action (see [LS1]). The *strongly G-regular* classes of elements in $H(\overline{F})$ are those mapping to the strongly regular elements in $G(\overline{F})$. We have then a simple notion of *image* for stable conjugacy classes of strongly G-regular elements in $H(F)$ using the *inverse* of the canonical map on classes. *Transfer* is again specified by (1.3), i.e., by

$$\Phi^{st}(\gamma_H, f^H) = \sum \Delta(\gamma_H, \gamma_G)\Phi(\gamma_G, f^G),$$

now for all strongly G-regular elements γ_H of $H(F)$, where $\Delta(\gamma_H, \gamma_G)$ is the transfer factor of [LS1]. For *local transfer at the identity* we require (1.3) only for those γ_H near the identity and we may replace $\Delta(\gamma_H, \gamma_G)$ by a locally defined term $\Delta_{loc}(\gamma_H, \gamma_G)$ which coincides with $\Delta(\gamma_H, \gamma_G)$ up to a constant.

To shorten the exposition we will assume F *nonarchimedean*. Suppose for now that G is *quasi-split*. There are various ways to describe $\Delta_{loc}(\gamma_H, \gamma_G)$. See for example, [H2], [LS2], [S2]. We follow [S2]. Given $\gamma_H \in H(F)$ strongly G-regular and sufficiently close to the identity we shall fix an admisssible embedding of $T_H = \mathrm{Cent}(\gamma_H, H)$ in G together with a set of a-data $\{a_\alpha\}$ for the image T_G of T_H. See [LS1, Sections (1.3), (2.2), (3.1)]. Let $\gamma_G \in T_G(F)$ be the image of γ_H under the embedding. As usual we parametrize the conjugacy classes in the stable conjugacy class of γ_G by $\mathcal{D}(T_G)$. Suppose $\gamma_G(\omega)$ is an element in the class attached to $\omega \in \mathcal{D}(T_G)$. Then

$$\Delta_{loc}(\gamma_H, \gamma_G(\omega)) = \kappa(\omega)\Delta_{loc}(\gamma_H, \gamma_G) \tag{2.1}$$

where κ is the character on $\mathcal{D}(T_G)$ determined by the endoscopic data underlying H [LS2]. We have to describe $\Delta_{loc}(\gamma_H, \gamma_G)$; it is the product of

(i) a root of unity determined by the embedding and a-data (Δ_I in [LS1]),
(ii) the usual discriminant function (Δ_{IV} in [LS1]), and
(iii) a term indexed by orbits of the Galois group Γ in the set R_G of roots of T_G in G.

An orbit $\mathcal{O}$ makes a nontrivial contribution only if it is symmetric, i.e., $\mathcal{O} = -\mathcal{O}$, *and* it consists of roots outside H, i.e., roots not lying in the image of R_H under the map induced by our embedding of T_H in G. Then take $\alpha \in \mathcal{O}$. For γ_H sufficiently near 1 the term $\alpha(\gamma_G)^{1/2} - \alpha(\gamma_G)^{-1/2}$ is defined in the usual way. Moreover,

$$\frac{\alpha(\gamma_G)^{1/2} - \alpha(\gamma_G)^{-1/2}}{a_\alpha}$$

lies in the fixed field $F_{\pm\alpha} \subset \overline{F}$ of the stabilizer of $\pm\alpha$ in Γ. Let F_α be the fixed field of the stabilizer of α in Γ. Because $\mathcal{O}$ is symmetric, F_α is a quadratic extension of $F_{\pm\alpha}$. Let χ_α be the attached quadratic character of $F_{\pm\alpha}$. Then the contribution to $\Delta_{loc}(\gamma_H, \gamma_G)$ from $\mathcal{O}$ is

$$\chi_\alpha\left(\frac{\alpha(\gamma_G)^{1/2} - \alpha(\gamma_G)^{-1/2}}{a_\alpha}\right).$$

For general G there is an additional contribution to $\Delta_{\text{loc}}(\gamma_H, \gamma_G)$. First, strongly G-regular γ_H may be the image of no γ_G in $G(F)$. In that case, $\Delta_{\text{loc}}(\gamma_H, \gamma_G) = 0$ for all strongly regular γ_G in $G(F)$. On the other hand, if γ_H is the image of some element γ_G then we may embed $T_G = \text{Cent}(\gamma_G, G)$ in G^*, the quasi-split inner form of G. Let γ_{G^*} be the image of γ_G under this embedding. Then γ_{G^*} is an image of γ_G in the sense of Section 1, and with H regarded as endoscopic for G^*, γ_H is an image of γ_{G^*}. The Local Hypothesis indicates how we are to write $\Delta_{\text{loc}}(\gamma_H, \gamma_G))$ in terms of $\Delta_{\text{loc}}(\gamma_H, \gamma_{G^*}))$. See [LS1], [LS2] for details.

To express local transfer at the identity in terms of Shalika germs, let $\mathcal{O}_G$ be a unipotent conjugacy class in $G(F)$ and define

$$\Gamma^{(H)}_{\mathcal{O}_G}(\gamma_H) = \sum_{\gamma_G} \Delta_{\text{loc}}(\gamma_H, \gamma_G)\Gamma_{\mathcal{O}_G}(\gamma_G)$$

with the usual notational conventions. Then local transfer at the identity amounts to showing:

(2.2) *for each unipotent conjugacy class $\mathcal{O}_G$ in $G(F)$, the transferred "κ-germ" $\Gamma^{(H)}_{\mathcal{O}_G}$ is a linear combination of the stable germs $\Gamma^{\text{st}}_{\mathcal{O}_H}$ for H.*

Property (2.2) is known for $\mathcal{O}_G$ regular [LS1], $\mathcal{O}_G$ trivial (an easy consequence of well-known results) or, in many cases, $\mathcal{O}_G$ subregular [H1]. There are also complete results for specific groups, e.g., $SU(3)$ [LS2], [H2] and $GSp(4)$ [H3].

Transition to globally defined transfer factors, i.e., factors defined on all strongly G-regular elements, is quite subtle. First we introduce a term (Δ_2 in [LS1]) which is a quasi-character on the Cartan subgroup $T_H(F)$ containing γ_H. This character is an analogue of the "ρ-shift" for real groups and, similarly, uses classification of the embeddings of the L-groups of Cartan subgroups in ${}^LG$. Data from this analysis also allows us to replace (iii) above with a globally defined term. Moreover, we do not use (2.1) and the Local Hypothesis separately but instead introduce a new unified term (Δ_1 in [LS1]). Now, however, only a *relative transfer factor* $\Delta(\gamma_H, \gamma_G; \overline{\gamma}_H, \overline{\gamma}_G)$ is well-defined (and canonical), for γ_H, $\overline{\gamma}_H$ any two strongly G-regular elements, with $\overline{\gamma}_H$ an image of $\overline{\gamma}_G$. But then we fix some such $\overline{\gamma}_H$ and $\overline{\gamma}_G$, specify $\Delta(\overline{\gamma}_H, \overline{\gamma}_G)$ arbitrarily, and set

$$\Delta(\gamma_H, \gamma_G) = \Delta(\gamma_H, \gamma_G; \overline{\gamma}_H, \overline{\gamma}_G)\Delta(\overline{\gamma}_H, \overline{\gamma}_G).$$

See [LS1, Sections 3.7, 4.1]. This normalization fits well with adelic considerations [LS1, Section 6] and the problem of handling several inner forms of G simultaneously.

With $\Delta(\gamma_H, \gamma_G)$ defined we continue the program of Section 1. Let semisimple ϵ_H in $H(F)$ be an image [LS1] of, say, ϵ_G in $G(F)$. Again we may assume H_{ϵ_H} quasi-split. Then H_{ϵ_H} is endoscopic for G_{ϵ_G} [LS3, Section 1.4]. We assume the analogue of (1.5), i.e., local transfer at the identity for all pairs $(G_{\epsilon_G}, H_{\epsilon_H})$. To obtain local transfer at ϵ_H we need only observe the following property of transfer factors under translation by central elements. We shall state it for (G, H). The center Z_G of $G(F)$ is canonically embedded in the center of $H(F)$. If $z \in Z_G$ and γ_H is an image of γ_G then $z\gamma_H$ is an image of $z\gamma_G$. According to [LS1, Lemma 4.4.A] there is a character λ on Z_G such that

$$\Delta(z\gamma_H, z\gamma_G) = \lambda(z)\Delta(\gamma_H, \gamma_G)$$

for all z, γ_H. The proof of this requires a detailed analysis of the term Δ_2 in the transfer factor (see [LS3, Sections 3, 4]).

We now assume the analogue of (1.6), i.e., each pair $(G_{\epsilon_G}, H_{\epsilon_H})$ admits local transfer around ϵ_H. For $f^G \in C_c^\infty(G(F))$, consider the "normalized κ-orbital integral"

$$\Phi(\gamma_H, f^G) = \sum_{\gamma_G} \Delta(\gamma_H, \gamma_G)\Phi(\gamma_G, f^G)$$

for γ_H near ϵ_H. If ϵ_H is not an image then $\Phi(\gamma_H, f^G)$ vanishes near ϵ_H. Otherwise fix ϵ_G with ϵ_H as an image. Each orbital integral $\Phi(\gamma_G, f^G)$ is, by Harish-Chandra descent, locally an orbital integral on $G_{\epsilon'_G}(F)$ for some suitable ϵ'_G stably conjugate to ϵ_G. We also need descent for the transfer factor $\Delta(\gamma_H, \gamma_G)$. The main theorem of [LS3] is that

$$\Delta_G(\gamma_H, \gamma_G) = (\text{const})\Delta_{G_{\epsilon'_G}}(\gamma_H, \gamma_G)$$

for γ_H near ϵ_H, γ_G near ϵ'_G. (For F archimedean, this is true in the limit). The result is better stated with relative transfer factors. Take $\overline{\gamma}_H, \overline{\gamma}_G$ near ϵ_H, ϵ'_G respectively, with $\overline{\gamma}_H$ an image of $\overline{\gamma}_G$, and set $\Theta = \Delta_G/\Delta_{G_{\epsilon'_G}}$. Then

$$\Theta(\gamma_H, \gamma_G; \overline{\gamma}_H, \overline{\gamma}_G) = 1. \tag{2.3}$$

The proof of (2.3) in [LS3] proceeds as follows. First we reduce easily to the case G quasi-split and $\epsilon'_G = \epsilon_G$ (see Section 3.1). Then we analyse in detail the cohomologically defined factors Θ_I, Θ_2 of Θ (Sections 3, 4). The only other factor in Θ possibly making a nontrivial contribution is Θ_{II} which is given by a simple explicit expression indexed by Galois orbits of roots. See (5.1); Θ_{II} is the right side of (5.1.1) and the final formulas for Θ_I, Θ_2 are also given. We find that only orbits of roots outside both G_{ϵ_G} and H may make a nontrivial contribution to Θ_I, Θ_2 or Θ_{II}. At the same time a long reduction argument allows us to assume all roots in G have the same length and that they take only the values ± 1 on ϵ_G while the coroots take only these values on the endoscopic datum s defining H (Section 5). Under such conditions we are able to compute the product of Θ_I and Θ_2 explicitly (see Theorem 6.3.C) and so compare orbit by orbit with Θ_{II}. We have, in particular, a product formula over all places which allows us to assume odd residual characteristic. The rest is a number theoretical computation (see Section 6.6).

We may now finish. The normalized κ-orbital integral $\Phi(\gamma_H, f)$ is a linear combination of such integrals for the groups $G_{\epsilon'_G}$ (see Sections 1.5, 1.7 of [LS3] for more precise information) and so we follow the argument for the stable case in Section 1. *Our conclusion is that what remains to prove transfer is the local problem (2.2) for Shalika germs.*

This long analysis has further consequences. We mention two examples. From the study of the regular unipotent contribution we obtain in the p-adic case a formula for regular unipotent germs [S2] and in the real case a proof that the transfer factors defined here coincide with those introduced earlier in [S1]. From descent we verify some conjectures of Kottwitz [K] about extending transfer factors and the matching of orbital integrals from the "most regular" classes to all equisingular semisimple classes [LS3, §2].

REFERENCES

[CD] L. Clozel and P. Delorme, *Le théorème de Paley-Wiener invariant pour les groupes de Lie réductifs*, Inv. Math. **77** (1984), 427–453.

[H1] T. Hales, *The subregular germ of orbital integrals*, Thesis, Princeton University.

[H2] ———, *Orbital integrals on* $U(3)$, to appear.

[H3] ———, *Shalika germs on* $GSp(4)$, Astérisque, Nos. 171–172 (1989), 195–256.

[K] R. Kottwitz, *Stable trace formula: elliptic singular terms*, Math. Ann. **275** (1986), 365–399.

[KS] R. Kottwitz and D. Shelstad, *Twisted endoscopy*, in preparation.

[LS1] R. Langlands and D. Shelstad, *On the definition of transfer factors*, Math. Ann. **278** (1987), 219–271.

[LS2] ———, *Orbital integrals on forms of* $SL(3)$, *II*, Can. J. Math. **XLI** (1989), 480–507.

[LS3] ———, *Descent for transfer factors*, The Grothendieck Festschrift, vol II, Birkhäuser, Boston, 1991, pp. 485–563.

[S1] D. Shelstad, *L-indistinguishability for real groups*, Mathematische Annalen **259** (1982), 385–430.

[S2] ———, *A formula for regular unipotent germs*, Astérisque **171–172** (1989), 275–277.

DEPARTMENT OF MATHEMATICS AND COMPUTER SCIENCE, RUTGERS UNIVERSITY, NEWARK, NEW JERSEY 07102

ON JACQUET MODULES OF INDUCED REPRESENTATIONS OF p–ADIC SYMPLECTIC GROUPS

MARKO TADIĆ

University of Zagreb

1. INTRODUCTION

We fix a reductive p-adic group G. One very useful tool in the representation theory of reductive p-adic groups is the Jacquet module. Let us recall the definition of the Jacquet module. Let (π, V) be a smooth representation of G and let P be a parabolic subgroup of G with a Levi decomposition $P = MN$. The Jacquet module of V with respect to N is

$$V_N = V/\operatorname{span}_{\mathbb{C}}\{\pi(n)v - v; n \in N, v \in V\}.$$

Here M acts in a natural way on V_N. We twist this action by $\delta_P^{-1/2}$, and fix such an action of M. The Jacquet functor is exact and left adjoint to the functor of parabolic induction, i.e., Frobenius reciprocity holds

$$\operatorname{Hom}_G(V, \operatorname{Ind}_P^G(\sigma)) \cong \operatorname{Hom}_M(V_N, \sigma),$$

for each smooth representation σ of M. Here $\operatorname{Ind}_P^G(\sigma)$ denotes the parabolically induced representation of G by σ from P (the induction we consider is normalized, i.e., unitarizable representations are carried to unitarizable representations).

Frobenius reciprocity indicates the importance of knowledge of Jacquet modules of representations. But the importance goes far beyond Frobenius reciprocity. The techniques of Jacquet modules are especially convenient for the analysis of parabolically induced representations. In this case it can be very hard—or even impossible—to understand the complete structure of the Jacquet module. One has a slightly weaker understanding of the Jacquet modules here: for a parabolically induced representation $\operatorname{Ind}_{P_1}^G(\sigma)$ and a parabolic subgroup $P_2 = M_2N_2$ there exists a filtration of $\left(\operatorname{Ind}_{P_1}^G(\sigma)\right)_{N_2}$,

$$\{0\} = U_0 \subseteq U_1 \subseteq \ldots \subseteq U_m = \left(\operatorname{Ind}_{P_1}^G(\sigma)\right)_{N_2}$$

This paper was written while the author was a visitor at the University of Utah.

as an M_2-representation, such that one can express U_i/U_{i-1}, $i = 1, \ldots, m$, as certain induced representations of M_2 by suitable Jacquet modules of σ. The above filtration is related to the Bruhat decomposition $P_1 \backslash G / P_2$. This description has been given by J. Bernstein and A. V. Zelevinsky ([BZ]), W. Casselman ([C]), and Harish-Chandra.

Such a description is less than we would like to have, but these formulas can still be very useful. The reason is that one can study Jacquet modules for different parabolic subgroups, and then compare them. In this way it is possible to get fairly explicit information about the induced representation. We will explain these ideas in more detail in a forthcoming paper.

The formulas referred to above are fairly complicated to apply in the analysis of induced representations, especially when one studies whole families of groups. So one may pose the problem of giving a more explicit description of Jacquet modules of induced representations. For a longtime, I have been interested in developing a better understanding of these formulas. (J. Bernstein, P. Deligne and D. Kazhdan, in their paper on the trace Paley-Wiener theorem for reductive p-adic groups ([BDK]), also raise the question of better understanding of the combinational structure related to describing Jacquet modules of induced representations.)

In this paper, we are going to explain one possible approach to the above problem for the case of symplectic groups. Joint work with P. Sally ([ST]), related more to $GSp(2)$, shows the usefulness of such an approach.

The structure describing semi-simplifications of Jacquet modules of induced representation for $GL(n)$ was obtained by Bernstein and Zelevinsky ([Z1]). This is a Hopf algebra structure. Since we need some notation from the GL-case, we will review this example in section two. This case is also good motivation for the symplectic case.

In section six we give one application of this approach to the construction of square integrable representations. Section seven contains an application to reducibility problems.

Complete proofs of results presented in this paper will appear elsewhere.

2. General linear groups

By F we will denote a p-adic field. We will assume that characteristic of F is different from two. Usually we will write $GL(n)$ for $GL(n, F)$.

For two smooth representations, π of $GL(n)$ and τ of $GL(m)$, we write

$$\pi \times \tau = \mathrm{Ind}_P^{GL(n+m)}(\pi \otimes \tau)$$

(see ([BZ])). Here P denotes the parabolic subgroup

$$P = \left\{ \left[\begin{array}{c|c} g & * \\ \hline 0 & h \end{array}\right] : g \in GL(n), h \in GL(m) \right\}$$

The Grothendieck group of the category of all finite length, smooth representations of $GL(m)$ is denoted by R_m. Recall that R_m is a free $\mathbb{Z}$-module over the set of equivalence classes of all irreducible smooth representations of $GL(m)$. In a natural way we can define a $\mathbb{Z}$-bilinear mapping on the basis by

$$\times : R_n \times R_m \to R_{n+m}$$
$$(\pi, \tau) \to s.s.(\pi \times \tau),$$

where $s.s.(\pi \times \tau)$ denotes the semi-simplification of $\pi \times \tau$. Set $R = \oplus_{m \geq 0} R_m$. Then we have graded ring structure on R. We define an additive mapping $m : R \otimes R \to R$, determined by the formula

$$m(\pi \otimes \tau) = \pi \times \tau.$$

While the above operation comes from induction, the following operation comes from Jacquet modules. Let σ be a finite length, smooth representation of $GL(n)$ and $0 \leq p \leq n$. Then

$$r_{(p,n-p),n}(\sigma)$$

will denote the Jacquet module of σ with respect to the parabolic subgroup given by

$$\left\{ \left[\begin{array}{c|c} g & * \\ \hline 0 & k \end{array} \right] : g \in GL(p), k \in GL(n-p) \right\}.$$

We may consider, in a natural way,

$$s.s.(r_{(p,n-p),n}(\sigma)) \in R_p \otimes R_{n-p}.$$

Set

$$m^*(\sigma) = \sum_{p=0}^{n} s.s.(r_{(p,n-p),n}(\sigma)) \in R \otimes R.$$

One extends m^* $\mathbb{Z}$-linearly to $m^* : R \to R \otimes R$.

In the proof that (R, m, m^*) is a Hopf algebra, the central property is that

$$m^* : R \to R \otimes R$$

is multiplicative when, on the right hand side, we define

$$(\pi_1 \otimes \pi_2) \times (\tau_1 \otimes \tau_2) = (\pi_1 \times \tau_1) \otimes (\pi_2 \times \tau_2).$$

This gives us a simple rule for computing the semisimplification of a Jacquet module of an induced representation in terms of the Jacquet module of the representation from which we are inducing. Let me mention two related examples at this point.

Finite fields. In place of F put a finite field $\mathbf{F}$. We obtain, in the same way, a Hopf algebra structure on R. Here m^* computes the Jacquet modules exactly since they are semisimple. A similar claim is true for m.

One introduces in a natural way a scalar product on R such that the irreducible representations form an orthonormal basis. Frobenius reciprocity then says that m^* is an adjoint operator to m. Also, if one introduces in a natural way a cone of positive elements in R, then the operations m and m^* are positive.

A. V. Zelevinsky ([Z2]) obtained a classification of irreducible representations of $GL(n, \mathbf{F})$ modulo cuspidal representations as a structure theory of the positive Hopf algebra R. He showed that

$$R = \bigotimes_{\rho \text{ cuspidal}} R(\rho)$$

as positive Hopf algebras. The irreducible representations in $R(\rho)$ are parametrized by partitions. For details of this elegant theory one should consult Zelevinsky's book.

Division algebras. In place of F put a central division F-algebra A; again we obtain a Hopf algebra structure on R. In our paper ([T]), we developed some important parts of the representation theory of $GL(n, A)$ using the Hopf algebra structure. We started in ([T]) from the results of P. Deligne, D. Kazhdan and M-F. Vigneras ([DKV]).

We now turn to the symplectic case.

3. Symplectic groups

J_n will denote the $n \times n$ matrix

$$J_n = \begin{bmatrix} 0 & 0 & \dots & 0 & 1 \\ 0 & 0 & \dots & 1 & 0 \\ \vdots & \vdots & \ddots & \vdots & \vdots \\ 1 & 0 & \dots & 0 & 0 \end{bmatrix},$$

and $Sp(n, F)$ will denote the group of all $(2n) \times (2n)$ matrices over F, satisfying

$${}^tS \begin{bmatrix} 0 & J_n \\ -J_n & 0 \end{bmatrix} S = \begin{bmatrix} 0 & J_n \\ -J_n & 0 \end{bmatrix}$$

(${}^tS$ denotes the transposed matrix of S).

Take a smooth representation π of $GL(n, F)$ and σ of $Sp(m, F)$. Set

$$\pi \rtimes \sigma = \operatorname{Ind}_P^{Sp(n+m)}(\pi \otimes \sigma),$$

where P is the parabolic subgroup in $Sp(n+m)$ consisting of the elements

$$\begin{bmatrix} g & * & * \\ 0 & h & * \\ 0 & 0 & {}^\tau g^{-1} \end{bmatrix}$$

(${}^\tau g$ denotes the transposed matrix of g with respect to the second diagonal). Here $\pi \otimes \sigma$ maps the above element to $\pi(g) \otimes \sigma(h)$.

Note that this type of multiplication of representations was introduced by Faddeev in ([F]) for the finite field case.

Let $R_n(S)$ be the Grothendieck group of the category of finite length, smooth representations of $Sp(n)$. Set

$$R(S) = \bigoplus_{n\geq 0} R_n(S).$$

Now we have

$$\rtimes : R_n \times R_m(S) \to R_{n+m}(S),$$

and further,

$$\rtimes : R \times R(S) \to R(S).$$

$\rtimes$ also factors to give

$$\mu : R \otimes R(S) \to R(S), \quad \mu(\pi \otimes \sigma) = s.s.(\pi \rtimes \sigma).$$

For a finite length representation σ of $Sp(n)$, and $0 \leq k \leq n$, we denote by

$$s_{(k),(0)}(\sigma)$$

the Jacquet module for the following parabolic subgroup in $Sp(n)$:

$$\left\{ \begin{bmatrix} g & * & * \\ 0 & h & * \\ 0 & 0 & {}^{\tau}g^{-1} \end{bmatrix} : g \in GL(k), h \in Sp(n-k) \right\}$$

Here the Levi factor is naturally isomorphic to $GL(k) \times Sp(n-k)$. We therefore consider

$$s.s.(s_{(k),(0)}(\sigma)) \in R_k \otimes R_{n-k}(S)$$

and define

$$\mu^*(\sigma) = \sum_{k=0}^{n} s.s.(s_{(k),(0)}\sigma)) \in R \otimes R(S).$$

μ^* contains semisimplifications of all Jacquet modules for maximal parabolic subgroups.

To have Jacquet modules of induced representations, it is enough to compute

$$\mu^*(\pi \rtimes \sigma)$$

for π a representation of GL and σ of Sp. We will now describe a formula for $\mu^*(\pi \rtimes \sigma)$.

Define a multiplication

$$\begin{aligned} \rtimes : \quad & (R \otimes R) \times (R \otimes R(S)) \to R \otimes R(S) \\ & ((\pi_1 \otimes \pi_2), (\pi_3 \otimes \sigma)) \quad \to (\pi_1 \times \pi_3) \otimes (\pi_2 \rtimes \sigma) \end{aligned}$$

A long calculation gives:

Theorem 1. *Let π be a finite length, smooth representation of GL, and σ a similar representation of Sp. Put*

$$M^*(\pi) = (m \otimes 1) \circ (\sim \otimes m^*) \circ s \circ m^*(\pi),$$

where $s(x \otimes y) = y \otimes x$ and $\sim: R \rightarrow R$ is given on the basis $\pi \longmapsto \tilde{\pi}$ (here $\tilde{\pi}$ denotes the contragredient representation of π, $\circ$ denotes the composition). Then

$$\mu^*(\pi \rtimes \sigma) = M^*(\pi) \rtimes \mu^*(\sigma).$$

4. $GSp(n, F)$

By definition, $GSp(n, F)$ is a group of all $(2n) \times (2n)$ matrices over F satisfying

$${}^tS \begin{bmatrix} 0 & J_n \\ -J_n & 0 \end{bmatrix} S = \mu(S) \begin{bmatrix} 0 & J_n \\ -J_n & 0 \end{bmatrix}$$

for some $\mu(S) \in F^\times$. We further define $GSp(0)$ to be $F^\times$. Standard maximal parabolic subgroups are

$$\left\{ \begin{bmatrix} g & * & * \\ 0 & h & * \\ 0 & 0 & \mu(h)^\tau g^{-1} \end{bmatrix} : g \in GL(k), k \in GSp(n-k) \right\},$$

$0 \leq k \leq n$. Again, Levi factors are naturally isomorphic to $GL(k) \times GSp(n-k)$. Therefore, as was done for Sp-groups, we may define a representation $\pi \rtimes \sigma$ of $GSp(n)$ if π is a representation of $GL(k)$ and σ is a representation of $GSp(n-k)$. The Grothendieck group of the category of finite length representations we denote by $R_n(G)$. Set $R(G) = \underset{n \geq 0}{\oplus} R_n(G)$.

In the same way as before we define

$$\rtimes : R \times R(G) \rightarrow R(G).$$

Analogously we define

$$\mu^* : R(G) \rightarrow R \otimes R(G).$$

Now a technical variation of the formula for $Sp(n)$ gives a formula for

$$\mu^*(\pi \rtimes \sigma)$$

for the GSp-case. We are not going to write this formula here.

For a character ω of $F^\times$, we use the same letter to denote the character $\omega \circ \mu$ of $GSp(n)$.

5. An application to square integrable representations of $GSp(n)$

Let ρ be a cuspidal representation of $GL(m)$. For a non-negative integer n, the set

$$[\rho, \nu^n \rho] = \{\rho, \nu\rho, \ldots, \nu^n \rho\}$$

is called a segment in the cuspidal representation. Here

$$\nu = |\det|.$$

The representation

$$\rho \times \nu\rho \times \ldots \times \nu^n \rho$$

contains a unique essentially square-integrable subquotient. We denote this subquotient by $\rho([\rho, \nu^n \rho])$. This classification was obtained by J. Bernstein.

Write $\rho = \nu^\alpha \rho_o$, where ρ_o is unitary and $\alpha \in \mathbb{R}$. Take a cuspidal representation σ of $GSp(r)$. Suppose that $\rho \rtimes \sigma$ reduces. Then

$$\rho_o \cong \tilde{\rho}_o \quad \text{and} \quad \omega_{\rho_o}\sigma \cong \sigma.$$

Here ω_{ρ_o} denotes the central character of ρ_o.

Suppose that $(\nu\rho_o) \rtimes \sigma$ reduces. Then this representation has a unique essentially square-integrable subquotient which we denote by $\delta(\{\nu\rho_o\}, \sigma)$. We can now define the representations $\delta([\nu\rho_o, \nu^n\rho_o], \sigma)$ recursively. There exists a unique subquotient of

$$(\nu^n \rho_o) \times (\nu^{n-1}\rho_o) \times \ldots \times (\nu\rho_o) \rtimes \sigma$$

denoted by $\delta([\nu\rho_o, \nu^n\rho_o], \sigma)$, such that

$$\begin{aligned}\mu^*(\delta([\nu\rho_o, \nu^n\rho_o], \sigma)) = 1 \otimes \delta([\nu\rho_o, \nu^n\rho_o], \sigma) + \nu^n\rho_o \otimes \delta([\nu\rho_o, \nu^{n-1}\rho_o], \sigma) \\ + \delta([\nu^{n-1}\rho_o, \nu^n\rho_o]) \otimes \delta([\nu\rho_o, \nu^{n-2}\rho_o], \sigma) + \ldots \\ \ldots + \delta([\nu^2\rho_o, \nu^n\rho_o]) \otimes \delta(\{\nu\rho_o\}, \sigma) \\ + \delta([\nu\rho_o, \nu^n\rho_o]) \otimes \sigma.\end{aligned}$$

Moreover, each $\delta([\nu\rho_o, \nu^n\rho_o], \sigma)$ is an essentially square integrable representation. These representations are obvious generalizations of the Steinberg representation.

Remark 1. If $(\nu^\alpha\rho_o) \rtimes \sigma$ reduces with $\alpha \neq 0, 1, -1$, then it is also possible to define a series of square integrable representations in a similar way. This situation of $\alpha \neq 0, 1, -1$ does occur, although the only example known to us was found by F. Shahidi ([S1], [S2]). We want to thank him for communicating this result to us and for the letter in which he supplied the details of the proof. In the Shahidi example, α is $1/2$, ρ_o is a unitary cuspidal representation of $GL(2)$, and σ is a character of $GSp(0)$. We use this result in ([ST]). F. Shahidi gave a general approach to such reducibility questions; one description can be found in ([S1]).

Now suppose $\rho_o \cong \tilde{\rho}_o$ but $\omega_{\rho_o}\sigma \not\cong \sigma$. Denote $\delta(\{\rho_o\}, \sigma) = \rho_o \rtimes \sigma$. This is not an essentially square integrable representation. However, the representation $\nu\rho_o \times \rho_o \rtimes \sigma$ contains a unique essentially square integrable subquotient, denoted by $\delta([\rho_o, \nu\rho_o], \sigma)$. We then define $\delta([\rho_o, \nu^n\rho_o], \sigma)$ recursively: there exists a unique subquotient $\delta([\rho_o, \nu^n\rho_o], \sigma)$ of

$$(\nu^n\rho_o) \times (\nu^{n-1}\rho_o) \times \ldots \times \rho_o \rtimes \sigma$$

such that

$$\begin{aligned}\mu^*(\delta([\rho_o,\nu^n\rho_o],\sigma)) = 1\otimes\delta([\rho_o,\nu^n\rho_o],\sigma) + \nu^n\rho_o\otimes\delta([\rho_o,\nu^{n-1}\rho_o],\sigma)\\ +\delta([\nu^{n-1}\rho_o,\nu^n\rho_o])\otimes\delta([\rho_o,\nu^{n-2}\rho_o],\sigma)+\dots\\ \dots+\delta([\nu^2\rho_o,\nu^n\rho_o])\otimes\delta([\rho_o,\nu\rho_o],\sigma)\\ +\delta([\nu\rho_o,\nu^n\rho_o])\otimes\delta(\{\rho_o\},\sigma)\\ +\delta([\rho_o,\nu^n\rho_o])\otimes(\sigma+\omega_{\rho_o}\sigma).\end{aligned}$$

The representations $\delta([\rho_o,\nu^n\rho_o],\sigma)$ are essentially square-integrable for $n \geq 1$. F. Rodier pointed out some of the above representations in [R1], in the case where ρ_o is a character of order 2, σ is a character, and $n = 1$.

Now we have the "mixed" case.

Theorem 2. *Let $\rho_1,\dots,\rho_n,\tau_1,\dots,\tau_m$ be mutually inequivalent unitary cuspidal representations of GL-groups (possibly $m = 0$ or $n = 0$). Let σ be a cuspidal representation of GSp, and X the group generated by central characters of $\rho_1,\rho_2,\dots,\rho_n$. Suppose that*

(i) $\rho_i \cong \tilde\rho_i$, $1 \le i \le n$, *and* $\tau_j \cong \tilde\tau_j$, $1 \le j \le m$
(the first condition implies Card $X \le 2^n$).

(ii) *Card $X = 2^n$ and $\omega\sigma \ncong \sigma$ for any non-trivial $\omega \in X$.*

(iii) $\nu\tau_j \rtimes \sigma$ *reduces for* $1 \le j \le m$.

Let p_i $(1 \le i \le n)$ and q_j $(1 \le j \le m)$ be positive integers. Set

$$\Delta_i = [\rho_i,\nu^{p_i}\rho_i],\ \Gamma_j = [\nu\tau_j,\nu^{q_j}\tau_j],\ 1\le i\le n,\ 1\le j\le m.$$

Then the representation $\delta(\Delta_1)\times\dots\times\delta(\Delta_n)\times\delta(\Gamma_1)\times\dots\times\delta(\Gamma_m)\rtimes\sigma$ contains a unique irreducible subquotient, denoted by $\delta(\Delta_1,\dots,\Delta_n,\Gamma_1,\dots,\Gamma_m,\sigma)$, which has

$$\delta(\Delta_1)\times\dots\times\delta(\Delta_n)\times\delta(\Gamma_1)\times\dots\times\delta(\Gamma_m)\otimes\sigma$$

for a subquotient of (a suitable) Jacquet module. This subquotient is essentially square integrable.

Remarks 2. (i) Suppose ρ_i,τ_j, and σ in Theorem 2 are all characters. In this case it can be shown that the list of representations in Theorem 2 is equal to the collection of essentially square integrable representations described by F. Rodier in ([R1]).

(ii) There is the problem of determining the reducibility points of

$$\nu^\alpha\rho_o\rtimes\sigma$$

where $\alpha\in\mathbb{R}$, ρ_o is a unitary cuspidal representation of GL, and σ a cuspidal representation of GSp such that $\tilde\rho_o\cong\rho$ and $\omega_{\rho_o}\sigma\cong\sigma$. In addition to the well-known case of $GSp(1) = GL(2)$, this problem has been solved completely by $J.-L.$ Waldspurger for one intermediate parabolic subgroup of $GSp(2)$ ([W]). For the other intermediate parabolic subgroup this has been done by F. Shahidi ([S1], [S2])—in fact, Shahidi's paper ([S1]) includes Waldspurger's case. More generally, Shahidi's paper studies general quasi-split reductive groups over F and contains one description of the reducibility points of induced representations from maximal parabolic subgroups by cuspidal representations (see Remark 1).

(iii) If, instead of condition *(ii)* in Theorem 2, one has that $\nu^{\alpha_j}\tau_j \rtimes \sigma$ reduces, $\alpha_j > 0$, $1 \leq j \leq m$, then it is possible to construct in a similar way essentially square integrable representations. We hope that such constructed representations will exhaust all essentially square integrable representations of GSp-groups which are subquotients of induced representations by regular cuspidal representations.

(iv) A. Moy pointed out to me that for GSp-groups there may be more than one irreducible square integrable subquotient in the same unramified principal series representation. Therefore, for GSp-groups there will exist more square-integrable representations than those described in Theorem 2 and its modification mentioned in *(iii)*. Moreover, Moy's remark motivated us to construct an example of a non-regular square integrable representation which is not supported in the minimal parabolic subgroup.

6. Reducibility points

We conclude this paper with examples showing how the formula obtained for $\mu^* \circ \mu$ can be very useful for determining the reducibility of induced representations.

Examples. Let χ and σ be characters of $F^\times$

(i) For $GSp(n+1)$ the representation $\chi \rtimes \rho([\nu 1_{F^\times}, \nu^n 1_{F^\times}], \sigma)$ is reducible if and only if

$$\chi \in \left\{1_{F^\times}, \nu^{\pm,(n+1)} 1_{F^\times}\right\}$$

($1_{F^\times}$ denotes the trivial character of $F^\times$).

(ii) The representation

$$\delta([\chi, \nu\chi]) \rtimes \delta([\nu 1_{F^\times}, \nu^n 1_{F^\times}], \sigma)$$

is reducible if and only if $[\chi, \nu\chi]$ is an element of

$$\{[\psi, \nu\psi], [\nu, \nu^2], [\nu^n, \nu^{n+1}], [\nu^{n+1}, \nu^{n+2}]\} \text{ or}$$
$$\{[\nu^{-1}\psi, \psi], [\nu^{-2}, \nu^{-1}], [\nu^{-(n+1)}, \nu^{-n}], [\nu^{-(n+2)}, \nu^{-(n+1)}]\},$$

where $\psi^2 = 1_{F^\times}$.

(iii) In the following example the representation is not supported in the minimal parabolic subgroup. Let ρ be a cuspidal unitary representation of $GL(m)$, $m > 1$, such that $\rho \cong \tilde{\rho}$ and $\omega_\rho \neq 1_{F^\times}$. Suppose that $\chi \neq \omega_\rho$. Then $\chi \rtimes \delta([\rho, \nu^n \rho], \sigma)$ is reducible if and only if $\chi \in \{\nu^{\pm 1}\}$.

Remark 3. R. Gustafson determined the reducibility points for representations of $Sp(n)$

$$(\sigma \circ det_m) \rtimes 1$$

where σ is an unramified character ([G]).

REFERENCES

[BDK] J. Bernstein, P. Deligne, and D. Kazhdan, *Trace Paley-Wiener theorem for reductive p-adic groups*, J. Analyse Math **47** (1986), 180–192.

[BZ] J. Bernstein and A.V. Zelevinsky, *Induced representations of reductive p-adic groups I*, Ann. Sci. École Norm. Sup. **10** (1977), 441–472.

[C] W. Casselman, *Intro. to the theory of admissible representations*, preprint.

[DKV] P. Deligne, D. Kazhdan, M.-F. Vigneras, *Représentations des algèbres centrales simples p-adiques*, Représentations des Groupes Réductifs sur un Corps Local, Hermann, Paris, 1984.

[F] D. K. Faddeev, *On multiplication of representations of classical groups over a finite field with representations of the full linear group*, Vestnik Lenigradskogo Universiteta **13** (1976), 35–40. (Russian)

[G] R. Gustafson, *The degenerate principal series for* $Sp(2n)$, Memoirs of the AMS **248** (1981).

[R1] F. Rodier, *Décomposition de la série principale des groupes réductifs p-adiques*, Non-Commutative Harmonic Analysis, Lecture Notes in Mathematics, Vol. 880, Springer-Verlag, Berlin-Heidelberg-New York, 1981.

[R2] ———, *Sur les représentations non ramifiées des groupes réductifs p-adiques; l'example de* $GSp(4)$, Bull. Soc. Math. France **116** (1988), 15–42.

[ST] P. J. Sally, M. Tadić, *On representations of non-archimedean* $GSp(2)$, manuscript.

[S1] F. Shahidi, *A proof of Langlands conjecture on Plancherel measure; complementary series for p-adic groups*, Annals of Math (2) **132** (1990), 273–330.

[S2] ———, *Langlands' conjecture on Plancherel measures for p-adic groups*, these proceedings.

[T] M. Tadić, *Induced representations of* $GL(n, A)$ *for p-adic division algebras* A, J. reine angew. Math. **405** (1990), 48–77.

[W] J.-L. Waldspurger, *Un exercice sur* $GSp(4, F)$ *et les représentations de Weil*, Bull. Soc. Math.France **115** (1987), 35–69.

[Z1] A. V. Zelevinsky, *Induced representations of reductive p-adic groups II, On irreducible representations of* $GL(n)$, Ann. Sci. École Norm. Sup. **13** (1980), 165–210.

[Z2] ———, *Representations of Finite Classical Groups, A Hopf Algebra Approach, Lecture Notes in Math., Vol. 869*, Springer-Verlag, Berlin-Heidelberg-New York, 1981.

DEPARTMENT OF MATHEMATICS, UNIVERSITY OF ZAGREB, P. O. BOX 187, 41001, ZAGREB, YUGOSLAVIA

ASSOCIATED VARIETIES AND UNIPOTENT REPRESENTATIONS

DAVID A. VOGAN, JR.

Massachusetts Institute of Technology

1. INTRODUCTION.

Suppose $G_{\mathbb{R}}$ is a semisimple Lie group. The philosophy of coadjoint orbits, as propounded by Kirillov and Kostant, suggests that unitary representations of $G_{\mathbb{R}}$ are closely related to the orbits of $G_{\mathbb{R}}$ on the dual $\mathfrak{g}_{\mathbb{R}}^*$ of the Lie algebra $\mathfrak{g}_{\mathbb{R}}$ of $G_{\mathbb{R}}$. One knows how to attach representations to semisimple orbits, but the methods used (which rely on the existence of nice "polarizing subalgebras" of $\mathfrak{g}$) cannot be applied to most nilpotent orbits.

One notion that *is* available for all representations is that of "associated variety." Let $K_{\mathbb{R}}$ be a maximal compact subgroup of $G_{\mathbb{R}}$, and K its complexification. Write $\mathfrak{g}$ for the complexification of $\mathfrak{g}_{\mathbb{R}}$. Attached to any admissible representation of $G_{\mathbb{R}}$ (for example, to any irreducible unitary representation) is a Harish-Chandra module X, which carries an algebraic action of K and a Lie algebra representation of $\mathfrak{g}$. If the original representation has finite length, as we assume from now on, then X is finitely generated as a $U(\mathfrak{g})$-module. Choose a finite-dimensional K-invariant generating subspace X_0 of X, and set

$$X_n = U_n(\mathfrak{g}) \cdot X_0 \tag{1.1)(a}$$

Here as usual $U_n(\mathfrak{g})$ is the (finite-dimensional) subspace of $U(\mathfrak{g})$ spanned by products of at most n elements of $\mathfrak{g}$. This defines a K-invariant increasing filtration on X, which is compatible with the standard filtration of $U(\mathfrak{g})$ in the sense that

$$U_p(\mathfrak{g}) \cdot X_q \subset X_{p+q} \tag{1.1)(b}$$

It follows that the associated graded space $\operatorname{gr} X$ is a module over $\operatorname{gr} U(\mathfrak{g})$. By the Poincaré-Birkhoff-Witt theorem, this last ring is naturally isomorphic to the symmetric algebra $S(\mathfrak{g})$. Because the filtration of X is K-invariant, K acts on $\operatorname{gr} X$ as well. Because of the compatibility of the K and $\mathfrak{g}$ actions on X, the action of the Lie algebra $\mathfrak{k}$ also preserves the filtration of X. It follows that the ideal generated by $\mathfrak{k}$ in $S(\mathfrak{g})$ annihilates

Supported in part by NSF grant DMS-8805665.

$\operatorname{gr} X$. Consequently $\operatorname{gr} X$ may be regarded as an $S(\mathfrak{g}/\mathfrak{k})$-module, equipped with a compatible action of K. Condition (1.1)(a) guarantees that $\operatorname{gr} X$ is generated by X_0; in particular, it is finitely generated.

Recall that the ***associated variety*** $\mathcal{V}(M)$ of a module M over a commutative ring R is defined to be the set of all prime ideals containing $\operatorname{Ann} M$. (When R is a finitely generated algebra over $\mathbb{C}$, $\mathcal{V}(M)$ is a closed subvariety of the affine algebraic variety $\operatorname{Spec} R$.) In the setting of the preceding paragraph, it is easy to check that the associated variety of $\operatorname{gr} X$ is independent of the choice of X_0; we will recall the argument in section 2. It is called the ***associated variety of*** X, and written $\mathcal{V}(X)$.

If V is any complex vector space, then the symmetric algebra $S(V)$ may be regarded as the algebra of polynomial functions on the dual vector space V^*. Evaluation at λ in V^* defines a homomorphism from $S(V)$ to $\mathbb{C}$, and therefore a maximal ideal in $S(V)$. All maximal ideals in $S(V)$ arise in this way, so the maximal spectrum of $S(V)$ (the closed points of $\operatorname{Spec} S(V)$) may be identified with V^*. Using these identifications (and speaking a little loosely), we may therefore write

$$\mathcal{V}(X) \subset \mathfrak{g}^* \tag{1.2)(a}$$

Explicitly,

$$\mathcal{V}(X) = \{\, \lambda \in \mathfrak{g}^* \mid p(\lambda) = 0 \text{ whenever } p \in \operatorname{Ann}(\operatorname{gr} X)\,\}. \tag{1.2)(b}$$

Because the elements of $\mathfrak{k}$ annihilate $\operatorname{gr} X$, they must be zero at elements of $\mathcal{V}(X)$. Consequently

$$\mathcal{V}(X) \subset (\mathfrak{g}/\mathfrak{k})^*. \tag{1.2)(c}$$

Because of the compatibility of the K action and the module structure on $\operatorname{gr} X$, $\mathcal{V}(X)$ is a K-invariant subvariety of $\mathfrak{g}^*$ (or of $(\mathfrak{g}/\mathfrak{k})^*$). It turns out in fact that $\mathcal{V}(X)$ is a union of a finite number of nilpotent orbits of K (Corollary 5.23).

We have therefore attached to any admissible representation of $G_{\mathbb{R}}$ of finite length a finite union of nilpotent K-orbits on $(\mathfrak{g}/\mathfrak{k})^*$. Our original intention was to relate representations to nilpotent orbits of $G_{\mathbb{R}}$ on $\mathfrak{g}_{\mathbb{R}}^*$. However, Sekiguchi has shown that there is a close formal relationship between these two kinds of orbits. We will recall his results in section 6, and related results of Schwartz in section 7. For now, it is sufficient to say that one can hope to pursue the philosophy of coadjoint orbits by investigating the relationship between a Harish-Chandra module and its associated variety. (A more precise formulation of this statement may be found in section 8.) Several natural questions arise at once.

1. Is $\mathcal{V}(X)$ the closure of a single orbit of K?
2. Is the closure of every nilpotent K-orbit on $(\mathfrak{g}/\mathfrak{k})^*$ the associated variety of an irreducible unitary representation of $G_{\mathbb{R}}$?
3. Is X uniquely determined by $\mathcal{V}(X)$?

4. Can the structure of X (global character, multiplicities of representations of K) be read off from $\mathcal{V}(X)$?

In a sense these questions are easy: the answer to each is no. The purpose of this paper is to establish some positive results along the lines suggested by the questions. Our inspiration comes from the (closely related) theory of characteristic varieties of D-modules. (We will have no need to be precise about what D is, or to define characteristic varieties; the expert reader can supply appropriate hypotheses.) There Kashiwara-Kawai, Gabber, and others have established deep and powerful results on what the characteristic variety of an irreducible regular holonomic D-module can look like, the extra structure it carries, and the extent to which a D-module can be recovered from its characteristic variety. For example, the characteristic variety of a simple D-module need not be irreducible. But Kashiwara and Kawai have shown that the irreducible components form a single equivalence class under the relation "intersect in codimension 1"; that is, they cannot be too far apart. We prove a weaker related result for associated varieties in section 4. Here is part of it.

Theorem 1.3. *Suppose X is an irreducible Harish-Chandra module, and $\mathcal{O}_\theta$ a K-orbit of maximal dimension in $\mathcal{V}(X)$. Suppose that the complement of $\mathcal{O}_\theta$ has codimension at least two in $\overline{\mathcal{O}_\theta}$. Then $\mathcal{V}(X) = \overline{\mathcal{O}_\theta}$.*

The proof uses a filtration of X different from (1.1)(a), but still satisfying (1.1)(b). Although it is entirely elementary, it is suggested by the noncommutative localization used by Gabber and Kashiwara-Kawai.

The codimension condition is satisfied in many interesting cases (for example, for representations attached to most "non-induced" orbits). In this paper, however, we will apply Theorem 1.3 only to prove the theorem of Borho-Brylinski and Joseph that the associated variety of a primitive ideal is irreducible (Corollary 4.7).

We turn now to the other questions listed above. In one way or another, they ask how X is related to $\mathcal{V}(X)$. To understand that, we must first understand how $\operatorname{gr} X$ is related to $\mathcal{V}(X)$. Very roughly speaking, a finitely generated module over a finitely generated $\mathbb{C}$-algebra looks like the space of sections of a vector bundle over its associated variety. That is, $\operatorname{gr} X$ is approximately the space of sections of a (K-equivariant) vector bundle on $\mathcal{V}(X)$. Now we have seen that $\mathcal{V}(X)$ is approximately a homogeneous space K/H. A K-equivariant vector bundle on K/H is (by passage to the isotropy action at the identity coset) the same thing as a representation of H. The corresponding space of sections is then an induced representation of K.

This suggests that we should be able to attach to X not only the variety $\mathcal{V}(X)$, but also representations of appropriate subgroups of K. This can be done without much difficulty (Definition 2.12). The resulting structure is analogous to the "characteristic cycle" for D-modules; the dimensions of the representations are the analogues of the multiplicities in the cycle.

Under appropriate hypotheses on the annihilator of X, it turns out to be possible to place very strong constraints on the possibilities for the representations (Theorem 8.7 below). This is analogous to the fact that the smooth part of the characteristic variety of a D-module carries a natural local system. In some cases (considered for example in [Sc]) these constraints cannot be satisfied at all; we get in this way a partial understanding of the negative answer to the second question above. In general we find that, as a representation of K, X must be (approximately) induced from a very special representation of a very special subgroup. This provides some information about the fourth question above. In particular, we formulate in section 12 a precise conjecture (together with strong evidence) about the restrictions to K of a large class of unipotent representations.

The theory of associated varieties of primitive ideals is perhaps more familiar to some readers, so we recall briefly how it is related to these ideas. More details may be found in section 4. Suppose I is any two-sided ideal in $U(\mathfrak{g})$. Then the quotient ring $U(\mathfrak{g})/I$ is a finitely generated $U(\mathfrak{g})$-module, so we may define an associated graded $S(\mathfrak{g})$-module $\operatorname{gr} U(\mathfrak{g})/I \simeq S(\mathfrak{g})/\operatorname{gr} I$. The annihilator of this module is obviously $\operatorname{gr} I$, so its associated variety (generally written $\mathcal{V}(I)$) is

$$\mathcal{V}(I) = \{\, \lambda \in \mathfrak{g}^* \mid p(\lambda) = 0 \text{ whenever } p \in \operatorname{gr} I) \,\}. \tag{1.4}$$

(Notice that this notation is inconsistent with that of (1.2): we should really call this $\mathcal{V}(U(\mathfrak{g})/I)$.) Because I is a two-sided ideal, the quotient ring $U(\mathfrak{g})/I)$ inherits the action of the adjoint group G_{ad} of $\mathfrak{g}$. It follows that $\mathcal{V}(I)$ is a G_{ad}-invariant subvariety of $\mathfrak{g}^*$.

Now suppose that I is the annihilator in $U(\mathfrak{g})$ of a Harish-Chandra module X. It is immediate from the definitions that

$$\operatorname{gr}\operatorname{Ann}(X) \subset \operatorname{Ann}(\operatorname{gr} X). \tag{1.5)(a}$$

Comparing (1.2) with (1.4) therefore gives

$$\mathcal{V}(X) \subset \mathcal{V}(\operatorname{Ann}(X)) \cap (\mathfrak{g}/\mathfrak{k})^*. \tag{1.5)(b}$$

It turns out that this containment is almost (but not quite) an equality; a precise statement appears in Theorem 8.4.

2. Associated varieties: elementary properties.

We continue now the discussion of associated varieties begun in the introduction. The results of this section are all easy and well-known, although in a few cases it is difficult to find good references. We have therefore included more proofs than the experts will need.

It is convenient and instructive to work in a slightly greater degree of generality. We continue to assume that $\mathfrak{g}$ is a complex reductive Lie algebra. Let K be an algebraic group equipped with an action Ad on $\mathfrak{g}$ (by automorphisms). We assume also that we are given an injective map on Lie algebras

$$i : \mathfrak{k} \rightarrow \mathfrak{g}$$

compatible with the differential of Ad. (We will use i to regard $\mathfrak{k}$ as a subalgebra of $\mathfrak{g}$, and generally drop it from the notation. The reader may wonder why we do not simply require K to be a subgroup of some algebraic group G with Lie algebra $\mathfrak{g}$. The reason is that in the setting of the introduction, this will not be possible if $G_{\mathbb{R}}$ is not a linear group.)

A module X for $\mathfrak{g}$ is called a $(\mathfrak{g}, K)$-*module* if it is equipped with an algebraic representation π of K satisfying the two conditions

$$\begin{aligned} \pi(k)(u \cdot x) &= (Ad(k)u) \cdot \pi(k)x \qquad (k \in K,\ u \in U(\mathfrak{g}),\ x \in X) \\ d\pi(Z)x &= Z \cdot x \qquad (Z \in \mathfrak{k},\ x \in X). \end{aligned} \tag{2.1(a)}$$

An increasing filtration of X indexed by $\mathbb{Z}$ is called *compatible* if it satisfies

$$\begin{aligned} U_p(\mathfrak{g}) \cdot X_q &\subset X_{p+q} \\ \pi(K)X_n &\subset X_n \end{aligned} \tag{2.1)(b}$$

The first condition allows one to define on $\operatorname{gr} X = \sum_{n \in \mathbb{Z}} X_n / X_{n-1}$ the structure of a graded $S(\mathfrak{g})$-module. (The nth summand will sometimes be called $\operatorname{gr}_n(X)$.) The second condition provides a graded algebraic action (still called π) of K on $\operatorname{gr} X$. These two structures satisfy

$$\begin{aligned} \pi(k)(p \cdot m) &= (Ad(k)p) \cdot \pi(k)m \qquad (k \in K,\ p \in S(\mathfrak{g}),\ m \in \operatorname{gr} X) \\ Z \cdot m &= 0 \qquad (Z \in \mathfrak{k},\ m \in \operatorname{gr} X). \end{aligned} \tag{2.1)(c}$$

An $S(\mathfrak{g})$-module carrying a representation of K satisfying (2.1)(c) is called an $(S(\mathfrak{g}), K)$-*module*. A compatible filtration of X is called *good* if $\cap_{n \in \mathbb{Z}} X_n = 0$, $\cup_{n \in \mathbb{Z}} X_n = X$, and $\operatorname{gr} X$ is a finitely generated $S(\mathfrak{g})$-module. This amounts to four conditions on the filtration:

$$\begin{aligned} X_{-n} &= 0 \qquad \text{(all } n \text{ sufficiently large);} \\ \cup_{n \in \mathbb{Z}} X_n &= X; \\ \dim X_n &< \infty; \\ U_p(\mathfrak{g}) \cdot X_q &= X_{p+q} \qquad \text{(all } q \text{ sufficiently large, all } p \geq 0). \end{aligned} \tag{2.1)(d}$$

The existence of a good filtration evidently implies that X is finitely generated (by X_q for large enough q, say.) Conversely, if X is finitely generated, then we can construct a good filtration of X as in (1.1). The first problem is that X will have many different good filtrations; in order to extract well-defined invariants from the $(S(\mathfrak{g}), K)$-module structure on $\operatorname{gr} X$, we must investigate the dependence of this structure on the filtration. Here and at many points below we will therefore need to consider several different filtrations at the same time, and the subscript notation for them becomes inconvenient. In these cases we may say that $\mathcal{F}$ is a filtration of X, and write $\mathcal{F}_n(X)$ and $\operatorname{gr}(X, \mathcal{F})$ instead of X_n and $\operatorname{gr} X$.

Proposition 2.2. *Suppose X is a $(\mathfrak{g}, K)$-module, and $\mathcal{F}$ and $\mathcal{G}$ are good filtrations of X.*

a) There are integers s and t so that for every integer p,

$$\mathcal{G}_{p-s}(X) \subset \mathcal{F}_p(X) \subset \mathcal{G}_{p+t}(X).$$

b) There are finite filtrations

$$0 = \mathrm{gr}(X, \mathcal{F})_{-1} \subset \mathrm{gr}(X, \mathcal{F})_0 \subset \cdots \subset \mathrm{gr}(X, \mathcal{F})_{s+t} = \mathrm{gr}(X, \mathcal{F})$$
$$0 = \mathrm{gr}(X, \mathcal{G})_{-1} \subset \mathrm{gr}(X, \mathcal{G})_0 \subset \cdots \subset \mathrm{gr}(X, \mathcal{G})_{s+t} = \mathrm{gr}(X, \mathcal{G})$$

by graded $(S(\mathfrak{g}), K)$-submodules, with the property that the corresponding subquotients are isomorphic:

$$\mathrm{gr}(X, \mathcal{F})_j / \mathrm{gr}(X, \mathcal{F})_{j-1} \cong \mathrm{gr}(X, \mathcal{G})_j / \mathrm{gr}(X, \mathcal{G})_{j-1} \qquad (0 \leq j \leq s+t).$$

(The isomorphism shifts the grading of the jth subquotient by $j - s$.)

Proof. The assertion in (a) is a consequence of the properties of good filtrations listed in (2.1)(d). We first choose n so large that

$$\mathcal{F}_p(X) = U_{p-n}(\mathfrak{g}) \cdot \mathcal{F}_n(X),$$

for all $p \geq n$. Next, we choose t so large that $\mathcal{F}_p(X) \subset \mathcal{G}_{p+t}(X)$ for all $p \leq n$; there are essentially only finitely many values of p to consider. For $p \geq n$, we have

$$\mathcal{F}_p(X) = U_{p-n}(\mathfrak{g}) \cdot \mathcal{F}_n(X) \subset U_{p-n}(\mathfrak{g}) \cdot \mathcal{G}_{n+t}(X) \subset \mathcal{G}_{p+t}(X).$$

(The last step uses the compatibility condition (2.1)(b) on $\mathcal{G}$.) This establishes the second containment in (a); the first follows by reversing the roles of $\mathcal{F}$ and $\mathcal{G}$.

For (b), we will construct the required filtration very explicitly. We want to define a graded submodule $\mathrm{gr}(X, \mathcal{F})_j$. We define its component in degree n to be the image of $\mathcal{F}_n \cap \mathcal{G}_{n+j-s}$ in $\mathrm{gr}_n(X, \mathcal{F})$. That is,

$$\mathrm{gr}_n(X, \mathcal{F})_j = (\mathcal{F}_n \cap \mathcal{G}_{n+j-s}) / (\mathcal{F}_{n-1} \cap \mathcal{G}_{n+j-s}).$$

That this actually defines an $(S(\mathfrak{g}), K)$-submodule of $\mathrm{gr}(X, \mathcal{F})$ follows from the compatibility of the filtration $\mathcal{G}$. By (a), the submodule is zero if $j < 0$, and is all of $\mathrm{gr}(X, \mathcal{F})$ if $j \geq s + t$. Computing the subquotient modules is easy: the nth graded piece of $\mathrm{gr}(X, \mathcal{F})_j / \mathrm{gr}(X, \mathcal{F})_{j-1}$ is

$$(\mathcal{F}_n \cap \mathcal{G}_{n+j-s}) / (\mathcal{F}_{n-1} \cap \mathcal{G}_{n+j-s} + \mathcal{F}_n \cap \mathcal{G}_{n+j-s-1})$$

Except for the shift from n to $n + j - s$, this formula is symmetric in $\mathcal{F}$ and $\mathcal{G}$. This provides the isomorphism in (b). □

Recall that a map d from an abelian category to an abelian semigroup is called *additive* if whenever

$$0 \rightarrow A \rightarrow B \rightarrow C \rightarrow 0$$

is a short exact sequence, we have $d(B) = d(A) + d(C)$. If d is such a map and A is an object in the category with a finite filtration

$$0 = A_{-1} \subset A_0 \subset \cdots \subset A_m = A,$$

then $d(A) = \sum_{j=0}^{m} d(A_j/A_{j-1})$. If B is another object admitting a filtration with subquotients isomorphic to those of A, then $d(A) = d(B)$. Proposition 2.2 therefore guarantees that any additive map on the category of finitely generated $(S(\mathfrak{g}), K)$-modules will give a well-defined (in fact additive) map on the category of $(\mathfrak{g}, K)$-modules.

Here are two important examples. Any finitely generated $S(\mathfrak{g})$-module A has a *Krull dimension* $\dim A$ which is a non-negative integer. (It is often convenient to say that the zero module has Krull dimension -1.) For a short exact sequence as above, we have $\dim B = \max(\dim A, \dim B)$. This map is additive if we make the integers into a semigroup by defining the "sum" of two integers to be their maximum. We therefore get a corresponding invariant for a finitely generated $U(\mathfrak{g})$-modules; it is the *Gelfand-Kirillov dimension*, usually written $\operatorname{Dim} X$.

Next, recall from the introduction that the associated variety $\mathcal{V}(A)$ for an $S(\mathfrak{g})$-module is the set of prime ideals containing the annihilator of A. We can make the collection of sets of prime ideals in $S(\mathfrak{g})$ into a semigroup with the union operation; then $\mathcal{V}$ is an additive map. (The reason is that given a short exact sequence as above, we must have

$$(\operatorname{Ann} A)(\operatorname{Ann} C) \subset \operatorname{Ann} B \subset (\operatorname{Ann} A) \cap (\operatorname{Ann} C).$$

Any prime ideal containing $\operatorname{Ann} B$ must contain the first term, and therefore must contain either $\operatorname{Ann} A$ or $\operatorname{Ann} C$. Conversely, any prime ideal containing either $\operatorname{Ann} A$ or $\operatorname{Ann} C$ contains their intersection, and therefore contains $\operatorname{Ann} B$.) We can therefore define the associated variety $\mathcal{V}(X)$ for a finitely generated $U(\mathfrak{g})$-module, as was claimed in the introduction. By the Nullstellensatz, this amounts to the assertion that the radical $\sqrt{\operatorname{Ann}(\operatorname{gr} X)}$ is independent of the choice of good filtration.

We want now to introduce some refinements of these invariants. We begin by recalling a little commutative algebra. Suppose M is a module for the commutative ring R. In addition to the associated variety, there are two other important sets of prime ideals attached to M. The *support* $\operatorname{Supp} M$ is defined to be the set of all prime ideals P in R for which the localization M_P is non-zero. The set of *associated primes* $\operatorname{Ass} M$ consists of those prime ideals which are annihilators of elements of M. It is easy to see that

$$\mathcal{V}(M) \supset \operatorname{Supp} M \supset \operatorname{Ass} M.$$

If M is finitely generated, then $\mathcal{V}(M) = \operatorname{Supp} M$ (see for example [M, pp.25–26]). If in addition R is Noetherian, then $\operatorname{Ass} M$ is a finite set including the minimal elements of $\mathcal{V}(M)$ [M, Theorem 6.5]. It follows that $\mathcal{V}(M)$ is the Zariski closure of $\operatorname{Ass} M$.

Suppose now that R is Noetherian and M is finitely generated. Let $P_1, \ldots, P_r$ be the set of minimal elements in $\mathcal{V}(M)$; that is, the set of minimal primes containing the annihilator of M. (Each P_i corresponds to an irreducible component $\mathcal{V}(P_i)$ of $\mathcal{V}(M)$ regarded as an algebraic variety.) We are going to define the characteristic cycle of M to be a certain formal sum of these prime ideals (with positive integer multiplicities). Roughly speaking, the coefficient of $\mathcal{V}(P_i)$ will measure how many copies of R/P_i are contained in M.

Theorem 6.4 of [M] guarantees that we can find a finite filtration of M by R-submodules so that each subquotient M_j/M_{j-1} is isomorphic to R/Q_j, with Q_j in $\mathcal{V}(M)$ a prime ideal. We will define the coefficient of $\mathcal{V}(P_i)$ to be the number of values of j for which $P_i = Q_j$. We have to check that this definition is independent of the choice of the filtration of M. This requires a little care. The set of prime ideals occurring among the Q_j *does* depend on the choice of filtration; but the minimal elements have well-defined multiplicities. To see this, we can compare each of two filtrations of M to a common refinement of them, using formal arguments and the following easy lemma.

Lemma 2.3. *Suppose P is a prime ideal in a commutative ring R. Regard $A = R/P$ as an R-module, and fix a finite filtration of A with subquotients A_j/A_{j-1} of the form R/Q_j, with Q_j a prime ideal in R. Then $Q_1 = P$, and every other Q_j properly contains P.*

We leave the remaining details to the reader. We have now shown that the following definition makes sense.

Definition 2.4. Let R be a commutative Noetherian ring and M a finitely generated R-module. Let $P_1, \ldots, P_r$ be the minimal prime ideals containing the annihilator of M. The *characteristic cycle* of M is the formal sum

$$\operatorname{Ch}(M) = \sum_{i=1}^{r} m(P_i, M) P_i,$$

where $m(P_i, M)$ is a positive integer defined as follows. Choose a finite filtration of M so that each subquotient M_j/M_{j-1} is of the form R/Q_j, with Q_j a prime ideal in R. Then $m(P_i, M)$ is the number of values of j for which $P_i = Q_j$.

Sometimes it is useful to define $m(Q, M)$ when Q is a prime ideal not among the $\{P_i\}$. If Q contains the annihilator of M but not minimally, we set $m(Q, M) = \infty$. If Q does not contain the annihilator of M, then $m(Q, M) = 0$. In terms of the local ring R_P at a prime ideal P, and the stalk $M_P = M \otimes_R R_P$ of M at a prime ideal P, it is not difficult to check that in every case

$$m(P, M) = \text{ length of } M_P \text{ as an } R_P\text{-module.}$$

Finally, we want to see that Ch is an additive map. To do that, we have to define a semigroup structure on its range. We take the range to be the set of finite formal sums $\sum_{i=1}^{r} m_i P_i$ (of prime ideals with positive integer coefficients) subject to the condition that there should be no containments among the P_i. To add two such expressions, we first throw away terms for ideals properly contained in other ideals, then add coefficients. Now the additivity is elementary.

As a consequence, we can define the *characteristic cycle* $\mathrm{Ch}(X)$ of a $(\mathfrak{g}, K)$-module; it consists of a positive integer weight attached to each irreducible component of $\mathcal{V}(X)$. Obviously this invariant refines the associated variety. Since the Gelfand-Kirillov dimension is just the dimension of $\mathcal{V}(X)$, the characteristic cycle contains that information as well.

One weakness of the characteristic cycle is that it contains no information about the action of K. In order to remedy this, we need some other ways to calculate the multiplicities in the characteristic cycle. Although our main results can be formulated strictly in terms of modules and ideals, it is very convenient to use the language of sheaves of modules along the way. For this we refer to [H, chapter II], or [Sh, Chapter VI]. In particular, we use (for a commutative ring R) the equivalence between the category of R-modules and the category of quasi-coherent sheaves of modules on $\mathrm{Spec}\, R$.

Proposition 2.5. *Suppose P is a prime ideal in the commutative Noetherian ring R, and M is a finitely generated R-module annihilated by P. Then there is an element f of R, not belonging to P, with the property that the localization M_f is a free $(R/P)_f$-module. (That is, M is free on the open set $f \neq 0$ in $\mathrm{Spec}\, R/P$). Its rank is the multiplicity $m(P, M)$ of P in the characteristic cycle of M.*

The first assertion may be found for example in [Sh, Proposition VI.3.1]. The second is an immediate consequence of Definition 2.4 and the exactness of localization.

Corollary 2.6. *In the setting of Proposition 2.5, suppose Q is any prime ideal containing P but not containing f. Then the stalk M_Q of M at Q is a free R_Q/PR_Q-module of rank equal to $m(P, M)$. In particular, this multiplicity is equal to the dimension of the vector space M_Q/QM_Q over the quotient field R_Q/QR_Q of R/Q.*

Corollary 2.7. *In the setting of Proposition 2.5, suppose $\mathfrak{m}$ is any maximal ideal containing P but not containing f. Then*

$$m(P, M) = \dim M/\mathfrak{m}M;$$

the dimension is taken over the field $R/\mathfrak{m}$.

Corollary 2.8. *Suppose R is a finitely generated commutative algebra over $\mathbb{C}$, and M is a finitely generated R-module. Choose a finite filtration* *of M so that each subquotient is annihilated by some prime ideal in $\mathcal{V}(M)$. Fix one of the minimal primes P in $\mathcal{V}(M)$. Then there is an open dense subset U of* Spec R/P *(which we regard as a subset of* Spec R*) so that if $\mathfrak{m}$ is any maximal ideal in U, the multiplicity of P is*

$$m(P, M) = \sum_j \dim_{\mathbb{C}} M_j/(\mathfrak{m}M_j + M_{j-1}).$$

It is tempting to try to omit the filtration of M in the last corollary; the formula would still make sense, and it would give the right answer for $M = R/Q$ whenever Q is prime. The difficulty is that the resulting function on modules is not additive. A simple example is $R = \mathbb{C}[x]$ and $M = R/\langle x^2 \rangle$. Then $\mathcal{V}(M)$ consists of just the maximal ideal $\mathfrak{m} = \langle x \rangle$ in R, which has multiplicity two. Nevertheless, one calculates immediately that $\dim M/\mathfrak{m}M = 1$.

Despite the possibility of such problems, we are going to need to calculate multiplicities using filtrations satisfying somewhat weaker conditions than the one in Corollary 2.8. This can be done using the following elementary extension of Proposition 2.5.

Proposition 2.9. *Suppose P is a prime ideal in the commutative Noetherian ring R, and M is a finitely generated R-module. The following conditions are equivalent.*

- *a) There is an ideal I of R, not contained in P, such that M is annihilated by IP.*
- *b) There is an element g of R, not contained in P, such that $g \cdot M$ is annihilated by P.*
- *c) There is an element h of R, not contained in P, such that M_h is annihilated by P.*
- *d) There is an element f of R, not belonging to P, with the property that M_f is annihilated by P, and defines a free $(R/P)_f$-module. (That is, M is free on the open set $f \neq 0$ in* Spec R/P*).*

In the setting (d), the rank of M_f is the multiplicity $m(P, M)$ of P in the characteristic cycle of M.

When the conditions (a)–(d) in the proposition are satisfied, we say that M is *generically reduced along* P. (Analogously, we might say that M is *reduced along* P when P annihilates M.) Notice that these are conditions only on the annihilator $\mathfrak{a}$ of M. We may also say that the ideal $\mathfrak{a}$ is *generically reduced along* P if the R-module $R/\mathfrak{a}$ is. This notion in the case of ideals is considered further in Lemmas 10.16 and 10.17.

Proof. We will show that (a) $\Rightarrow$ (b) $\Rightarrow$ (c) $\Rightarrow$ (d) $\Rightarrow$ (a). Assume (a), and choose g in I not in P; then (b) is immediate. Next, suppose g is as in (b); then (c) follows with $h = g$. Suppose (c) holds. Apply Proposition 2.5 to the prime ideal PR_h in R_h, obtaining an element $f_0 = h^{-n} f_1$ of R_h; here f_1 is in R. It is easy to verify that the element $f = f_1 h$ has the properties we require (since R_f is naturally isomorphic to $(R_h)_{f_0}$.

Suppose f is as in (d). Choose generators $(m_1, \ldots, m_r)$ for M and $(p_1, \ldots, p_s)$ for P. The assumption that P annihilates M_f means that for each i, j there is a positive integer $N(i,j)$ so that $f^{N(i,j)} \cdot (p_i \cdot m_j) = 0$. Taking N to be the maximum of the $N(i,j)$, we get $f^N \cdot (P \cdot M) = 0$. We can take I to be the ideal generated by f^N in (a).

Finally, we prove the assertion about multiplicities. Fix f as in (d), and let S be the submodule of M annihilated by P. As we just saw, the action of f^N maps M into S. It follows that P does not contain the annihilator of M/S, so $m(P, M/S) = 0$, and consequently $m(P,M) = m(P,S)$. By the exactness of localization, $M_f = S_f$. Proposition 2.5 applied to S therefore gives the result. □

Corollary 2.10. *Suppose R is a finitely generated commutative algebra over $\mathbb{C}$, and M is a finitely generated R-module. Fix one of the minimal primes P in $\mathcal{V}(M)$. Choose a finite filtration of M so that each subquotient is generically reduced along P. Then there is an open dense subset U of* $\operatorname{Spec} R/P$ *(which we regard as a subset of* $\operatorname{Spec} R$*) so that if $\mathfrak{m}$ is any maximal ideal in U, the multiplicity of P is*

$$m(P,M) = \sum_j \dim_{\mathbb{C}} M_j/(\mathfrak{m}M_j + M_{j-1}).$$

We are going to define a refinement of the characteristic cycle that reflects some of the action of K. We will make no attempt to define it in very great generality. Recall that a *virtual character* of an algebraic group H is a finite formal integer combination of irreducible (finite-dimensional algebraic) representations of H. The set of all virtual characters is a free abelian group with basis the set of irreducible representations of H. A character is called *genuine* if the coefficients are non-negative; that is, if it is the character of a (finite-dimensional reducible algebraic) representation π of H. We write $\Theta(\pi)$ for the character of π. The connection with more classical terminology in representation theory is that we may identify $\Theta(\pi)$ with the function on H (also denoted $\Theta(\pi)$) sending h to $\operatorname{tr} \pi(h)$. Passing to differences, we can attach such a function to any virtual character. This identifies the lattice of virtual characters of H with a certain space of conjugation-invariant regular functions on H. Because the unipotent radical U of H acts by unipotent operators in any algebraic representation, these functions must be constant on U. Consequently they descend to H/U. The theory of characters is really therefore a theory about reductive groups.

In order to define the refined characteristic cycle, we need one more fact.

Lemma 2.11. *Suppose M is a finitely generated $(S(\mathfrak{g}), K)$-module. Then there is a finite filtration of M by $(S(\mathfrak{g}), K)$-submodules with the property that every subquotient is generically reduced along every minimal prime in $\mathcal{V}(M)$.*

Proof. If M is zero, $\mathcal{V}(M)$ is empty and there is nothing to prove. Otherwise we proceed by induction on the dimension of $\mathcal{V}(M)$, and then by induction on the sum of the multiplicities of the components of largest dimension. So fix a minimal prime ideal P_1 in $\mathcal{V}(M)$, with R/P_1 of maximal dimension. The action of K on $S(\mathfrak{g})$ evidently permutes the prime ideals in $\mathcal{V}(M)$, and therefore the (finitely many) minimal primes. List the minimal primes in $\mathcal{V}(M)$ and the K orbit of P_1 as $P_1, \ldots, P_s$. Define M_1 to be the submodule of M annihilated by the intersection J of all of these ideals. Because J is K-invariant, M_1 is an $(S(\mathfrak{g}), K)$-submodule. Since P_1 is an associated prime of M, P_1 has non-zero multiplicity in M_1; so the inductive hypothesis applies to the quotient M/M_1. So we need only prove that M_1 is generically reduced along every minimal prime P in $\mathcal{V}(M)$. If P is not one of the P_i, then P does not contain J; so condition (a) of Proposition 2.9 may be satisfied (in a trivial way) by taking I equal to J. If P is one of the P_i, take I equal to the intersection of the remaining ones. Then IP is contained in J, which annihilates M_1; so condition (a) of Proposition 2.9 is again satisfied. □

Definition 2.12. Suppose M is a finitely generated $(S(\mathfrak{g}), K)$-module, and P is a minimal prime ideal in $\mathcal{V}(M)$ (so that $\mathcal{V}(P)$ is a component of $\mathcal{V}(M)$). Assume that for some $\lambda \in \mathfrak{g}^*$ the K-orbit of λ contains a dense open subset of $\mathcal{V}(P)$. Write $K(\lambda)$ for the isotropy group of the action of K at λ, and $\mathfrak{m}(\lambda)$ for the maximal ideal in $S(\mathfrak{g})$ corresponding to λ. We want to attach to M a finite-dimensional representation of $K(\lambda)$. If N is any $(S(\mathfrak{g}), K)$-module, then $K(\lambda)$ acts algebraically on the finite-dimensional vector space $N/\mathfrak{m}(\lambda)N$. Choose a finite filtration of M by $(S(\mathfrak{g}), K)$-submodules M_i, so that the subquotients are generically reduced along P (Lemma 2.11). The *character* of M at λ is the (genuine) virtual character

$$\chi(\lambda, M) = \sum_j M_j/(\mathfrak{m}(\lambda)M_j + M_{j-1})$$

of $K(\lambda)$. (We will explain in a moment why this is independent of choices.) By Corollary 2.10,

$$\dim \chi(\lambda, M) = m(P, M).$$

There is no difficulty in extending this definition to the setting of a coherent sheaf of modules with K-action over an algebraic variety Z on which K acts. If $Z = K/H$ is a homogeneous space, then such a sheaf M must be the sheaf of sections of a vector bundle on which K acts. Such a vector bundle in turn is given by a representation π of H. In this case $\chi(eH, M)$ is the virtual representation of H represented by π. The definition (like that of the characteristic cycle) gives precision to the idea that a module is more or less the space of sections of a vector bundle over its support.

To see that $\chi(\lambda, M)$ is well-defined, one can imitate the proof of Proposition 2.2. At various points in the argument one encounters short exact sequences

$$0 \to A \to B \to C \to 0.$$

These give rise to right exact sequences

$$A/\mathfrak{m}(\lambda)A \to B/\mathfrak{m}(\lambda)B \to C/\mathfrak{m}(\lambda)C \to 0$$

which must be shown to be exact. But the last formula in Definition 2.12 will guarantee the additivity of dimensions in this sequence, and the exactness follows. (This argument is the reason we needed Corollary 2.10.) Once we know that $\chi(\lambda, M)$ is well-defined, its additivity is obvious.

If the element λ of Proposition 2.2 is replaced by some $k \cdot \lambda$, with $k \in K$, then $K(\lambda)$ and $\chi(\lambda, M)$ are replaced by their conjugates under k. In this sense the choice of λ is immaterial. We could in fact even manage without the hypothesis that some K orbit meet $\mathcal{V}(P)$ in a dense set: in any case there is a dense open set in $\mathcal{V}(P)$ on which the reductive parts of the isotropy groups of K belong to a single conjugacy class, and (on a slightly smaller open set) the same definition as above will yield a well-defined conjugacy class of genuine virtual representations of this class of subgroups. We will have no need for this generality, however.

The next theorem summarizes some of what we have established in this section.

Theorem 2.13. *Suppose X is a finitely generated $(\mathfrak{g}, K)$-module, and $\mathcal{V}(X) \subset \mathfrak{g}^*$ is its associated variety. Assume that K has a finite number of orbits on $\mathcal{V}(X)$. List the maximal orbits (those not contained in the closures of others) as $\mathcal{O}_1, \ldots, \mathcal{O}_s$, and choose a representative $\lambda_i \in \mathcal{O}_i$ for each orbit. Write K_i for the isotropy group of K at λ_i. Then attached to X is a non-zero genuine virtual representation $\chi_i = \chi(\lambda_i, X)$ of K_i.*

In light of the remarks after Definition 2.12, this theorem says that X has something to do with sections of the vector bundles $K \times_{K_i} \chi_i$. We will make this more precise in Theorem 4.2.

3. Associated varieties: microlocal properties.

In this section we will consider properties of associated varieties that are suggested by the "microlocalization" techniques of Sato-Kashiwara-Kawai and Gabber (see [Gi], [Ga], [Sp], and [SKK]). First we frame some of the definitions of section 2 more generally, beginning with a filtered algebra A (with a unit) over $\mathbb{C}$. (Often A will be the universal enveloping algebra $U(\mathfrak{g})$ or a quotient of it.) This means that A is equipped with an increasing filtration by subspaces indexed by $\mathbb{Z}$:

$$\cdots \subset A_{-1} \subset A_0 \subset A_1 \subset \cdots, \qquad A_p A_q \subset A_{p+q}. \tag{3.1)(a}$$

(Occasionally it will be convenient to index a filtration by $b\mathbb{Z}$ for some fraction b; this causes no difficulties.) We can then define an associated graded ring

$$R = \operatorname{gr} A = \sum_{n\in\mathbb{Z}} R^n, \qquad R^n = A_n/A_{n-1}. \tag{3.1)(b}$$

We will assume for convenience that R is commutative, although some of the preliminary formalism requires much less. Of course if A is $U(\mathfrak{g})$, then R is $S(\mathfrak{g})$. Suppose X is an A-module. A ***compatible filtration*** on X is an increasing family of subspaces of X indexed by $\mathbb{Z}$ (or, occasionally, by $a + b\mathbb{Z}$), satisfying

$$A_p \cdot X_q \subset X_{p+q}. \tag{3.1)(c}$$

In this case we can define an associated graded module

$$M = \operatorname{gr} X = \sum_{n\in\mathbb{Z}} M^n, \qquad M^n = X_n/X_{n-1}, \tag{3.1)(d}$$

a graded R-module. When we want the notation to allow for several filtrations, we write $\mathcal{F}_n(X)$ and $\operatorname{gr}(X, \mathcal{F})$ instead of X_n and $\operatorname{gr} X$. We will need our filtrations to be ***exhaustive***:

$$\bigcup_n A_n = A, \qquad \bigcup_n X_n = X. \tag{3.1)(e}$$

(The dual requirements that $\bigcap_n A_n = 0$ and $\bigcap_n X_n = 0$ will appear eventually, but they are not needed for the general formal development.)

The notation of (3.1) will be in force throughout this section.

The notion of good filtration is a little subtle in general (see [Gi, Proposition 1.2.2]). We will need it only in the special case when $A_{-1} = 0$. Then it is given by the obvious analogue of (2.1)(d), and Proposition 2.2 carries over immediately.

The theory of microlocalization constructs certain (noncommutative) filtered localizations A_S of A related to graded localizations of R. Tensoring with these localizations gives a localization theory for A-modules; that is, a way of mapping an A-module X into larger, "smoother" objects X_S. A compatible filtration $\mathcal{F}$ on X induces one $\mathcal{F}_S$ on X_S as well. We can then pull $\mathcal{F}_S$ back to X, getting a new and "smoother" compatible filtration $\mathcal{F}(S)$ on X. From the existence of these improved filtrations, one can hope to deduce interesting results about X. One problem with this program is that it appears to require an understanding of microlocalization. What we propose to do is construct the new filtrations directly, without explicit use of A_S. The price is of course a loss in conceptual power; the present paragraph is intended to alleviate that loss.

What we cannot avoid is commutative localization, and we recall now a little about that. Recall that R is a graded commutative ring. A *closed cone* in $\operatorname{Spec} R$ is any subvariety defined by a homogeneous ideal in R, and an *open cone* is the complement of a closed cone. If f is in R, write $D(f) \subset \operatorname{Spec} R$ for the set of prime ideals not containing f. An open cone is a union of such sets, for various homogeneous elements f. Let us make this explicit in the case of $S(\mathfrak{g})$. Suppose U is an open cone in $\mathfrak{g}^*$. The complement of U is a closed cone, which is therefore the set of simultaneous zeros of a finite set S of homogeneous polynomials. The usual "identification" of $\operatorname{Spec} S(\mathfrak{g})$ with $\mathfrak{g}^*$ identifies $D(f)$ with the subset of $\mathfrak{g}^*$ at which the polynomial f does not vanish. Hence

$$U = \{\lambda \in \mathfrak{g}^* \mid f(\lambda) \neq 0 \text{ for some } f \text{ in } S\} = \bigcup_{f \in S} D(f).$$

In general, whenever S is a subset of a commutative ring R, we will write

$$U_S = \bigcup_{f \in S} D(f) \tag{3.2)(a}$$

for the set of prime ideals not containing S, and

$$Z_S = \{P \in \operatorname{Spec} R \mid S \subset P\} \tag{3.2)(b}$$

for its complement. Of course Z_S is just the variety of S, a closed subset of $\operatorname{Spec} R$.

Suppose M is an R-module. Recall (say from [H]) that M defines a sheaf of modules (which we denote by $\tilde{M}$) on $\operatorname{Spec} R$. By definition, $\tilde{M}(D(f))$ is the localization M_f of M at f. A typical element of this localization is of the form $f^{-n}m$, with n a non-negative integer and $m \in M$. Two such elements $f^{-n}m$ and $f^{-n'}m'$ are equal if $f^{N-n}m$ is equal to $f^{N-n'}m'$ in M for all large N. If f is homogeneous of degree p, we can therefore grade M_f by defining

$$M_f^n = \sum_{k \in \mathbb{N}} f^{-k} M^{n+kp}. \tag{3.2)(c}$$

Notice that this grading may have negative terms even if the one on M does not.

Next, we want to compute the value of the sheaf $\tilde{M}$ on the open cone U_S defined by a set S of homogeneous elements. If f and g belong to S, then there is a natural map $\phi_{f,g}$ from M_f to M_{fg} (sending $f^{-n}m$ to $(fg)^{-n}(g^n m)$). The module $\tilde{M}(U_S)$ is contained in the direct product over S of the various localizations M_f; it is defined to be

$$\tilde{M}(U_S) = \left\{ m = (m_f) \in \prod_{f \in S} M_f \;\middle|\; \begin{array}{c} \phi_{f,g}(m_f) = \phi_{g,f}(m_g) \\ \text{for all } f,\, g \in S \end{array} \right\}. \tag{3.2)(d}$$

The maps $\phi_{f,g}$ preserve degrees in the gradings of the previous paragraph. If R is noetherian (so that S may be taken to be finite) then it follows that $\tilde{M}(U_S)$ is spanned by elements (m_f) in which all coordinates have the same degree:

$$\tilde{M}(U_S)^n = \left\{ m = (m_f) \in \prod_{f \in S} M_f^n \;\middle|\; \begin{array}{c} \phi_{f,g}(m_f) = \phi_{g,f}(m_g) \\ \text{for all } f,\, g \in S \end{array} \right\}. \tag{3.2)(e}$$

In this way $\tilde{M}(U_S)$ acquires a grading; again it may have negative terms. As an immediate consequence of this description, we get the following lemma.

Lemma 3.3. *Suppose R is a graded commutative ring, S is a homogeneous subset of R, and U_S is the open cone (3.2)(a). Then the kernel of the natural map $M \to \tilde{M}(U_S)$ is*

$$\{ m \in M \mid \text{ for all } f \in S \text{ there is some } N = N(f,m) \text{ so that } f^N m = 0 \}.$$

The support of this kernel is contained in the closed cone Z_S defined by S.

Definition 3.4. In the setting of (3.1), suppose $\mathcal{F}$ is a compatible filtration of the A-module X, and S is a homogeneous subset of $R = \operatorname{gr} A$. List the elements of S as $\{f_i\}_{i \in I}$; say f_i has degree p_i. Choose a set $\Sigma = \{\phi_i\}_{i \in I}$ of representatives of S in A:

$$\phi_i \in A_{p_i}, \qquad \operatorname{gr} \phi_i = f_i.$$

Suppose $J = (i_1, \dots, i_N) \in I^N$ is an ordered N-tuple of elements of I. Define

$$p_J = \sum_{j=1}^{N} p_{i_j}, \qquad f_J = \prod_{j=1}^{N} f_{i_j} \in R^{p_J}, \qquad \phi_J = \prod_{j=1}^{N} \phi_{i_j} \in A_{p_J}.$$

Notice that ϕ_J depends on the order in which the product is taken.

The *S-localization of $\mathcal{F}$* is a new filtration $\mathcal{F}(S)$ on X, defined as follows:

$$\mathcal{F}(S)_n(X) = \left\{ x \in X \;\middle|\; \begin{array}{l} \text{for all } N \text{ sufficiently large, and} \\ \text{for all } J \in I^N,\ \phi_J \cdot x \in \mathcal{F}_{n+p_J}(X) \end{array} \right\}.$$

When the set S and the filtration $\mathcal{F}$ are understood, it will be convenient to write $X_{[n]}$ in place of $\mathcal{F}(S)_n(X)$. We will then write $[\operatorname{gr}] X$ for the associated graded object.

In order to work conveniently with this definition, we need to extend its notation somewhat.

Definition 3.5. In the setting of (3.1)(a) and (3.1)(b), fix a collection $\Sigma = \{\phi_i\}_{i \in I}$ of elements of A; say $\phi_i \in A_{p_i}$. Define ϕ_J, p_J as in Definition 3.4. By a *product of type* (N, r, Σ) we will mean (roughly) a product π of elements of A of degrees adding up to r, with at least N of the factors in Σ. (When Σ is understood, we may simply say a product of type (N, r).) A little more precisely, we mean that there are elements $(b_0, \dots, b_N)$ of A, with $b_j \in A_{q_j}$, and a sequence $J = (i_j) \in I^N$, so that

$$\pi = b_0 \phi_{i_1} b_1 \cdots b_{N-1} \phi_{i_N} b_N, \qquad r = p_J + \sum_{j=0}^{N} q_j.$$

Notice that a product of type (N, r) belongs to A_r.

Lemma 3.6. *In the setting of Definitions 3.4 and 3.5, $X_{[n]} = \mathcal{F}(S)_n(X)$ is a vector subspace of X containing $X_n = \mathcal{F}_n(X)$. Suppose that x is an element of $X_{[n]}$. Then there is a positive integer N_0 (depending on x) having the following property: if π is a product of type (N, r) and $N \geq N_0$, then $\pi \cdot x \in X_{n+r}$.*

If we could rearrange the product π to put all the terms in Σ on the right, then the result would be obvious. By Definition 3.4, the ϕ_J factor would take x into X_{n+p_J} (if N is large enough); and the remaining factor (which lies in A_{r-p_J}) would take X_{n+p_J} into X_{n+r}. The difficulty is that A is not commutative, so we are not completely free to rearrange the product. What saves us is the commutativity of $R = \operatorname{gr} A$. This says that we can rearrange products up to error terms of lower order: if $\phi \in A_p$ and $b \in A_q$, then there is a $c \in A_{p+q-1}$ such that $\phi b = b\phi + c$. Repeating this argument, we see that any product π of type (N, r) may be rewritten as

$$\pi = b\phi_J + (\text{sum of products of type } (N-1, r-1)) \tag{3.7}$$

with $b \in A_{r-p_J}$. (What is critical for this is that $\operatorname{gr} A$ be commutative, or at least that the elements $\operatorname{gr} \phi_i$ be central in it.)

Lemma 3.8. *In the setting of Definitions 3.4 and 3.5, fix integers N, s, and t, with $0 \leq s, t \leq N$, and $s + t \leq N + 1$. Then any product π of type (N, r) is equal to a sum of terms of two forms. The first form is $c\phi_J$, with $J \in I^s$ and c a product of type $(N-s-m, r-p_J-m)$ (with $0 \leq m \leq t-1$). The second form is just products of type $(N-t, r-t)$.*

Notice that a term of the first form in the lemma is a special kind of product of type $(N - m, r - m)$. The simplest case of the lemma is $N = 1, s = t = 1$. In that case it says that if $\phi \in A_p, b \in A_q$, and $b' \in A_{q'}$ (so that the r of the lemma is $p+q+q'$), then $b'\phi b = c\phi + c'$, with $c \in A_{q+q'}$, and $c' \in A_{p+q+q'-1}$. The term $c\phi$ is of the first form in the lemma (with $m = 0 = t - 1$) and the term c' is of the second form. Of course we may take $c = b'b, c' = b'(\phi b - b\phi)$.

Proof of Lemma 3.8. We proceed by induction on N. If $t = 0$, then the result is trivial (since π is already of the second form allowed in the conclusion). If $t = 1$, then the conclusion of the lemma follows from (3.7). So suppose $t > 1$. It follows that N is positive, and that $s \leq N - 1$. Therefore we can find $b \in A_q, \phi \in A_p \cap \Sigma$, and a product π' of type $(N-1, r-p-q)$, so that $\pi = b\phi\pi'$. We apply the inductive hypothesis to $(\pi', N-1, s, t-1)$, and multiply the resulting expansion of π by $b\phi$. The only difficulty arises from terms π'' of type $((N-1)-(t-1), r-p-q-t+1)$ in the expansion of π'. After multiplication by $b\phi$, such a term is of type $(N-t+1, r-t+1)$. Applying the inductive hypothesis to $(b\phi\pi'', N-t+1, s, 1)$ gives the conclusion of the lemma. □

Proof of Lemma 3.6. The first assertion of the lemma is obvious. For the rest, choose s so large that for all all $J \in I^s$, $\phi_J \cdot x \in X_{n+p_J}$ (Definition

3.4). Since the filtration of X is exhaustive, there is a positive integer t such that $x \in X_{n+t}$. Set $N_0 = s + t - 1$. Suppose $N \geq N_0$, and π is a product of type (N, r). We must show that $\pi \cdot x \in X_{n+r}$. By Lemma 3.8, we may replace π by a sum of terms of two forms. If $c\phi_J$ is a term of the first form, then $\phi_J \cdot x \in X_{n+p_J}$ and $c \in A_{r-p_J-m}$. It follows that $c\phi_J \cdot x \in X_{n+r-m}$ (with $m \geq 0$). If π' is of the second form, then it belongs to A_{r-t}. Since $x \in X_{n+t}$, it follows that $\pi' \cdot x \in X_{(n+t)+(r-t)}$. □

Corollary 3.9. *In the setting of Definition 3.4, the localized filtration $\mathcal{F}(S)$ may be described as*

$$\mathcal{F}(S)_n(X) = \left\{ x \in X \;\middle|\; \begin{array}{c} \text{for all } N \text{ large and all } \pi \text{ of type } (N,r), \\ \pi \cdot x \in \mathcal{F}_{n+r}(X) \end{array} \right\}.$$

It is a compatible exhaustive filtration of X, depending only on the set S (and not on the choice of representatives Σ in A).

Proof. Write $X_{[n]}$ for $\mathcal{F}(S)_n(X)$. The description of $X_{[n]}$ is immediate from Lemma 3.6. Since $X_n \subset X_{[n]}$, the localized filtration is exhaustive (cf. (3.1)(e)). To check the compatibility, suppose $x \in X_{[n]}$ and $b \in A_q$. We want to show that $b \cdot x \in X_{[n+q]}$. We use the description just established for the filtration. If π is a product of type (N, r), then $\pi \cdot b$ is a product of type $(N, r+q)$. It follows that if N is large enough, $\pi \cdot (b \cdot x) \in X_{n+r+q}$. That is, $b \cdot x$ satisfies our new criterion for belonging to $X_{[n+q]}$.

To see that $\mathcal{F}(S)$ is independent of the choice of Σ, suppose $\Sigma' = \{\phi'_i\}_{i \in I}$ is another such set. We can write $\phi'_i = \phi_i + v_i$, with $v_i \in A_{p_i - 1}$. If $J \in I^N$, then clearly ϕ'_J is equal to ϕ_J plus a sum of products of various types $(N - s, p_J - s)$, with $1 \leq s \leq N$; these are obtained from ϕ_J by replacing s of the ϕ_i factors by v_i. Now it follows from Lemma 3.6 that if $x \in X_{[n]}$ and N is large enough, then $\phi'_J \cdot x \in X_{[n+p_J]}$. This shows that the filtration defined using Σ' contains the one defined using Σ. □

A similar argument shows that $\mathcal{F}(S)$ is unaffected by adding to S any finite homogeneous subset of the ideal generated by S. A consequence is

Corollary 3.10. *In the setting of Definition 3.4, suppose S is finite, and that S' is another finite homogeneous set generating the same ideal in R. Then $\mathcal{F}(S) = \mathcal{F}(S')$.*

With a little more effort, one sees that only the radical of the ideal matters. Since we will not use this fact, we omit a detailed proof.

Proposition 3.11. *In the setting of Definition 3.4, write U_S for the open cone in* Spec R *whose complement is defined by S. Then there is a natural map of R-modules*

$$\sigma(S) : \mathrm{gr}\,(X, \mathcal{F}(S)) \to \tilde{M}(U_S)$$

giving rise to a commutative diagram

$$\begin{array}{ccc} \mathrm{gr}\,(X, \mathcal{F}) & \longrightarrow & \mathrm{gr}\,(X, \mathcal{F}(S)) \\ \downarrow & & \downarrow \\ M & \longrightarrow & \tilde{M}(U_S). \end{array}$$

The first vertical arrow is the isomorphism defining M, the upper horizontal arrow comes from the inclusion $\mathcal{F}_n(X) \subset \mathcal{F}(S)_n(X)$ of Lemma 3.6, and the lower horizontal arrow is restriction of sections (cf. Lemma 3.3).

If S is finite, then $\sigma(S)$ is injective.

Proof. Write $\sigma_n : X_n \to M^n$ for the quotient map defining M. We may construct $\sigma(S)$ as a family of (symbol) maps

$$\sigma(S)_n : \mathcal{F}(S)_n(X) \to \tilde{M}(U_S)^n \tag{3.12)(a}$$

trivial on $\mathcal{F}(S)_{[n-1]}(X)$. As usual it is convenient to let $X_{[n]}$ denote $\mathcal{F}(S)_n(X)$. Then we write $\sigma_{[n]}$ for $\sigma(S)_n$:

$$\sigma_{[n]} : X_{[n]} \to \tilde{M}(U_S)^n. \tag{3.12)(a)'$$

Fix an element $x \in X_{[n]}$. Choose N_0 as in Lemma 3.6, and fix $N > N_0$. We want to define an element

$$\sigma_{[n]}(x) = m = (m_i)_{i \in I} \in \tilde{M}(U_S) \tag{3.12)(b}$$

(cf. (3.2)(d)), with $m_i \in M^n_{f_i}$. By Lemma 3.6, $(\phi_i)^N \cdot x \in X_{n+Np_i}$. Write

$$m_i' = \sigma_{n+Np_i}((\phi_i)^N \cdot x), \qquad m_i = (f_i)^{-N} m_i'. \tag{3.12)(c}$$

The first problem is to show (m_i) actually belongs to $\tilde{M}(U_S)$. By (3.2)(d), this is equivalent to showing that for $i, j \in I$ we have $f_j^N \cdot m_i' = f_i^N \cdot m_j'$ in the localized module $M_{f_i f_j}$. Of course it suffices to prove the equality in M. By inspection of the definitions, we see that it amounts to

$$(\phi_j)^N (\phi_i)^N \cdot x = (\phi_i)^N (\phi_j)^N \cdot x \pmod{X_{n+Np_i+Np_j-1}}. \tag{*}$$

Now since $\operatorname{gr} A$ is commutative, it is easy to check that $(\phi_j)^N(\phi_i)^N$ is equal to $(\phi_i)^N(\phi_j)^N$ modulo products of type $(2N-2, Np_i + Np_j - 1)$. Lemma 3.6 guarantees that such products map x into $X_{n+Np_i+Np_j-1}$, proving $(*)$. A similar argument shows that $\sigma(S)$ respects the action of R. If $x \in X_n$, then m_i' is equal to $(f_i)^N \cdot (x + X_{n-1})$. It follows that m is the natural image of $\sigma(x)$ in $\tilde{M}(U_S)$, proving the commutativity of the diagram in the proposition.

It remains to prove (assuming S is finite) the injectivity of $\sigma(S)$; that is, that the kernel of $\sigma_{[n]}$ is precisely $X_{[n-1]}$. So suppose $x \in X_{[n]}$, and $\sigma_{[n]}(x) = 0$. This means that for every $i \in I$, the element m_i of M_{f_i} must be zero. By the remarks before (3.2)(c), this means that $f_i^N \cdot m_i' = 0$ (as an element of M) for all large N. By the definition of m_i', this means that there is a positive integer N_i such that

$$(\phi_i)^{N_i} \cdot x \in X_{n+N_ip_i-1}. \tag{**}$$

We want to deduce from this that $x \in X_{[n-1]}$. We use the criterion of Definition 3.4. Suppose N is larger than the cardinality of S times the maximum of the various N_i, and also larger than N_0; and that $J \in I^N$. We claim that $\phi_J \cdot x \in X_{n+p_J-1}$. This will complete the proof. Clearly there is an i that occurs at least N_i times among the ϕ_{i_j}. Write J' for what is left after N_i i's are removed from J (so that $p_J = p_{J'} + N_i p_i$). Then ϕ_J is equal to $\phi_{J'}\phi_i^{N_i}$ plus a sum of products of type $(N-1, p_J - 1)$. Since $N-1$ is at least N_0, the second kind of product maps x into X_{n+p_J-1}. By (**), the first term has this property as well. □

To see that the localized filtration $\mathcal{F}(S)$ still captures much of the structure of X, we need a simple definition and a condition on X.

Definition 3.13. In the setting of (3.1), set

$$A_{-\infty} = \bigcap_n A_n, \qquad X_{-\infty} = \bigcap_n X_n.$$

When we wish to emphasize the filtration, we may write $\mathcal{F}_{-\infty}(X)$. In the simplified notation at the end of Definition 3.4, we write $X_{[-\infty]}$ for $\mathcal{F}(S)_{-\infty}(X)$.

Notice that $X_{-\infty}$ is automatically an A-submodule of X. (This uses only the compatibility of the filtration and the assumption that the filtration of A is exhaustive.) Combining Proposition 3.11 with Lemma 3.3, we get

Corollary 3.14. *In the setting of Definition 3.4, suppose S is finite. Then the restriction of the original filtration $\mathcal{F}$ to $\mathcal{F}(S)_{-\infty}(X) = X_{[-\infty]}$ defines an injection of* gr $(X_{[-\infty]})$ *into a submodule of M supported on the closed cone Z_S defined by S (cf. (3.2)(b).*

Suppose in particular that $A_{-1} = 0$, R is Noetherian, and $\mathcal{F}$ is a good filtration of X. Then the characteristic variety of $X_{[-\infty]}$ is contained in Z_S.

When X is irreducible and Z_S does not contain its characteristic variety (or even under various weaker conditions) Corollary 3.14 will allow us to deduce that $X_{[-\infty]}$ must be zero, and therefore that the localized filtration "sees" all of the structure of X. The reason we cannot easily get a great deal of information from this is that the localized module $\tilde{M}(U_S)$ need not be finitely generated (over R), and therefore the localized filtration need not be good. In the next section we will see how it is sometimes possible to circumvent this problem in the case of $(\mathfrak{g}, K)$-modules, using the extra rigidity imposed by the action of K.

4. Microlocalization for $(\mathfrak{g}, K)$-modules

In this section we will apply the results of section 3 to associated varieties of $(\mathfrak{g}, K)$-modules. The main problem will be to find conditions under which the module $\tilde{M}(U_S)$ of (3.2) is finitely generated over $S(\mathfrak{g})$. Before considering this, we give a simpler application. Recall from after Corollary 2.10 the notion of virtual characters of an algebraic group H. Suppose Θ

is such a virtual character, and π is a finite-dimensional representation of H. We can find a finite set of distinct irreducible representations $\{\rho_i\}$ of H and integers m_i so that $\Theta = \sum m_i\rho_i$, and $\pi = \sum n_i\rho_i$ (as a virtual representation). Define the *quotient multiplicity of* Θ *in* π to be

$$[\Theta : \pi]_{H,\mathrm{quo}} = \sum m_i \dim \mathrm{Hom}_H(\pi, \rho_i). \tag{4.1)(a}$$

(We may drop the H if this causes no confusion.) Similarly, the *submodule multiplicity of* Θ *in* π is

$$[\Theta' : \pi]_{H,\mathrm{sub}} = \sum m_i \dim \mathrm{Hom}_H(\rho_i, \pi). \tag{4.1)(b}$$

Finally, the *multiplicity of* Θ *in* π is

$$[\Theta : \pi]_H = \sum m_i n_i. \tag{4.1)(c}$$

If π is completely reducible (for example, if H is reductive) then the three definitions coincide; we may drop the subscripts sub and quo from the notation in that case. These definitions may be extended to the case when π is any rational representation (possibly infinite-dimensional); in that case we need to require either that Θ be genuine, or that π have finite multiplicities, to avoid the appearance of $\infty - \infty$. If Θ is the character of some representation τ, then

$$[\Theta : \pi] \geq [\Theta : \pi]_{\mathrm{quo}} \geq \dim \mathrm{Hom}_H(\pi, \tau) \tag{4.1)(d}$$

with equality whenever π and τ are completely reducible. Of course there is an analogous inequality for submodule multiplicities.

Theorem 4.2. *Suppose X is an irreducible $(\mathfrak{g}, K)$-module (cf. (2.1)), $\lambda \in \mathfrak{g}^*$, and that the K-orbit*

$$\mathcal{O} = K \cdot \lambda \simeq K/K(\lambda)$$

of λ is dense in some irreducible component of the associated variety $\mathcal{V}(X)$. Define a genuine virtual representation $\chi(\lambda, X)$ of $K(\lambda)$ as in section 2. If τ is any representation of K, then

$$\dim \mathrm{Hom}_K(\tau, X) \leq [\chi(\lambda, X) : \tau]_{K(\lambda),\mathrm{quo}}.$$

Because of Frobenius reciprocity, this theorem can be regarded as a precise version of the remarks after Theorem 2.13. (The proof should make this clearer.) We are most interested in the case when K is reductive and τ is irreducible. Then the left side is the multiplicity of τ as a K-type of X. A somewhat surprising feature of the result is that one needs only a single component of $\mathcal{V}(X)$ to control all of the K-types of X.

Proof. Fix a good filtration $\mathcal{F}$ of X, and write $M = \mathrm{gr}\, X$ as in sections 2 and 3. Necessarily $\mathcal{O}$ is open in $\mathcal{V}(X)$, so its complement in $\mathcal{V}(X)$ is a closed cone in $\mathfrak{g}^*$. Choose a finite homogeneous set S in $S(\mathfrak{g})$ defining this complement: in the notation of (3.2)(b),

$$\mathcal{V}(X) = \mathcal{O} \cup Z_S,$$

a disjoint union. We may assume that the ideal generated by S (or even the linear span of S) is K-invariant. The open cone U_S (cf. (3.2)(a)) meets $\mathcal{V}(X)$ precisely in $\mathcal{O}$. Let $\mathcal{F}(S)$ be the localized filtration of X (Definition 3.4); this will be K-invariant by Corollary 3.10. (The proof of that corollary becomes almost trivial when S and S' have the same linear span, which is the only case we need.) Then $X_{[-\infty]}$ is a $(\mathfrak{g}, K)$-submodule of X, with characteristic variety contained in Z_S (Corollary 3.14). Since $\mathcal{V}(X)$ is not contained in Z_S, $X_{[-\infty]}$ is a proper submodule; so it is zero by the irreducibility of X. Proposition 3.11 gives an embedding $[\mathrm{gr}]\, X \hookrightarrow \tilde{M}(U_S)$, and it follows that

$$\begin{aligned} \dim \mathrm{Hom}_K(\tau, X) &\le \dim \mathrm{Hom}_K(\tau, [\mathrm{gr}]\, X) \\ &\le \dim \mathrm{Hom}_K(\tau, \tilde{M}(U_S)). \end{aligned} \tag{4.3)(a}$$

Now choose a finite filtration of M as in Lemma 2.11. The functor taking a module N to $\tilde{N}(U_S)$ is left exact, as is clear from (3.2) and the exactness of localization. Consequently

$$\dim \mathrm{Hom}_K(\tau, \tilde{M}(U_S)) \le \sum_i \dim \ \mathrm{Hom}_K(\tau, (\widetilde{M_i/M_{i-1}})(U_S)). \tag{4.3)(b}$$

But U_S meets $\mathcal{V}(X)$ only in the orbit $\mathcal{O}$, which may be identified with the homogeneous space $K/K(\lambda)$. The assumption that M_i/M_{i-1} is generically reduced along $\mathcal{O}$ implies by K-invariance that it is reduced everywhere on $\mathcal{O}$; so the restriction of the sheaf $\widetilde{M_i/M_{i-1}}$ to U_S may be identified with a K-equivariant sheaf of modules on the homogeneous space $K/K(\lambda)$. Such a sheaf of modules is necessarily the equivariant vector bundle induced by the (fiber) representation of $K(\lambda)$ on $E_i = M_i/(\mathfrak{m}(\lambda)M_i + M_{i-1})$. By Frobenius reciprocity,

$$\dim \mathrm{Hom}_K(\tau, (\widetilde{M_i/M_{i-1}})(U_S)) = \dim \ \mathrm{Hom}_{K(\lambda)}(\tau, E_i). \tag{4.3)(c}$$

By (4.1)(d), this is bounded above by the quotient multiplicity of E_i in τ.

Now the sum of the various E_i is $\chi(\lambda, X)$ (Definition 2.12); so the inequalities in (4.3) give the conclusion of the theorem. □

In order to decide when a localized filtration is good, we need to know when the R-module $\tilde{M}(U_S)$ is finitely generated.

Theorem 4.4 ([Gr, Proposition 5.11.1]). *Suppose R is a finitely generated commutative algebra over a field k, M is a finitely generated R-module, and U is an open set in* Spec R. *Write Z for the complement of U in*

Spec R. *Then the* R*-module* $\tilde{M}(U)$ *is finitely generated if and only if for every prime ideal* $P \in U \cup \operatorname{Ass} M$, *the closure* Y *of* P *in* Spec R *satisfies*

$$Y \cap Z \text{ has codimension at least 2 in } Y.$$

Needless to say, Grothendieck actually proves a much more general result.

To check the condition in this theorem, we will use the following easy lemma.

Lemma 4.5. *Suppose* M *is a finitely generated* $(S(\mathfrak{g}), K)$*-module. Then* K *(through its action on* $S(\mathfrak{g})$*) permutes the set of associated primes of* M*; the identity component* K_0 *preserves each associated prime.*

Proof. The first claim is obvious. Since Ass M is finite, the stabilizer in K of each associated prime must be a closed subgroup of K of finite index. Such a subgroup contains the identity component. □

Theorem 4.6. *In the setting of (2.1), suppose* X *is a finitely generated* $(\mathfrak{g}, K)$*-module. Fix* $\lambda \in \mathfrak{g}^*$ *belonging to* $\mathcal{V}(X)$*, and assume that*

i) the closure of $\mathcal{O} = K \cdot \lambda$ *contains an irreducible component of* $\mathcal{V}(X)$*; and*

ii) if X' *is a non-zero* $(\mathfrak{g}, K)$*-submodule of* X*, then* $\mathcal{V}(X') \supset \mathcal{O}$.

Define $\partial\mathcal{O}$ *to be the complement of* $\mathcal{O}$ *in its closure* $\overline{\mathcal{O}}$*. Then there are two (non-exclusive) possibilities: either*

a) $\mathcal{V}(X)$ *is equal to* $\overline{\mathcal{O}}$*; or*

b) $\partial\mathcal{O}$ *has codimension one in* $\overline{\mathcal{O}}$.

Proof. Choose a good filtration $\mathcal{F}$ of X, and write $M = \operatorname{gr} X$ as usual. Define Z to be the complement of $\mathcal{O}$ in $\mathcal{V}(X)$. Then Z is a closed cone in $\mathfrak{g}^*$, and $Z \cap \overline{\mathcal{O}} = \partial\mathcal{O}$. Choose a finite homogeneous subset S of $S(\mathfrak{g})$ defining Z, and construct the localized filtration $\mathcal{F}(S)$ as in section 3. By Corollary 3.9, this is a compatible exhaustive filtration of X. By Corollary 3.14 and hypothesis (ii) of the theorem, $X_{[-\infty]} = 0$.

Suppose for the rest of the proof that conclusion (b) of the theorem fails; that is, that $Z \cap \overline{\mathcal{O}}$ has codimension at least two in $\overline{\mathcal{O}}$. Theorem 4.4 implies that $\tilde{M}(U_S)$ is finitely generated. To see that, suppose P is an associated prime of M not belonging to Z. By Lemma 4.5, the associated variety Y of P meets $\mathcal{O}$ in a K_0-invariant subset Y_0. Consequently Y_0 is open in $\mathcal{O}$, so $Z \cap Y \subset Z \cap \overline{\mathcal{O}}$ has codimension at least two in Y.

Now Proposition 3.11 implies that $\operatorname{gr}(X, \mathcal{F}(S))$ is finitely generated. By the definition at (2.1), it follows that $\mathcal{F}(S)$ is a good filtration of X. Now it is easy to check that the associated primes of $\tilde{M}(U)$ must belong to U; so in our case the associated primes of $\operatorname{gr}(X, \mathcal{F}(S))$ must correspond to the connected components of $\mathcal{O}$. Since the associated primes are dense in the characteristic variety, we get $\mathcal{V}(X) = \overline{\mathcal{O}}$. □

Corollary 4.7 ([BB], [J]). *Let $\mathfrak{g}$ be a reductive Lie algebra and $I \subset U(\mathfrak{g})$ a primitive ideal. Then the associated variety $\mathcal{V}(I)$ (defined, using the Poincaré-Birkhoff-Witt theorem, to be the variety in $\mathfrak{g}^*$ corresponding to the ideal $\operatorname{gr} I$ in $S(\mathfrak{g})$) is the closure of a single nilpotent coadjoint orbit in $\mathfrak{g}^*$.*

Proof. The expert will recognize this as almost an immediate consequence of Theorem 4.6; but then the expert already knew the result. For the benefit of other readers, we will give a more complete argument, including sketches of proofs for some standard intermediate results. To begin, we need some auxiliary definitions.

Suppose G is a connected reductive algebraic group with Lie algebra $\mathfrak{g}$. The adjoint action Ad of G on $\mathfrak{g}$ extends to an algebraic action (still denoted Ad) of G on $U(\mathfrak{g})$ by algebra automorphisms. The differential of this action, denoted ad, sends an element $Y \in \mathfrak{g}$ to the derivation

$$(\operatorname{ad} Y)(u) = Yu - uY \qquad (u \in U(\mathfrak{g})). \tag{4.8)(a}$$

Because I is a two-sided ideal, $\operatorname{ad}(\mathfrak{g})$ preserves I. Since G is connected, $\operatorname{Ad}(G)$ preserves I as well. We therefore get an action (still called Ad) of G on $U(\mathfrak{g})/I$ by algebra automorphisms. The differential of this action is given by a formula like (4.8)(a).

On the other hand, the Lie algebra $\mathfrak{g} \times \mathfrak{g}$ acts on $U(\mathfrak{g})$ by

$$(Y_1, Y_2) \cdot u = Y_1 u - uY_2 \qquad (u \in U(\mathfrak{g}), Y_i \in \mathfrak{g}). \tag{4.8)(b}$$

The defining relations of $U(\mathfrak{g})$ show that this action is a Lie algebra representation. Again the fact that I is a two-sided ideal implies that the action factors to $U(\mathfrak{g})/I$. The diagonal embedding of $\mathfrak{g}$ in $\mathfrak{g} \times \mathfrak{g}$ now provides the structure considered in (2.1), and (4.8)(a) and (b) make $U(\mathfrak{g})/I$ into a $(\mathfrak{g} \times \mathfrak{g}, G)$-module. We express this by saying that $U(\mathfrak{g})/I$ is a *Harish-Chandra bimodule.* We sometimes write $\mathfrak{g}_\Delta \subset \mathfrak{g} \times \mathfrak{g}$ for the diagonal subalgebra, which will play the role of $\mathfrak{k}$.

Restriction of linear functionals to the first factor provides a G-equivariant isomorphism

$$(\mathfrak{g} \times \mathfrak{g}/\mathfrak{g}_\Delta)^* \to \mathfrak{g}^* \tag{4.8)(c}$$

which we use to identify $\mathcal{V}(U(\mathfrak{g})/I)$ (defined by (1.2)) with the associated variety $\mathcal{V}(I)$ defined in the statement of the corollary. (The main point is that the standard filtration of $U(\mathfrak{g})$ defines by passage to the quotient a good filtration of the $(\mathfrak{g} \times \mathfrak{g}, G)$-module $U(\mathfrak{g})/I$.)

Write

$$\mathfrak{Z}(\mathfrak{g}) = \text{ center of } U(\mathfrak{g}); \tag{4.9}$$

this is the subalgebra of $U(\mathfrak{g})$ on which the adjoint action of G is trivial. Because I is primitive, it contains a maximal ideal $\mathcal{I}$ in $\mathfrak{Z}(\mathfrak{g})$. It follows at once that the $(\mathfrak{g} \times \mathfrak{g}, G)$-module $U(\mathfrak{g})/I$ is annihilated by the maximal ideal

$$\mathfrak{Z}(\mathfrak{g}) \otimes \mathcal{I} + \mathcal{I} \otimes \mathfrak{Z}(\mathfrak{g}) \subset \mathfrak{Z}(\mathfrak{g} \times \mathfrak{g}).$$

Consequently the Harish-Chandra bimodule $U(\mathfrak{g})/I$ is finitely generated and annihilated by an ideal of finite codimension in the center of the enveloping algebra. By one of Harish-Chandra's basic finiteness theorems, it follows that $U(\mathfrak{g})/I$ has finite length as a bimodule. In particular, it satisfies ACC and DCC on two-sided ideals. Following [D], we can therefore choose a minimal two-sided ideal J properly containing I. Because I is prime, J is unique. It follows that every non-zero submodule X' of $U(\mathfrak{g})/I$ must contain J/I. In particular, we must have

$$\mathcal{V}(X') \supset \mathcal{V}(J/I). \tag{4.10}$$

We will use this in a moment to deduce hypothesis (ii) in Theorem 4.6.

The finiteness under the center of the enveloping algebra also guarantees that (for any subquotient X of $U(\mathfrak{g})/I$) $\mathcal{V}(X)$ consists of nilpotent elements (see Corollary 5.4 below). Since G has only a finite number of nilpotent coadjoint orbits (Theorem 5.8 below) $\mathcal{V}(X)$ must be a finite union of these, of dimensions bounded above by $\operatorname{Dim} X$. Now the additivity of associated varieties implies that

$$\mathcal{V}(U(\mathfrak{g})/I) = \mathcal{V}(U(\mathfrak{g})/J) \cup \mathcal{V}(J/I).$$

On the other hand, J properly contains the prime ideal I. By a theorem in [BK], it follows that $\operatorname{Dim} U(\mathfrak{g})/I > \operatorname{Dim} U(\mathfrak{g})/J$. Assembling these observations, we find that $\mathcal{V}(J/I)$ must contain a G-orbit $\mathcal{O}$ of dimension equal to the Gelfand-Kirillov dimension of $U(\mathfrak{g})/I$, and that the closure of $\mathcal{O}$ will in that case be an irreducible component of $\mathcal{V}(U(\mathfrak{g})/I)$. In conjunction with (4.10), this establishes the hypotheses for Theorem 4.6.

To complete the argument, recall that any coadjoint orbit of a Lie group carries a natural symplectic structure, and is therefore even-dimensional. The boundary of $\mathcal{O}$ (as a finite union of nilpotent coadjoint orbits) must therefore have even codimension in the closure of $\mathcal{O}$. This rules out conclusion (b) of Theorem 4.6; and (a) is what we wished to show. □

As a final application, we return to the problem of K-multiplicities.

Theorem 4.11. *In the setting of (2.1), suppose X is a finitely generated $(\mathfrak{g}, K)$-module. Fix $\lambda \in \mathfrak{g}^*$ belonging to $\mathcal{V}(X)$, and define*

$$\mathcal{O} = K \cdot \lambda = K/K(\lambda), \qquad \partial\mathcal{O} = \overline{\mathcal{O}} - \mathcal{O}.$$

Assume that

i) $\overline{\mathcal{O}}$ contains an irreducible component of $\mathcal{V}(X)$;
ii) if X' is a non-zero $(\mathfrak{g}, K)$-submodule of X, then $\mathcal{V}(X') \supset \mathcal{O}$; and
iii) $\partial\mathcal{O}$ has codimension at least two in $\overline{\mathcal{O}}$.

Define a genuine virtual representation $\chi(\lambda, X)$ of $K(\lambda)$ as in section 2, and choose a completely reducible representation $V(\lambda, X)$ of $K(\lambda)$ representing $\chi(\lambda, X)$. Then there is a finitely generated $(S(\mathfrak{g}), K)$-module Q supported on $\partial\mathcal{O}$, with the property that

$$X = Ind_{K(\lambda)}^{K}(V(\lambda, X)) - Q$$

as virtual representations of K. If K is reductive, this means that the multiplicity in X of any irreducible representation τ of K is

$$\dim \operatorname{Hom}_K(\tau, X) = \dim \operatorname{Hom}_{K(\lambda)}(\tau\mid_{K(\lambda)}, V(\lambda, X)) - \dim \operatorname{Hom}_K(\tau, Q).$$

This follows from the argument given for Theorem 4.2, in conjunction with Theorem 4.4. Because the details are a little delicate, and we will use this result (in section 12) only as evidence for a conjecture, we omit the details.

5. Associated varieties for $(\mathfrak{g}, K)$-modules

In this section we recall those basic facts about associated varieties for $(\mathfrak{g}, K)$-modules that depend on the structure theory of reductive Lie groups. The main result is Corollary 5.23; Corollary 5.20 will also be crucial for the proof of Theorem 8.7.

For this section, we will assume that

$$\mathfrak{g} \text{ is an algebraic Lie algebra.} \tag{5.1)(a}$$

As in (4.8) and (4.9), it is often convenient to fix a connected algebraic group G with Lie algebra $\mathfrak{g}$. Then G acts by Ad on the algebra $S(\mathfrak{g})$ of polynomial functions on $\mathfrak{g}^*$; we write

$$S(\mathfrak{g})^G \tag{5.1)(b}$$

for the algebra of $\operatorname{Ad}(G)$-invariant polynomials, and

$$\mathfrak{Z}(\mathfrak{g}) = U(\mathfrak{g})^G \tag{5.1)(c}$$

for the algebra of G-invariants in $U(\mathfrak{g})$. (If G is connected, $\mathfrak{Z}(\mathfrak{g})$ is the center of $U(\mathfrak{g})$.)

Lemma 5.2. *Filter $\mathfrak{Z}(\mathfrak{g})$ by the restriction of the standard filtration of $U(\mathfrak{g})$. Then*

$$\operatorname{gr} \mathfrak{Z}(\mathfrak{g}) \simeq S(\mathfrak{g})^G.$$

Proof. Obviously the symbol maps

$$\sigma_n : U_n(\mathfrak{g})/U_{n-1}(\mathfrak{g}) \to S^n(\mathfrak{g})$$

of the Poincaré-Birkhoff-Witt theorem restrict to an inclusion

$$\sigma : \operatorname{gr} \mathfrak{Z}(\mathfrak{g}) \hookrightarrow S(\mathfrak{g})^G.$$

We must prove that σ is surjective. The symmetrization map β provides a degree-preserving $\operatorname{Ad}(G)$-equivariant map from $S(\mathfrak{g})$ to $U(\mathfrak{g})$. When restricted to homogeneous polynomials, β is a one-sided inverse for the symbol maps. That is, if p is a homogeneous polynomial of degree n, then $\sigma_n(\beta(p)) = p$. Since β respects the adjoint action, it maps $S(\mathfrak{g})^G$ into $\mathfrak{Z}(\mathfrak{g})$. The surjectivity of σ follows immediately. □

Corollary 5.3. *Suppose $\mathcal{I}$ is a proper ideal of finite codimension in $\mathfrak{Z}(\mathfrak{g})$. Then $\operatorname{gr}\mathcal{I}$ is a proper graded ideal of finite codimension in $S(\mathfrak{g})^G$. Its radical is the ideal $S^+(\mathfrak{g})^G$ of invariant polynomials without constant term.*

Proof. Only the last assertion requires comment. The radical will be the intersection of the maximal ideals containing $\operatorname{gr}\mathcal{I}$. These must form a cone (since $\operatorname{gr}\mathcal{I}$ is graded) and a finite set (since $\operatorname{gr}\mathcal{I}$ has finite codimension). A set of maximal ideals in an $\mathbb{N}$-graded $\mathbb{C}$-algebra with these two properties corresponds to maximal ideals in the degree zero subalgebra. In our case this subalgebra is reduced to the constants, so the only maximal ideal involved is $S^+(\mathfrak{g})^G$. □

Corollary 5.4. *Suppose X is a $\mathfrak{g}$-module of finite length. Then the associated variety $\mathcal{V}(X)$ is contained in the cone $\mathcal{N}^*$ defined by $S^+(\mathfrak{g})^G$:*

$$\mathcal{N}^* = \{\, \lambda \in \mathfrak{g}^* \mid p(\lambda) = 0, \text{ all } p \in S^+(\mathfrak{g})^G \,\}.$$

Proof. Any irreducible $\mathfrak{g}$-module is annihilated by a maximal ideal in $\mathfrak{Z}(\mathfrak{g})$; so any $\mathfrak{g}$-module of finite length is annihilated by the product of a finite number of maximal ideals. Such a product is of finite codimension in $\mathfrak{Z}(\mathfrak{g})$. Now apply Corollary 5.3 and (1.2). □

Because of Corollary 5.4, we turn our attention now to the cone $\mathcal{N}^*$. Write Ad^* for the coadjoint action of G on $\mathfrak{g}^*$. For any $\lambda \in \mathfrak{g}^*$, define

$$G(\lambda) = \{\, g \in G \mid \mathrm{Ad}^*(g)(\lambda) = \lambda \,\}, \qquad \mathfrak{g}(\lambda) = \mathrm{Lie}(G(\lambda)). \qquad (5.5)(a)$$

More generally, if H is a Lie group acting by automorphisms on $\mathfrak{g}$, we will define $H(\lambda)$ analogously; this notation will be applied particularly with $H = K$ in the setting of (2.1), and with $H = G_{\mathbb{R}}$ (a real Lie group with complexified Lie algebra $\mathfrak{g}$) in the setting of the introduction. It is an elementary exercise to verify that

$$\mathfrak{g}(\lambda) = \{\, x \in \mathfrak{g} \mid \text{ for all } y \in \mathfrak{g}, \lambda([x,y]) = 0 \,\}. \qquad (5.5)(b)$$

We say that λ is *nilpotent* if its restriction to the subalgebra $\mathfrak{g}(\lambda)$ is zero; that is, if

$$\lambda([x,\mathfrak{g}]) = 0 \Rightarrow \lambda(x) = 0. \qquad (5.5)(c)$$

When G is reductive, we are going to see that $\mathcal{N}^*$ is precisely the cone of nilpotent elements (and that this definition of nilpotent agrees with more familiar ones). Here is a preliminary step.

Lemma 5.6. *In the setting of (5.5), identify the tangent space $T_\lambda(G\cdot\lambda)$ at with a subspace of the ambient vector space $\mathfrak{g}^*$. Then*

$$T_\lambda(G\cdot\lambda) = \{\, \mu \in \mathfrak{g}^* \mid \mu|_{\mathfrak{g}(\lambda)} = 0 \,\}.$$

Proof. By general results on homogeneous spaces, the tangent space in question may be obtained by applying the differentiated (coadjoint) action of G to λ. A typical element is therefore of the form $\mathrm{ad}^*(x)(\lambda)$. We check that such an element satisfies the condition in the lemma. Fix y in $\mathfrak{g}(\lambda)$. Then

$$\mathrm{ad}^*(x)(\lambda)(y) = -\lambda(\mathrm{ad}(x)(y)) = \lambda([y,x]) = 0,$$

the last equality coming from (5.5)(b). On the other hand, both spaces in the lemma have dimension equal to the codimension of $\mathfrak{g}(\lambda)$ in $\mathfrak{g}$. The containment we have just proved therefore implies their equality. □

Theorem 5.7. *Suppose G is a complex connected algebraic group with Lie algebra $\mathfrak{g}$, and $\lambda \in \mathfrak{g}^*$. The following conditions are equivalent.*

a) *λ is nilpotent; that is, $\lambda|_{\mathfrak{g}(\lambda)} = 0$.*
b) *$\lambda \in T_\lambda(G \cdot \lambda)$*
c) *There is an element $x \in \mathfrak{g}$ such that $\mathrm{ad}^*(x)\lambda = \lambda$*
d) *For all non-zero complex numbers t, $t\lambda \in G \cdot \lambda$.*
e) *For infinitely many complex numbers t, $t\lambda \in G \cdot \lambda$.*

If in addition G is reductive, these are also equivalent to

f) *There is a Borel subalgebra $\mathfrak{b}$ of $\mathfrak{g}$ such that $\lambda|_{\mathfrak{b}} = 0$*
g) *$0 \in \overline{G \cdot \lambda}$*
h) *$\lambda \in \mathcal{N}^*$; that is, every G-invariant polynomial without constant term vanishes at λ.*

Proof. We first prove the equivalence of (a)–(e). The equivalence of (a) and (b) is Lemma 5.6, and that of (b) and (c) is formal (see the proof of Lemma 5.6). Exponentiating (c) gives

$$\mathrm{Ad}^*(\exp(sx))\lambda = e^s \cdot \lambda,$$

which implies (d); and (e) follows from (d) . Conversely, assume (e). The set of complex numbers for which the condition in (e) holds is automatically a subgroup of $\mathbb{C}^\times$. It therefore contains a sequence converging to 1, and (c) follows.

Next, we show that (c) implies (f). By the definition of ad^*, (c) is equivalent to

$$\lambda|_{\mathrm{im}(1+\mathrm{ad}\,x)} = 0.$$

The image in question (call it $\mathfrak{s}$) contains all of the generalized eigenspaces of $\mathrm{ad}\,x$ except that for the eigenvalue -1. On the other hand, the sum of all the generalized eigenspaces of $\mathrm{ad}\,x$ corresponding to eigenvalues with non-negative real part is a parabolic subalgebra of $\mathfrak{g}$. (Here we are using for the first time the assumption that $\mathfrak{g}$ is reductive.) It follows that $\mathfrak{s}$ contains a parabolic subalgebra, and hence a Borel subalgebra. This is (f).

To see that (f) implies (g), suppose B is the Borel subgroup of G corresponding to $\mathfrak{b}$. There is an element h of B such that $\mathrm{Ad}^*(h)$ has only real eigenvalues strictly smaller than one on $(\mathfrak{g}/\mathfrak{b})^*$. It follows that $\lim_{n\to\infty} \mathrm{Ad}(h^n)(\lambda) = 0$, which implies (g). That (g) implies (h) is easy (since G-invariant polynomials are constant on orbit closures).

To complete the proof, it is enough to show that (h) implies (e). This seems to be deeper than the rest of the argument. We will use the following result of Kostant.

Theorem 5.8 ([K]). *If G is reductive, the cone $\mathcal{N}^*$ (Corollary 5.4) is the union of a finite number of orbits of G.*

Assume now that $\lambda \in \mathcal{N}^*$. Since $\mathcal{N}^*$ is a cone, all multiples of λ belong to it as well. By Theorem 5.8, we can find infinitely many in a single G-orbit. If one of these is $t\lambda$, then multiplying by t^{-1} gives infinitely many multiples of λ in the orbit of λ, as required. □

Although we will make every effort to avoid doing so, it is occasionally convenient to use an identification of $\mathfrak{g}^*$ with $\mathfrak{g}$.

Lemma 5.9. *Suppose G is a reductive Lie group with Lie algebra $\mathfrak{g}$. Then there is a non-degenerate symmetric G-invariant bilinear form $\langle , \rangle$ on $\mathfrak{g}$. If $\mathfrak{p}$ is any parabolic subalgebra of $\mathfrak{g}$, with nil radical $\mathfrak{n}$, then*

$$\mathfrak{p}^{\perp} = \textit{radical of } \langle , \rangle |_{\mathfrak{p}} = \mathfrak{n}.$$

This is standard and easy: one can add to the Killing form any non-degenerate symmetric form on the center. Such a form gives a G-equivariant identification

$$\mathfrak{g}^* \simeq \mathfrak{g}, \qquad \lambda \mapsto x_\lambda. \tag{5.10}$$

Because G is an algebraic group, there is a notion of semisimple and nilpotent elements in $\mathfrak{g}$ (and a Jordan decomposition). In any algebraic Lie algebra, an element x is nilpotent if and only if it belongs to the nil radical of some Borel subalgebra. (This is not quite the definition, but it follows immediately from the fact every element belongs to a Borel subalgebra.)

Corollary 5.11. *In the setting of Theorem 5.7, suppose G is reductive. Then the isomorphism of (5.10) identifies $\mathcal{N}^*$ with the cone $\mathcal{N}$ of nilpotent elements in $\mathfrak{g}$.*

Proof. Suppose $\lambda \in \mathfrak{g}^*$ corresponds to $x_\lambda \in \mathfrak{g}$. If $\mathfrak{b}$ is a Borel subalgebra (or indeed any subspace of $\mathfrak{g}$) then $\lambda |_{\mathfrak{b}} = 0$ if and only if $x_\lambda \in \mathfrak{b}^{\perp}$. Now apply condition (f) of Theorem 5.7, Lemma 5.9, and the remarks preceding the corollary. □

Except under special additional hypotheses on the module X (as in the proof of Corollary 4.7, for example) there is no reason for the group G to act on an associated variety $\mathcal{V}(X)$. The basic finiteness result in Theorem 5.8 is therefore insufficient for us. We first recall the additional structure used in section 2.

Definition 5.12. A *pair* is a pair (G, K) of algebraic groups endowed with
i) an inclusion of Lie algebras $i : \mathfrak{k} \to \mathfrak{g}$; and
ii) an algebraic action Ad of K on $\mathfrak{g}$ by automorphisms, compatible with i.

Since the conditions (i) and (ii) refer only to $\mathfrak{g}$, we may also speak of the pair $(\mathfrak{g}, K)$. Define the *nilpotent cone for the pair* by

$$\begin{aligned}\mathcal{N}_{\mathfrak{k}}^* &= \{\lambda \in \mathcal{N}^* \mid \lambda|_{\mathfrak{k}} = 0\}\\ &= \mathcal{N}^* \cap (\mathfrak{g}/\mathfrak{k})^*.\end{aligned}$$

From Corollary 5.4 and (1.2)(c) we get at once

Corollary 5.13. *Suppose X is a $(\mathfrak{g}, K)$-module of finite length. Then the associated variety $\mathcal{V}(X)$ is a union of K-orbits contained in the cone $\mathcal{N}_{\mathfrak{k}}^*$ of Definition 5.12.*

We must therefore study the orbits of K on $\mathcal{N}_{\mathfrak{k}}^*$. Many of our results are most conveniently phrased in terms of the symplectic structure on a coadjoint orbit, which we therefore recall. For details the reader may consult for example [GS]. Fix a G-orbit

$$\mathcal{O} \subset \mathfrak{g}^*. \tag{5.14)(a}$$

For each $\lambda \in \mathcal{O}$, define a skew-symmetric bilinear form ω_λ on $\mathfrak{g}$ by

$$\omega_\lambda(x, y) = \lambda([x, y]). \tag{5.14)(b}$$

The radical of ω_λ is $\mathfrak{g}(\lambda)$ (by (5.5)(b)), so ω_λ defines a non-degenerate symplectic form

$$\omega_\lambda \text{ on } \mathfrak{g}/\mathfrak{g}(\lambda) \simeq T_\lambda(\mathcal{O}). \tag{5.14)(c}$$

Evidently these forms fit together to define a smooth algebraic 2-form ω on $\mathcal{O}$. One checks fairly easily that ω is closed, so $\mathcal{O}$ is in a natural G-invariant way a complex symplectic manifold. (In particular, the dimension of $\mathcal{O}$ is even; recall that this was critical to the proof of Corollary 4.7.) Write $\mathrm{Sp}(\omega_\lambda)$ for the group of linear transformations of $T_\lambda(\mathcal{O})$ preserving the form ω_λ. Then the isotropy action at λ gives a natural homomorphism

$$G(\lambda) \to \mathrm{Sp}(\omega_\lambda). \tag{5.14)(d}$$

A somewhat different construction of ω_λ, emphasizing the "Poisson structure," is implicit in sections 10 and 11 (see (11.13)).

A submanifold Y of a symplectic manifold (Z, ω) is called *isotropic* if the symplectic form restricts to zero on each tangent space; that is, if

$$T_y(Y)^\perp \supset T_y(Y) \tag{5.15)(a}$$

for every $y \in Y$. (Here we identify $T_y(Y)$ with a subspace of $T_y(Z)$, and form the perpendicular with respect to the symplectic form ω_y. We say that Y is *coisotropic* if

$$T_y(Y)^\perp \subset T_y(Y), \tag{5.15)(b}$$

and *Lagrangian* if it is both isotropic and coisotropic; that is, if

$$T_y(Y)^\perp = T_y(Y). \tag{5.15)(c}$$

If Z is algebraic, we could make these definitions for arbitrary subvarieties, or even subschemes, using the Zariski tangent space. This leads to a conflict with standard terminology, however: a subvariety is usually called Lagrangian if its smooth locus is Lagrangian. Such a subvariety will satisfy (5.15)(b) but not (5.15)(c) at singular points y. We will therefore try to avoid the terminology in the singular case.

Proposition 5.16. *Suppose (G, K) is a pair (Definition 5.12), $\lambda \in (\mathfrak{g}/\mathfrak{k})^*$, and $\mathcal{O} = G \cdot \lambda$ is a coadjoint orbit of dimension $2n$. Define*

$$\mathcal{O}_{\mathfrak{k}} = \mathcal{O} \cap (\mathfrak{g}/\mathfrak{k})^*,$$

which we regard as a subscheme of the algebraic variety $\mathcal{O}$.

a) *The orbit $K \cdot \lambda$ is a smooth isotropic subvariety of $\mathcal{O}$ (cf. (5.15)(a)), contained in $\mathcal{O}_{\mathfrak{k}}$. In particular, $\dim K \cdot \lambda \leq n$.*
b) *The subscheme $\mathcal{O}_{\mathfrak{k}}$ of $\mathcal{O}$ is coisotropic, in the sense that its Zariski tangent space at each (closed) point λ satisfies (5.15)(b). In particular, the dimension of each such tangent space is at least n.*

In (b), we do not claim that $\mathcal{O}_{\mathfrak{k}}$ has dimension at least n; this scheme could be non-reduced, and so could have no smooth points. I know of no example where this happens, however.

Proof. The orbit $K \cdot \lambda$ is smooth because it is homogeneous. Its tangent space at λ is $\mathfrak{k}/\mathfrak{k}(\lambda)$. If x and y are elements of $\mathfrak{k}$ (representing tangent vectors) then $[x, y]$ also belongs to $\mathfrak{k}$. Since λ is assumed to vanish on $\mathfrak{k}$, it follows that

$$\omega_\lambda(x, y) = \lambda([x, y]) = 0.$$

This is (a).

For (b), we compute first that

$$T_\lambda(K \cdot \lambda)^\perp = \{\, y + \mathfrak{g}(\lambda) \in \mathfrak{g}/\mathfrak{g}(\lambda) \mid \omega_\lambda(y, \mathfrak{k}) = 0 \,\}.$$

.Lemma 5.6 allows us to identify this with a subspace of $\mathfrak{g}^*$, namely

$$\begin{aligned} &\{\, \mu \in \mathfrak{g}^* \mid \mu|_{\mathfrak{k}+\mathfrak{g}(\lambda)} = 0 \,\} \\ &\qquad = \{\, \mu \in \mathfrak{g}^* \mid \mu|_{\mathfrak{g}(\lambda)} = 0 \,\} \cap \{\, \mu \in \mathfrak{g}^* \mid \mu|_{\mathfrak{k}} = 0 \,\} \\ &\qquad = T_\lambda(G \cdot \lambda) \cap T_\lambda((\mathfrak{g}/\mathfrak{k})^*). \end{aligned}$$

Now the Zariski tangent space of an intersection is the intersection of the tangent spaces. (Here it is essential to take scheme-theoretic intersection; this assertion would not be true in general if we considered only the underlying variety.) We have therefore shown that

$$T_\lambda(K \cdot \lambda)^\perp = T_\lambda\left(G \cdot \lambda \cap (\mathfrak{g}/\mathfrak{k})^*\right). \tag{5.17}$$

Now (b) follows from (a). □

Corollary 5.18. *In the setting of Prop. 5.16, the following are equivalent.*

a) $\dim K \cdot \lambda = n$.

b) *$K \cdot \lambda$ is a Lagrangian subvariety of $G \cdot \lambda$.*

c) *The intersection $G \cdot \lambda \cap (\mathfrak{g}/\mathfrak{k})^*$ is reduced at λ, and $K \cdot \lambda$ is open in it.*

Proposition 5.19 ([KR]). *In the setting of Proposition 5.16, assume that $\mathfrak{k}$ is the algebra of fixed points of an involutive automorphism θ of $\mathfrak{g}$. Then the conditions of Corollary 5.18 are satisfied.*

Proof. Write $\mathfrak{p}$ for the -1 eigenspace of θ on $\mathfrak{g}$, and $\mathfrak{p}(\lambda)$ for its intersection with $\mathfrak{g}(\lambda)$. If x and y belong to $\mathfrak{p}$, then

$$\theta([x,y]) = [\theta(x),\theta(y)] = [-x,-y] = [x,y];$$

so $[x,y]$ belongs to $\mathfrak{k}$. The argument for Proposition 5.16(a) shows that $\mathfrak{p}/\mathfrak{p}(\lambda)$ is an isotropic subspace of $T_\lambda(G \cdot \lambda)$. Since $\mathfrak{g} = \mathfrak{k} + \mathfrak{p}$, $T_\lambda(G \cdot \lambda)$ is spanned by the two isotropic subspaces $\mathfrak{k}/\mathfrak{k}(\lambda)$ and $\mathfrak{p}/\mathfrak{p}(\lambda)$. By linear algebra, the sum must be direct and the subspaces of dimension n. □

Corollary 5.20 ([KR]). *Suppose (G,K) is a pair (Definition 5.12), and $\lambda \in \mathfrak{g}^*$. Assume that $\mathfrak{k}$ is the algebra of fixed points of an involutive automorphism θ of $\mathfrak{g}$. Then the intersection $G \cdot \lambda \cap (\mathfrak{g}/\mathfrak{k})^*$ is a finite union of K-orbits. It is a smooth reduced Lagrangian subvariety of $G \cdot \lambda$.*

Only the finiteness of the number of K-orbits requires comment; and this follows from the fact that each is open in the intersection.

Definition 5.21. A *reductive symmetric pair* is a pair (G,K) of reductive algebraic groups as in Definition 5.12, endowed in addition with

iii) an involutive automorphism θ of $\mathfrak{g}$, commuting with $\mathrm{Ad}\, K$, with fixed point set $\mathfrak{k}$.

We say that (G,K) is of *Harish-Chandra class* if the automorphisms in $\mathrm{Ad}\, K$ are inner. (This is automatic if K is connected.) We may again speak of the pair $(\mathfrak{g},K)$.

Corollary 5.22 ([KR]). *Suppose $(\mathfrak{g},K)$ is a reductive symmetric pair. Then the cone $\mathcal{N}_\mathfrak{k}^*$ of Definition 5.12 is a finite union of K-orbits.*

Corollary 5.23. *Suppose $(\mathfrak{g},K)$ is a reductive symmetric pair, and X is a $(\mathfrak{g},K)$-module of finite length. Then the associated variety $\mathcal{V}(X)$ is a finite union of K-orbits in $\mathcal{N}_\mathfrak{k}^*$.*

6. Connection with real nilpotent orbits.

In this section we recall results of Kostant-Rallis and Sekiguchi relating the nilpotent K orbits considered in section 5 with real nilpotent orbits. The ultimate goal, about which we will say more in section 7, is to make some philosophical connections between associated varieties and the method of coadjoint orbits.

Definition 6.1. A *real reductive Lie group* $G_{\mathbb{R}}$ is one with the following three properties: $G_{\mathbb{R}}$ has a finite number of connected components; the Lie algebra $\mathfrak{g}_{\mathbb{R}}$ is reductive; and the center of the derived group of the identity component is finite.

Such a group has a maximal compact subgroup $K_{\mathbb{R}}$, unique up to conjugacy by $G_{\mathbb{R}}$, and a Cartan involution θ with fixed point group $K_{\mathbb{R}}$. The complexification K of $K_{\mathbb{R}}$ is a complex algebraic group, which acts on the complexification $\mathfrak{g}$ of $\mathfrak{g}_{\mathbb{R}}$. Consequently $(\mathfrak{g}, K)$ is a reductive symmetric pair. Conversely, it can be shown that every reductive symmetric pair arises in this manner from a real reductive group.

Given a real reductive group $G_{\mathbb{R}}$, we identify $\mathfrak{g}^*$ with the space of $\mathbb{R}$-linear maps from $\mathfrak{g}_{\mathbb{R}}$ to $\mathbb{C}$. This is a complex vector space containing as a real form the space $\mathfrak{g}_{\mathbb{R}}^*$ of real-valued linear functionals on $\mathfrak{g}_{\mathbb{R}}$. Taking the real part defines a restriction map

$$\mathrm{Re} : \mathfrak{g}^* \to \mathfrak{g}_{\mathbb{R}}^*$$

analogous to the restriction map from $\mathfrak{g}^*$ to $\mathfrak{k}^*$ used in section 5. The analogue of $(\mathfrak{g}/\mathfrak{k})^*$ is then the space $\mathfrak{g}_{i\mathbb{R}}^*$ of purely imaginary-valued linear functionals on $\mathfrak{g}_{\mathbb{R}}$. The analogue of $\mathcal{N}_{\mathfrak{k}}^*$ is the *imaginary nilpotent cone*

$$\mathcal{N}_{i\mathbb{R}}^* = \{\lambda \in \mathcal{N}^* \mid \mathrm{Re}\,\lambda = 0\} = \mathcal{N}^* \cap \mathfrak{g}_{i\mathbb{R}}^*.$$

Of course multiplication by i defines a $G_{\mathbb{R}}$-equivariant isomorphism from (for example) $\mathfrak{g}_{i\mathbb{R}}^*$ onto $\mathfrak{g}_{\mathbb{R}}^*$. We could therefore equally well consider the real nilpotent cone, and it is traditional to do this. The aesthetic advantages of $\mathcal{N}_{i\mathbb{R}}^*$ (such as the improved analogy with $\mathcal{N}_{\mathfrak{k}}^*$) were pointed out to me by H. Matumoto. For the moment, notice only that the $G_{\mathbb{R}}$-orbits on $\mathfrak{g}_{i\mathbb{R}}^*$ are a very natural setting for the method of coadjoint orbits.

Theorem 6.2 ([Se]). *Suppose that $G_{\mathbb{R}}$ is a real reductive group (Definition 6.1) with maximal compact subgroup $K_{\mathbb{R}}$. Then there is a natural one-to-one correspondence between the (finite) set of $G_{\mathbb{R}}$-orbits on the imaginary nilpotent cone $\mathcal{N}_{i\mathbb{R}}^*$ and the K-orbits on $\mathcal{N}_{\mathfrak{k}}^*$ (Definition 5.12), implemented by Theorem 6.4 below. Suppose that in this correspondence the orbit of $\lambda_{i\mathbb{R}} \in \mathcal{N}_{i\mathbb{R}}^*$ corresponds to that of $\lambda_{\mathfrak{k}} \in \mathcal{N}_{\mathfrak{k}}^*$. Let G be any complex group with Lie algebra $\mathfrak{g}$.*

a) The G-orbits of $\lambda_{\mathfrak{k}}$ and $\lambda_{i\mathbb{R}}$ coincide.

b) $\dim_{\mathbb{R}} G_{\mathbb{R}} \cdot \lambda_{i\mathbb{R}} = 2 \cdot \dim_{\mathbb{C}} K \cdot \lambda_{\mathfrak{k}} = \dim_{\mathbb{C}} G \cdot \lambda_{\mathfrak{k}}$.

c) The maximal compact subgroups of the isotropy groups $K(\lambda_{\mathfrak{k}})$ and $G_{\mathbb{R}}(\lambda_{i\mathbb{R}})$ are isomorphic (canonically, up to inner automorphism).

We recall the outline of Sekiguchi's argument (since we need most of it just to write down the correspondence). Unfortunately we must begin by choosing a non-degenerate symmetric real bilinear form on $\mathfrak{g}_{\mathbb{R}}$, which is invariant under $G_{\mathbb{R}}$ and θ, negative definite on $\mathfrak{k}_{\mathbb{R}}$, and positive definite on the -1-eigenspace $\mathfrak{p}_{\mathbb{R}}$ of θ. Such a form exists and is unique up to a positive real scalar on each simple factor of $\mathfrak{g}_{\mathbb{R}}$. We use it to identify $\mathcal{N}_{i\mathbb{R}}^*$ with $\mathcal{N}_{i\mathbb{R}}$ (the cone of purely imaginary nilpotent elements of the complexified Lie algebra) and $\mathcal{N}_{\mathfrak{p}}^*$ with $\mathcal{N}_{\mathfrak{p}}$ (the cone of nilpotent elements in the -1-eigenspace of θ). In this way the theorem becomes one about nilpotent Lie algebra elements. (The identification of elements made here depends on the choice of the form, but the identification of *orbits* does not. The reason is that multiplication by a positive real scalar on each simple factor sends a nilpotent element to a conjugate one.)

Write σ for the complex conjugation on $\mathfrak{g}$ defining the real form $G_{\mathbb{R}}$. Then σ commutes with θ, and the involution $\tau = \sigma\theta$ of $\mathfrak{g}$ is a Cartan involution for the complex Lie algebra. After the reduction of the previous paragraph, Theorem 6.2 describes a relationship between nilpotent elements in the -1-eigenspaces of σ and θ on $\mathfrak{g}$. We are going to define it by reduction to the case of $\mathfrak{sl}(2,\mathbb{R})$. It is convenient to consider the three (commuting) involutive automorphisms of $\mathfrak{sl}(2)$ analogous to θ, σ, and τ. Recall first that $\mathfrak{sl}(2)$ consists of the two by two complex matrices of trace zero. Then for $x \in \mathfrak{sl}(2)$, set

$$\theta_0 x = -{}^t x \qquad \sigma_0 x = \bar{x} \qquad \tau_0 x = -{}^t\bar{x}. \tag{6.3}$$

Then θ_0 is a complexified Cartan involution for the real form $\mathfrak{sl}(2,\mathbb{R})$; σ_0 is the complex conjugation for $\mathfrak{sl}(2,\mathbb{R})$; and τ_0 is a Cartan involution for $\mathfrak{sl}(2)$. (Equivalently, τ_0 is the complex conjugation for a compact real form of $\mathfrak{sl}(2)$. With this notation, we may reformulate the bijection in Sekiguchi's result as follows.

Theorem 6.4. *Suppose $G_{\mathbb{R}}$ is a real reductive group (Definition 6.1) with Cartan involution θ and maximal compact subgroup $K_{\mathbb{R}}$. Write σ for the complex conjugation on $\mathfrak{g}$ defining $\mathfrak{g}_{\mathbb{R}}$, and K for the complexification of $K_{\mathbb{R}}$. Then the following sets are in natural one-to-one correspondence.*

a) *$G_{\mathbb{R}}$-orbits on the cone $\mathcal{N}_{i\mathbb{R}}$ of purely imaginary nilpotent elements in $\mathfrak{g}$.*
b) *$G_{\mathbb{R}}$-conjugacy classes of homomorphisms $\phi_{i\mathbb{R}}$ from $\mathfrak{sl}(2)$ to $\mathfrak{g}$, intertwining σ_0 with σ. (That is, we require $\sigma(\phi_{i\mathbb{R}}(x)) = \phi_{i\mathbb{R}}(\sigma_0(x))$.)*
c) *$K_{\mathbb{R}}$-conjugacy classes of homomorphisms $\phi_{i\mathbb{R},\mathfrak{p}}$ from $\mathfrak{sl}(2)$ to $\mathfrak{g}$, intertwining σ_0 with σ and θ_0 with θ.*
d) *K-conjugacy classes of homomorphisms $\phi_{\mathfrak{p}}$ from $\mathfrak{sl}(2)$ to $\mathfrak{g}$, intertwining θ_0 with θ.*
e) *K-orbits on the cone $\mathcal{N}_{\mathfrak{p}}$ of nilpotent elements in the -1-eigenspace of θ.*

Here the bijections from (c) to (b) and (d) are given by the obvious inclusions; that from (b) to (a) sends (the conjugacy class of) $\phi_{i\mathbb{R}}$ to (the orbit of) $\phi_{i\mathbb{R}}\begin{pmatrix} 0 & i \\ 0 & 0 \end{pmatrix}$; and that from (d) to (e) sends $\phi_{\mathfrak{p}}$ to $\phi_{\mathfrak{p}}\begin{pmatrix} 1/2 & -i/2 \\ -i/2 & -1/2 \end{pmatrix}$.

Sketch of proof. Fix a nilpotent element $x_{i\mathbb{R}} \in \mathcal{N}_{i\mathbb{R}}$. By the Jacobson-Morozov theorem (applied to the real Lie algebra $\mathfrak{g}_{\mathbb{R}}$) we can find elements $y_{i\mathbb{R}} \in \mathcal{N}_{i\mathbb{R}}$ and $h_{i\mathbb{R}} \in \mathfrak{g}_{\mathbb{R}}$ satisfying

$$[h_{i\mathbb{R}}, x_{i\mathbb{R}}] = 2x_{i\mathbb{R}}, \quad [h_{i\mathbb{R}}, y_{i\mathbb{R}}] = -2y_{i\mathbb{R}}, \quad [x_{i\mathbb{R}}, y_{i\mathbb{R}}] = h_{i\mathbb{R}} \tag{6.5)(a}$$

$$\sigma x_{i\mathbb{R}} = -x_{i\mathbb{R}}, \quad \sigma y_{i\mathbb{R}} = -y_{i\mathbb{R}}, \quad \sigma h_{i\mathbb{R}} = h_{i\mathbb{R}}. \tag{6.5)(b}$$

The elements $y_{i\mathbb{R}}$ and $h_{i\mathbb{R}}$ are unique up to conjugation by the centralizer of $x_{i\mathbb{R}}$ in $G_{\mathbb{R}}$. The three elements span the complexification of a real subalgebra $\mathfrak{s}_{\mathbb{R}}$ of $\mathfrak{g}_{\mathbb{R}}$, which is obviously a homomorphic image of $\mathfrak{sl}(2,\mathbb{R})$. Explicitly, we can define a homomorphism $\phi_{i\mathbb{R}}$ by

$$\phi_{i\mathbb{R}}\begin{pmatrix} a & b \\ c & -a \end{pmatrix} = ah_{i\mathbb{R}} - ibx_{i\mathbb{R}} + icy_{i\mathbb{R}}. \tag{6.5)(c}$$

Because of (6.5)(b), the homomorphism $\phi_{i\mathbb{R}}$ intertwines the complex conjugation σ_0 for $\mathfrak{sl}(2,\mathbb{R})$ with σ. This gives the correspondence between (a) and (b) in Theorem 6.4.

To go from (b) to (c), we must conjugate $\phi_{i\mathbb{R}}$ by an element of $G_{\mathbb{R}}$ to make it intertwine Cartan involutions. The standard Cartan involution θ_0 for $\mathfrak{sl}(2,\mathbb{R})$ (negative transpose) is mapped to a certain automorphism θ_0' of the image. On the other hand, it is known that any Cartan involution of a semisimple subalgebra of $\mathfrak{g}_{\mathbb{R}}$ must extend to one on all of $\mathfrak{g}_{\mathbb{R}}$. Consequently θ_0' is the restriction of some Cartan involution θ' of $\mathfrak{g}_{\mathbb{R}}$. Clearly $\phi_{i\mathbb{R}}$ intertwines θ_0 with θ'. Now (by the uniqueness of Cartan involutions) θ' must differ from θ by conjugation by some element g of (the identity component of) $G_{\mathbb{R}}$. Write

$$\phi_{i\mathbb{R},\mathfrak{k}} = \mathrm{Ad}(g) \circ \phi_{i\mathbb{R}}. \tag{6.6}$$

Then $\phi_{i\mathbb{R},\mathfrak{k}}$ intertwines θ_0 with θ and σ_0 with σ. Sekiguchi shows [Se, Lemma 1.5] that this property (together with the specified $G_{\mathbb{R}}$-conjugacy class of $\phi_{i\mathbb{R}}$) determines $\phi_{i\mathbb{R},\mathfrak{k}}$ up to $K_{\mathbb{R}}$-conjugacy. This gives the correspondence from (b) to (c).

Next, we show how to go from (e) to (d). We begin with a nilpotent element $x_{\mathfrak{k}} \in \mathcal{N}_{\mathfrak{k}}$. As is shown in [KR], we can find elements $y_{\mathfrak{k}} \in \mathcal{N}_{\mathfrak{k}}$ and $h_{\mathfrak{k}} \in \mathfrak{k}$ so that

$$[h_{\mathfrak{k}}, x_{\mathfrak{k}}] = 2x_{\mathfrak{k}}, \quad [h_{\mathfrak{k}}, y_{\mathfrak{k}}] = -2y_{\mathfrak{k}}, \quad [x_{\mathfrak{k}}, y_{\mathfrak{k}}] = h_{\mathfrak{k}}. \tag{6.7)(a}$$

$$\theta x_{\mathfrak{k}} = -x_{\mathfrak{k}}, \quad \theta y_{\mathfrak{k}} = -y_{\mathfrak{k}}, \quad \theta h_{\mathfrak{k}} = h_{\mathfrak{k}}. \tag{6.7)(b}$$

Again these three elements determine a homomorphism $\phi_{\mathfrak{k}}$ from $\mathfrak{sl}(2)$ to $\mathfrak{g}$, by the requirements

$$\phi_{\mathfrak{k}}\begin{pmatrix} 0 & +i \\ -i & 0 \end{pmatrix} = h_{\mathfrak{k}}, \quad \phi_{\mathfrak{k}}\begin{pmatrix} +1/2 & -i/2 \\ -i/2 & -1/2 \end{pmatrix} = x_{\mathfrak{k}}, \quad \phi_{\mathfrak{k}}\begin{pmatrix} 1/2 & i/2 \\ i/2 & -1/2 \end{pmatrix} = y_{\mathfrak{k}}. \tag{6.7)(c}$$

The homomorphism $\phi_{\mathfrak{k}}$ intertwines θ_0 with θ. This establishes the correspondence from (e) to (d).

Suppose now that we are given $\phi_{\mathfrak{k}}$ as in (d) of Theorem 6.4. We wish to modify $\phi_{\mathfrak{k}}$ by conjugation by an element of K so that it intertwines σ_0 with σ. Of course it is equivalent to have τ_0 intertwined with τ. Now τ_0 is mapped by $\phi_{\mathfrak{k}}$ to a Cartan involution τ_0' of the image. Now we use a slight refinement of the result about extending Cartan involutions from subalgebras: that if the subalgebra is preserved by a fixed involutive automorphism θ, and the given Cartan involution on the subalgebra commutes with θ, then the extension may be chosen to commute with θ as well [vanD, Proposition 2]. We conclude that τ_0' may be extended to a Cartan involution τ' of $\mathfrak{g}$ commuting with θ. Now τ' and τ are two Cartan involutions commuting with θ. Consequently they differ by conjugation by some element k of (the identity component of) $K_{\mathbb{C}}$ [L, Chapter IV, Theorem 2.1]. Write

$$\phi_{i\mathbb{R},\mathfrak{k}} = \mathrm{Ad}(k) \circ \phi_{\mathfrak{k}}. \tag{6.8}$$

Then $\phi_{i\mathbb{R},\mathfrak{k}}$ intertwines θ_0 with θ and τ_0 with τ; so it also intertwines σ_0 with σ. This is the correspondence from (d) to (c); again we refer to [Se] for the proof that it is well-defined. □

We turn now to the rest of the proof of Theorem 6.2. That $G \cdot \lambda_{\mathfrak{k}}$ is equal to $G \cdot \lambda_{i\mathbb{R}}$ is clear from the construction of the bijection in Theorem 6.4. We have already seen (Corollary 5.20) that $K \cdot \lambda_{\mathfrak{k}}$ may be regarded as a complex Lagrangian subvariety of $G \cdot \lambda_{\mathfrak{k}}$; this proves the second equality of dimensions in (b). Suppose now that we regard $G \cdot \lambda_{i\mathbb{R}}$ as a *real* symplectic manifold, by considering only the real part of $\omega_{\lambda_{i\mathbb{R}}}$ (cf. (5.14)). Then the assumption that $\lambda_{i\mathbb{R}}$ is imaginary-valued forces the real submanifold $G_{\mathbb{R}} \cdot \lambda_{i\mathbb{R}}$ to be isotropic; and the proof of Proposition 5.19 (with σ replacing θ) shows that it is actually Lagrangian. This gives the first equality of dimensions in (b).

For Theorem 6.2(c), we need some additional notation based on (6.5). Write

$$G_{\mathbb{R}}(x_{i\mathbb{R}}) = \text{ centralizer in } G_{\mathbb{R}} \text{ of } x_{i\mathbb{R}}, \tag{6.9)(a}$$

$$G_{\mathbb{R}}(\phi_{i\mathbb{R}}) = \text{ centralizer in } G_{\mathbb{R}} \text{ of } \phi_{i\mathbb{R}}(\mathfrak{sl}(2)). \tag{6.9)(b}$$

The element $h_{i\mathbb{R}}$ has integral eigenvalues in its adjoint action on $\mathfrak{g}$, and preserves $\mathfrak{g}_{\mathbb{R}}(x_{i\mathbb{R}})$. Whenever $\mathfrak{s}$ is an $\mathrm{ad}(h_{i\mathbb{R}})$-stable subspace of $\mathfrak{g}$, we write

$$\mathfrak{s}[k; h_{i\mathbb{R}}] = \mathfrak{s}[k] = \{\, s \in \mathfrak{s} \mid [h_{i\mathbb{R}}, s] = ks \,\}. \tag{6.9)(c}$$

By the representation theory of $\mathfrak{sl}(2)$), the eigenvalues of $\mathrm{ad}(h_{i\mathbb{R}})$ on $\mathfrak{g}_{\mathbb{R}}(x_{i\mathbb{R}})$ are non-negative, and $\mathfrak{g}_{\mathbb{R}}(\phi_{i\mathbb{R}})$ is precisely the zero eigenspace:

$$\mathfrak{g}_{\mathbb{R}}(\phi_{i\mathbb{R}}) = \mathfrak{g}_{\mathbb{R}}(x_{i\mathbb{R}})[0]. \tag{6.9)(d}$$

Define

$$\mathfrak{u}_{\mathbb{R}}(x_{i\mathbb{R}}) = \sum_{k>0} \mathfrak{g}_{\mathbb{R}}(x_{i\mathbb{R}})[k], \tag{6.9)(e}$$

the sum of the positive eigenspaces of $\mathrm{ad}(h_{i\mathbb{R}})$ on $\mathfrak{g}_{\mathbb{R}}(x_{i\mathbb{R}})$. Then $\mathfrak{g}_{\mathbb{R}}(x_{i\mathbb{R}})$ is evidently the semidirect product of the reductive subalgebra $\mathfrak{g}_{\mathbb{R}}(\phi_{i\mathbb{R}})$ by the nilpotent ideal $\mathfrak{u}_{\mathbb{R}}(x_{i\mathbb{R}})$. We want to get this decomposition on the group level. By putting $h_{i\mathbb{R}}$ in a Cartan subalgebra of $\mathfrak{g}_{\mathbb{R}}$, we see that $\mathfrak{u}_{\mathbb{R}}(x_{i\mathbb{R}})$

is contained in a maximal nilpotent subalgebra of $\mathfrak{g}_{\mathbb{R}}$. It follows that the corresponding connected subgroup

$$U_{\mathbb{R}}(x_{i\mathbb{R}}) = \exp(\mathfrak{u}_{\mathbb{R}}(x_{i\mathbb{R}})) \tag{6.9)(f}$$

is a simply connected unipotent Lie group.

Lemma 6.10 (cf. [BV, section 2] and [K]). *In the setting of (6.9), the centralizer $G_{\mathbb{R}}(x_{i\mathbb{R}})$ is the semidirect product of the reductive group $G_{\mathbb{R}}(\phi_{i\mathbb{R}})$ by the unipotent normal subgroup $U_{\mathbb{R}}(x_{i\mathbb{R}})$. In particular, suppose that the image of $\phi_{i\mathbb{R}}$ is preserved by θ. Then the restriction of θ to $G_{\mathbb{R}}(\phi_{i\mathbb{R}})$ is a Cartan involution, so*

$$K_{\mathbb{R}} \cap G_{\mathbb{R}}(\phi_{i\mathbb{R}})$$

is a maximal compact subgroup of $G_{\mathbb{R}}(\phi_{i\mathbb{R}})$ or of $G_{\mathbb{R}}(x_{i\mathbb{R}})$.

The main point in the proof is to show that, given $x_{i\mathbb{R}}$, any two choices of $y_{i\mathbb{R}}$ satisfying (6.5)(a)-(b) must be conjugate by a (unique) element of $U_{\mathbb{R}}(x_{i\mathbb{R}})$. We omit the argument.

We can define exactly parallel notation for K based on (6.7). In that case we are working with algebraic groups, and the analogue of Lemma 6.10 is precisely a Levi decomposition.

Lemma 6.11. *In the setting of (6.7) (and with notation analogous to (6.9)) the centralizer $K(x_{\mathfrak{k}})$ is the semidirect product of the reductive group $K(\phi_{\mathfrak{k}})$ by the unipotent normal subgroup $U(x_{\mathfrak{k}})$. In particular, suppose that the image of $\phi_{\mathfrak{k}}$ is preserved by the involution τ (defined before (6.3)). Then the restriction of τ to $K(\phi_{\mathfrak{k}})$ is a Cartan involution, so*

$$K_{\mathbb{R}} \cap K(\phi_{\mathfrak{k}})$$

is a maximal compact subgroup of $K(\phi_{\mathfrak{k}})$ or of $K(x_{\mathfrak{k}})$.

Corollary 6.12. *In the setting of Theorem 6.4, suppose ϕ is a homomorphism from $\mathfrak{sl}(2)$ to $\mathfrak{g}$, intertwining σ_0 with σ and θ_0 with θ. Define*

$$x_{i\mathbb{R}} = \phi\begin{pmatrix} 0 & i \\ 0 & 0 \end{pmatrix}, \qquad x_{\mathfrak{k}} = \phi\begin{pmatrix} 1/2 & -i/2 \\ -i/2 & -1/2 \end{pmatrix}.$$

Then the centralizer in $K_{\mathbb{R}}$ of the image of ϕ is a maximal compact subgroup of $G_{\mathbb{R}}(x_{i\mathbb{R}})$ and of $K(x_{\mathfrak{k}})$.

In light of Theorem 6.4, this establishes Theorem 6.2(c).

7. Admissible orbits.

In this section we recall Duflo's notion of "admissible" (imaginary) coadjoint orbits. We then present a result of J. Schwartz (Theorem 7.14) describing which nilpotent $K_{\mathbb{C}}$ orbits on $\mathcal{N}_{\mathfrak{k}}^*$ correspond (via Theorem 6.2) to admissible imaginary orbits.

Suppose to begin that $G_{\mathbb{R}}$ is any real Lie group, and

$$\lambda_{i\mathbb{R}} \in \mathfrak{g}_{i\mathbb{R}}^* \tag{7.1)(a}$$

is a purely imaginary-valued linear functional. Write $\mathcal{O}_{i\mathbb{R}}$ for the $G_{\mathbb{R}}$-orbit of $\lambda_{i\mathbb{R}}$. As in (5.14) we can define an (imaginary-valued) symplectic form $\omega_{\lambda_{i\mathbb{R}}}$ on the tangent space

$$T_{\lambda_{i\mathbb{R}}}(\mathcal{O}_{i\mathbb{R}}) \simeq \mathfrak{g}_{\mathbb{R}}/\mathfrak{g}_{\mathbb{R}}(\lambda_{i\mathbb{R}}). \tag{7.1)(b}$$

Write $Sp(\omega_{\lambda_{i\mathbb{R}}})$ for the group of (real linear) symplectic linear transformations of this tangent space. Then the isotropy action gives a natural homomorphism

$$G_{\mathbb{R}}(\lambda_{i\mathbb{R}}) \xrightarrow{j} Sp(\omega_{\lambda_{i\mathbb{R}}}). \tag{7.1)(c}$$

On the other hand, the real symplectic group has a natural two-fold covering group, the *metaplectic group*:

$$1 \longrightarrow \{1, \epsilon\} \longrightarrow Mp(\omega_{\lambda_{i\mathbb{R}}}) \xrightarrow{p} Sp(\omega_{\lambda_{i\mathbb{R}}}) \longrightarrow 1. \tag{7.1)(d}$$

This covering may be pulled back via the homomorphism (7.1)(c) to give the *metaplectic double cover* of the isotropy group:

$$1 \longrightarrow \{1, \epsilon\} \longrightarrow \tilde{G}_{\mathbb{R}}(\lambda_{i\mathbb{R}}) \xrightarrow{p(\lambda_{i\mathbb{R}})} G_{\mathbb{R}}(\lambda_{i\mathbb{R}}). \tag{7.1)(e}$$

Explicitly, this covering group is defined by

$$\tilde{G}_{\mathbb{R}}(\lambda_{i\mathbb{R}}) = \{\, (g, m) \in G_{\mathbb{R}}(\lambda_{i\mathbb{R}}) \times Mp(\omega_{\lambda_{i\mathbb{R}}}) \mid j(g) = p(m) \,\}. \tag{7.1)(f}$$

Definition 7.2 ([D]). In the setting of (7.1), a representation χ of $\tilde{G}_{\mathbb{R}}(\lambda_{i\mathbb{R}})$ is called *genuine* if $\chi(\epsilon) = -I$. (In [V2, chapter 10], such representations are called *metaplectic*. Notice that if χ is irreducible, $\chi(\epsilon)$ is necessarily $+I$ or $-I$.) We say that χ is *admissible* if it is genuine, and the differential of χ is a multiple of $\lambda_{i\mathbb{R}}$; that is, if

$$\chi(\exp x) = \exp(\lambda_{i\mathbb{R}}(x)) \cdot I$$

for all $x \in \mathfrak{g}_{\mathbb{R}}(\lambda_{i\mathbb{R}})$. (Here the exponential map on the left is the one for $G_{\mathbb{R}}$, and the one on the right is for complex numbers.) Notice that if $G_{\mathbb{R}}(\lambda_{i\mathbb{R}})$ has a finite number of connected components, an irreducible admissible representation of $\tilde{G}_{\mathbb{R}}(\lambda_{i\mathbb{R}})$ is necessarily unitarizable.

If admissible representations exist, we say that $\lambda_{i\mathbb{R}}$ (or the orbit $\mathcal{O}_{i\mathbb{R}}$) is *admissible*. A pair $(\lambda_{i\mathbb{R}}, \chi)$ consisting of an element of $\mathfrak{g}_{i\mathbb{R}}^*$ and an irreducible admissible representation of $\tilde{G}_{\mathbb{R}}(\lambda_{i\mathbb{R}})$ is called an *admissible $G_{\mathbb{R}}$-orbit datum*. Two such are called equivalent if they are conjugate by $G_{\mathbb{R}}$.

Admissible orbit data are the raw material of the orbit method. Here is a rough version of what one expects. (The words "nice" and "usually" below reflect my ignorance; I do not know a more precise statement that is correct.)

Desideratum 7.3. *Suppose $G_{\mathbb{R}}$ is a nice type I Lie group, and $\mathcal{D}=(\lambda_{i\mathbb{R}},\chi)$ is an admissible orbit datum for $G_{\mathbb{R}}$ (Definition 7.2). Attached to the equivalence class of $\mathcal{D}$ there should be a unitary representation $\pi(\mathcal{D})$ of $G_{\mathbb{R}}$. This representation should be a direct sum of a finite number (possibly zero) of irreducibles; and usually $\pi(\mathcal{D})$ itself should be irreducible.*

For more information about what can be proved in this direction, the reader may consult for example [D]; the case of reductive groups is discussed in [V2, chapter 10]. Of course our primary concern here is with the case of nilpotent orbit data for reductive groups. The condition of admissibility is very simple in this case.

Observation 7.4. *In the setting of Definition 7.2, assume that $\lambda_{i\mathbb{R}}$ is nilpotent (cf. (5.5)). Then a representation χ of $\tilde{G}_{\mathbb{R}}(\lambda_{i\mathbb{R}})$ is admissible if and only if it is trivial on the identity component, and $\chi(\epsilon) = -I$. (Here ϵ is the non-trivial element of the kernel of the covering map $p(\lambda_{i\mathbb{R}})$.) In particular, $\lambda_{i\mathbb{R}}$ is admissible if and only if the preimage (under $p(\lambda_{i\mathbb{R}})$) of the identity component $G_{\mathbb{R}}(\lambda_{i\mathbb{R}})_0$ is disconnected.*

Example 7.5. Suppose $G_{\mathbb{R}}$ is the group $SO(3)$. The Lie algebra $\mathfrak{g}_{\mathbb{R}}$ may be identified with skew-symmetric three by three real matrices. Fix a non-zero real number t, and let $\lambda_{i\mathbb{R}}(t)$ denote the linear functional whose value at a matrix x is it times the $(1,2)$ entry of x. Then the isotropy group $G_{\mathbb{R}}(\lambda_{i\mathbb{R}}(t))$ is $SO(2)$, embedded in $G_{\mathbb{R}}$ as the upper left two by two block. The characters of $SO(2)$ are parametrized by $\mathbb{Z}$; we can arrange the parametrization so that the nth character χ_n has differential $\lambda_{i\mathbb{R}}(n)$ (restricted to $\mathfrak{so}(2)$). The complexified isotropy action on $\mathfrak{g}/\mathfrak{g}(\lambda_{i\mathbb{R}})$ has the two weights corresponding to $+1$ and -1. The symplectic group for a two-dimensional vector space is just $SL(2,\mathbb{R})$. It follows that the isotropy representation maps $SO(2)$ isomorphically onto a maximal compact subgroup of $SL(2,\mathbb{R})$. Consequently the metaplectic double cover of $SO(2)$ is the connected double cover; its genuine representations are again one-dimensional characters χ_j parametrized by $j \in \mathbb{Z}+1/2$. We have therefore found admissible orbit data $\mathcal{D}(j) = (\lambda_{i\mathbb{R}}(j),\chi_j)$ for each $j \in \mathbb{Z}+1/2$. It turns out that the $\mathcal{D}(j)$ and $\mathcal{D}(-j)$ are conjugate, and that every admissible orbit datum except $\mathcal{T} = (0,\mathbb{C})$ is conjugate to some $\mathcal{D}(j)$. (Here $\mathcal{T}$ stands for "trivial"; this orbit datum exists for every $G_{\mathbb{R}}$, and $\pi(\mathcal{T})$ is the trivial representation.) There are various ways to attach representations to the other orbit data; the most natural all make $\pi(\mathcal{D}(j))$ the irreducible representation of $SO(3)$ of dimension $2j$. Notice in particular that the trivial representation is attached both to $\mathcal{T}$ and to $\mathcal{D}(1/2)$.

In this example, only the orbit datum $\mathcal{T}$ is nilpotent. Before we consider some interesting examples of nilpotent orbit data, it will be helpful to have a little more machinery.

Lemma 7.6. *Suppose G is a real Lie group having a finite number of connected components, K is a maximal compact subgroup of G, and ϵ is* *a central element of G of finite order m. Fix an mth root of unity ζ. Say that a representation of χ of G is admissible if it is trivial on the identity component of G, and $\chi(\epsilon) = \zeta I$; and define admissible representations of K analogously. Then restriction to K induces a bijection from admissible representations of G to admissible representations of K.*

This is an immediate consequence of Mostow's result that G is topologically the product of K with a vector space. Because of this result (and Observation 7.4), the question of admissibility for nilpotent orbit data can be studied on the level of maximal compact subgroups. Suppose V is a real vector space carrying a non-degenerate imaginary-valued symplectic form ω. We can choose a complex structure and positive-definite Hermitian form h on V so that ω is the imaginary part of h. Write $U(h)$ for the unitary group of h; this is a maximal compact subgroup of $Sp(\omega)$. The complex determinant defines a one-dimensional character

$$\det : U(h) \to \mathbb{C}^{\times}. \tag{7.7)(a}$$

Using this homomorphism, we can pull back the connected double cover of $\mathbb{C}^{\times}$ to a double cover of $U(h)$:

$$\tilde{U}(h) = \{ (g, z) \in U(h) \times \mathbb{C}^{\times} \mid \det(g) = z^2 \} \tag{7.7)(b}$$

This is called the *square root of the determinant covering*, because projection on the second factor defines a homomorphism

$$\det{}^{1/2} : \tilde{U}(h) \to \mathbb{C}^{\times}. \tag{7.7)(c}$$

whose square is precisely det. It turns out that this (delightfully simple) covering is precisely the one induced by the (delightfully complicated) metaplectic covering $Mp(\omega)$. The next lemma shows how to compute with such coverings.

Lemma 7.8. *Suppose G is a real Lie group, and γ is a one-dimensional character of G. Let $\tilde{G}$ denote the square root of γ covering of G (cf. (7.7)), and ϵ the non-trivial element of the kernel of the covering map. Define an admissible representation χ of $\tilde{G}$ to be one trivial on the identity component, satisfying $\chi(\epsilon) = -1$. Define a γ-admissible representation χ_0 of G to be one whose differential is half the differential of γ:*

$$\chi_0(\exp(x)) = \gamma(\exp(x/2)) \cdot I \qquad (x \in \mathfrak{g})$$

Then there is a natural bijection between admissible representations of $\tilde{G}$ and γ-admissible representations of G.

Proof. We use the one-dimensional character $\gamma^{1/2}$ of $\tilde{G}$ (cf. (7.7)(c)). Tensoring with $\gamma^{1/2}$ sends admissible representations of $\tilde{G}$ to (the pullbacks to $\tilde{G}$ of) γ-admissible representations of G. $\square$

Corollary 7.9 ([Sc]). *Suppose $G_{\mathbb{R}}$ is a real reductive Lie group, and $\lambda_{i\mathbb{R}} \in \mathcal{N}_{i\mathbb{R}}^*$ is a purely imaginary nilpotent linear functional. Fix a maximal compact subgroup H of the isotropy group $G_{\mathbb{R}}(\lambda_{i\mathbb{R}})$. Choose an H-invariant complex structure and hermitian form h on the symplectic vector space $\mathfrak{g}_{\mathbb{R}}/\mathfrak{g}_{\mathbb{R}}(\lambda_{i\mathbb{R}})$ as in (7.7); this is possible since the compact subgroup $p(\lambda_{i\mathbb{R}})(H)$ of the symplectic group must be contained in some maximal compact subgroup. Now define γ to be the corresponding complex determinant character of H. Then the admissible representations of $\tilde{G}_{\mathbb{R}}(\lambda_{i\mathbb{R}})$ (Definition 7.2) are in natural one-to-one correspondence with the γ-admissible representations of H (Lemma 7.8). In particular, $\lambda_{i\mathbb{R}}$ is admissible if and only if the restriction of γ to the identity component H_0 of H is the square of another character of H_0.*

This follows at once from Lemmas 7.6 and 7.8, and Observation 7.4. It is still short of complete information about admissibility, for the character γ is difficult to compute explicitly from this description. Nevertheless we can treat some illustrative examples.

Example 7.10 (see [Sc]). Suppose $G_{\mathbb{R}}$ is the symplectic group $Sp(2n, \mathbb{R})$ for the standard symplectic form ω on $\mathbb{R}^{2n} = \mathbb{R}^n \times \mathbb{R}^n$: in terms of the usual dot product, this is

$$\omega((x,y),(x',y')) = x \cdot y' - y \cdot x'.$$

Then $G_{\mathbb{R}}$ consists of $2n \times 2n$ matrices $\begin{pmatrix} A & B \\ C & D \end{pmatrix}$ such that A^tB and D^tC are symmetric, and $A^tD - B^tC = I$. Its Lie algebra consists of matrices $\begin{pmatrix} X & Y \\ Z & -{}^tX \end{pmatrix}$ with Y and Z symmetric.

Fix non-negative integers p, q, and r such that $p + q + r = n$. We are going to define a nilpotent linear functional $\lambda_{i\mathbb{R}}(p,q,r)$. With obvious notation for matrix entries, it is

$$\lambda_{i\mathbb{R}}(p,q,r)\begin{pmatrix} X & Y \\ Z & -{}^tX \end{pmatrix} = i\sum_{k=1}^{p} y_{kk} - i\sum_{k=p+1}^{p+q} y_{kk}.$$

Using the explicit description of the group given above, it is not too hard to compute the isotropy group $G_{\mathbb{R}}(\lambda_{i\mathbb{R}}(p,q,r))$ explicitly. The answer is most conveniently expressed in terms of the semidirect product decomposition (Lemma 6.10). The reductive factor is $O(p,q) \times Sp(2r, \mathbb{R})$ (embedded in $Sp(2n, \mathbb{R})$ in an "obvious" way which we encourage the reader to untangle). The unipotent normal subgroup is two-step nilpotent; its Lie algebra $\mathfrak{u}_{\mathbb{R}}(\lambda_{i\mathbb{R}}(p,q,r))$ consists of matrices of the form $\begin{pmatrix} A & 0 \\ C & -{}^tA \end{pmatrix}$, subject to the conditions

$$a_{kl} = 0 \text{ unless } k > p+q, l \leq p+q, C = {}^tC, \text{ and } c_{kl} = 0 \text{ if } k,l > p+q.$$

Consequently the maximal compact subgroup H of the isotropy group is $O(p) \times O(q) \times U(r)$.

To compute the character γ, we need to understand something about the symplectic vector space

$$V(p,q,r) = \mathfrak{g}_{\mathbb{R}}/\mathfrak{g}_{\mathbb{R}}(\lambda_{i\mathbb{R}}(p,q,r))$$

The dimension of $V(p,q,r)$ is easily computed from the description of the isotropy subgroup; it is $2r(p+q)+(p+q)(p+q+1)$. This formula suggests a description of $V(p,q,r)$ as a symplectic representation of $O(p,q)\times Sp(2r,\mathbb{R})$, which is not too difficult to verify:

$$V(p,q,r) \simeq (\mathbb{R}^{2r}\otimes\mathbb{R}^{p+q}) \oplus (S^2(\mathbb{R}^{p+q}) \oplus S^2(\mathbb{R}^{p+q})^*) .$$

Here the first summand carries the tensor product of the symplectic form on the first factor with the orthogonal form on the second; this is a symplectic form preserved by the product of the groups of the small forms. The second summand is the sum of a group representation and its dual; it therefore carries a natural group-invariant symplectic form. (The $Sp(2r,\mathbb{R})$ factor acts trivially on the second summand.)

We can now express the character γ of H (Corollary 7.9) in terms of the standard determinant characters of the factors $U(r)$, $O(p)$, and $O(q)$:

$$\gamma = (\det_{U(r)})^{p+q} \otimes (\det_{O(p)})^{n-1} \otimes (\det_{O(q)})^{n-1}.$$

On the identity component only the first factor is non-trivial. We conclude that $\lambda_{i\mathbb{R}}(p,q,r)$ is admissible if and only if either $p+q$ is even or $r=0$. In this case (Corollary 7.9 again) the number of irreducible admissible representations – that is, the number of inequivalent orbit data – is 4 if p and q are both non-zero, 2 if exactly one is non-zero, and 1 if both are zero. (This last case corresponds to the trivial orbit datum.)

It turns out that all the linear functionals $\lambda_{i\mathbb{R}}(p,q,r)$ are admissible for the metaplectic covering $Mp(2n,\mathbb{R})$. Schwartz gives many examples of nilpotent orbits for $Sp(2n,\mathbb{R})$ that are inadmissible for all coverings, however.

The next theorem relates the character γ of Corollary 7.9 to the corresponding K-orbit on $\mathcal{N}_{\mathfrak{k}}^*$ (Theorem 6.2).

Theorem 7.11 ([Sc]). *In the setting of Theorem 6.2, suppose that the element $\lambda_{\mathfrak{k}} \in \mathcal{N}_{\mathfrak{k}}^*$ corresponds to $\lambda_{i\mathbb{R}} \in \mathcal{N}_{i\mathbb{R}}^*$. Write $\gamma_{i\mathbb{R}}$ for the character of the maximal compact subgroup of $G_{\mathbb{R}}(\lambda_{i\mathbb{R}})$ (Corollary 7.9). Define a character $\gamma_{\mathfrak{k}}$ of $K(\lambda_{\mathfrak{k}})$ by*

$$\gamma_{\mathfrak{k}}(k) = \det(\mathrm{Ad}(k))\,|_{(\mathfrak{k}/\mathfrak{k}(\lambda))^*} .$$

Then the restriction of $\gamma_{\mathfrak{k}}$ to the maximal compact subgroup is identified (by the isomorphism of Theorem 6.2) with $\gamma_{i\mathbb{R}}$.

The character $\gamma_{\mathfrak{k}}$ gives the action of $K(\lambda_{\mathfrak{k}})$ on top degree differential forms on $K\cdot\lambda_{\mathfrak{k}}$, at the point $\lambda_{\mathfrak{k}}$. Because of the symplectic structure, this is dual to the action on the top exterior power of the conormal bundle $T^*_{K\cdot\lambda_\mathfrak{k}}(G\cdot\lambda_\mathfrak{k})$ of $K\cdot\lambda_\mathfrak{k}$ in the (complex symplectic) variety $G\cdot\lambda_\mathfrak{k}$, at the point $\lambda_\mathfrak{k}$. By the proof of Proposition 5.19, this can be phrased as

$$\gamma_\mathfrak{k}(k) = \det(\mathrm{Ad}(k))\,|_{(\mathfrak{p}/\mathfrak{p}(\lambda))}\,. \tag{7.12}$$

Before giving the proof of Theorem 7.11, we record a consequence. It is convenient to make a definition parallel to Definition 7.2.

Definition 7.13. Suppose $(\mathfrak{g}, K)$ is a reductive symmetric pair (Definition 5.21), and $\lambda_\mathfrak{k} \in \mathcal{N}^*_\mathfrak{k}$. Define a character $\gamma_\mathfrak{k}$ of $K(\lambda_\mathfrak{k})$ as in Theorem 7.11 and (7.12). An algebraic representation χ of $K(\lambda_\mathfrak{k})$ is called *admissible* if its differential is half the differential of $\gamma_\mathfrak{k}$; that is, if

$$\chi(\exp(x)) = \gamma_\mathfrak{k}(\exp(x/2))\cdot I$$

for all $x \in \mathfrak{k}(\lambda_\mathfrak{k})$. If admissible representations exist, we say that $\lambda_\mathfrak{k}$ (or the orbit $K\cdot\lambda_\mathfrak{k}$) is admissible. A pair $(\lambda_\mathfrak{k}, \chi)$ consisting of an element of $\mathcal{N}^*_\mathfrak{k}$ and an irreducible admissible representation of $K(\lambda_\mathfrak{k})$ is called a *nilpotent admissible K-orbit datum.* Two such are equivalent if they are conjugate by K.

Theorem 7.14 ([Sc]). *Suppose $G_\mathbb{R}$ is a real reductive Lie group, $K_\mathbb{R}$ is a maximal compact subgroup, and K is its complexification. Then there is a natural bijection between equivalence classes of nilpotent admissible $G_\mathbb{R}$-orbit data (Definition 7.2) and equivalence classes of nilpotent admissible K-orbit data (Definition 7.13).*

Proof. Fix a nilpotent linear functional $\lambda_{i\mathbb{R}} \in \mathcal{N}^*_{i\mathbb{R}}$, and let $\lambda_\mathfrak{k} \in \mathcal{N}^*_\mathfrak{k}$ be a corresponding element (Theorem 6.2). We want to associate to each admissible representation $\chi_{i\mathbb{R}}$ of $\tilde{G}_\mathbb{R}(\lambda_{i\mathbb{R}})$ an admissible representation $K(\lambda_\mathfrak{k})$. Let H be a maximal compact subgroup of $G_\mathbb{R}(\lambda_{i\mathbb{R}})$. By Theorem 6.2, we may as well assume that H is also a maximal compact subgroup of $K(\lambda)$. Let γ be the character of H constructed in Corollary 7.9 (or, by Theorem 7.11, the restriction of $\gamma_\mathfrak{k}$ to H). By Lemma 7.8, restriction to the preimage of H and twisting by the "square root of γ" defines a bijection from admissible representations of $\tilde{G}_\mathbb{R}(\lambda_{i\mathbb{R}})$ to γ-admissible representations of H. An even simpler fact about algebraic groups (essentially Weyl's "unitarian trick") guarantees that restriction to H is a bijection from admissible representations of $K(\lambda_\mathfrak{k})$ to γ-admissible representations of H. (An admissible representation of $K(\lambda_\mathfrak{k})$ is automatically trivial on the unipotent radical, so this is really just a statement about reductive groups.) The theorem follows. □

Proof of Theorem 7.11. Fix a nondegenerate symmetric real bilinear form b on $\mathfrak{g}_\mathbb{R}$, invariant under $G_\mathbb{R}$ and θ, negative definite on $\mathfrak{k}_\mathbb{R}$, and positive definite on $\mathfrak{p}_\mathbb{R}$, as in the proof of Theorem 6.2. Then the bilinear form

$$b_\theta(u,v) = b(\theta u, v) = b(u, \theta v) \tag{7.15)(a}$$

is negative definite on $\mathfrak{g}_{\mathbb{R}}$. Define elements $x_{i\mathbb{R}}$ and $x_{\mathfrak{k}}$ of $\mathfrak{g}$ by the requirements

$$\lambda_{i\mathbb{R}}(y) = b(x_{i\mathbb{R}}, y), \qquad \lambda_{\mathfrak{k}}(y) = b(x_{\mathfrak{k}}, y) \tag{7.15)(b}$$

for all $y \in \mathfrak{g}$. By Theorem 6.4, we may as well assume that there is a homomorphism ϕ from $\mathfrak{sl}(2)$ to $\mathfrak{g}$, intertwining σ_0 with σ and θ_0 with θ, and satisfying

$$\phi\begin{pmatrix} 0 & i \\ 0 & 0 \end{pmatrix} = x_{i\mathbb{R}}, \qquad \phi\begin{pmatrix} 1/2 & -i/2 \\ -i/2 & -1/2 \end{pmatrix} = x_{\mathfrak{k}}. \tag{7.15)(c}$$

In terms of ϕ, we can give two descriptions of the compact group H:

$$H = \{\, g \in G_{\mathbb{R}} \mid \mathrm{Ad}(g)(x_{i\mathbb{R}}) = x_{i\mathbb{R}}, \theta g = g \,\}; \tag{7.15)(d}$$

$$H = \{\, k \in K \mid \mathrm{Ad}(k)(x_{\mathfrak{k}}) = x_{\mathfrak{k}}, \sigma k = k \,\}. \tag{7.15)(e}$$

The symplectic structure $\omega = \omega_{x_{i\mathbb{R}}}$ on $V = \mathfrak{g}_{\mathbb{R}}/\mathfrak{g}_{\mathbb{R}}(x_{i\mathbb{R}})$ is defined by

$$\omega(u, v) = b(x_{i\mathbb{R}}, [u, v]). \tag{7.16)(a}$$

In order to calculate the character $\gamma_{i\mathbb{R}}$, we must (according to Corollary 7.9) construct a complex structure J on V; that is, a linear transformation satisfying $J^2 = -I$. In addition, J must commute with the action of H, and satisfy

$$\omega(Ju, v) = \omega(Jv, u) \tag{7.16)(b}$$

$$(1/i)\omega(Ju, u) \geq 0. \tag{7.16)(c}$$

This makes V into a complex vector space, and H acts by complex-linear transformations. (Although we will not use it, the hermitian form h of (7.7) is

$$h(u, v) = (1/i)\omega(Ju, v) + \omega(u, v). \tag{7.16)(d}$$

The action of H is unitary for this hermitian form.) The character $\gamma_{i\mathbb{R}}$ is the determinant of the action of H on this complex vector space. Equivalently, it is the determinant of the action of H on the $+i$-eigenspace of J in the space $V_{\mathbb{C}} = \mathfrak{g}/\mathfrak{g}(x_{i\mathbb{R}})$.

We begin now the construction of J. Define

$$E = \phi\begin{pmatrix} 0 & 1 \\ 0 & 0 \end{pmatrix}, \qquad F = \phi\begin{pmatrix} 0 & 0 \\ 1 & 0 \end{pmatrix} = -\theta E. \tag{7.17)(a}$$

These are elements of $\mathfrak{g}_{\mathbb{R}}$, and $iE = x_{i\mathbb{R}}$. If T is a linear transformation of $\mathfrak{g}_{\mathbb{R}}$, write T^{θ} for its adjoint with respect to the negative definite form b_{θ} of (7.15)(a):

$$b_{\theta}(Tu, v) = b_{\theta}(u, T^{\theta} v).$$

Obviously θ is self-adjoint, and

$$\mathrm{ad}(x)^{\theta} = -\mathrm{ad}(\theta x). \tag{7.17)(b}$$

Our first approximation to J is the linear transformation

$$Q = \theta \circ \operatorname{ad} E = -\operatorname{ad} F \circ \theta. \tag{7.17)(c}$$

Obviously the kernel of Q is precisely $\mathfrak{g}_{\mathbb{R}}(x_{i\mathbb{R}})$, and it follows from (7.17)(b) that Q is skew-adjoint with respect to b_θ: $Q^\theta = -Q$. A first consequence is that Q defines an invertible linear transformation $\overline{Q}$ on V. A second is that $R = -Q^2$ is a non-negative self-adjoint linear transformation on $\mathfrak{g}_{\mathbb{R}}$. By (7.17)(c),

$$R = \operatorname{ad} F \circ \operatorname{ad} E. \tag{7.17)(d}$$

Clearly Q commutes with the action of H. By (7.16),

$$\begin{aligned} \omega(Qu, v) &= b(iE, [\theta[E,u], v]) \\ &= -ib(\theta[E,u], [E,v]) \\ &= -ib_\theta([E,u],[E,v]). \end{aligned} \tag{7.18)(a}$$

From the last expression it follows that

$$\omega(Qu, v) = \omega(Qv, u), \qquad (1/i)\omega(Qu, u) \geq 0. \tag{7.18)(b}$$

These assertions correspond to the requirements (7.16)(b) and (c) for J. The only difficulty is that $\overline{Q}^2$ is not $-I$, but only the negative operator $-\overline{R}$. We correct this using a square root in the usual way. Let S denote the non-negative self-adjoint square root of R. Then S defines an invertible linear transformation $\overline{S}$ on V, and we set

$$J = (\overline{S})^{-1}\overline{Q}. \tag{7.18)(c}$$

The correction factor commutes with all operators commuting with $\overline{Q}$, including $\overline{Q}$ itself and the adjoint action of H; so $J^2 = -I$, and J commutes with H. Now (7.18)(b) implies that $\omega(-Q^2u, v) = \omega(u, -Q^2v)$. Considering this equation on each eigenspace of $-Q^2$, we deduce that

$$\omega(Su, v) = \omega(u, Sv).$$

Now a trivial formal argument gives (7.16)(b) and (c) from (7.18)(b).

In light of the remarks after (7.16)(d), and the construction just given for J, we find that the character $\gamma_{i\mathbb{R}}$ of H may be described as the determinant of the adjoint action on the span of the positive eigenspaces of

$$(1/i)Q = -\theta \circ ad(x_{i\mathbb{R}}); \tag{7.19}$$

this operator may be taken to act on all of $\mathfrak{g}$, or just on the quotient $V_{\mathbb{C}} = \mathfrak{g}/\mathfrak{g}(x_{i\mathbb{R}})$. (It is a hermitian operator with respect to the hermitian form obtained from b_θ; this is another way to understand the fact that it is diagonalizable, with real eigenvalues.) Our next task is to relate these positive eigenspaces to $\mathfrak{k}/\mathfrak{k}(x_\mathfrak{k})$.

We begin by constructing a Cayley transform. Even though the homomorphism ϕ may not exponentiate to SL(2), its composition with ad does.

Every element g of $SL(2,\mathbb{C})$ therefore gives rise to an automorphism of $\mathfrak{g}$, which we write as $\mathrm{Ad}(\phi(g))$ (even though $\phi(g)$ by itself is undefined). These automorphisms commute with the action of H. Set

$$c = (1/\sqrt{2})\begin{pmatrix} i & -1 \\ 1 & -i \end{pmatrix}. \qquad (7.20)(a)$$

By calculation in SL(2), we find that

$$\mathrm{Ad}(\phi(c))(x_{i\mathbb{R}}) = x_{\mathfrak{k}}. \qquad (7.20)(b)$$

Define

$$T = \mathrm{Ad}(\phi(c)) \circ (1/i)Q \circ \mathrm{Ad}(\phi(c))^{-1}. \qquad (7.20)(c)$$

Then T may be regarded as a linear transformation on $\mathfrak{g}$ or on $\mathfrak{g}/\mathfrak{g}(x_{\mathfrak{k}})$. It is diagonalizable with real eigenvalues, and the character $\gamma_{i\mathbb{R}}$ of H is the determinant of the adjoint action on the span of the positive eigenspaces. To calculate T, we need to know how θ acts on $\mathrm{Ad}(\phi(c))$. This can be computed from the fact that ϕ intertwines θ_0 and θ. Using (7.20)(b) and (7.19), we find (after a little calculation in SL(2))

$$T = -\theta \circ \mathrm{Ad}(\phi\begin{pmatrix} 0 & -i \\ -i & 0 \end{pmatrix}) \circ ad(x_{\mathfrak{k}}) = \mathrm{Ad}(\phi\begin{pmatrix} 0 & i \\ i & 0 \end{pmatrix}) \circ ad(x_{\mathfrak{k}}) \circ \theta. \quad (7.20)(d)$$

Write $W = \mathfrak{g}/\mathfrak{g}(x_{\mathfrak{k}})$, W^+ for the sum of the positive eigenspaces of T on W, and W^- for the sum of the negative eigenspaces. We have

$$W = W^+ \oplus W^- = (W \cap \mathfrak{p}) \oplus (W \cap \mathfrak{k}). \qquad (7.21)(a)$$

The theorem we are trying to prove says that the determinants of the actions of H on W^+ and $W \cap \mathfrak{k}$ agree. Obviously it suffices to prove that

$$W^+ \simeq W \cap \mathfrak{p} \qquad (7.21)(b)$$

as representations of H. In order to do this, we need another decomposition of W. Define $\zeta = \mathrm{Ad}(\phi(-I))$, an involutive automorphism of $\mathfrak{g}$. Then ζ commutes with θ, T, and the image of ϕ. It therefore lifts to a linear transformation of order 2 on W (still called ζ), commuting with everything else. Write

$$W = W_e \oplus W_o \qquad (7.21)(c)$$

for the decomposition into the +1 ("even") and -1 ("odd") eigenspaces of ζ. We use analogous notation for other spaces and operators; thus for example W_o^+ is the sum of the positive eigenspaces of T_o. The desired isomorphism (7.21)(b) would follow from two separate isomorphisms

$$W_e^+ \simeq W_e \cap \mathfrak{p}, \qquad (7.21)(d)$$

and

$$W_o^+ \simeq W_o \cap \mathfrak{p}. \qquad (7.21)(e)$$

These seem to require rather different treatments, and we will prove them separately.

First we consider (7.21)(d). It follows from (7.20)(d) that T and θ anticommute on W_e:

$$T_e\theta_e = -\theta_e T_e. \tag{7.22)(a}$$

Consequently θ_e interchanges W_e^+ and W_e^-, and T_e interchanges $W_e \cap \mathfrak{k}$ and $W_e \cap \mathfrak{p}$. It follows immediately that all four representations of H are isomorphic:

$$W_e^+ \simeq W_e^- \simeq W_e \cap \mathfrak{k} \simeq W_e \cap \mathfrak{p}. \tag{7.22)(b}$$

The first isomorphism is given by θ_e; the second by restricting to W_e^- the projection $(I+\theta_e)/2$; and the third by T_e. In particular, we get (7.21)(d).

For (7.21)(e), we need to decompose W further using the action $\mathrm{ad}\circ\phi$ of $\mathfrak{sl}(2)$. Fix a non-negative integer N (the highest weight), and write $S(N)$ for the irreducible representation of $\mathfrak{sl}(2)$ of dimension $N+1$ (realized say on homogeneous polynomials of degree N in two variables). Define

$$\mathfrak{g}(N) = \text{ sum of all copies of } S(N) \text{ in } \mathfrak{g}, \tag{7.23)(a}$$

the corresponding isotypic subspace of $\mathfrak{g}$. We have

$$\mathfrak{g}_o = \sum_{N\ odd} \mathfrak{g}(N) \tag{7.23)(b}$$

Recall from (6.7)(c) the element

$$h_{\mathfrak{k}} = \phi\begin{pmatrix} 0 & -i \\ i & 0 \end{pmatrix} \in \mathfrak{k}.$$

The eigenvalues of $\left(\begin{smallmatrix} 0 & -i \\ i & 0 \end{smallmatrix}\right)$ on $S(N)$ are $N, N-2, \ldots, -N+2, -N$, each occurring once. Write $S(N)_m$ for the m-eigenspace, and

$$\mathfrak{g}(N)_m = \ m\text{-eigenspace of } h_{\mathfrak{k}} \text{ in } \mathfrak{g}(N). \tag{7.23)(c}$$

We have

$$\mathrm{ad}(x_{\mathfrak{k}}) : \mathfrak{g}(N)_m \to \mathfrak{g}(N)_{m+2}\,; \tag{7.23)(d}$$

this map is an isomorphism unless $m = N$, in which case it is zero. Because it also interchanges $\mathfrak{k}$ and $\mathfrak{p}$, and commutes with the action of H, we deduce that

$$\mathfrak{g}(N)_m \cap \mathfrak{k} \simeq \mathfrak{g}(N)_{m+2} \cap \mathfrak{p} \qquad (m \neq N), \tag{7.23)(e}$$

and similarly with $\mathfrak{k}$ and $\mathfrak{p}$ exchanged. These facts pass at once to W. Writing $W(N)_m$ for the image of $\mathfrak{g}(N)_m$ in W, we find that

$$W = \sum_{N \neq m} W(N)_m, \qquad W(N)_m \simeq \mathfrak{g}(N)_m \quad (N \neq m). \tag{7.23)(f}$$

All of these spaces are preserved by θ, and

$$T : W(N)_m \to W(N)_{-m-2}. \tag{7.23)(g}$$

In particular, $W(N)$ is T-invariant, so we may speak of $W(N)^+$ and $W(N)^-$. The isomorphism (7.21)(e) will follow if we can show that

$$W(N)^+ \simeq W(N) \cap \mathfrak{p} \tag{7.23)(h}$$

for every odd non-negative integer N.

So fix such an integer $N = 2j+1$. We first compute the right side of (7.23)(h). Define

$$A(N) = W(N)_{-1} \cap \mathfrak{k}, \tag{7.24)(a}$$

a representation of H. The bilinear form b induces a natural isomorphism

$$(\mathfrak{g}(N)_{-1} \cap \mathfrak{p})^* \simeq \mathfrak{g}(N)_1 \cap \mathfrak{p}. \tag{7.24)(b}$$

Because of (7.23)(e), it follows that

$$A(N)^* \simeq W(N)_{-1} \cap \mathfrak{p}, \tag{7.24)(c}$$

In the same way we may compute all the spaces $W(N)_m \cap \mathfrak{k}$ and $W(N)_m \cap \mathfrak{p}$ in terms of $A(N)$ and $A(N)^*$:

$$W(N)_m \cap \mathfrak{k} \simeq A(N) \quad \text{and} \quad W(N)_m \cap \mathfrak{p} \simeq A(N)^* \quad \text{when } m \equiv -1 \pmod 4 \tag{7.24)(d}$$

$$W(N)_m \cap \mathfrak{k} \simeq A(N)^* \quad \text{and} \quad W(N)_m \cap \mathfrak{p} \simeq A(N) \quad \text{when } m \equiv 1 \pmod 4 \tag{7.24)(e}$$

In particular,

$$\begin{aligned} W(N) \cap \mathfrak{p} &\simeq (A(N) \oplus A(N)^*)^j \oplus A(N)^* \quad (N \equiv 1 \pmod 4) \\ &\simeq (A(N) \oplus A(N)^*)^j \oplus A(N) \quad (N \equiv 3 \pmod 4) \end{aligned} \tag{7.24)(f}$$

Next, we compute the left side of (7.23)(h). Fix $m > -1$. Because of (7.23)(g), the sum of the positive eigenspaces of T on $W(N)_m \oplus W(N)_{-m-2}$ is isomorphic (by projection on the first summand) to $W(N)_m$. By (7.24), this space is isomorphic to $A(N) \oplus A(N)^*$. Consequently

$$W(N)^+ \simeq (A(N) \oplus A(N)^*)^j \oplus W(N)^+_{-1}. \tag{7.25)(a}$$

Recall now the formula (7.20)(d) for T. Comparing it with (7.24)(f) and (7.25)(a), we find that the desired isomorphism (7.23)(h) follows from the fact that

$$(-1)^j \operatorname{Ad}\left(\phi\begin{pmatrix} 0 & -i \\ -i & 0 \end{pmatrix}\right) \circ ad(x_\mathfrak{k})$$

acts by a positive scalar on $W(N)_{-1}$. (This is an assertion about finite-dimensional representations of $\mathfrak{sl}(2)$, and is easily verified by direct calculation.) This proves (7.23)(h), and therefore (7.21)(e), and therefore (7.21)(b), and therefore the theorem. □

The proof has some useful consequences. Because W_e and W_o are perpendicular with respect to the symplectic form $\omega_{x_{\mathfrak{k}}}$ on W, each of them separately is a symplectic vector space. This structure is preserved by H, so the determinant of H acting on W_e is one. The isomorphisms (7.22)(b) imply that this determinant is the square of the determinant on (say) $W_e \cap \mathfrak{k}$. This last determinant is therefore ± 1:

$$\det(\mathrm{Ad}(x)) \text{ on } (W_e \cap \mathfrak{p}) = \pm 1. \tag{7.26)(a}$$

In particular, this determinant is trivial on the identity component of H. On the odd part of W, the determinant characters of H on $A(N)$ and $A(N)^*$ differ by the square of a character. It therefore follows from (7.24)(f) that there is a character γ_1 of H such that (as characters of H)

$$\det \circ \mathrm{Ad} \text{ on } (W_o \cap \mathfrak{p}) = [\det \circ \mathrm{Ad} \text{ on } (W_{-1} \cap \mathfrak{p})][\gamma_1]^2. \tag{7.26)(b}$$

In light of (7.26)(a), we can drop the subscript o on the identity component of H:

$$\gamma_{\mathfrak{k}} |_{H_0} = [\det \circ \mathrm{Ad} \text{ on } (W_{-1} \cap \mathfrak{p})][\gamma_1]^2. \tag{7.26)(c}$$

Corollary 7.27 ([Sc]). *In the setting of (6.7), let H be a maximal compact subgroup of the centralizer $K(x_{\mathfrak{k}})$. Then the nilpotent element $x_{\mathfrak{k}}$ is admissible (Defn. 7.13) if and only if the determinant of the action of H_0 on the -1-eigenspace of $h_{\mathfrak{k}}$ on $\mathfrak{p}$ is the square of another character of H_0.*

Recall that a nilpotent element is called *even* if the corresponding semisimple element $h_{\mathfrak{k}}$ (or, equivalently, $h_{i\mathbb{R}}$) has only even eigenvalues. This means that the representation $\mathrm{ad} \circ \phi$ of $\mathfrak{sl}(2)$ has only odd-dimensional irreducible constituents. In the classical groups, a nilpotent element is even if and only if all its Jordan block sizes have the same parity. For an even nilpotent the space W_{-1} is zero, so we deduce

Corollary 7.28 ([Sc]). *Every even nilpotent element is admissible. For such an element, the character of Theorem 7.11 takes values in $\{\pm 1\}$.*

8. Representations attached to admissible orbits.

Recall that one of our goals is a better understanding of what it means for a representation of a reductive group $G_{\mathbb{R}}$ to be attached to a nilpotent imaginary orbit $G_{\mathbb{R}} \cdot \lambda_{i\mathbb{R}}$. We have asserted in section 7 that one should actually try to attach a representation to a $G_{\mathbb{R}}$-admissible orbit datum $(\lambda_{i\mathbb{R}}, \chi_{i\mathbb{R}})$. In section 6, we saw that $\lambda_{i\mathbb{R}}$ corresponds to a nilpotent element $\lambda_{\mathfrak{k}} \in (\mathfrak{g}/\mathfrak{k})^*$. In section 7, we saw that $\chi_{i\mathbb{R}}$ corresponds to a representation $\chi_{\mathfrak{k}}$ of the isotropy group $K(\lambda_{\mathfrak{k}})$, having the following property. Write $\gamma_{\mathfrak{k}}$ for the determinant character of $K(\lambda_{\mathfrak{k}})$ on

$$T^*_{\lambda_{\mathfrak{k}}}(K \cdot \lambda_{\mathfrak{k}}) \simeq (\mathfrak{k}/\mathfrak{k}(\lambda_{\mathfrak{k}}))^* \simeq \mathfrak{p}/\mathfrak{p}(\lambda_{\mathfrak{k}}). \tag{8.1)(a}$$

Then we require

$$\chi_{\mathfrak{k}}(x^2) = \gamma_{\mathfrak{k}}(x) \cdot I \tag{8.1)(b}$$

for all x in the identity component of $K(\lambda_{\mathfrak{k}})$. On the other hand, suppose X is a $(\mathfrak{g}, K)$-module of finite length. By Theorem 2.13 (and Corollary 5.23) one can associate to X a certain finite set of K-orbits

$$K \cdot \lambda_1, \ldots, K \cdot \lambda_s \subset \mathcal{N}_{\mathfrak{k}}^* \tag{8.1)(c}$$

and (genuine virtual) representations χ_i of $K(\lambda_i)$. It is therefore natural to impose the following requirement on the (still undefined) process of attaching representations to nilpotent orbits.

Desideratum 8.2. *Suppose $G_{\mathbb{R}}$ is a real reductive group, with complexified maximal compact subgroup K. Suppose $\mathcal{D} = (\lambda_{i\mathbb{R}}, \chi_{i\mathbb{R}})$ is a nilpotent admissible $G_{\mathbb{R}}$-orbit datum, and $(\lambda_{\mathfrak{k}}, \chi_{\mathfrak{k}})$ is a corresponding nilpotent admissible K-orbit datum. If X is a $(\mathfrak{g}, K)$-module attached to $\mathcal{D}$, then $(\lambda_{\mathfrak{k}}, \chi_{\mathfrak{k}})$ should be (up to K-conjugacy) one of the pairs (λ_i, χ_i) attached to X by Theorem 2.13 (cf. (8.1)(c)).*

This is a rather weak requirement. In the case of the principal nilpotent orbit for a quasisplit group, it allows whole translation families of fundamental series representations. (According to Desideratum 7.3, there should be at most finitely many representations attached to an orbit datum.) On the other hand, if $G_{\mathbb{R}}$ is semisimple, this requirement alone correctly attaches only the trivial representation to the trivial orbit datum $(0, \mathbb{C})$. In general the requirement is stronger for smaller orbits. Since it is the representations attached to small orbits that are the most troublesome technically, it is worthwhile to pursue Desideratum 8.2. We are therefore led to the problem that is the main concern of this section.

Problem 8.3. *Given an orbit $K \cdot \lambda \subset \mathcal{N}_{\mathfrak{k}}^*$, find conditions on a Harish-Chandra module X guaranteeing that*

a) *$K \cdot \lambda$ contains a component of $\mathcal{V}(X)$, and*
b) *the corresponding isotropy representation $\chi(\lambda, X)$ (Theorem 2.13) is admissible (Definition 7.13).*

Towards part (a) of Problem 8.3, we will prove only the following result.

Theorem 8.4. *Suppose (G, K) is a reductive symmetric pair of Harish-Chandra class (Definition 5.21), and X is an irreducible $(\mathfrak{g}, K)$-module. Write*

$$J = \operatorname{Ann} X \subset U(\mathfrak{g}),$$

a primitive ideal in $U(\mathfrak{g})$. Let $\mathcal{O} \subset \mathcal{V}(J) \subset \mathfrak{g}^*$ be the dense nilpotent G-orbit (Corollary 4.7).

a) *$\mathcal{V}(X) \subset \mathcal{V}(J) \cap (\mathfrak{g}/\mathfrak{k})^*$.*
b) *$\mathcal{O} \cap (\mathfrak{g}/\mathfrak{k})^*$ is the union of a finite number of K-orbits $\mathcal{O}_1, \ldots, \mathcal{O}_r$, each of which has dimension equal to half the dimension of $\mathcal{O}$.*
c) *Some of the $\mathcal{O}_i$ are contained in $\mathcal{V}(X)$; they are precisely the K-orbits of maximal dimension in $\mathcal{V}(X)$.*

Proof. Part (a) is an immediate consequence of the definitions: elements of $\operatorname{gr} J$ (defined using the standard filtration of $U(\mathfrak{g})$) obviously annihilate

$\operatorname{gr} X$, and therefore constrain $\mathcal{V}(X)$ according to (1.2)(a). Part (b) follows from Corollary 5.20. Now (c) is evidently equivalent to the assertion that the Gelfand-Kirillov dimension of X (section 2) is at least half that of $U(\mathfrak{g})/J$. Such a relationship holds for any faithful module for a primitive algebra; a proof of the equality in this special case appears in [V1, Lemma 3.4]. □

Part (a) of Problem 8.3 is therefore just slightly stronger than asking that

$$\mathcal{V}(\operatorname{Ann}(X)) = \overline{G \cdot \lambda}. \tag{8.5}$$

A great deal is known about how to check such a condition, but we will not review it here.

We turn now to part (b) of Problem 8.3. To formulate a result, we need to recall the *transpose anti-automorphism* $u \mapsto {}^t u$ of $U(\mathfrak{g})$. This is a linear map, characterized by the properties

$${}^t x = -x \quad (x \in \mathfrak{g}), \qquad {}^t(uv) = {}^t v\, {}^t u \quad (u, v \in U(\mathfrak{g})). \tag{8.6}$$

Theorem 8.7. *Suppose (G, K) is a reductive symmetric pair of Harish-Chandra class (Definition 5.21), and X is an irreducible $(\mathfrak{g}, K)$-module. Write*

$$J = \operatorname{Ann} X \subset U(\mathfrak{g}),$$

a primitive ideal in $U(\mathfrak{g})$. Let $\mathcal{O} \subset \mathcal{V}(J) \subset \mathfrak{g}^*$ be the dense nilpotent G-orbit (Corollary 4.7). Assume that

i) The $S(\mathfrak{g})$-module $S(\mathfrak{g})/\operatorname{gr} J$ is generically reduced along $\mathcal{O}$ (Proposition 2.9; this is automatic if $\operatorname{gr} J$ is prime, or more generally if $\mathcal{O}$ has multiplicity one in the characteristic cycle of $S(\mathfrak{g})/\operatorname{gr} J$).

ii) The ideal J is preserved by the transpose antiautomorphism of $U(\mathfrak{g})$ (cf. (8.6)).

Fix $\lambda \in \mathcal{O} \cap (\mathfrak{g}/\mathfrak{k})^*$, and write H for the corresponding isotropy subgroup of K. Then the character $\chi(\lambda, X)$ of X at λ (Theorem 2.13) is admissible (Definition 7.13).

The proof will occupy the next three sections. We will conclude this section with some easy differential geometry intended to provide motivation for the proof. (Since the results will not be applied directly, we will feel free to omit any inconvenient proofs.) The problem is to understand what kind of natural condition can force χ to be admissible. Suppose for a moment that instead of admissibility (which specifies the differential of χ in a slightly complicated way) we were seeking conditions that force the differential of χ to be zero. The following well-known result provides such conditions; we formulate it in the smooth context for the sake of familiarity.

Proposition 8.8. *Suppose $H \subset K$ are Lie groups, and (χ, V) is a finite-dimensional representation of H. Write $\mathcal{V}$ for the corresponding vector bundle on K/H. Then the following conditions are equivalent.*

a) The differential of χ is zero.
b) $\mathcal{V}$ has a K-invariant structure of local coefficient system on K/H.
c) Attached to every vector field ξ on K/H there is a differential operator L_ξ on the space $\Gamma\mathcal{V}$ of sections of $\mathcal{V}$. This correspondence is complex-linear, and satisfies
  i) $L_\xi(f \cdot \sigma) = (\partial_\xi f) \cdot \sigma + f \cdot (L_\xi \sigma) \quad (f \in C^\infty(K/H), \sigma \in \Gamma\mathcal{V})$.
  ii) $L_{f\xi}(\sigma) = f \cdot (L_\xi \sigma) \quad (f \in C^\infty(K/H), \sigma \in \Gamma\mathcal{V})$.
  iii) Suppose $x \in \mathfrak{k}$, and $\xi(x)$ is the corresponding vector field on K/H. Then $L_{\xi(x)}$ is the natural action of x on sections of $\mathcal{V}$.

In (c), condition (i) simply says that L_ξ is a first-order differential operator with symbol ξ (times an appropriate identity operator). This is a consequence of (ii) and (iii), but we include it to clarify the nature of L.

Condition (b) has been included only for motivation; we are not going to define local coefficient system carefully here. The expert reader will easily supply a proof of its equivalence with (a) or (c). We will prove only the equivalence of (a) and (c).

Proof. Suppose that (c) holds. Write $\mathfrak{m}$ for the ideal in $C^\infty(K/H)$ of functions vanishing at eH. There is a natural isomorphism

$$\Gamma\mathcal{V}/\mathfrak{m}\Gamma\mathcal{V} \simeq V;$$

this takes the restriction to $\mathfrak{h}$ of the natural action of $\mathfrak{k}$ to the differential of χ. If $x \in \mathfrak{h}$, then the vector field $\xi(x)$ vanishes at eH, and may therefore be written in the form

$$\xi(x) = \sum f_i \xi_i$$

with $f_i \in \mathfrak{m}$. If $v \in V$ is represented by a section $\sigma \in \Gamma\mathcal{V}$, then (by (iii)) $\chi(x)v$ is represented by

$$L_{\xi(x)}\sigma = \sum L_{f_i\xi_i}\sigma = \sum f_i L_{\xi_i}\sigma.$$

(Here we have used condition (ii).) This last expression belongs to $\mathfrak{m}\Gamma\mathcal{V}$, and therefore represents zero in V. Therefore the differential of χ is zero, which is (a).

The other direction is very easy, and we will be sketchy. Suppose (a) holds. Write H_0 for the identity component of H. Then K/H_0 is a covering of K/H, and the pullback $\mathcal{V}_0$ of $\mathcal{V}$ to this covering is the vector bundle corresponding to the representation $\chi|_{H_0}$ of H_0. By assumption this last bundle is trivial (in a K-invariant way): its sections are just functions on K/H_0 with values in V. It follows immediately that $\mathcal{V}_0$ has the structure required in (c). (L_ξ acts on vector-valued functions by acting on each coordinate separately.) But the structure in (c) is purely local; so its existence on the covering space K/H_0 immediately implies its existence on K/H. □

Next, we prove an analogue of Proposition 8.8 for the case of admissible representations of H.

Proposition 8.9. *Suppose $H \subset K$ are Lie groups, and (χ, V) is a finite-dimensional representation of H. Write $\mathcal{V}$ for the corresponding vector bundle on K/H. Define γ to be the determinant character of H acting on $(\mathfrak{k}/\mathfrak{h})^*$ (the cotangent space at eH to K/H. Fix a complex number k. Then the following conditions are equivalent.*

a) *The differential of χ is equal to k times the differential of γ.*

b) *Attached to every vector field ξ on K/H there is a differential operator L_ξ on the space $\Gamma\mathcal{V}$ of sections of $\mathcal{V}$. This correspondence is complex-linear, and satisfies*

   i) $L_\xi(f \cdot \sigma) = (\partial_\xi f) \cdot \sigma + f \cdot (L_\xi \sigma) \quad (f \in C^\infty(K/H),\ \sigma \in \Gamma\mathcal{V})$.

   ii) $L_{f\xi}(\sigma) = f \cdot (L_\xi \sigma) + k \cdot (\partial_\xi f) \cdot \sigma \quad (f \in C^\infty(K/H),\ \sigma \in \Gamma\mathcal{V})$.

   iii) *Suppose $x \in \mathfrak{k}$, and $\xi(x)$ is the corresponding vector field on K/H. Then $L_{\xi(x)}$ is the natural action of x on sections of $\mathcal{V}$.*

I do not know a good name for the kind of structure considered in (b); of course the second term in (ii) means that it is *not* a connection. The L is intended to suggest "Lie derivative;" the Lie derivative action of vector fields on volume forms on a manifold satisfies the conditions in (b) with $k = 1$.

Proof. Suppose that (b) holds. Write $\mathcal{T}$ for the tangent bundle of K/H, so that $\Gamma\mathcal{T}$ is the space of vector fields. Exactly as in the proof of Proposition 8.8, we have

$$\Gamma\mathcal{T}/\mathfrak{m}\Gamma\mathcal{T} \simeq T_{eH}(K/H). \tag{8.10)(a}$$

The commutator of two vector fields vanishing at eH again vanishes at eH. Consequently there is a well-defined action of vector fields vanishing at eH on the tangent space at eH:

$$A : \mathfrak{m}\Gamma\mathcal{T} \to \mathrm{End}(T_{eH}(K/H)), \quad A(\tau)(\xi(eH)) = [\tau, \xi](eH). \tag{8.10)(b}$$

If x belongs to $\mathfrak{h}$, then $\xi(x)$ vanishes at eH, and the action of $\xi(x)$ just described is just the adjoint action on $\mathfrak{k}/\mathfrak{h}$. Its trace is therefore the negative of the differential of γ at x:

$$\gamma(x) = -\mathrm{tr}\, A(\xi(x)) \quad (x \in \mathfrak{h}). \tag{8.10)(c}$$

To compute this trace, choose a set of vector fields so that $\xi_1(eH), \ldots, \xi_n(eH)$ is a basis of $T_{eH}(K/H)$. If τ is any vector field vanishing at eH, we can write (at least near eH)

$$\tau = \sum f_i \xi_i \tag{8.11)(a}$$

for some functions f_i vanishing at eH. Then

$$\begin{aligned} A(\tau)(\xi_j(eH)) &= \sum_i [f_i\xi_i, \xi_j](eH) \\ &= \sum_i (f_i\xi_i\xi_j - \xi_j f_i \xi_i)(eH) \\ &= \sum_i \left(f_i(\xi_i\xi_j - \xi_j\xi_i) - (\partial_{\xi_j} f_i)\xi_i\right)(eH) \\ &= \sum_i \left(f_i(eH)[\xi_i, \xi_j](eH) - (\partial_{\xi_j} f_i)(eH)\xi_i(eH)\right) \\ &= \sum_i (\partial_{\xi_j} f_i)(eH)\xi_i(eH) \end{aligned} \tag{8.11)(b}$$

Consequently

$$\operatorname{tr} A(\tau) = -\sum_i (\partial_{\xi_i} f_i)(eH). \tag{8.11)(c}$$

Now assume that (c) holds. We calculate the differential of χ as in the proof of Proposition 8.5, by identifying V with $\Gamma\mathcal{V}/\mathfrak{m}\Gamma\mathcal{V}$. Suppose that an element $v \in V$ is represented by a section σ, and that $x \in \mathfrak{h}$. Then

$$\chi(x)v \text{ is represented by } L_{\xi(x)}(\sigma). \tag{8.12)(a}$$

Write $\xi(x) = \sum_i f_i \xi_i$ as in (8.11)(a), with $f_i(eH) = 0$. Then condition (ii) shows that

$$L_{\xi(x)}(\sigma) = \sum_i f_i L_{\xi_i}(\sigma) + k \cdot (\partial_{\xi_i} f_i) \cdot \sigma. \tag{8.12)(b}$$

Evaluating at eH gives

$$\chi(x)v = k \cdot \left(\sum_i (\partial_{\xi_i} f_i)(eH)\right) \cdot v. \tag{8.12)(c}$$

By (8.11), the sum is $\gamma(x)$, proving (a).

Since we will not use the converse, we omit a detailed proof. One approach is to pass to a covering as in the proof of Proposition 8.8, and relate $\mathcal{V}$ to the "kth symmetric power" of the volume form bundle. (This is straightforward if k is an integer.) We have already remarked that the Lie derivative action on volume forms satisfies the conditions in (b) with k=1, and one can proceed from there. □

With this proposition as a guide, we can now formulate some analogous algebraic results. We will start in a purely commutative setting (Proposition 9.9); then show how the commutative structure can arise from "Poisson algebras" (Proposition 10.7); and finally show (sometimes) how to get the necessary Poisson algebra structure from a $(\mathfrak{g}, K)$-module (section 11).

9. Proof of Theorem 8.7: commutative algebra.

Suppose C is a graded commutative $\mathbb{C}$-algebra. We are looking for a version of Proposition 8.9, in which this algebra will correspond to the smooth functions on K/H. Recall that a *derivation* of C is a linear endomorphism δ of C satisfying

$$\delta(cc') = c\delta(c') + \delta(c)c'$$

The derivations of C form a graded C-module $\mathrm{Der}(C)$: we say that δ has degree p if it carries C^n to C^{n+p-1}. If δ and δ' are derivations of degrees p and p', then the commutator $[\delta, \delta']$ is a derivation of degree $p+p'-1$. For our purposes it will be convenient not to work with $\mathrm{Der}(C)$ directly.

Definition 9.1. A *module of derivations* of C is a graded C-module D endowed with a degree-preserving C-module map

$$\partial : D \to \mathrm{Der}(C), \quad \xi \mapsto \partial_\xi,$$

and a $\mathbb{C}$-bilinear skew-symmetric bracket

$$\{,\} : D \times D \to D$$

of degree -1. We require these to satisfy

$$\{\xi, c \cdot \xi'\} = c \cdot \{\xi, \xi'\} + \partial_\xi(c) \cdot \xi' \qquad (9.1)(a)$$

$$\partial_{\{\xi,\xi'\}} = [\partial_\xi, \partial_{\xi'}] \qquad (9.1)(b)$$

$$\{\xi, \{\xi', \xi''\}\} = \{\{\xi, \xi'\}, \xi''\} + \{\xi', \{\xi, \xi''\}\}. \qquad (9.1)(c)$$

(The last condition is the Jacobi identity.) All three conditions are easily verified for the case $D = \mathrm{Der}(C)$. The module D will replace the vector fields in Proposition 8.9.

Definition 9.2. Suppose D is a module of derivations of C, and M is a graded C-module. A *Lie derivative* on M is a graded $\mathbb{C}$-linear map

$$L : D \to \mathrm{Hom}_{\mathbb{C}}(M, M), \quad \xi \mapsto L_\xi$$

satisfying

$$L_\xi(c \cdot m) = c \cdot L_\xi(m) + \partial_\xi(c) \cdot m. \qquad (9.2)(a)$$

It is called *torsion-free* if L maps the bracket to commutator of operators:

$$L_{\{\xi,\xi'\}} = [L_\xi, L_{\xi'}]. \qquad (9.2)(b)$$

(Notice that this condition makes M a representation of the Lie algebra D.) It is said to be of *k-form type* (for $k \in \mathbb{C}$) if

$$L_{c\cdot\xi}(m) = c \cdot L_\xi(m) + k \cdot \partial_\xi(c) \cdot m. \qquad (9.2)(c)$$

(By "k-forms" we understand here not differential forms of degree k, but rather k-th powers of the volume form. Ultimately we will be concerned with the case $k = 1/2$, where the "half-form bundle" considered in distribution theory is of 1/2-form type in the present sense. This is the origin of the terminology.) Of course such a module M will replace the space $\Gamma\mathcal{V}$ of sections of $\mathcal{V}$ in Proposition 8.9.

Notice that we can define a torsion-free Lie derivative on D itself by $L_\xi(\xi') = \{\xi, \xi'\}$. It is not in general of k-form type for any k, however. The map $L = \partial$ is a torsion-free Lie derivative of 0-form type for C (regarded as a module over itself).

We can now introduce the analogue of the adjoint action of $\mathfrak{h}$ on $\mathfrak{k}/\mathfrak{h}$, as in (8.10).

Lemma 9.3. *In the setting of Definition 9.1, suppose $\mathfrak{m}$ is a maximal ideal in C, corresponding to a homomorphism*

$$\lambda : C \to \mathbb{C}.$$

Then $\mathfrak{m}D$ is a Lie subalgebra of the Lie algebra $(D, \{,\})$. Write

$$A_\mathfrak{m} : \mathfrak{m}D \to \mathrm{End}(D/\mathfrak{m}D), \quad A_\mathfrak{m}(\tau)(\xi + \mathfrak{m}D) = \{\tau, \xi\} + \mathfrak{m}D$$

for the action induced by the adjoint action. Then every endomorphism $A_\mathfrak{m}(\tau)$ has finite rank; so we can define a one-dimensional representation γ of $\mathfrak{m}D$ by

$$\gamma_\mathfrak{m}(\tau) = -\mathrm{tr}\, A_\mathfrak{m}(\tau).$$

Explicitly, suppose $\{\xi_i\}$ is a finite subset of D, and $a_i \in \mathfrak{m}$. Then

$$\gamma_\mathfrak{m}\left(\sum a_i \xi_i\right) = \sum \lambda(\partial_{\xi_i}(a_i)).$$

Suppose that M is a C-module and L is a Lie derivative on M (Definition 9.2). Then the submodule $\mathfrak{m}M$ is invariant under the action of $\mathfrak{m}D$, so there is a natural representation

$$\chi_\mathfrak{m} : \mathfrak{m}D \to \mathrm{End}(M/\mathfrak{m}M), \quad \chi_\mathfrak{m}(\tau)(m + \mathfrak{m}M) = L_\xi(m) + \mathfrak{m}M$$

If L is of k-form type (Definition 9.2), then

$$\chi_\mathfrak{m}(\tau) = k \cdot \gamma_\mathfrak{m}(\tau) \cdot I \quad (\tau \in \mathfrak{m}D).$$

The proof is a formal translation of that of Proposition 8.9, and we omit it. (It is worth observing that $\mathfrak{m}^2 D$ is a Lie ideal in $\mathfrak{m}D$, and that the representations $A_\mathfrak{m}$, $\gamma_\mathfrak{m}$, and $\chi_\mathfrak{m}$ all factor to $\mathfrak{m}D/\mathfrak{m}^2 D$.)

We need to fit this structure together with group actions. So suppose K is a complex algebraic group, acting by degree-preserving algebraic automorphisms of the algebra C: we write

$$\mathrm{Ad} : K \to \mathrm{Aut}(C). \qquad (9.4)(a)$$

(When ambiguity might arise, we may write instead Ad_C.) The differential of the K action defines a Lie algebra homomorphism from $\mathfrak{k}$ to the (degree-preserving!) derivations of degree 1 of C:

$$\mathrm{ad} : \mathfrak{k} \rightarrow \mathrm{Der}^1(C). \tag{9.4)(b}$$

A degree-preserving algebraic action π of K on a C-module M is called *compatible* if it satisfies

$$\pi(g)(c \cdot m) = (\mathrm{Ad}(g)(c)) \cdot (\pi(g)(m)) \tag{9.4)(c}$$

(cf. (2.1)(c)). (Generally we will write such actions on modules with a dot ("module notation") rather than choose a name for the representation.) In this case we call M a *compatible* (C, K)*-module.*

Definition 9.5. In the setting of (9.4)(a), a module D of derivations of C is called *compatible* if we are given an action of K on D by Lie algebra automorphisms, and a K-equivariant Lie algebra homomorphism

$$i : \mathfrak{k} \rightarrow D^1$$

satisfying the following conditions:

a) D is a compatible (C, K)-module;
b) for all $x \in \mathfrak{k}$, we have $\partial_{i(x)} = ad_C(x)$; and
c) the differential of the action of K on D is the map $\mathrm{ad}_D \circ i$ (from $\mathfrak{k}$ to $\mathrm{End}(D)$).

Definition 9.6. Suppose D is a compatible module of derivations of C, and M is a compatible graded C-module. A Lie derivative L on M is called *compatible* if

a) $L_{g \cdot \xi}(g \cdot m) = g \cdot (L_\xi(m))$; and
b) the differential of the action of K on M is $L \circ i$.

Notice that the Lie derivatives defined on C and on D after Definition 9.2 are automatically compatible.

Suppose now that D is a compatible module of derivations of C, and that $\mathfrak{m}$ is a maximal ideal in C, corresponding to a homomorphism

$$\lambda : C \rightarrow \mathbb{C}. \tag{9.7)(a}$$

The group K acts on the set of such λ; define H to be the isotropy subgroup. Its Lie algebra is easily computed to be

$$\mathfrak{h} = \{ x \in \mathfrak{k} \mid \mathrm{ad}(x)(\mathfrak{m}) \subset \mathfrak{m} \}. \tag{9.7)(b}$$

Because of (9.5)(b), this contains as a subalgebra

$$\mathfrak{h}_1 = \{ x \in \mathfrak{k} \mid i(x) \in \mathfrak{m}D \}. \tag{9.7)(c}$$

The group H acts on $D/\mathfrak{m}D$ and (if M is a compatible (C, K)-module) on $M/\mathfrak{m}M$: we write

$$A_H : H \rightarrow \mathrm{End}(D/\mathfrak{m}D), \qquad \chi_H : H \rightarrow \mathrm{End}(M/\mathfrak{m}M). \tag{9.7)(d}$$

The definitions of compatibility have been arranged to guarantee that the differentials of these representations agree on $\mathfrak{h}_1$ with the Lie algebra representations of Lemma 9.3:

$$A_H(x) = A_{\mathfrak{m}}(i(x)), \quad \chi_H(x) = \chi_{\mathfrak{m}}(i(x)) \quad (x \in \mathfrak{h}_1) \tag{9.7)(e}$$

(In fact the representations $A_{\mathfrak{m}}$ and $\chi_{\mathfrak{m}}$ may be extended from $\mathfrak{m}D$ to the larger Lie algebra

$$\{\tau \in D \mid \partial_\tau(\mathfrak{m}) \subset \mathfrak{m}\},$$

which contains $i(x)$; and (9.7)(e) remains valid. The reason we have not done this is that the final — and most interesting — conclusion of Lemma 9.3 appears *not* to extend.)

Finally, we need something like the character γ for H. The Zariski tangent space to $\operatorname{Spec} C$ at $\mathfrak{m}$ is

$$T_{\mathfrak{m}}(\operatorname{Spec} C) = \operatorname{Hom}(\mathfrak{m}/\mathfrak{m}^2, C/\mathfrak{m}). \tag{9.8)(a}$$

Any derivation of C sends $\mathfrak{m}^2$ into $\mathfrak{m}$, and so defines an element of this tangent space; so we get a natural map

$$\partial_{\mathfrak{m}} : D/\mathfrak{m}D \to T_{\mathfrak{m}}(\operatorname{Spec} C). \tag{9.8)(b}$$

We can also define

$$i_{\mathfrak{m}} : \mathfrak{k} \to D/\mathfrak{m}D. \tag{9.8)(c}$$

By inspection of the definitions,

$$\mathfrak{h} = \ker \partial_{\mathfrak{m}} \circ i_{\mathfrak{m}}, \qquad \mathfrak{h}_1 = \ker i_{\mathfrak{m}}. \tag{9.8)(d}$$

(This suggests already the importance of the case when $\partial_{\mathfrak{m}}$ is an isomorphism.) Finally, we define a character γ_H of H by

$$\gamma_H(h) = \det A_H(h)^{-1} \tag{9.8)(e}$$

at least when $D/\mathfrak{m}D$ is finite-dimensional. As in (9.7)(e), this definition is obviously compatible with the one in Lemma 9.3:

$$\gamma_H(x) = \gamma_{\mathfrak{m}}(i(x)), \quad (x \in \mathfrak{h}_1). \tag{9.8)(f}$$

These considerations and Lemma 9.3 prove

Proposition 9.9. *Suppose K is an algebraic group of automorphisms of the graded commutative algebra C, D is a compatible module of derivations of C (Definition 9.6), and M is a compatible (C,K)-module of k-form type (Definition 9.2). Fix a maximal ideal $\mathfrak{m}$ in C, and let H denote its stabilizer in K. Define $\mathfrak{h}_1$ as in (9.7)(c). Let χ_H be the representation of H on $M/\mathfrak{m}M$, and let γ_H be the inverse of the determinant character of H on $D/\mathfrak{m}D$ (assuming this to be finite-dimensional). Then for $x \in \mathfrak{h}_1$, we have*

$$\chi_H(x) = k \cdot \gamma_H(x) \cdot I.$$

Suppose in addition that the map $\partial_{\mathfrak{m}}$ of (9.8)(b) is an isomorphism. Then $\mathfrak{h}_1$ is all of $\mathfrak{h}$, and γ_H is equal to the determinant of the action of H on the Zariski cotangent space $T^*_{\mathfrak{m}}(\operatorname{Spec} C)$.

10. Proof of Theorem 8.7: Poisson algebras.

Our next goal is to show how the structures required in Proposition 9.9 can arise from the theory of Poisson algebras.

Definition 10.1. A *graded Poisson algebra* is a graded commutative algebra R endowed with a Lie algebra structure

$$\{,\} : R \times R \to R$$

of degree -1, such that the bracket with each element of R is a derivation of the commutative algebra structure:

$$\{r, st\} = \{r, s\}t + s\{r, t\}.$$

An ideal $Q \subset R$ is called a *Poisson ideal* if $\{Q, R\} \subset Q$; in this case R/Q is again a Poisson algebra. An ideal I is called *integrable* if $\{I, I\} \subset I$; we will see in a moment how to exploit this weaker condition.

Example 10.2. Suppose G is a Lie group with Lie algebra $\mathfrak{g}$. Then the symmetric algebra $S(\mathfrak{g})$ carries a Poisson algebra structure, determined by the property

$$\{x, y\} = [x, y] \qquad (x, y \in \mathfrak{g}).$$

More generally

$$\{x, p\} = \text{ad}(x)(p) \qquad (x \in \mathfrak{g}, p \in S(\mathfrak{g})).$$

Since ad is the differential of the adjoint action of G, it follows that any $\text{Ad}(G)$-invariant subspace of $S(\mathfrak{g})$ is preserved by bracket with $\mathfrak{g}$; so any $\text{Ad}(G)$-invariant ideal is preserved by Poisson bracket with anything. If X is an $\text{Ad}(G)$-invariant subset of $\mathfrak{g}^*$, then the ideal of zeros $Q(X)$ is a Poisson ideal; so $S(\mathfrak{g})/Q(X)$ is a Poisson algebra. It will be graded if X is homogeneous—for example, if X is a nilpotent coadjoint orbit.

Integrable ideals are even easier to construct in this example. Suppose Q is a Poisson ideal in $S(\mathfrak{g})$, and $\mathfrak{k}$ is a subalgebra of $\mathfrak{g}$. Then $Q + \mathfrak{k}S(\mathfrak{g})$ is an integrable ideal. It will be graded if Q is.

Lemma 10.3. *Suppose R is a graded Poisson algebra, and I is a graded integrable ideal. Set*

$$C = R/I, \qquad D = I/I^2.$$

Then C is a commutative algebra, and D is a graded C-module. D has the structure of a module of derivations of C (Definition 9.1), by

$$\partial_{a+I^2}(b + I) = \{a, b\}, \quad \{a + I^2, a' + I^2\} = \{a, a'\} + I^2 \quad (a, a' \in I, b \in R).$$

We omit the straightforward verification.

Next, we consider the Poisson algebra structure that will give rise to Lie derivatives.

Definition 10.4. Suppose R is a graded Poisson algebra, and M is a graded R-module. A *first-order structure* on M is a graded (complex-linear) map of degree -1

$$a_1 : R \times M \to M$$

having the following property:

$$a_1(r, s \cdot m) + r a_1(s, m) = a_1(rs, m) + 1/2\{r, s\} \cdot m \qquad (r, s \in R, m \in M).$$

This structure is said to be *torsion-free* if whenever r and s both annihilate M, we have

$$a_1(\{r, s\}, m) = a_1(r, a_1(s, m)) - a_1(s, a_1(r, m)).$$

This definition (which comes from Gerstenhaber's theory of deformations) is certainly difficult to motivate directly. We will see later how first-order structures arise from $\mathfrak{g}$-modules. For the moment, we offer only a simpler analogy. If M is a complex vector space, we could define a *zeroth-order structure* on M to be a map a_0 of degree 0 from $R \times M$ to M, satisfying

$$a_0(r, a_0(s, m)) = a_0(rs, m).$$

This is nothing but an R-module structure. We could call it *torsion-free* if whenever $a_0(r, m)$ and $a_0(s, m)$ are both zero, then $a_0(\{r, s\}, m)$ is also zero. This just says that the annihilator of M is an integrable ideal. Definition 10.4 is in a technical sense a natural refinement of these notions (of R-module and integrable ideal).

Lemma 10.5. *Suppose R is a graded Poisson algebra, I is a graded integrable ideal, and M is a graded R-module annihilated by I. Define C, D, ∂, and $\{,\}$ as in Lemma 10.3. Then a first-order structure a_1 on M induces a Lie derivative*

$$L : D \to \operatorname{End}(M)$$

of half-form type (Definition 9.2), by the formula

$$L_{s+I^2}(m) = a_1(s, m) \qquad (s \in I, m \in M).$$

If a_1 is torsion-free, then so is L.

Proof. Again this is a routine verification from the definitions. Consider for example the "half-form type" condition. So suppose $r \in R$ and $s \in I$. Then by definition

$$L_{rs+I^2}(m) = a_1(rs, m) = r a_1(s, m) + a_1(r, sm) - 1/2\{r, s\}m.$$

Since s annihilates M, the second term on the right is zero; and the others are

$$(r + I)L_{s+I^2}(m) + 1/2\partial_{s+I^2}(r)m,$$

as required. We leave the other details to the reader. □

We need a way to understand the map $\partial_{\mathfrak{m}}$ of (9.8) in terms of the Poisson structure. Our goal is Proposition 10.9 below. Suppose R is a Poisson algebra and $\mathfrak{m}$ is a maximal ideal in R, corresponding to a homomorphism

$$\lambda : R \to \mathbb{C}, \qquad \ker \lambda = \mathfrak{m}. \tag{10.6)(a}$$

We define a (possibly degenerate) symplectic form $\phi_{\mathfrak{m}}$ (or ϕ_λ) on the Zariski *cotangent* space $\mathfrak{m}/\mathfrak{m}^2$ by

$$\phi_{\mathfrak{m}}(f + \mathfrak{m}^2, g + \mathfrak{m}^2) = \lambda(\{f, g\}) \qquad (f, g \in \mathfrak{m}). \tag{10.6)(b}$$

Of course a bilinear form on the cotangent space is the same as a linear map from the cotangent space to the tangent space; we write

$$\Phi_{\mathfrak{m}} : T^*_{\mathfrak{m}}(\operatorname{Spec} R) \to T_{\mathfrak{m}}(\operatorname{Spec} R). \tag{10.6)(c}$$

Then $\phi_{\mathfrak{m}}$ is non-degenerate if and only if $\Phi_{\mathfrak{m}}$ is an isomorphism.

Suppose now that $I \subset \mathfrak{m}$ is an ideal. Then I spans a subspace of $\mathfrak{m}/\mathfrak{m}^2$ that we call the *conormal space to R/I at* $\mathfrak{m}$:

$$T^*_{R/I,\mathfrak{m}}(\operatorname{Spec} R) = I/(\mathfrak{m}^2 \cap I). \tag{10.7)(a}$$

By definition, the Zariski tangent space $T_{\mathfrak{m}}(R/I)$ consists of linear functionals on the quotient $\mathfrak{m}/(\mathfrak{m}^2 + I)$. This may be identified with linear functionals on $\mathfrak{m}/\mathfrak{m}^2$ vanishing on the conormal space to R/I:

$$T_{\mathfrak{m}}(\operatorname{Spec} R/I) = \left(T^*_{R/I,\mathfrak{m}}(\operatorname{Spec} R)\right)^{\perp}. \tag{10.7)(b}$$

Suppose in addition that I is integrable. Then it is immediate from the definitions that

$$\phi_{\mathfrak{m}} = 0 \text{ on } T^*_{R/I,\mathfrak{m}}(\operatorname{Spec} R) \quad (I \text{ integrable}). \tag{10.7)(c}$$

Consequently $\Phi_{\mathfrak{m}}$ restricts to a map

$$\Phi_{\mathfrak{m},I} : I/(\mathfrak{m}^2 \cap I) \to T_{\mathfrak{m}}(\operatorname{Spec} R/I). \tag{10.7)(d}$$

It is clear from the definitions that $\partial_{\mathfrak{m}}$ is the composition of $\Phi_{\mathfrak{m},I}$ with the natural projection

$$I/\mathfrak{m}I \to I/(\mathfrak{m}^2 \cap I). \tag{10.7)(e}$$

Definition 10.8. Suppose R is a Poisson algebra (over $\mathbb{C}$) and $\mathfrak{m}$ is a maximal ideal in R, with $R/\mathfrak{m} = \mathbb{C}$. $\operatorname{Spec} R$ is *symplectic at* $\mathfrak{m}$ if

i) it is regular at $\mathfrak{m}$ (that is, if $R_{\mathfrak{m}}$ is a regular local ring); and
ii) the form $\phi_{\mathfrak{m}}$ is non-degenerate.

(It is not hard to see that the first condition is in fact a consequence of the second.) Now suppose $I \subset \mathfrak{m}$ is an integrable ideal. We say that R/I is *Lagrangian at* $\mathfrak{m}$ if

i) R is symplectic at $\mathfrak{m}$;
ii) R/I is regular at $\mathfrak{m}$; and
iii) the subspace $T^*_{R/I,\mathfrak{m}}(\operatorname{Spec} R)$ of $T^*_\mathfrak{m}(\operatorname{Spec} R)$ is maximally isotropic for the form $\phi_\mathfrak{m}$.

Again it is fairly easy to see that the second condition is a consequence of the other two.

Here is our result about $\partial_\mathfrak{m}$.

Proposition 10.9. *Suppose R is a Noetherian Poisson algebra, $I \subset R$ is an integrable ideal, and $\mathfrak{m} \supset I$ is a maximal ideal. Assume that R/I is Lagrangian at $\mathfrak{m}$. Then the map $\partial_\mathfrak{m}$ (Lemma 10.3 and (9.8)) is an isomorphism.*

Before starting the proof of Proposition 10.9, we consider the underlying linear algebra.

Lemma 10.10. *Suppose (V, ϕ) is a finite-dimensional vector space with a (possibly degenerate) symplectic form, and $S \subset V$ is an isotropic subspace. Write $S^\perp \subset V^*$ for the linear functionals vanishing on S. Then the natural map $S \rightarrow S^\perp$ defined by ϕ is*
i) surjective if and only if S is maximal isotropic, and
ii) injective if and only if S meets the radical of ϕ only in $\{0\}$.

We leave the elementary proof to the reader. As a consequence of Lemma 10.10 and the description of $\partial_\mathfrak{m}$ in (10.7), we get immediately the following result.

Lemma 10.11. *In the setting of Lemma 10.3, fix a maximal ideal $\mathfrak{m} \supset I$. Then $\partial_\mathfrak{m}$ (cf. (9.8)) is an isomorphism if and only if the following two conditions are satisfied:*
i) the subspace $T^*_{R/I,\mathfrak{m}}(\operatorname{Spec} R)$ is a maximal isotropic subspace;
ii) the symplectic form $\phi_\mathfrak{m}$ is non-degenerate; and
iii) the projection $I/\mathfrak{m}I \rightarrow I/(\mathfrak{m}^2 \cap I)$ is an isomorphism.

Conditions (i) and (ii) here are obvious consequences of the Lagrangian hypothesis in Proposition 10.9. For (iii), we use the following lemma.

Lemma 10.12. *Suppose R is a commutative Noetherian ring, $\mathfrak{m}$ is a maximal ideal in R, and $I \subset \mathfrak{m}$ is any ideal. Assume that R is regular at $\mathfrak{m}$. Then R/I is regular at $\mathfrak{m}$ if and only if*

$$I \cap \mathfrak{m}^2 = \mathfrak{m}I.$$

Proof. (We follow [H, p.178], where the result is proved for varieties.) Write n for the dimension of the local ring $R_\mathfrak{m}$. Then R is regular at $\mathfrak{m}$ if and only if the Zariski cotangent space $\mathfrak{m}/\mathfrak{m}^2$ has dimension n (as a vector space over $R/\mathfrak{m}$) [AM, Theorem 11.22]. For the cotangent space to R/I, we have the exact sequence

$$0 \to I/(I \cap \mathfrak{m}^2) \to \mathfrak{m}/\mathfrak{m}^2 \to \mathfrak{m}/(\mathfrak{m}^2 + I) \to 0; \qquad (10.13)(a)$$

that is,

$$0 \to I/(I \cap \mathfrak{m}^2) \to T^*_{\mathfrak{m}}(\text{ Spec } R) \to T^*_{\mathfrak{m}}(\text{Spec } R/I) \to 0. \qquad (10.13)(b)$$

Suppose first that the condition in the lemma is satisfied. Write r for the dimension of $I/\mathfrak{m}I$. Then by Nakayama's lemma, I is generated by r elements near $\mathfrak{m}$ (that is, after inverting a finite number of elements not in $\mathfrak{m}$). It follows that $(R/I)_{\mathfrak{m}}$ has dimension at least $n-r$. On the other hand, (10.13)(b) shows that the cotangent space at $\mathfrak{m}$ to Spec R/I has dimension equal to $n-r$. This implies that the dimension of $(R/I)_{\mathfrak{m}}$ is at most $n-r$ [AM, Corollary 11.15], with equality if and only if R/I is regular at $\mathfrak{m}$.

Conversely, suppose R/I is regular of dimension d at $\mathfrak{m}$. By (10.13)(b), $I/(I \cap \mathfrak{m}^2)$ has dimension $n-d$. Choose elements $x_1, \ldots, x_{n-d}$ of I whose images form a basis for this quotient, and define J to be the ideal they generate in R. Obviously $J \subset I$, and by construction

$$J/(J \cap \mathfrak{m}^2) = I/(I \cap \mathfrak{m}^2),$$

a space of dimension $n-d$. On the other hand, $J/\mathfrak{m}J$ has dimension at most $n-d$ (since J has $n-d$ generators); so the natural projection from $J/\mathfrak{m}J$ onto $J/(J \cap \mathfrak{m}^2)$ must be an isomorphism. So J satisfies the condition in the lemma. By the first half of the proof, $(R/J)_{\mathfrak{m}}$ is a regular local ring of dimension d. On the other hand $(R/I)_{\mathfrak{m}}$ is a quotient of $(R/J)_{\mathfrak{m}}$. Since these are regular local rings (hence integral domains) of the same dimension, it follows that $I_{\mathfrak{m}} = J_{\mathfrak{m}}$. It follows at once that $I/\mathfrak{m}I = J/\mathfrak{m}J$; and we have already seen that the right-hand side is isomorphic to $I/(I \cap \mathfrak{m}^2)$. □

Proof of Proposition 10.9. Because of Lemma 10.12, the Lagrangian hypothesis in the proposition implies all three of the conditions given in Lemma 10.11 for $\partial_{\mathfrak{m}}$ to be an isomorphism. □

Finally, we can bring a group into the picture. Suppose the algebraic group K acts algebraically by graded automorphisms on the graded Poisson algebra R:

$$\text{Ad} : K \to \text{Aut}(R). \qquad (10.14)(a)$$

We say that this action is ***Hamiltonian*** if there is a K-equivariant Lie algebra homomorphism

$$i : \mathfrak{k} \to R^1 \qquad (10.14)(b)$$

with the property that the differential of Ad at x is given by Poisson bracket with $i(x)$. An ideal I is called ***K-integrable*** if it is integrable and K-invariant, and

$$I \supset i(\mathfrak{k}). \qquad (10.14)(c)$$

Proposition 10.15. *In the setting of Lemma 10.5, assume also that R carries a Hamiltonian action of K (cf. (10.14)); that M is a compatible (R, K)-module; that I is K-integrable; and that the first-order structure a_1 is K-equivariant. Then the structures C, D, ∂, $\{,\}$, and L of Lemmas 10.3 and 10.5 all carry compatible K-actions.*

We leave this to the reader.

The conditions appearing in Proposition 10.9 are phrased in terms of Zariski cotangent spaces. It will be convenient for us to relate such conditions to the notion of "generically reduced" (cf. Proposition 2.9). This is accomplished by Lemma 10.17. below.

Lemma 10.16. *Suppose P is a prime ideal in the commutative Noetherian ring R, and $Q \subset R$ is any other ideal. Then Q is generically reduced along P (Proposition 2.9) if and only if there is an element f of R, not belonging to P, such that Q_f is equal either to P_f or to R_f. The second possibility occurs exactly when Q is not contained in P.*

Proof. Suppose first that Q is generically reduced. By Proposition 2.9(c), there is an f not in P so that $(R/Q)_f$ is a free $(R/P)_f$-module. In particular, $(R/Q)_f$ is annihilated by P, so it is a quotient of $(R/P)_f$. Any proper quotient of $(R/P)_f$ has a larger annihilator, and so cannot be free unless it is zero. The required statement now follows from the exactness of localization. The argument reverses trivially. □

Lemma 10.17. *Suppose R is a finitely generated commutative algebra over $\mathbb{C}$, P is a prime ideal in R, and $Q \subset P$ is any ideal. Write d for the dimension of R/P. Then Q is generically reduced along P if and only if there is an element f of R, not belonging to P, with the following property: for every maximal ideal $\mathfrak{m}$ of R containing P but not containing f, the Zariski cotangent space*

$$\mathfrak{m}/(\mathfrak{m}^2 + Q)$$

to Spec R/Q *at $\mathfrak{m}$ has dimension d. In this case we have*

$$T^*_{\mathfrak{m}}(\mathrm{Spec}\, R/Q) \simeq T^*_{\mathfrak{m}}(\mathrm{Spec}\, R/P)$$

for such $\mathfrak{m}$.

Proof. Suppose Q is generically reduced along P. By Lemma 10.16, there is an h not in P so that $(R/Q)_h = (R/P)_h$. Consequently the Zariski tangent spaces coincide away from the zeros of h. Choose g not in P vanishing on the singular locus of Spec R/P; then $f = gh$ satisfies the requirement of the lemma.

Conversely, suppose such an f exists. Fix a maximal ideal $\mathfrak{m}$ containing P but not f, and consider the local ring $(R/Q)_{\mathfrak{m}}$. Since $Q \subset P$, the dimension of Q is at least d. The hypothesis therefore guarantees that $(R/Q)_{\mathfrak{m}}$ is a regular local ring [AM, Theorem 11.22] of dimension d. Just as in the proof of Lemma 10.12, it follows that $P_{\mathfrak{m}} = Q_{\mathfrak{m}}$. If we let g be a common denominator for the images in $Q_{\mathfrak{m}}$ of a finite set of generators of P, then $gP \subset Q$. Hence $P_g = Q_g$. The element g does not belong to $\mathfrak{m}$, so it does not belong to P either. By Lemma 10.16, Q is generically reduced along P. □

11. Proof of Theorem 8.7: representation theory.

We can now present the setting in which we will construct first-order structures, Lie derivatives of half-form type, and eventually admissible representations of isotropy groups. Suppose (G, K) is a reductive symmetric pair of Harish-Chandra class (Definition 5.21; we assume G is connected, although K need not be). Let X be a $(\mathfrak{g}, K)$-module of finite length. Define

$$J = \operatorname{Ann} X \subset U(\mathfrak{g}), \qquad (11.1)(a)$$

a primitive ideal in $U(\mathfrak{g})$. Define

$$Q = \operatorname{gr} J \subset S(\mathfrak{g}), \qquad (11.1)(b)$$

the associated graded ideal. Because J is $\operatorname{Ad}(G)$-invariant, Q is as well; so Q is a Poisson ideal (Example 10.2). Its associated variety is the union of the closures of several nilpotent G-orbits (just one if X is irreducible). Fix one of these $\mathcal{O}$, which we assume contains a component of $\mathcal{V}(J)$:

$$\mathcal{V}(J) = \mathcal{V}(Q) \supset \overline{\mathcal{O}}. \qquad (11.1)(c)$$

Our Poisson algebra is

$$R = S(\mathfrak{g})/Q. \qquad (11.1)(d)$$

Of course we have a Hamiltonian action Ad of K on R; the map i is induced by the inclusion of $\mathfrak{k}$ in $S(\mathfrak{g})$. Define

$$I = Q + \mathfrak{k}S(\mathfrak{g}). \qquad (11.1)(e)$$

This is an integrable ideal by Example 10.2; it is obviously K-integrable by the definition (cf. (10.6)(c)). We can therefore introduce

$$C = R/I, D = I/I^2, \partial \qquad (11.1)(f)$$

and so on as in Lemma 10.3. The set of maximal ideals in $\operatorname{Spec} C$ is the associated variety of I; that is (on the level of points)

$$\operatorname{Spec} C = \mathcal{V}(I) = \mathcal{V}(Q) \cap (\mathfrak{g}/\mathfrak{k})^* = \overline{\mathcal{O}} \cap (\mathfrak{g}/\mathfrak{k})^*. \qquad (11.1)(g)$$

(Here we have used the obvious fact that the associated variety of $\mathfrak{k}S(\mathfrak{g})$ is $(\mathfrak{g}/\mathfrak{k})^*$.) To get the module M, choose any finite-dimensional K-stable generating subspace X_0 of X, and define a filtration as in (1.1). Set

$$M = \operatorname{gr} X; \qquad (11.1)(h)$$

obviously M is a compatible (R, K)-module annihilated by I.

We can now state a crucial technical lemma.

Lemma 11.2. *In the setting of (11.1), assume that*

Q is generically reduced along $\mathcal{O}$

(cf. Proposition 2.9; the condition is automatically fulfilled if Q is prime, or more generally if $\mathcal{O}$ has multiplicity one in the characteristic cycle of $S(\mathfrak{g})/Q$). Fix a weight $\lambda \in \mathcal{O} \cap (\mathfrak{g}/\mathfrak{k})^*$, and write $\mathcal{O}_\mathfrak{k}$ for its orbit under K. Let $\mathfrak{m}$ denote the maximal ideal corresponding to evaluation at λ (in $S(\mathfrak{g})$, R, or C). Define H to be the stabilizer of λ in K.

a) The Poisson algebra R is symplectic at $\mathfrak{m}$ (Definition 10.8).
b) The ideal I and the C-module M are generically reduced along $\mathcal{O}_\mathfrak{k}$, so the representation χ_H of H on $M/\mathfrak{m}M$ (cf. (9.7)(d)) coincides with $\chi(\lambda, M)$ as defined in Definition 2.12.
c) The quotient R/I is Lagrangian at $\mathfrak{m}$ (Definition 10.8).
d) The Zariski tangent space $T_\mathfrak{m}(\operatorname{Spec} C)$ is naturally isomorphic to $\mathfrak{k}/\mathfrak{h}$; so the character γ_H of H defined in (9.8)(e) coincides with $\gamma_\mathfrak{k}$ as defined in Theorem 7.11.

We postpone the proof (which is a straightforward application of Proposition 5.20) to the end of the section.

In addition, we need a way to define a first-order structure on M.

Lemma 11.3. *In the setting of (11.1), assume the ideal $J \subset U(\mathfrak{g})$ is preserved by the transpose antiautomorphism of (8.6). Then the R-module M admits a K-equivariant torsion-free first-order structure (Definition 10.4).*

Again we postpone the proof for a moment.

Proof of Theorem 8.7. We introduce the structure and notation of (11.1). By Lemma 11.2 and hypothesis (i), what we must show is that the differential of the representation χ_H of H (cf. (9.7)(d)) is half the differential of γ_H (cf. (8.17)(e)). By Proposition 9.9, it suffices to find a Lie derivative on M (as a C-module) of half-form type. By Lemma 10.5, it suffices to find a K-equivariant first-order structure on M as an R-module. By Lemma 11.3, the existence of such a structure follows from hypothesis (ii). □

Proof of Lemma 11.3. The idea of the proof is that the module structure on M captures X as an A-module "to order zero;" the first order structure captures slightly more of the A-module structure. The implementation we give of this idea is borrowed from [Ge] and [BFFLS]; see also [V3, §3].

We begin with a simple fact about the transpose map. Recall the symmetrization map β from $S(\mathfrak{g})$ to $U(\mathfrak{g})$, and the symbol maps σ_n (see the proof of Lemma 5.2). If f is a homogeneous polynomial of degree n in $S(\mathfrak{g})$, then one checks easily that

$$^t(\beta(f)) = (-1)^n \beta(f). \tag{11.4}$$

Write

$$A = U(\mathfrak{g})/J, \quad J_n = J \cap U_n(\mathfrak{g}), \quad A_n = U_n(\mathfrak{g})/J_n \tag{11.5)(a}$$

Then A is a filtered algebra, and the adjoint action of G on $U(\mathfrak{g})$ factors to A. The isomorphism σ from $\operatorname{gr} A$ onto R is implemented by (surjective) symbol maps

$$\sigma_n : A_n \to R^n \tag{11.5)(b}$$

with kernel precisely A_{n-1}. These maps respect the adjoint action of G. Because G is reductive, we can choose G-equivariant cross-sections

$$\alpha_n : R^n \to A_n, \qquad \sigma_n \circ \alpha_n = \text{ identity}. \tag{11.5)(c}$$

These maps can be added over n, giving a filtered linear isomorphism

$$\alpha : R \to A. \tag{11.5)(d}$$

(When confusion may arise, we will write these maps as σ^A and α^A.) By the hypothesis on J, the transpose antiautomorphism factors to a filtered antiautomorphism of A. By (11.4), the associated graded (anti)automorphism of R acts by $(-1)^n$ on R^n. Now the transpose is involutive, so A is the direct sum of its $+1$ and -1 eigenspaces. In particular, there is a G-invariant complement for A_{n-1} in A_n on which the transpose acts by $(-1)^n$. If we use such a complement to define α, we get

$$^t\alpha_n(r) = (-1)^n \alpha_n(r). \tag{11.5)(e}$$

(It is easy to check that that this requirement determines α_n modulo $U_{n-2}(\mathfrak{g})$.)

Because α is a linear isomorphism, we may use it to pull back the algebra structure from A to R, obtaining a new multiplication that we write as $*$:

$$r * s = \alpha^{-1}(\alpha(r)\alpha(s)). \tag{11.6)(a}$$

This multiplication respects the filtration of R by degree, so we may write it as an infinite sum of bilinear maps m_n of degree $-n$. This means that

$$m_n : R^p \times R^q \to R^{p+q-n} \tag{11.6)(b}$$

and that

$$r * s = \sum_{n=0}^{\infty} m_n(r, s). \tag{11.6)(c}$$

There is no difficulty about convergence: for fixed r and s (say homogeneous of degrees p and q) we will have $m_n(r, s) = 0$ whenever $n > p+q$ (since the gradation of R has no negative terms). The sum in (11.6)(c) is therefore finite. By the definition of associated graded algebra,

$$m_0(r, s) = rs \tag{11.6)(d}$$

In order to prove the lemma, we will need to calculate m_1 as well. Notice first that if a and b belong to A, then

$$^t(ab + ba) = ab + ba, \qquad ^t(ab - ba) = -(ab - ba). \tag{11.7)(a}$$

By (11.5)(e), it follows that

$$r * s + s * r \in R^{even}, \qquad r * s - s * r \in R^{odd} \tag{11.7)(b}$$

(with obvious notation). In terms of the expansion (11.6)(c), this says that

$$m_{2n+1}(r,s) = -m_{2n+1}(s,r), \qquad m_{2n}(r,s) = m_{2n}(s,r). \tag{11.7)(c}$$

If $u \in U_p(\mathfrak{g})$ and $v \in U_q(\mathfrak{g})$, then it is well-known that $uv - vu \in U_{p+q-1}(\mathfrak{g})$. Writing σ^U for the symbol maps in $U(\mathfrak{g})$, we obtain

$$\sigma^U_{p+q-1}(uv - vu) = \{\sigma^U_p(u), \sigma^U_q(v)\}. \tag{11.7)(d}$$

(This is easily proved by induction on p and q; when $p = q = 1$, it amounts to the defining relation of $U(\mathfrak{g})$.) Of course the analogous formula relates commutators in A to the Poisson structure in R. Translated into a statement about the product $*$ on R, (11.7)(d) is

$$m_1(r,s) - m_1(s,r) = \{r,s\}.$$

Now use (11.7)(c) with $n = 0$ to get

$$m_1(r,s) = 1/2\{r,s\}. \tag{11.7)(e}$$

We now make analogous constructions relating X and M. By definition of M, there are K-equivariant symbol maps

$$\sigma_n : X_n \to M^n \tag{11.8)(a}$$

with kernel precisely X_{n-1}. Because K is reductive, we can choose K-equivariant cross-sections

$$\alpha_n : M^n \to X_n, \qquad \sigma_n \circ \alpha_n = \text{ identity}. \tag{11.8)(b}$$

These maps can be added over n, giving a filtered K-equivariant linear isomorphism

$$\alpha : M \to X. \tag{11.8)(c}$$

(Again we will sometimes write σ^X and α^X to avoid confusion.) This isomorphism can be used to pull the A-module structure on X back to an $(R,*)$-module structure on M, which we again denote with a $*$:

$$r * m = (\alpha^X)^{-1}\left(\alpha^A(r) \cdot \alpha^X(m)\right). \tag{11.8)(d}$$

Just as in the case of the product $*$ on R, this action respects the filtrations by degrees, so it may be expanded as a sum of K-equivariant bilinear maps a_n of degree $-n$:

$$a_n : R^p \times M^q \to M^{p+q-n}, \qquad r * m = \sum_{n=0}^{\infty} a_n(r,m). \tag{11.8)(e}$$

By the definition of $\operatorname{gr} M$,

$$a_0(r,m) = r \cdot m \qquad (11.8)(f).$$

The map a_1 will turn out to be the first-order structure on M that we are seeking. We have provided no normalization of α "up to order $n-2$" analogous to (11.5)(e), so we cannot expect to find a closed formula for a_1 analogous to (11.7)(e). To get information about a_1, we use the fact that X is an A-module, and therefore that M is an $(R,*)$-module. This means that

$$r*(s*m) = (r*s)*m \qquad (r,s \in R, m \in M). \qquad (11.9)(a)$$

Written in terms of the a_n and m_n, this becomes

$$\sum_p a_p\left(r, \sum_q a_q(s,m)\right) = \sum_p a_p\left(\sum_q m_q(r,s), m\right). \qquad (11.9)(b)$$

Now suppose r, s, and m are homogeneous of degrees i, j, and k respectively. Then a typical term on either side is homogeneous of degree $i+j+k-p-q$. Collecting terms of the same degree, we obtain finally

$$\sum_{p+q=n} a_p(r, a_q(s,m)) = \sum_{p+q=n} a_p(m_q(r,s),m)). \qquad (11.9)(c)$$

We examine these identities one at a time, beginning with $n=0$:

$$a_0(r, a_0(s,m)) = a_0(m_0(r,s),m).$$

Using (11.6)(d) and (11.8)(f), we can rewrite this as

$$r\cdot(s\cdot m) = (rs)\cdot m:$$

that is, M is an R-module. This is an important fact, but hardly new to us. For $n=1$, we get

$$a_0(r, a_1(s,m)) + a_1(r, a_0(s,m)) = a_0(m_1(r,s),m) + a_1(m_0(r,s),m).$$

Using (11.6)(d), (11.7)(e), and (11.8)(f), we get

$$r\cdot a_1(s,m) + a_1(r, s\cdot m) = 1/2\{r,s\}\cdot m + a_1(rs,m). \qquad (11.9)(d).$$

This is precisely the requirement for a first-order structure.

To see that a_1 is torsion free, we consider (11.9)(c) with $n=2$. This is

$$\begin{aligned} r\cdot a_2(s,m) &+ a_1(r, a_1(s,m)) + a_2(r, s\cdot m) \\ &= m_2(r,s)\cdot m + a_1(1/2\{r,s\},m) + a_2(rs,m). \end{aligned} \qquad (11.10)(a)$$

This property is difficult to use, since we do not know m_2 explicitly. However, we do know that m_2 is symmetric (cf. (11.7)(c)). Skew-symmetrizing

(11.10)(a) in the variables r and s therefore eliminates the first and third terms on the right, leaving

$$\begin{aligned}(r \cdot a_2(s,m) - s \cdot a_2(r,m)) + (a_1(r,s\cdot m) - a_1(s,r\cdot m)) \\ + (a_2(r,s\cdot m) - a_2(s,r\cdot m)) \\ = a_1(\{r,s\},m).\end{aligned} \tag{11.10)(b}$$

If r and s both annihilate M, then the first and third terms on the left vanish, leaving

$$a_1(r,s\cdot m) - a_1(s,r\cdot m) = a_1(\{r,s\},m) \qquad (r,s \in \operatorname{Ann} M). \tag{11.10)(c}$$

This is the definition of torsion-free. □

Proof of Lemma 11.2. Write P for the prime ideal in R defined by $\mathcal{O}$. By the hypothesis and Lemma 10.17,

$$T^*_\lambda(\operatorname{Spec} R) \simeq T^*_\lambda(\operatorname{Spec} R/P) \simeq T^*_\lambda(\mathcal{O}). \tag{11.11)(a}$$

(The set of λ for which this holds is Zariski dense and G-invariant, and therefore includes all of $\mathcal{O}$.) We need the same fact for R/I. Choose g not in P so that $P_g = Q_g$ (Lemma 10.16); by the G-invariance of P and Q we may assume that $g \notin \mathfrak{m}$. Write P' for the prime ideal in R defined by the component of $\mathcal{O}_\mathfrak{k}$ containing λ, and consider the three ideals

$$P' \supset P + \mathfrak{k}R \supset I = Q + \mathfrak{k}R. \tag{11.11)(b}$$

By Proposition 5.20, the first two coincide after we localize at some element f not vanishing at λ. By hypothesis, P coincides with Q after localization at some element g, which may also be chosen not to vanish at λ. So all three coincide after localization at fg. It follows that I is generically reduced along $\mathcal{O}_\mathfrak{k}$. This is part (b) of the lemma (since I annihilates M). By Lemma 10.17,

$$T^*_\lambda(\operatorname{Spec} R/I) \simeq T^*_\lambda(\mathcal{O}_\mathfrak{k}). \tag{11.11)(c}$$

Part (d) of the lemma follows.

To continue, we need to better understand the symplectic form ϕ_λ on $T^*_\lambda(\operatorname{Spec} R)$ (cf. (10.6)). Each function f in R defines an element

$$df = (f - \lambda(f)) + \mathfrak{m}^2 \in T^*_\lambda(\mathcal{O}). \tag{11.12)(a}$$

This function also defines a tangent vector $\partial_\lambda(f)$ by the requirement

$$\partial_\lambda(f)(dg) = \lambda(\{f,g\}). \tag{11.12)(b}$$

The map Φ_λ of (10.6) is

$$\Phi_\lambda(df) = \partial_\lambda(f). \tag{11.12)(c}$$

Suppose $x \in \mathfrak{g}$; write $i(x)$ for the restriction to $\mathcal{O}$ of the (linear) function x on $\mathfrak{g}^*$, and $\operatorname{ad}^*(x)$ for the vector field on $\mathcal{O}$ induced by the coadjoint action of G. Then (because the coadjoint action is Hamiltonian)

$$\partial_\lambda(i(x)) = \operatorname{ad}^*(x)(\lambda). \tag{11.12)(d}$$

Now every tangent vector to $\mathcal{O}$ at λ comes from $\mathfrak{g}$; so this implies that the map Φ_λ is surjective. Consequently ϕ_λ is non-degenerate, proving part (a) of the lemma.

Because Φ_λ is an isomorphism, we can use it to transfer the symplectic form ϕ_λ to a symplectic form ϕ_λ^* on the tangent space. Now we already have a non-degenerate symplectic form ω_λ on $T_\lambda(\mathcal{O})$ (cf. (5.14)(b)). Of course we would like to know that this it form coincides with ϕ_λ^*: that is, that

$$\omega_\lambda(\Phi_\lambda(df), \Phi_\lambda(dg)) = \phi_\lambda(df, dg). \tag{11.13)(a}$$

Because of (11.12)(c), this is equivalent to

$$\lambda(\{f, g\}) = \omega_\lambda(\partial_\lambda(f), \partial_\lambda(g)). \tag{11.13)(b}$$

Each side is a bilinear form vanishing on $\mathfrak{m}^2$ and on constants, so it suffices to prove (11.13)(b) for $f = i(x)$ and $g = i(y)$, with x and y in $\mathfrak{g}$. Then both sides are just $\lambda([x, y])$ (cf. Example 10.2 and (5.14)(b)).

Part (c) of the lemma follows from (11.13) and Corollary 5.20. □

12. K-MULTIPLICITIES IN UNIPOTENT REPRESENTATIONS.

In this section we present a conjectural description of the K-multiplicities in certain unipotent representations. Since unipotent representations have not been defined in general, the reader may prefer to regard this as a desideratum rather than a conjecture; but see the remarks after the statement.

Conjecture 12.1. *Suppose $(\mathfrak{g}, K)$ is a reductive symmetric pair of Harish-Chandra class, and $\mathcal{O} \subset \mathfrak{g}^*$ is a nilpotent coadjoint orbit. Assume that $\partial\mathcal{O}$ has codimension at least four in $\overline{\mathcal{O}}$. Suppose X is an irreducible unipotent $(\mathfrak{g}, K)$-module attached to $\mathcal{O}$. Then there is*

a) an element $\lambda \in \mathcal{O} \cap (\mathfrak{g}/\mathfrak{k})^*$, and

b) an admissible representation χ of the stabilizer $K(\lambda)$ (Definition 7.13)

such that, as a representation of K,

$$X \simeq Ind_{K(\lambda)}^K(\chi).$$

It is possible to replace the "unipotent $(\mathfrak{g}, K)$-modules attached to $\mathcal{O}$" in this conjecture by a precisely defined class. Here is one way to do that. One can find in [V3, Definition 5.5] a definition of "unipotent Dixmier algebra attached to $\mathcal{O}$." (It is not known that algebras satisfying the requirements there exist, but it is conjectured that they do.) Such an algebra A is equipped with a map $U(\mathfrak{g}) \to A$, of which the kernel is a primitive ideal J_A satisfying $\mathcal{V}(\operatorname{gr} J_A) = \overline{\mathcal{O}}$. We can call J_A a ***unipotent primitive ideal attached to*** $\mathcal{O}$. A unipotent representation attached to $\mathcal{O}$ should have annihilator equal to some J_A. (It seems to be a bad idea to try to use this as a definition of unipotent representation — there are some non-unitary representations annihilated by a J_A, and they should probably not be called unipotent.) At any rate, the conjecture above should hold for any irreducible X with $\operatorname{Ann} X = J_A$ (always assuming the codimension condition on $\partial\mathcal{O}$).

Let us consider how close we are to proving this more precise conjecture. Of course the idea is to apply Theorem 8.7. The ideal J_A will always satisfy ${}^tJ_A = J_A$ (as a consequence of the definition of Dixmier algebra). It will *not* in general satisfy the first hypothesis of Theorem 8.7 (that $\operatorname{gr} J_A$ is generically reduced); but it will satisfy an analogous condition. This analogous condition (which is too involved to restate here) allows the proof of Theorem 8.7 to be repeated almost without change. We thus find a $\lambda \in \mathcal{O} \cap (\mathfrak{g}/\mathfrak{k})^*$ so that $\chi = \chi(\lambda, X)$ is a non-zero admissible representation of $K(\lambda)$. Now Theorem 4.4 shows that as representations of K

$$X \simeq Ind_{K(\lambda)}^{K}(\chi) - E, \tag{12.2}$$

with an "error term" E related to $\partial\mathcal{O}$. (The "codimension at least four" condition in the conjecture gives the "codimension at least two" hypothesis in Theorem 4.4 because of Corollary 5.20.) The K-multiplicities in E are of a lower order of magnitude than those in X, because of the condition on the support of E. Therefore (12.2) says that Conjecture 12.1 is approximately correct. For the rest of this section, we will consider some examples offering various kinds of support or illumination for Conjecture 12.1.

Example 12.3. Suppose $G_{\mathbb{R}}$ is $SL(2,\mathbb{R})$, and $\mathcal{O}$ is the principal nilpotent orbit. This orbit does *not* satisfy the codimension condition of Conjecture 12.1; we want to see how the conclusion of the conjecture fails. The group K is isomorphic to $\mathbb{C}^\times$, so its representations are characters τ_n parametrized by the integers. There are two orbits of K on $\mathcal{O} \cap (\mathfrak{g}/\mathfrak{k})^*$; each has $K(\lambda) = \{\pm 1\}$. Thus $K(\lambda)$ has two admissible representations χ_0 and χ_1, both admissible; we have

$$Ind_{K(\lambda)}^{K}(\chi_a) = \sum_{n \equiv a \,(\mathrm{mod}\, 2)} \tau_n.$$

By contrast, Arthur attaches three (special) unipotent representations to $\mathcal{O}$ (the constituents of the unitary principal series with continuous parameter 0). Their K-characters are

$$\sum_{k=-\infty}^{\infty} \tau_{2k}, \quad \sum_{k=0}^{\infty} \tau_{2k+1}, \quad \sum_{k=0}^{\infty} \tau_{-2k-1}.$$

The first of these is given by a formula as in Conjecture 12.1, but the last two are not.

This particular example suggests that certain combinations of representations might obey the multiplicity formula in Conjecture 12.1 without the codimension hypothesis. This is false, as one sees by examining the principal orbit in $SU(2,1)$.

Example 12.4. Suppose $G_{\mathbb{R}}$ is the double cover of $SL(3,\mathbb{R})$, and $\mathcal{O}$ is the minimal non-zero nilpotent orbit. Then $\dim \mathcal{O} = 4$ (and $\partial\mathcal{O} = \{0\}$ so the codimension condition is satisfied). K is isomorphic to $SL(2,\mathbb{C})$, which has one irreducible representation τ_n of each dimension n. Up to K conjugacy there is only one possibility for λ, namely a highest weight vector in the representation (isomorphic to τ_5) of K on $\mathfrak{g}/\mathfrak{k})^*$. Consequently

$$K(\lambda) = \left\{ \begin{pmatrix} t & x \\ 0 & t^{-1} \end{pmatrix} \mid t^4 = 1 \right\}.$$

The admissible representations of $K(\lambda)$ are those trivial on the identity component. There are four irreducible ones, given by

$$\chi_a \begin{pmatrix} t & x \\ 0 & t^{-1} \end{pmatrix} = t^a \qquad (a = 0, 1, 2, 3).$$

The corresponding induced representations are

$$Ind_{K(\lambda)}^{K}(\chi_a) = \sum_{n \equiv a+1 \,(\mathrm{mod}\, 4)} \tau_n.$$

For $a = 0$, $a = 1$, and $a = 2$ there are unitary representations attached to $\mathcal{O}$ having these K-types (see [T]). For $a = 3$ there is no $(\mathfrak{g}, K)$-module with exactly these K-types annihilated by a unipotent primitive ideal. (There *are* two irreducible $(\mathfrak{g}, K)$-modules with these K-types, but they are not unitary. Their primitive ideals are interchanged by the transpose antiautomorphism.)

The point of this example is to show that we cannot insist on the existence of unipotent representations attached to all admissible χ. A good explanation of why χ_3 is not allowed in this case would be extremely valuable.

References

[AM] M. Atiyah and I. MacDonald, *Introduction to Commutative Algebra*, Addison-Wesley, Reading, Massachusetts, 1969.

[BV] D. Barbasch and D. Vogan, *Unipotent representations of complex semisimple Lie groups*, Ann. of Math. **121** (1985), 41–110.

[BFFLS] F. Bayen, M. Flato, C. Fronsdal, A. Lichnerowicz, and D. Sternheimer, *Deformation theory and quantization I: Deformation of symplectic structures*, Ann. Physics **111** (1978), 61–110.

[BB] W. Borho and J.-L. Brylinski, *Differential operators on homogeneous spaces III*, Inventiones Math. **80** (1985), 1–68.

[BK] W. Borho and H. Kraft, *Über Bahnen und deren Deformationen bei linearen Aktionen reduktiver Gruppen*, Comment. Math. Helvetici **54** (1979), 61–104.

[D] M. Duflo, *Théorie de Mackey pour les groupes de Lie algèbriques*, Acta Math. **149** (1982), 153–213.

[Ga] O. Gabber, *The integrability of the characteristic variety*, Amer. J. Math. **103** (1981), 445–468.

[Ge] M. Gerstenhaber, *On the deformation of rings and algebras*, Ann. of Math. **79** (1964), 59–103.

[Gi] V. Ginsburg, *Characteristic varieties and vanishing cycles*, Inventiones math. **84** (1986), 327–402.

[Gr] A. Grothendieck (rédigés avec la collaboration de J. Dieudonné), *Eléments de Géométrie Algébrique IV. Etude locale des schémas et des morphismes de schémas (Seconde Partie)*, Publications Mathématiques **24**, Institut des Hautes Etudes Scientifiques, Le Bois-Marie, Bures-sur-Yvette, France, 1965.

[GS] V. Guillemin and S. Sternberg, *Geometric Asymptotics*, Math Surveys **14**, American Mathematical Society, Providence, Rhode Island, 1978.

[H] R. Hartshorne, *Algebraic Geometry*, Springer-Verlag, New York-Heidelberg-Berlin, 1977.

[J] A. Joseph, *On the associated variety of a primitive ideal*, J. Algebra **93** (1985), 509–523.

[K] B. Kostant, *The principal three-dimensional subgroup and the Betti numbers of a complex simple Lie group*, Amer. J. Math. **81** (1959), 973–1032.

[KR] B. Kostant and S. Rallis, *Orbits and representations associated with symmetric spaces*, Amer. J. Math. **93** (1971), 753–809.

[L] O. Loos, *Symmetric Spaces*, Vol I, Benjamin, New York-Amsterdam, 1969.

[M] H. Matsumura, *Commutative Ring Theory*, translated by M. Reid, Cambridge University Press, Cambridge, 1986.

[Sc] J. Schwartz, *The determination of the admissible orbits in the real classical groups*, Ph.D. dissertation, Massachusetts Institute of Technology, Cambridge, Massachusetts.

[Se] J. Sekiguchi, *Remarks on nilpotent orbits of a symmetric pair*, J. Math. Soc. Japan **39** (1987), 127–138.

[Sh] I. Shafarevich, *Basic Algebraic Geometry*, translated by K. Hirsch, Springer-Verlag, Berlin-Heidelberg-New York, 1977.

[SKK] M. Sato, M. Kashiwara, and T. Kawai, *Hyperfunctions and pseudodifferential equations*, Lecture Notes in Mathematics, vol. 287, Springer-Verlag, Berlin-Heidelberg-New York, 1973, pp. 265-529.

[Sp] T.A. Springer, *Microlocalisation algébrique*, Séminaire d'algébre Paul Dubreil et Marie-Paul Malliavin, 36ème année (Paris, 1983-1984), Lecture Notes in Mathematics, vol. 1146, Springer-Verlag, Berlin-New York, 1985, pp. 299–316.

[T] P. Torasso, *Quantification geometrique, operateurs d'entrelacement et representations unitaires de $SL_3(\mathbb{R})$*, Acta Math. **150** (1983), 153-242.

[vanD] G. van Dijk, *Invariant eigendistributions on the tangent space of a rank one semisimple symmetric space*, Math. Ann. **268** (1984), 405–416.

[V1] D. Vogan, *Gelfand-Kirillov dimension for Harish-Chandra modules*, Inventiones Math. **48** (1978), 75–98.

[V2] ———, *Unitary Representations of Reductive Lie Groups*, Annals of Mathematics Studies, Princeton University Press, Princeton, New Jersey, 1987.

[V3] ———, *Noncommutative algebras and unitary representations*, The Mathematical Heritage of Hermann Weyl Proceedings of Symposia in Pure Mathematics, Vol. 48 (R. O. Wells, Jr., eds.), American Mathematical Society, Providence, Rhode Island, 1988.

DEPARTMENT OF MATHEMATICS, MASSACHUSETTS INSTITUTE OF TECHNOLOGY, CAMBRIDGE, MA 02139

E-mail: dav @ math.mit.edu

70 ADOLPHSON/CONREY/GHOSH/YAGER. Analytic Number Theory and Diophantine Problems: Proceedings of a Conference at Oklahoma State University

71 GOLDSTEIN. Séminaire de Théorie des Nombres, Paris 1985–86

72 VAISMAN. Symplectic Geometry and Secondary Characteristic Classes

73 MOLINO. Riemannian Foliations

74 HENKIN/LEITERER. Andreotti-Grauert Theory by Integral Formulas

75 GOLDSTEIN. Séminaire de Théorie des Nombres, Paris 1986–87

76 COSSEC/DOLGACHEV. Enriques Surfaces I

77 REYSSAT. Quelques Aspects des Surfaces de Riemann

78 BORHO /BRYLINSKI/MACPHERSON. Nilpotent Orbits, Primitive Ideals, and Characteristic Classes

79 MCKENZIE/VALERIOTE. The Structure of Decidable Locally Finite Varieties

80 KRAFT/PETRIE/SCHWARZ. Topological Methods in Algebraic Transformation Groups

81 GOLDSTEIN. Séminaire de Théorie des Nombres, Paris 1987–88

82 DUFLO/PEDERSEN/VERGNE. The Orbit Method in Representation Theory: Proceedings of a Conference held in Copenhagen, August to September 1988

83 GHYS/DE LA HARPE. Sur les Groupes Hyperboliques d'après Mikhael Gromov

84 ARAKI/KADISON. Mappings of Operator Algebras: Proceedings of the Japan-U.S. Joint Seminar, University of Pennsylvania, Philadelphia, Pennsylvania, 1988

85 BERNDT/DIAMOND/HALBERSTAM/ HILDEBRAND. Analytic Number Theory: Proceedings of a Conference in Honor of Paul T. Bateman

86 CARTIER/ILLUSIE/KATZ/LAUMON/ MANIN/RIBET. The Grothendieck Festschrift: A Collection of Articles Written in Honor of the 60th Birthday of Alexander Grothendieck. Vol. I

87 CARTIER/ILLUSIE/KATZ/LAUMON/ MANIN/RIBET. The Grothendieck Festschrift: A Collection of Articles Written in Honor of the 60th Birthday of Alexander Grothendieck. Volume II

88 CARTIER/ILLUSIE/KATZ/LAUMON/ MANIN/RIBET. The Grothendieck Festschrift: A Collection of Articles Written in Honor of the 60th Birthday of Alexander Grothendieck. Volume III

89 VAN DER GEER/OORT / STEENBRINK. Arithmetic Algebraic Geometry

90 SRINIVAS. Algebraic K-Theory

91 GOLDSTEIN. Séminaire de Théorie des Nombres, Paris 1988–89

92 CONNES/DUFLO/JOSEPH/RENTSCHLER. Operator Algebras, Unitary Representations, Envelopping Algebras, and Invariant Theory. A Collection of Articles in Honor of the 65th Birthday of Jacques Dixmier

93 AUDIN. The Topology of Torus Actions on Symplectic Manifolds

94 MORA/TRAVERSO (eds.) Effective Methods in Algebraic Geometry

95 MICHLER/RINGEL (eds.) Represen- ation Theory of Finite Groups and Finite Dimensional Algebras

96 MALGRANGE. Equations Différentielles à Coefficients Polynomiaux

97 MUMFORD/NORI/NORMAN. Tata Lectures on Theta III

98 GODBILLON. Feuilletages, Etudes géométriques

99 DONATO /DUVAL/ELHADAD/TUYNMAN. Symplectic Geometry and Mathematical Physics. A Collection of Articles in Honor of J.-M. Souriau

100 TAYLOR. Pseudodifferential Operators and Nonlinear PDE

101 Barker/Sally. Harmonic Analysis on Reductive Groups